Linear Algebra

Linear Algebra From the Beginning

ERIC A. CARLEN
Georgia Institute of Technology

MARIA CONCEIÇÃO CARVALHO
University of Lisbon

W. H. Freeman and Company

New York

Publisher: Craig Bleyer
Executive Editor: Ruth Baruth
Senior Acquisitions Editor: Terri Ward
Marketing Manager: Victoria Anderson
Senior Media Editor: Roland Cheyney
Project Editor: Vivien Weiss
Assistant Editor: Christine Ondreicka
Design Manager: Blake Logan
Text Designer: Patrice Sheridan
Illustration Coordinator: Bill Page
Illustrations: Network Graphics
Production Coordinator: Paul W. Rohloff
Composition: MacMillan India
Printing: RR Donnelley

Library of Congress Control Number: 2006936847

ISBN-13: 978-0-7167-4894-6
ISBN-10: 0-7167-4894-0

W. H. Freeman and Company
41 Madison Avenue
New York, NY 10010

Houndmills, Basingstoke RG21 6XS, England

www.whfreeman.com

Contents

Contents   ix

Preface

Linear Algebra: From the Beginning is an introductory text designed to prepare students for the many and increasingly widespread uses of linear algebra in science and engineering. At the same time, it is written for beginners and does not assume any mathematical sophistication on the part of the reader beyond what would be acquired in a basic single-variable calculus course.

Linear Algebra: From the Beginning is focused on problem solving. It contains a wealth of examples and exercises, and we are sure that any student who worked through them all would master the subject and be ready to apply its methods in almost any context. Being able to do this requires *perspective*, which the text, examples, and exercises provide. To apply the methods of linear algebra successfully, one must understand the ideas on which they are based *and how those ideas fit together*. For this reason, we provide complete proofs. But more than that, we have paid close attention to the narrative line: We set up a context in which each new idea, definition, and theorem is as natural as possible. One thing leads to the next, as the story progresses from solving linear systems of equations, to finding least squares solutions when they cannot be solved exactly, to finding nice bases that simplify the presentation of transformations and the data on which they operate.

Some teachers first judge the quality of a linear algebra text book by looking up the definition of linear independence to see how soon it is introduced—the earlier the better. We do not subscribe to this criterion. We have found that if students *first* learn how to compute one-to-one parameterizations of the solution set of a linear system in the form

$$\mathbf{x}_0 + t_1\mathbf{v}_1 + \cdots + t_k\mathbf{v}_k,$$

one can *then* introduce linear independence as the property that makes such a parameterization one-to-one. This puts it into the practical context of solving equations. Students meet the linear independence not as a new and challenging abstract concept but as a name for something quite familiar. The beautiful abstraction, which we too cherish, is all here. But it is always introduced in a clear problem-solving context.

Because many science and engineering students will work with linear algebra using computers to manipulate large matrices, we explain several approaches to many problems and discuss which approaches are suitable for numerical computation, which would be prone to round-off errors, and why. There are computer exercises included to further develop this. However, *Linear Algebra: From the Beginning* is not a book on numerical linear algebra per se. Our goal is to give students a solid knowledge base in the core concepts of linear algebra and to develop the theoretical understanding and mathematical perspective needed to make full use of modern software for doing large scale computation. We do not, however, explain the use of any particular software package, hence the book can easily be used as the text for a course in which computers will play no role, skipping over a few subsections and exercises that would not be relevant to such a course. But the support is also there for a course syllabus that *would* make extensive use of computers.

The first five chapters cover the essential core of the subject, concerning linear transformations and linear systems of equations in $\mathbb{R}^n$ or $\mathbb{C}^n$. The initial focus in these chapters is on solving systems of linear equations or finding least squares solutions when that is not possible. Then, toward the end of Chapter 3, the focus begins to shift: More and more it is on finding nice bases in which linear transformations, and the data on which they operate, can be represented in a clear and useful way. This part of the story culminates, in Chapter 5, with an introduction to eigenvectors and diagonalization.

Almost every student in a modern science or engineering curriculum should know the material covered in the first five chapters. The last two chapters develop the story further, going into more specialized topics that will still be core topics for some students but not others. Chapter 6 discusses, among other things, the discrete Fourier transform and the diagonalization of large matrices, and it provides an introduction to iterative methods.

Chapter 7 introduces abstract linear algebra. In making this chapter the last in the book, we have taken the point of view that "abstract" is an active verb as well as an adjective. Students are good at abstracting concrete experience—once they have acquired the concrete experience in the first place. For this reason, our book starts out working with vectors in $\mathbb{R}^n$—lists of data—and transformations from $\mathbb{R}^n$ to $\mathbb{R}^m$. Once students have gained concrete computational experience in this setting, they can easily make the leap from thinking about vectors as lists of numbers to, say, functions on the real line, which is essential in order to apply linear algebra techniques to solve differential equations, for example. Indeed, the whole point of abstraction is to apply ideas and strategies that were developed for one purpose in a wider range of situations. To reveal the *full power* of the abstract approach in linear algebra, one has to show it at work in other fields, which cannot *really* be done without going beyond the bounds of our subject. However, one can make a good case for the utility of abstraction within linear algebra itself, and we do so here. For example, the abstract ideas introduced in Chapter 7 are very helpful in proving that for any $n \times n$ matrix A, there is a basis of $\mathbb{C}^n$ consisting of generalized eigenvectors, which is why the treatment of the Jordan canonical form appears in this chapter.

Again, let us emphasize that *Linear Algebra: From the Beginning* does not assume any mathematical sophistication on the part of the reader, beyond what would be acquired in a basic single variable calculus course. However, it does aim to develop it! As we have explained, to make full use of the powerful software available for solving large scale linear algebra problems, one must have a certain amount of mathematical perspective and understanding. There are many choices to be made about which software tools to use and how. But beyond this practical consideration, there is another point: *In our experience, linear algebra provides the ideal introduction to rigorous mathematical reasoning.* It would be a shame to neglect this aspect of the subject, and we have not. There is much here for anyone who wants to learn about mathematical proofs and how to go about constructing them.

This is especially true of Chapter 7. A course syllabus that included the content of Chapter 7 in detail would make an ideal "bridge course" for math majors to take between their calculus courses and their upper level analysis courses.

Indeed, the proofs in calculus are often quite delicate, depending on the fact that one is working with the real number system, not the rational number system. This is subtle. Things often get swept under the rug, and good students who really try to grasp the logic are often left unsatisfied. In linear algebra, the logic shines right through. For the most part, the proofs do not rely on the completeness property of the real numbers or other such subtleties. The logic of the proofs can be grasped and appreciated by all students who are inclined to try. Though linear algebra is sometimes considered to be a subject that requires "mathematical maturity," we believe instead that it is the ideal subject—much better than calculus—with which to develop mathematical maturity from the start.

1. Instructor's Manual with Solutions. This manual, written by Eric Carlen, contains helpful material for new teachers of the linear algebra course in addition to the solutions to all exercises in the textbook.
ISBN: 1-4292-0427-3

2. Student Solutions Manual. This manual, written by Eric Carlen, contains the solutions to all odd-numbered exercises in the textbook.
ISBN: 1-4292-0428-1

Acknowledgments

Many colleagues and students in the mathematical community have made valuable contributions and suggestions since the development of *Linear Algebra: From the Beginning* began. We are very thankful to a number of colleagues at Georgia Institute of Technology for making helpful suggestions—especially Fred Andrew, Rena Brakebill, John Elton, Jeff Geronimo, Michael Loss, and Enid Steinbart—as well as the many other colleagues and students who have emailed us about unclear points, typos, and errors.

We are also very grateful to the following instructors who provided detailed manuscript reviews.

David Meredith	San Francisco State University
Samar Khatiwala	Columbia University
Oleg Gleizer	University of California–Los Angeles
Alex Martsinkovsky	Northeastern University
Herman Gollwitzer	Drexel University
Andreas Seeger	University of Wisconsin–Madison
Thomas Judson	Harvard University
Richard Barshinger	Pennsylvania State University–Worthington–Scranton
Avinash Sathaye	University of Kentucky
Harry Allen	Ohio State University
Magdalena Daniela Toda	Texas Tech University
Vakhtang Putkaradze	University of New Mexico
Carlos Cabrelli	Georgia Institute of Technology
Luckhana Rena Brakebill	Georgia Institute of Technology
Ursula Molter	Georgia Institute of Technology
Laszlo Erdos	Georgia Institute of Technology

In writing this book, we have benefited from the advice and suggestions of Frank Purcell, to whom we are grateful for helping us keep the needs of first-year students in mind at every stage. His suggestions and analyses of reviews have helped us stick to our mission—and better accomplish it. We are thankful to have had the chance to work with him.

In addition, we appreciate the enthusiastic and helpful support of Craig Bleyer, Ruth Baruth, Terri Ward, Christine Ondreicka, Vivien Weiss, and Paul Rohloff at W. H. Freeman and Company.

SOME VERY IMPORTANT FORMULAS

V.I.F.1: Let $A = [v_1, v_2, \ldots, v_n]$ be an $m \times n$ matrix with columns $v_1, v_2, \ldots, v_n$, and let $x = \begin{bmatrix} x_1 \\ x_2 \\ \vdots \\ x_n \end{bmatrix}$ by any vector in $\mathbb{R}^n$. Then

$$Ax = [v_1, v_2, \ldots, v_n] \begin{bmatrix} x_1 \\ x_2 \\ \vdots \\ x_n \end{bmatrix} = x_1 v_1 + x_2 v_2 + \cdots + x_n v_n.$$

V.I.F.2: Let A be any $m \times n$ matrix with rows $r_1, r_2, \ldots, r_m$ so that

$$A = \begin{bmatrix} r_1 \\ r_2 \\ \vdots \\ r_m \end{bmatrix}.$$

Let $\mathbf{x}$ be any vector in $\mathbb{R}^n$. Then

$$A\mathbf{x} = \begin{bmatrix} r_1 \cdot \mathbf{x} \\ r_2 \cdot \mathbf{x} \\ \vdots \\ r_m \cdot \mathbf{x} \end{bmatrix}$$

V.I.F.3: Let $A = [v_1, v_2, \ldots, v_n]$ be an $m \times n$ matrix with columns $v_1, v_2, \ldots, v_n$. Let B be any $p \times m$ matrix. Then

$$BA = [B\mathbf{v}_1, B\mathbf{v}_2, \ldots, B\mathbf{v}_n].$$

V.I.F.4: Let A be any $m \times n$ matrix, and let B be any $p \times m$ matrix. Then

$$(BA)_{i,j} = (\text{row } i \text{ of } B) \cdot (\text{column } j \text{ of } A).$$

V.I.F.5: Let A be any $m \times n$ matrix, $\mathbf{x}$ any vector in $\mathbb{R}^m$, and $\mathbf{y}$ any vector in $\mathbb{R}^n$. Then

$$\mathbf{x} \cdot (A\mathbf{y}) = (A^t\mathbf{x}) \cdot \mathbf{y}.$$

Vectors, Matrices, and Linear Transformations

Overview of Chapter 1

This book is about functions that take lists of data as input and return lists of data as output, and about solving equations involving these "list in, list out" functions.

Such functions are common and probably familiar. For example, a weather service may give you a list of numerical data: air pressure, temperature, humidity, and so forth (the output list) as a function of your position as specified by a list of coordinates—say latitude and longitude (the input list). In mathematical terminology, "list variables" are called *vector variables*, and functions that transform one vector into another are called *vector functions* or *vector transformations*.

Here we focus attention on a special class of vector functions: The class of *linear transformations*.* This class deserves special attention for two reasons. First, linear transformations come up in many applications, so an ability to work effectively with them is esential in many

---

*You are not expected to be familiar with the definition of any technical terms in the overview; please read on. It helps to know the cast of main characters from the beginning.

quantitative fields. Second, the method of "linearization" coming from the differential calculus allows linear transformation methods to be applied in great generality to equations that involve nonlinear transformations.

You may have already encountered vectors, considered as elements of two- or three-dimensional Euclidean space, and seen vectors defined as "quantities with both magnitude and direction." In this nineteenth-century approach, vectors are *geometric* quantities from the beginning. This approach had its advantages as long as people were mainly concerned with vectors in two- or three-dimensional Euclidean space.

In a more modern approach, the starting point is algebra: We define certain algebraic operations on vectors, such as multiplying a vector by a number, adding two vectors, and the "dot product" of two vectors. Since the definition of these operations is purely algebraic and independent of our intuitive understanding of two- and three-dimensional geometry, it extends right away to arbitrary dimensions, that is, to arbitrarily long lists of numerical data. If you have done some computer programming, you are likely to have encountered "list" or "array" variables and certain "methods" or "procedures" defined on them. If so, the developments here are likely to be quite familiar.

Once the algebra is developed, we use it to "leverage" our intuitive understanding of geometry in two and three dimensions to arbitrary dimensions. Geometry has an important role in almost everything we do here, but unless one has a direct intuitive understanding of geometry in arbitrary dimensions, the path to a geometric understanding of vector problems proceeds through algebra. Fortunately, the path is quite short, and we come to n-dimensional geometry very soon, in the fourth section of the first chapter.

Without further ado, we start by examining some examples of functions that take vectors as input and return vectors as output.

SECTION 1 | Vectors and Multivariable Transformations

1.1 Some examples

Many functions considered in mathematics and science take *lists of data* as input and return *lists of data* as output. Here are some examples.

EXAMPLE 1   **(Watching your weight)** Suppose you are watching your weight. At the beginning of each month, you record your weight. After four months, you have a list of data consisting of four weights: w_1, w_2, w_3, and w_4. Let us record this list in a column of data:

$$\begin{bmatrix} w_1 \\ w_2 \\ w_3 \\ w_4 \end{bmatrix}.$$

This is *raw data* concerning your weight. To better understand these data, we can *transform* them into a more intelligible form. The transformation process produces a new list of data whose entries are of greater interest.

One interesting quantity that we can extract from this list is your *average recorded weight*. Let us call this quantity x_1:

$$x_1 = \frac{w_1 + w_2 + w_3 + w_4}{4}. \tag{1.1}$$

Other quantities of interest are your *changes of weight from month to month*:

$$x_2 = w_2 - w_1, \qquad x_3 = w_3 - w_2, \qquad \text{and} \qquad x_4 = w_4 - w_3.$$

Finally, you might well be interested in the total change over the course of the experiment. Define

$$x_5 = w_4 - w_1.$$

We now have a new list of data,

$$\begin{bmatrix} x_1 \\ x_2 \\ x_3 \\ x_4 \\ x_5 \end{bmatrix}.$$

We can summarize the relationship between this new list of transformed data and the list of raw data by

$$\begin{bmatrix} x_1 \\ x_2 \\ x_3 \\ x_4 \\ x_5 \end{bmatrix} = \begin{bmatrix} (w_1 + w_2 + w_3 + w_4)/4 \\ w_2 - w_1 \\ w_3 - w_2 \\ w_4 - w_3 \\ w_4 - w_1 \end{bmatrix}. \tag{1.2}$$

To focus on the *transformation of lists* aspect of this example, let **w** denote the list of raw data and let **x** denote the list of transformed data. Let f denote the transformation process itself so that we can write

$$\mathbf{x} = f(\mathbf{w}). \tag{1.3}$$

That is,

$$\begin{bmatrix} x_1 \\ x_2 \\ x_3 \\ x_4 \\ x_5 \end{bmatrix} = f\left(\begin{bmatrix} w_1 \\ w_2 \\ w_3 \\ w_4 \end{bmatrix} \right) = \begin{bmatrix} (w_1 + w_2 + w_3 + w_4)/4 \\ w_2 - w_1 \\ w_3 - w_2 \\ w_4 - w_3 \\ w_4 - w_1 \end{bmatrix}.$$

These equations define the function f as a transformation from the "set of 4-entry lists" to the "set of 5-entry lists." For example,*

$$f\left(\begin{bmatrix} 102 \\ 98 \\ 96 \\ 100 \end{bmatrix} \right) = \begin{bmatrix} 99 \\ -4 \\ -2 \\ 4 \\ -2 \end{bmatrix}.$$

Now suppose that you give somebody **x**, the list of processed data. Can they reconstruct your list of raw data from it? The problem posed to them is to find a "4-entry list" **w** so that

$$f(\mathbf{w}) = \begin{bmatrix} 99 \\ -4 \\ -2 \\ 4 \\ -2 \end{bmatrix}. \tag{1.4}$$

The *single* list "list equation," (1.4), can be written as a *system* of equations for the numerical variables w_1, w_2, w_3, and w_4:

$$\begin{aligned} \frac{w_1}{4} + \frac{w_2}{4} + \frac{w_3}{4} + \frac{w_4}{4} &= 99 \\ w_2 - w_1 &= -4 \\ w_3 - w_2 &= -2 \\ w_4 - w_3 &= 4 \\ w_4 - w_1 &= -2. \end{aligned} \tag{1.5}$$

This system is made up of five equations for four unknown quantities. To determine the unknowns, we eliminate variables. That is, we manipulate the system to eliminate w_1 from all but one of the equations, then w_2 from all but one of the remaining equations, and so forth.

---

*Choose your units—kilograms, pounds, whatever—to suit yourself.

With this goal in mind, multiply the first equation through by 4, and ignore the last equation—for now:

$$\begin{aligned} w_1 + w_2 + w_3 + w_4 &= 396 \\ -w_1 + w_2 &= -4 \\ -w_2 + w_3 &= -2 \\ -w_3 + w_4 &= 4. \end{aligned}$$

Adding equals to equals produces equals. Adding (each side of) the first equation to (the corresponding side of) the second equation, our system becomes

$$\begin{aligned} w_1 + w_2 + w_3 + w_4 &= 396 \\ 2w_2 + w_3 + w_4 &= 392 \\ -w_2 + w_3 &= -2 \\ -w_3 + w_4 &= 4. \end{aligned}$$

Notice that the variable w_1 appears only in the first equation; before it figured in both the first and the second equations. This is progress. We can make further progress by eliminating w_2 from the third equation. Multiplying the third equation through by 2 and adding (each side of) the second equation to (the corresponding side of) the third equation, our system becomes

$$\begin{aligned} w_1 + w_2 + w_3 + w_4 &= 396 \\ 2w_2 + w_3 + w_4 &= 392 \\ 3w_3 + w_4 &= 388 \\ -w_3 + w_4 &= 4. \end{aligned}$$

Multiplying the fourth equation through by 3, and adding (each side of) the third equation to (the corresponding side of) the fourth equation, our system becomes

$$\begin{aligned} w_1 + w_2 + w_3 + w_4 &= 396 \\ 2w_2 + w_3 + w_4 &= 392 \\ 3w_3 + w_4 &= 388 \\ 4w_4 &= 400. \end{aligned}$$

This is real progress. The *only* variable in the fourth equation is w_4. This equation says that $w_4 = 100$. Knowing this, we can reduce the third equation to

$$3w_3 = 288,$$

which means that $w_3 = 96$. If we know the values of w_3 and w_4, we can reduce the second equation to

$$2w_2 = 196,$$

which says that $w_2 = 98$. Knowing the values of w_2, w_3, and w_4, we can reduce the first equation to

$$w_1 = 102.$$

We have recovered the list of weights:

$$\begin{bmatrix} w_1 \\ w_2 \\ w_3 \\ w_4 \end{bmatrix} = \begin{bmatrix} 102 \\ 98 \\ 96 \\ 100 \end{bmatrix}.$$

Finally, notice that with $w_3 = 96$ and $w_4 = 100$, $w_4 - w_3 = 4$, so that the fifth equation in (1.5) is satisfied as well and hence so is (1.4).

There are at least two ways to describe what we have just done. One is to say that we have solved a system of four equations for four numerically valued variables, w_1, w_2, w_3, and w_4. Another is to say that we have solved an equation $f(\mathbf{w}) = \mathbf{x}$ relating two *list variables*.

The second mode of description turns out to be closely connected with a powerful way of thinking about relations between lists of data. This way of thinking has led to much beautiful mathematics and many beautiful mathematical applications. This book is about that way of thinking. But first, more examples will help us to see the way forward.

EXAMPLE 2

(Sorting lists of numbers) Let $(x_1, x_2, \ldots, x_n)$ be any list of n numbers. Here, we are writing our lists horizontally instead of vertically, as we did in Example 1.

Whether horizontal or vertical, our list will often more readily convey useful information if it is sorted. Very often, it is useful to sort lists so that the largest number is listed first, then the second largest, and so on.

For example, consider the list

$$(-3, 5, 0, 1, 2).$$

A *sorting transformation* would rearrange this list into

$$(5, 2, 1, 0, -3).$$

Again, what we have here is a mathematically well-defined transformation of one list into another. We define the *sorting function*, g_{sort}, on all finite lists of numbers to be the function whose output value at a given list is the corresponding sorted list. For example,

$$g_{\text{sort}}((-3, 5, 0, 1, 2)) = (5, 2, 1, 0, -3). \tag{1.6}$$

Consider the "list equation"

$$g_{\text{sort}}((x_1, x_2, x_3, x_4, x_5)) = (5, 2, 1, 0, -3). \tag{1.7}$$

The solution set of this equation is the set of all five entry lists that become $(5, 2, 1, 0, -3)$ on sorting. There are $5! = 120$ solutions of this equation; every permutation of the list $(5, 2, 1, 0, -3)$ is a solution.

Notice a difference between the equations considered in Examples 1 and 2. The definition of the transformation f in Example 1 involved algebraic operations—addition, subtraction, multiplication, and division—and so did our solution of (1.4). This is not the case in Example 2. There are all kinds of list transformations that one could consider. We will be mainly concerned with list transformations defined in terms of algebraic operations.

Here is another example.

EXAMPLE 3

(Voltages and currents) Here is a diagram of a simple electric circuit with two voltage sources and three resistors.

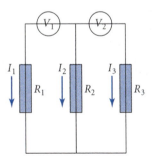

When the voltage sources are activated (for example, when batteries are put in place), electric current will flow. Let I_1, I_2, and I_3 denote the currents (measured in *amps*) flowing through the three resistors (whose resistance R is measured in *ohms*). We call the current positive if it flows in the direction of the corresponding arrow, and negative otherwise. Let the values of the three resistances be R_1, R_2, and R_3. To make our mathematical points in the simplest possible setting, let us suppose all three resistances are equal: $R_1 = R_2 = R_3 = R$. Finally, let the two voltages be V_1 and V_2, as indicated.

Regard the resistances as a fixed attribute of the circuit, while the voltages can be varied with voltage regulators or a change of batteries. Then there are two lists of variables to consider:

Voltage list: (V_1, V_2), the list of voltages.

Current list: (I_1, I_2, I_3), the list of currents.

Which is the input list and which is the output list? That depends on the question you ask. If you are given one list of data, you can figure out the other one by using certain physical laws.

Suppose we are given the list of currents and want to figure out the list of voltages. As we have said, this can be done using physical laws, namely *Ohm's law* and *Kirchhoff's rules*. We will not explain exactly what they say because that is not our point. We just jump to the conclusion and give the rule for obtaining the voltage list from the current list as an *example of a list-to-list function,* which is our point right now:

$$\begin{aligned} V_1 &= I_2 R - I_1 R \\ V_2 &= I_3 R - I_2 R. \end{aligned} \tag{1.8}$$

Equations (1.8) specify the functional dependence of the voltages V_1 and V_2 on the currents I_1, I_2, and I_3. Let us write this as

$$(V_1, V_2) = f(I_1, I_2, I_3), \tag{1.9}$$

which is just a shorthand notation for the computational rule expressed in (1.8). It is the function specifying how the voltage list depends on the current list.

Now let us consider another question. Suppose this time that the voltages are given, and we want to figure out what the currents are. Just to be concrete, let us ask:

- *What currents will flow in our circuit if we plug a 6-volt battery in place so that $V_1 = 6$ and a 9-volt battery in place so that $V_2 = 9$?*

To answer this question, we can try to solve (1.8) for I_1, I_2, and I_3. But that will not work: We have only two equations and there are three unknowns to determine.

We need another equation to solve for the currents, and physics provides it. Another of Kirchhoff's rules states:

$$I_1 + I_2 + I_3 = 0. \tag{1.10}$$

Equation (1.10) expresses the fact that electrical charge cannot be steadily flowing into or out of the node at the bottom of the circuit. It is a mathematical expression of the physical law of *electrical charge conservation*.

The mathematical significance of this equation is that not just any list of three numbers can be a list of the currents. No matter how the voltages are set, (1.10) must hold. So if we know any two values on the list of currents, we know the third. Therefore, we can *eliminate* any one of these current variables from the list as redundant information.

Let us eliminate I_2 from (1.8) using $I_2 = -(I_1 + I_3)$:

$$\begin{aligned} V_1 &= -(I_3 + 2I_1)R \\ V_2 &= (2I_3 + I_1)R. \end{aligned} \tag{1.11}$$

We get a second functional relationship,

$$(V_1, V_2) = g(I_1, I_3), \tag{1.12}$$

where g is a shorthand notation for the computational rules expressed in (1.11). This time there is no physical restriction on the input list. Any pair of values for I_1 and I_3 is physically admissible. (Except, of course, that in a real circuit, if either I_1 or I_3 gets *too* large, something will melt. But let us ignore this and consider an "ideal" circuit for which Ohm's law and Kirchhoff's rules hold no matter what.)

We can now *solve the equations* (1.11) to *determine the currents* as a function of the voltages: Multiplying the top equation in (1.11) through by 2 and adding it to the bottom equation,

$$2V_1 + V_2 = -3RI_1.$$

Multiplying the bottom equation in (1.11) through by 2 and adding it to the top equation,

$$V_1 + 2V_2 = 3RI_3.$$

Hence

$$I_1 = -\frac{2}{3R}V_1 - \frac{1}{3R}V_2$$
$$I_3 = \quad \frac{1}{3R}V_1 + \frac{2}{3R}V_2. \tag{1.13}$$

In view of (1.10),

$$I_2 = \frac{1}{3R}(V_1 - V_2). \tag{1.14}$$

Together, (1.13) and (1.14) give us a rule for determining the list of currents (I_1, I_2, I_3) given the list of voltages (V_1, V_2). As such, it is a function from the set of "2-entry lists of numbers" to the set of "3-entry lists of numbers." Let h denote this functional dependence so that

$$(I_1, I_2, I_3) = h(V_1, V_2).$$

For example, (1.13) and (1.14) tell us that if V_1 is 6 volts, V_2 is 9 volts, and R is measured in ohms, then $I_1 = -7/R$ amps, $I_2 = -1/R$ amps, and $I_3 = 8/R$ amps. That is,

$$\left(\frac{-7}{R}, \frac{-1}{R}, \frac{8}{R}\right) = h(6, 9).$$

This example has many important features to which we return later. What we want to focus on at this point are simply the several functional relationships between the "list of voltages" and the "list of currents" as examples of functions taking lists to lists.

We will see many more examples of such functions in what follows, but Examples 1–3 are enough to get going. Each of these examples has involved a function, or, in other words, a transformation of one list into another. There are two ways to think about such functions:

1. As a list of functions of several numerical variables, namely the numerical variables constituting the entries of the input list, as in (1.9)

2. As a single "list valued" function of a single "list" variable, as in (1.6)

In many instances, the second way is more natural, and computationally more convenient as well. If you have done some programming, you have probably run into "list" or "array" variables. If so, you have run into "methods" or "procedures" for working on array-type variables, transforming one such array into another. The sorting transformation in Example 2 is a case in point.

1.2 Basic definitions: Vectors and transformations

Before going any further, it will be good to set some terminology and notation once and for all. The term "list" has a clear meaning, which is why we have used it up to this point. But it is not the standard mathematical term, which is "vector."

DEFINITION	**(Vectors in $\mathbb{R}^n$)** A *vector* is an ordered list of n numbers $x_j, j = 1, 2, \ldots, n$ for some positive integer n, which is called the *dimension* of the vector. The integers $j = 1, 2, \ldots, n$ that order the list are called the *indices*, and the corresponding numbers x_j are called the *entries*. That is, for each $j = 1, 2, \ldots, n$, x_j is the jth entry on the list. The set of all n-dimensional vectors is denoted by $\mathbb{R}^n$.

The term *vector* comes from the Latin for "to carry." When the term was coined, people had in mind the description of motion by a list of velocities. For example, one might describe the instantaneous position of a particle being carried along in a stream of fluid by giving a list of coordinates as a function of the time t:

$$\begin{bmatrix} x(t) \\ y(t) \\ z(t) \end{bmatrix}.$$

Listing the derivatives of the coordinate functions gives

$$\begin{bmatrix} x'(t) \\ y'(t) \\ z'(t) \end{bmatrix},$$

the "velocity vector." This vector describes the instantaneous state of motion of the particle as it is carried along by the fluid.

This velocity vector expresses both the *direction* of motion and the *magnitude* of motion, which would be the speed at which the particle is moving, namely

$$\sqrt{(x'(t))^2 + (y'(t))^2 + (z'(t))^2}.$$

Hence, vectors are sometimes defined as "mathematical quantities having direction as well as magnitude." For some examples, such as velocity vectors, it is clear what "direction" means and what sort of "carrying" is going on. For other examples it is not.

It is not so obvious that the "carrying" idea is relevant for a vector whose entries are the three electrical currents I_1, I_2, and I_3 discussed in Example 3. What does "direction" even mean for a vector whose entries specify electrical currents? And what if our circuit had been more complicated, involving many more than three currents?

Perhaps surprisingly, there is a useful* notion of length and direction in any number of dimensions. We will explain all that soon enough, but at the beginning, *vectors are just lists of numbers*.

That explains what vectors are. Now, what sort of notation are we going to use to write about them? From here on, our standard practice is to denote vectors by boldface lowercase roman letters with, for example, **x** denoting the vector whose jth entry is x_j. We specify the

---

*By "useful," we mean useful for solving equations, among other things. In other words, useful in a practical sense, even in, say, eight dimensions.

vector in terms of its entries by listing them in *column form*:

$$\mathbf{x} = \begin{bmatrix} x_1 \\ x_2 \\ \vdots \\ x_n \end{bmatrix}.$$

For example,

$$\mathbf{x} = \begin{bmatrix} 2 \\ 4 \\ 1 \end{bmatrix}$$

is a vector in $\mathbb{R}^3$ for which $x_1 = 2$, $x_2 = 4$, and $x_3 = 1$.

DEFINITION

(Transformation from $\mathbb{R}^n$ to $\mathbb{R}^m$) Let m and n be positive integers. A *transformation* from $\mathbb{R}^n$ to $\mathbb{R}^m$ is simply a function f assigning a uniquely determined output vector $\mathbf{y}$ in $\mathbb{R}^m$ to each input vector $\mathbf{x}$ in $\mathbb{R}^n$, in which case we write

$$\mathbf{y} = f(\mathbf{x}).$$

In this book, the terms *transformation* and *function* are synonymous.

We have seen examples of transformations already. The transformation

$$f\left(\begin{bmatrix} w_1 \\ w_2 \\ w_3 \\ w_4 \end{bmatrix} \right) = \begin{bmatrix} (w_1 + w_2 + w_3 + w_4)/4 \\ w_2 - w_1 \\ w_3 - w_2 \\ w_4 - w_3 \\ w_4 - w_1 \end{bmatrix}$$

was considered in Example 1. Without changing any notation, we now see f as a transformation from $\mathbb{R}^4$ to $\mathbb{R}^5$.

Likewise, (1.8) can be written as a transformation from $\mathbb{R}^3$ to $\mathbb{R}^2$ if we just introduce the vectors

$$\begin{bmatrix} I_1 \\ I_2 \\ I_3 \end{bmatrix} \text{ and } \begin{bmatrix} V_1 \\ V_2 \end{bmatrix}.$$

Then (1.13) and (1.14) together give us a linear transformation from $\mathbb{R}^2$ to $\mathbb{R}^3$ in terms of the same vectors.

Before going further, we briefly recall some terminology and a number of definitions concerning *functions from one set to another*—for the most part, probably familiar. But because they are so fundamental, a careful review is worthwhile.

1.3 Functions: A brief review

Let X and Y be two sets. They could be sets of numbers, sets of lists, sets of triangles—it does not matter—just some pair of sets of mathematical objects. A function f from X to Y is a rule associating exactly one member of the set Y to each member of the set X.

The set X of inputs for which f is defined is usually called its *domain,* and the set Y of its possible outputs is called its *range.* Here are some kinds of functions:

1. *Algebraic functions of a real variable.* In this case, both sets X and Y are $\mathbb{R}$ or some subset of $\mathbb{R}$, and the function is given by an algebraic formula like $f(x) = \sqrt{1 + x^2}$.

2. *Transcendental functions of a real variable.* Again in this case, both sets X and Y are $\mathbb{R}$, or some subset of $\mathbb{R}$, but the function is not given by an algebraic formula. For example, $\sin(\theta)$ is not given by any algebraic formula. Instead, for $0 \leq \theta \leq \pi/2$, $\sin(\theta) = s/h$, where s is the length of the side of a right triangle opposite a vertex with angle θ and h is the length of the hypotenuse. This is a well-defined function specifying, by geometric means instead of algebraic means, an output number given an input number.

3. *Functions on finite sets.* Consider, for example, $X = \{1, 2, 3, 4, 5\}$ and $Y = \{2, 4, 6, 8\}$. That is, X is the set consisting of the first five natural numbers, and Y is the set consisting of the first four even natural numbers. In this case, since there are only finitely many inputs to consider, we can specify the action of the function by a *table*. A convenient way to do this is to list the outputs right beneath the inputs. For example, the table

$$
\begin{array}{ccccc}
1 & 2 & 3 & 4 & 5 \\
\downarrow & \downarrow & \downarrow & \downarrow & \downarrow \\
8 & 4 & 8 & 2 & 2
\end{array}
\tag{1.15}
$$

describes a function f from X to Y for which $f(1) = 8$, $f(2) = 2$, and so forth. The arrows emphasize the input–output relation.

The important thing to bear in mind is that functions need to be given not by a *formula* but by some rule for determining the output given the input. This rule might be given by a formula, but it might be given by a table as well. A table can specify a function f only if the domain X is a finite set (or else the table would be infinitely long, which is not very useful).

Also, although the "table" type of function may look artificial, the vector functions that we will work with are most conveniently studied in table terms.*

Given another function g from Y to some third set Z, we can form the *composition product* $g \circ f$, which is a function from X to Z through the rule

$$
g \circ f(x) = g(f(x))
$$

for all x in the domain of f. For example, if $f(x) = 1 + x^2$ and $g(y) = 1/y$ for $y \neq 0$, then

$$
g \circ f(x) = \frac{1}{1 + x^2}.
$$

Let us do another example where the functions are given by tables. Let f be the function given by the table in (1.15). Let $Z = \{2, 3, 5, 7\}$ be the set consisting of the first four prime numbers. Then with $Y = \{2, 4, 6, 8\}$ as earlier, define a function g from Y to Z by

$$
\begin{array}{cccc}
2 & 4 & 6 & 8 \\
\downarrow & \downarrow & \downarrow & \downarrow \\
5 & 3 & 2 & 7
\end{array}
\tag{1.16}
$$

Let us figure out what the table is for $g \circ f$. Since $f(1) = 8$, and $g(8) = 7$, $g \circ f(1) = g(f(1)) = g(8) = 7$. Since $f(2) = 4$, and $g(4) = 3$, $g \circ f(2) = g(f(2)) = g(4) = 3$. Continuing in this way, we find the table for $g \circ f$:

$$
\begin{array}{ccccc}
1 & 2 & 3 & 4 & 5 \\
\downarrow & \downarrow & \downarrow & \downarrow & \downarrow \\
7 & 3 & 7 & 5 & 5
\end{array}
\tag{1.17}
$$

The notation $g \circ f$ suggests a product, and in fact $g \circ f$ is sometimes called the "composition product" of f and g. Note, however, that it is not a commutative product. Indeed, $g \circ f$ is defined only if the range of f is contained in the domain of g. So it can be the case that $f \circ g$ is not even

---

*In fact, as we will see in the next section, "matrices" are just a way of writing certain vector functions in a sort of table form.

defined, though $g \circ f$ is: Let $g(y) = 0$ for all y, and let $f(x) = 1/x$ for $x \neq 0$. Then $g \circ f$ is defined for $x \neq 0$, but $g \circ f$ is not defined for any x.

However, the composition product is associative: If f, g, and h are any three functions with g defined on the range of f, and h defined on the range of g, then

$$(h \circ g) \circ f = h \circ (g \circ f).$$

In fact, by the definition, for any x in the domain of f,

$$((h \circ g) \circ f)(x) = (h \circ g)(f(x)) = h(g(f(x))),$$

and likewise,

$$(h \circ (g \circ f))(x) = h((g \circ f)(x)) = h(g(f(x))).$$

Either way, we get the same thing.

There is one more operation on functions, besides composition, that we are concerned with: *inverting functions*. To invert a function is to "run it backwards" so that the input is recovered from the output. Just as not every pair of functions can be composed, not every function can be inverted. Let us see what we require of a function if we are going to try to "run it backwards" and deduce the inputs from the outputs.

If every element in Y is an actual output value of f, that is, if for every y in Y there is *at least one* x in X so that $f(x) = y$, then we say f transforms X *onto* Y. For example, the function f defined in (1.15) does not transform X onto Y because 6, which belongs to Y, is not an output value of f. There is no answer to the question "Which input from X produces the output 6 in Y?" On the other hand, the function g defined in (1.16) does transform Y onto Z. Every element of $Z = \{2, 3, 5, 7\}$ is an output value of g.

One more property is required for a function f to be invertible: If f has the property that

$$f(x_1) = f(x_2) \Rightarrow x_1 = x_2$$

so that no two different inputs produce the same output, then we say that f is *one-to-one*. For example, the function f defined in (1.15) is not one-to-one, since it assigns the same output value, namely 8, to two different inputs, namely 1 and 3. (It also assigns the output 2 to both inputs 4 and 5, but once we have found one problem, that is it: The function is not one-to-one.) There are two answers to the question "Which input from X produces the output 8 in Y?" On the other hand, the function g defined in the table in (1.16) is one-to-one: Every input is assigned its own unique output value. There is just one answer, namely 4, to the question "Which input from Y produces the output 3 in Z?"

A function f from X to Y is *invertible* exactly when it is a one-to-one function from X onto Y. Since every y in Y is an actual output value of f, that is, $y = f(x)$ for some x in X, and since it is possible to determine (in principle at least) what the input x was if we are given the output y, we can define a new function f^{-1} from Y back to X by defining $f^{-1}(y)$ to be the unique x such that $f(x) = y$. Clearly, then, $f^{-1} \circ f$ is the *identity function* on X: $f^{-1} \circ f(x) = x$ for all x in X. The function f^{-1} is called the *inverse* of f. Invertible functions set up a one-to-one correspondence between their domains and ranges.

For example, the function g defined in (1.16) is invertible. (We have already checked that it is onto and one-to-one.) To specify g^{-1}, we just give a table that has the elements of Z as input values, and underneath we list the elements of Y that are associated to them by g:

$$\begin{array}{cccc} 2 & 3 & 5 & 7 \\ \downarrow & \downarrow & \downarrow & \downarrow \\ 6 & 4 & 2 & 8 \end{array} \qquad (1.18)$$

There are two points to bear in mind as a summary of the foregoing discussion:

- *The composition product of functions is not commutative in general, but it is associative, which is to say that $h \circ (g \circ f) = (h \circ g) \circ f$ whenever the domain of g contains the range of f and the domain of h contains the range of g.*

- *A function is invertible if and only if it is both onto and one-to-one.*

Now let us get back to functions whose domains and ranges are in $\mathbb{R}^m$ and $\mathbb{R}^n$ for some m and n.

EXAMPLE 4   **(A vector function with an algebraic formula)** Consider

$$f\left(\begin{bmatrix} x \\ y \end{bmatrix}\right) = \begin{bmatrix} x + 2y \\ y \end{bmatrix},$$

which defines a transformation f from $\mathbb{R}^2$ to $\mathbb{R}^2$. Here we have a mathematical formula, "built" out of algebraic operations applied to the individual variables x and y. If we let

$$\mathbf{u} = \begin{bmatrix} u \\ v \end{bmatrix}$$

denote the output vector, then

$$\mathbf{u} = f(\mathbf{x}), \tag{1.19}$$

or in long form,

$$\begin{bmatrix} u \\ v \end{bmatrix} = f\left(\begin{bmatrix} x \\ y \end{bmatrix}\right). \tag{1.20}$$

This single vector equation, written in either the short form (1.19) or the long form (1.20), is equivalent to the pair of equations

$$\begin{aligned} u &= x + 2y \\ v &= y \end{aligned} \tag{1.21}$$

relating the four numerical-valued variables x, y, u, and v.

Is this transformation from $\mathbb{R}^2$ to $\mathbb{R}^2$ invertible? Yes. To see that it is, let us fix a vector $\mathbf{u}$ in $\mathbb{R}^2$ and try to solve (1.19). The transformation f is onto if and only if (1.19) has a solution for *every* $\mathbf{u}$ in $\mathbb{R}^2$, and it is one-to-one if and only if (1.19) never has more than one solution for *any* $\mathbf{u}$ in $\mathbb{R}^2$. *Finding inverses is intimately connected with solving equations, which is why we are discussing the topic.*

Since presumably you are more familiar with numerical variables, we solve (1.19) through (1.21).

Solving (1.21) is easy: The second equation tells us that $y = v$, and substituting this into the first equation, we have $u = x + 2v$, or $x = u - 2v$. Notice that x and y are uniquely determined by u and v, so for every $\mathbf{u}$ there is exactly one $\mathbf{x}$ satisfying (1.19). In other words, f is both onto and one-to-one, and f^{-1} is given by

$$f^{-1}\left(\begin{bmatrix} u \\ v \end{bmatrix}\right) = \begin{bmatrix} u - 2v \\ v \end{bmatrix}.$$

Here we analyzed f by considering u and v separately. They can be thought of as algebraic functions of the numerical variables x and y, that is, as $u(x, y)$ and $v(x, y)$. This is the "multivariable" point of view. There is another way to look at this function in terms of algebraic operations on vector variables. We introduce this way of looking at things in the next section, and it turns out to be very advantageous.*

Exercises

1.1 Let f be the transformation from $\mathbb{R}^2$ to $\mathbb{R}^3$ given by

$$f\left(\begin{bmatrix} x \\ y \end{bmatrix}\right) = \begin{bmatrix} xy \\ x - y \\ x + y \end{bmatrix}$$

and let g be the transformation from $\mathbb{R}^3$ to $\mathbb{R}^2$ given by

$$g\left(\begin{bmatrix} x \\ y \\ z \end{bmatrix}\right) = \begin{bmatrix} z \\ y \end{bmatrix}.$$

(a) Compute $f\left(\begin{bmatrix} 1 \\ 2 \end{bmatrix}\right)$. **(b)** Compute $g\left(\begin{bmatrix} 2 \\ -1 \\ 3 \end{bmatrix}\right)$. **(c)** Compute $g \circ f\left(\begin{bmatrix} 1 \\ 2 \end{bmatrix}\right)$. **(d)** Compute a formula for the transformation $g \circ f$.

1.2 Define an appropriate transformation from $\mathbb{R}^n$ to $\mathbb{R}^m$, and write the following systems of equations as single-vector equations.

(a) $\quad 2x + 3y = 3$
$\qquad\quad x/y = 1$

(b) $\quad xy + yz + zx = 3$
$\qquad\qquad x + y = 1$
$\qquad\quad x - y + z = 2$

(c) $\quad x + y + x = 3$
$\qquad\quad y + z = 2$
$\qquad\quad x + z = 2$
$\qquad\quad x + y = 1$

1.3 Let f, g, and h be the functions from $\{1, 2, 3, 4, 5\}$ to itself given by

$$f: \begin{array}{ccccc} 1 & 2 & 3 & 4 & 5 \\ \downarrow & \downarrow & \downarrow & \downarrow & \downarrow \\ 2 & 4 & 3 & 5 & 1 \end{array} \qquad g: \begin{array}{ccccc} 1 & 2 & 3 & 4 & 5 \\ \downarrow & \downarrow & \downarrow & \downarrow & \downarrow \\ 2 & 2 & 4 & 4 & 2 \end{array} \qquad h: \begin{array}{ccccc} 1 & 2 & 3 & 4 & 5 \\ \downarrow & \downarrow & \downarrow & \downarrow & \downarrow \\ 5 & 4 & 3 & 2 & 1 \end{array}.$$

(a) Give the tables specifying $f \circ g$, $g \circ f$, $h \circ g$, and $h \circ h$.

(b) Give the tables specifying $h \circ (g \circ f)$ and $(h \circ g) \circ f$. Do your results agree with the associativity of the composition product?

(c) Which of the functions f, g, and h are one-to-one? Which are onto? Which are invertible? For those that are invertible, give the table specifying the inverses.

1.4 Let f, g, and h be the functions from $\{1, 2, 3, 4, 5\}$ to itself given by

$$f: \begin{array}{ccccc} 1 & 2 & 3 & 4 & 5 \\ \downarrow & \downarrow & \downarrow & \downarrow & \downarrow \\ 2 & 5 & 4 & 2 & 1 \end{array} \qquad g: \begin{array}{ccccc} 1 & 2 & 3 & 4 & 5 \\ \downarrow & \downarrow & \downarrow & \downarrow & \downarrow \\ 1 & 5 & 3 & 4 & 2 \end{array} \qquad h: \begin{array}{ccccc} 1 & 2 & 3 & 4 & 5 \\ \downarrow & \downarrow & \downarrow & \downarrow & \downarrow \\ 4 & 2 & 1 & 5 & 3 \end{array}.$$

(a) Give the tables specifying $f \circ g$, $g \circ f$, $h \circ g$, and $h \circ h$.

(b) Give the tables specifying $h \circ (g \circ f)$ and $(h \circ g) \circ f$. Do your results agree with the associativity of the composition product?

(c) Which of the functions f, g, and h are one-to-one? Which are onto? Which are invertible? For those that are invertible, give the table specifying the inverses.

*There would not be much of an advantage in the context of Example 4, which is so simple that *any* approach deals with it pretty quickly. But in more realistic problems with more numerical variables, the vector approach shows its strength.

1.5 Consider the function from $\mathbb{R}^2$ to $\mathbb{R}^2$ defined by

$$f\left(\begin{bmatrix} x \\ y \end{bmatrix}\right) = \begin{bmatrix} x - y \\ 2x + y \end{bmatrix}.$$

(a) Is f one-to-one? **(b)** Is f onto? **(c)** Is f invertible? If so, find a formula for the inverse. If not, explain why not.

1.6 Consider the function from $\mathbb{R}^2$ to $\mathbb{R}^2$ defined by

$$f\left(\begin{bmatrix} x \\ y \end{bmatrix}\right) = \begin{bmatrix} x^2 - y^2 \\ 2xy \end{bmatrix}.$$

(a) Is f one-to-one? **(b)** Is f onto? **(c)** Is f invertible? If so, find a formula for the inverse. If not, explain why not.

1.7 Consider the function from $\mathbb{R}^2$ to $\mathbb{R}^2$ defined by

$$f\left(\begin{bmatrix} x \\ y \end{bmatrix}\right) = \begin{bmatrix} x^2 - y^2 \\ x - 2 \end{bmatrix}.$$

(a) Find all solutions of

$$f\left(\begin{bmatrix} x \\ y \end{bmatrix}\right) = \begin{bmatrix} 0 \\ 0 \end{bmatrix}.$$

(b) Is f invertible? If so, find a formula for the inverse. If not, explain why not.

1.8 Consider the function from $\mathbb{R}^3$ to $\mathbb{R}^2$ defined by

$$f\left(\begin{bmatrix} x \\ y \\ z \end{bmatrix}\right) = \begin{bmatrix} 2x + y \\ x - z \end{bmatrix}.$$

(a) Compute

$$f\left(\begin{bmatrix} 0 \\ 0 \\ 0 \end{bmatrix}\right) \text{ and } f\left(\begin{bmatrix} 1 \\ -2 \\ 1 \end{bmatrix}\right).$$

(b) Is f invertible? If so, find a formula for the inverse. If not, explain why not.

1.9 Consider the function from $\mathbb{R}^2$ to $\mathbb{R}^3$ defined by

$$f\left(\begin{bmatrix} x \\ y \end{bmatrix}\right) = \begin{bmatrix} x - y \\ -2x \\ x + y \end{bmatrix}.$$

(a) Show that if

$$\begin{bmatrix} u \\ v \\ w \end{bmatrix} = f\left(\begin{bmatrix} x \\ y \end{bmatrix}\right),$$

then $u + v + w = 0$.

(b) Is f invertible? If so, find a formula for the inverse. If not, explain why not.

1.10 Consider the function from $\mathbb{R}^2$ to $\mathbb{R}^2$ defined by

$$f\left(\begin{bmatrix} x \\ y \end{bmatrix}\right) = \begin{bmatrix} x + ay \\ 2x + y \end{bmatrix}.$$

(a) Find all values of a, if any, for which f is one-to-one.

(b) Find all values of a, if any, for which f is onto.

1.11 Consider the function from $\mathbb{R}^2$ to $\mathbb{R}^2$ defined by

$$f\left(\begin{bmatrix} x \\ y \end{bmatrix}\right) = \begin{bmatrix} x^2 + ay^2 \\ x \end{bmatrix}.$$

(a) Find all values of a, if any, for which f is one-to-one.

(b) Find all values of a, if any, for which f is onto.

Transformations of polynomials

The next exercises concern polynomials. Polynomials can be specified by *listing* their coefficients. This allows us to think of polynomials as lists, that is, as vectors.

For example, the polynomial $p(x) = 2 + 4x - x^2$ can be specified by the vector

$$\begin{bmatrix} 2 \\ 4 \\ -1 \end{bmatrix},$$

where, by the convention we choose to follow, the first entry is the constant coefficient, the second entry is the coefficient multiplying x, and the third entry is the coefficient multiplying x^2. This convention establishes a one-to-one correspondence between second-degree polynomials and vectors in $\mathbb{R}^3$. In the same way, we identify the set of polynomials of degree (at most) k with $\mathbb{R}^{k+1}$.

By means of this correspondence, any transformation on the set of second-degree polynomials induces a transformation on $\mathbb{R}^3$. There are many interesting transformations on the set of polynomials. For example, if you differentiate a polynomial of degree (at most) k, you get a polynomial of degree (at most) $k - 1$. If you square a polynomial of degree (at most) k, you get a polynomial of degree (at most) $2k$.

The following exercises concern expressing a transformation of polynomials as a transformation on the list (vector) of its coefficients.

1.12 Define a transformation f from $\mathbb{R}^3$ to $\mathbb{R}^2$ as follows. Define

$$f\left(\begin{bmatrix} a \\ b \\ c \end{bmatrix}\right) = \begin{bmatrix} u \\ v \end{bmatrix},$$

where

$$u + vx = \frac{d}{dx}(a + bx + cx^2).$$

(a) Is f one-to-one? **(b)** Is f onto? **(c)** Is f invertible?

1.13 Define a transformation f from $\mathbb{R}^3$ to $\mathbb{R}^5$ as follows. Define

$$f\left(\begin{bmatrix} a \\ b \\ c \end{bmatrix}\right) = \begin{bmatrix} r \\ s \\ t \\ u \\ v \end{bmatrix},$$

where

$$r + sx + tx^2 + ux^3 + vx^4 = (a + bx + cx^2)^2.$$

(a) Is f one-to-one? **(b)** Is f onto? **(c)** Is f invertible?

1.14 Define a transformation f from $\mathbb{R}^3$ to $\mathbb{R}^3$ as follows. Define

$$f\left(\begin{bmatrix} a \\ b \\ c \end{bmatrix}\right) = \begin{bmatrix} t \\ u \\ v \end{bmatrix}$$

where

$$t + ux + vx^2 = \frac{\mathrm{d}}{\mathrm{d}x}(x(a + bx + cx^2)).$$

(The polynomial transformation is to multiply by x, then differentiate.)

(a) Is f one-to-one? **(b)** Is f onto? **(c)** Is f invertible?

1.15 Define a transformation f from $\mathbb{R}^3$ to $\mathbb{R}^3$ as follows. Define

$$f\left(\begin{bmatrix} a \\ b \\ c \end{bmatrix}\right) = \begin{bmatrix} t \\ u \\ v \end{bmatrix}$$

where

$$t + ux + vx^2 = \frac{\mathrm{d}^2}{\mathrm{d}x^2}(x^2(a + bx + cx^2)).$$

(The polynomial transformation is to multiply by x^2, then differentiate twice.)

(a) Is f one-to-one? **(b)** Is f onto? **(c)** Is f invertible?

SECTION 2 | Vector Operations and Linear Transformations

The most familiar sort of functions of a single real variable x, say $f(x) = x^2 + 1$, are "built up" out of certain algebraic operations that can be applied to a real variable. An important class of vector functions arises this way, too.

2.1 Vector algebra

DEFINITION

(**Scalar multiplication**) Given a number a in $\mathbb{R}$ and a vector

$$\mathbf{x} = \begin{bmatrix} x_1 \\ x_2 \\ \vdots \\ x_n \end{bmatrix},$$

define the product of a and $\mathbf{x}$, denoted $a\mathbf{x}$, by

$$a\mathbf{x} = \begin{bmatrix} ax_1 \\ ax_2 \\ \vdots \\ ax_n \end{bmatrix}.$$

For any vector $\mathbf{x}$, $-\mathbf{x}$ denotes the product of -1 and $\mathbf{x}$.

EXAMPLE 5 **(Multiplying numbers and vectors)** Here are several examples that should be clear without much discussion:*

$$2 \begin{bmatrix} 1 \\ -1 \\ 0 \end{bmatrix} = \begin{bmatrix} 2 \\ -2 \\ 0 \end{bmatrix}$$

$$\pi \begin{bmatrix} 1/2 \\ -1/2 \end{bmatrix} = \begin{bmatrix} \pi/2 \\ -\pi/2 \end{bmatrix} = \frac{1}{2} \begin{bmatrix} \pi \\ -\pi \end{bmatrix}$$

$$0 \begin{bmatrix} 1/2 \\ -1/2 \end{bmatrix} = \begin{bmatrix} 0 \\ 0 \end{bmatrix}.$$

DEFINITION

(Vector addition) Given two vectors $\mathbf{x}$ and $\mathbf{y}$ in $\mathbb{R}^n$ for some n, define their *vector sum*, $\mathbf{x} + \mathbf{y}$, by summing the corresponding entries:

$$\mathbf{x} + \mathbf{y} = \begin{bmatrix} x_1 + y_1 \\ x_2 + y_2 \\ \vdots \\ x_n + y_n \end{bmatrix}.$$

We define the *vector difference* of $\mathbf{x}$ and $\mathbf{y}$, $\mathbf{x} - \mathbf{y}$ by $\mathbf{x} - \mathbf{y} = \mathbf{x} + (-\mathbf{y})$.

Notice that vector addition does not mix up the entries of the vectors involved at all: The third entry, say, of the sum depends only on the third entries of the summands.

- *For this reason, vector addition inherits the commutative and associative properties of addition in the real numbers. It is just the addition of real numbers "done in parallel."*

EXAMPLE 6 **(Vector addition)**

$$\begin{bmatrix} 1 \\ -2 \\ -1 \end{bmatrix} + \begin{bmatrix} 1 \\ 1 \\ 1 \end{bmatrix} = \begin{bmatrix} 2 \\ -1 \\ 0 \end{bmatrix}$$

$$\begin{bmatrix} 1 \\ -2 \\ -1 \end{bmatrix} + \begin{bmatrix} -1 \\ 2 \\ 1 \end{bmatrix} = \begin{bmatrix} 0 \\ 0 \\ 0 \end{bmatrix}$$

$$\begin{bmatrix} 0 \\ 0 \\ 0 \end{bmatrix} + \begin{bmatrix} 1 \\ -2 \\ -1 \end{bmatrix} = \begin{bmatrix} 1 \\ -2 \\ -1 \end{bmatrix}.$$

There is a geometric way to think about vector addition in $\mathbb{R}^2$. If we identify the vector $\begin{bmatrix} x \\ y \end{bmatrix}$ in $\mathbb{R}^2$ with the point (x, y) in the Euclidean plane, we can then represent this vector geometrically by drawing an arrow with its tail at the origin and its head at (x, y). The following diagram shows three vectors represented this way:

$$\mathbf{x} = \begin{bmatrix} -1/2 \\ 1/2 \end{bmatrix}, \ \mathbf{y} = \begin{bmatrix} 3/2 \\ 1 \end{bmatrix}, \ \text{and their sum,} \ \mathbf{x} + \mathbf{y} = \begin{bmatrix} 1 \\ 3/2 \end{bmatrix}.$$

---

*Notice the smaller type used in the examples. The examples are more computational, and, like the problems, they are meant to be worked through as well as read. For this reason they are set in the same smaller type.

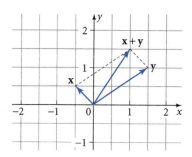

The vectors **x**, **y**, and **x**+**y** themselves are drawn in boldface. There are also two dashed lines: One is parallel to **x** and the other is parallel to **y**. The dashed lines complete the parallelogram whose vertices are the origin and the points corresponding to **x**, **y**, and **x** + **y**. As you see, the arrow representing **x** + **y** is the diagonal of this parallelogram that has its "tail end" at the origin.

A similar diagram could be drawn for any pair of vectors and their sum, and you see that we can think of vector addition in the plane as corresponding to the following operation:

- *Represent the vectors by arrows as in the diagram. Transport one arrow without turning it—that is, in a parallel motion—to bring its tail to the other arrow's head. The head of the transported arrow is now at the point corresponding to the sum of the vectors.*

EXAMPLE 7   **(Subtraction of vectors)** Let **x** and **y** be two vectors in the plane $\mathbb{R}^2$, and let **w** = **x** − **y**. Then, using the associative and commutative properties of vector addition,

$$\mathbf{x} = \mathbf{x} + (\mathbf{y} - \mathbf{y}) = (\mathbf{x} - \mathbf{y}) + \mathbf{y} = \mathbf{y} + \mathbf{w}.$$

Using the same diagram, with the arrows labeled a bit differently, we see that **w** = **x** − **y** is the arrow running from the head of **y** to the head of **x**, "parallel transported" so that its tail is at the origin.

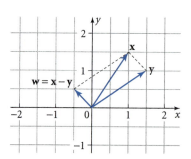

Having defined addition and subtraction of vectors, it might seem natural at this point to go on and define a vector product **x** × **y** by

$$\mathbf{x} \times \mathbf{y} = \begin{bmatrix} x_1 \times y_1 \\ x_2 \times y_2 \\ \vdots \\ x_n \times y_n \end{bmatrix}.$$

We *do not* do this. The reason is that there are very few examples of interesting vector transformations f for which, in the sense of this "definition," $f(\mathbf{x} \times \mathbf{y}) = f(\mathbf{x}) \times f(\mathbf{y})$. On the other hand, there are *many* interesting examples of vector transformations f with the property that $f(a\mathbf{x} + b\mathbf{y}) = af(\mathbf{x}) + bf(\mathbf{y})$ for all a, b, $\mathbf{x}$, and $\mathbf{y}$.

2.2 Linear transformations

DEFINITION	**(Linear transformation)** A transformation from $\mathbb{R}^n$ to $\mathbb{R}^m$ is a *linear transformation* if and only if $$f(a\mathbf{x} + b\mathbf{y}) = af(\mathbf{x}) + bf(\mathbf{y}) \qquad (2.1)$$ for all real numbers a and b and all vectors $\mathbf{x}$ and $\mathbf{y}$ in $\mathbb{R}^n$.

EXAMPLE 8 **(Checking for linearity)** We have seen many examples of transformations from $\mathbb{R}^n$ to $\mathbb{R}^m$ for various values of m and n in the previous section. Which of these are linear? Let us look at two examples. Rewrite (1.11) and (1.12) in vector form:

$$\begin{bmatrix} V_1 \\ V_2 \end{bmatrix} = g\left(\begin{bmatrix} I_1 \\ I_3 \end{bmatrix} \right) = \begin{bmatrix} -(2I_1 + I_3)R \\ (2I_3 + I_1)R \end{bmatrix}.$$

From the definition of g, we now check that for any $\begin{bmatrix} I_1 \\ I_3 \end{bmatrix}$ and $\begin{bmatrix} \tilde{I}_1 \\ \tilde{I}_3 \end{bmatrix}$,

$$g\left(a\begin{bmatrix} I_1 \\ I_3 \end{bmatrix} + b\begin{bmatrix} \tilde{I}_1 \\ \tilde{I}_3 \end{bmatrix} \right) = g\left(\begin{bmatrix} aI_1 + b\tilde{I}_1 \\ aI_3 + b\tilde{I}_3 \end{bmatrix} \right)$$

$$= \begin{bmatrix} -(2(aI_1 + b\tilde{I}_1) + (aI_3 + b\tilde{I}_3))R \\ (2(aI_3 + b\tilde{I}_3) + (aI_1 + b\tilde{I}_1))R \end{bmatrix}$$

$$= a\begin{bmatrix} -(2I_1 + I_3)R \\ (2I_3 + I_1)R \end{bmatrix} + b\begin{bmatrix} -(2\tilde{I}_1 + \tilde{I}_3)R \\ (2\tilde{I}_3 + \tilde{I}_1)R \end{bmatrix}$$

$$= ag\left(\begin{bmatrix} I_1 \\ I_3 \end{bmatrix} \right) + bg\left(\begin{bmatrix} \tilde{I}_1 \\ \tilde{I}_3 \end{bmatrix} \right).$$

Therefore, this transformation is linear.

On the other hand, consider the sorting function g_{sort}, defined in Example 2. Writing it as a vector function,

$$g_{\text{sort}}\left(\begin{bmatrix} 1 \\ 2 \\ 3 \\ 4 \\ 5 \end{bmatrix} + \begin{bmatrix} 5 \\ 4 \\ 3 \\ 2 \\ 1 \end{bmatrix} \right) = g_{\text{sort}}\left(\begin{bmatrix} 6 \\ 6 \\ 6 \\ 6 \\ 6 \end{bmatrix} \right) = \begin{bmatrix} 6 \\ 6 \\ 6 \\ 6 \\ 6 \end{bmatrix},$$

while

$$g_{\text{sort}}\left(\begin{bmatrix} 1 \\ 2 \\ 3 \\ 4 \\ 5 \end{bmatrix} \right) + g_{\text{sort}}\left(\begin{bmatrix} 5 \\ 4 \\ 3 \\ 2 \\ 1 \end{bmatrix} \right) = \begin{bmatrix} 5 \\ 4 \\ 3 \\ 2 \\ 1 \end{bmatrix} + \begin{bmatrix} 5 \\ 4 \\ 3 \\ 2 \\ 1 \end{bmatrix} = \begin{bmatrix} 10 \\ 8 \\ 6 \\ 4 \\ 2 \end{bmatrix}.$$

(Here we have taken $a = b = 1$.) Evidently our sorting transformation is not linear.

Another pair of definitions will help us "divide and conquer" questions about linearity.

DEFINITION

(Homogeneous and additive) A transformation from $\mathbb{R}^n$ to $\mathbb{R}^m$ is *homogeneous* if and only if for all $\mathbf{x}$ in $\mathbb{R}^n$ and all numbers a:

$$f(a\mathbf{x}) = af(\mathbf{x}). \qquad (2.2)$$

A transformation from $\mathbb{R}^n$ to $\mathbb{R}^m$ is *additive* if and only if for all $\mathbf{x}$ and $\mathbf{y}$ in $\mathbb{R}^n$:

$$f(\mathbf{x} + \mathbf{y}) = f(\mathbf{x}) + f(\mathbf{y}). \qquad (2.3)$$

If you think about it a bit, you will see that *a transformation from $\mathbb{R}^n$ to $\mathbb{R}^m$ is linear if and only if it is both homogeneous and additive.* In fact, in Example 7 we showed that g_{sort} was not linear by showing that it was not even additive.

We can make some more progress by bringing geometry into the game.

EXAMPLE 9 **(Linearity of rotations)** Consider the transformation f from $\mathbb{R}^2$ to $\mathbb{R}^2$ corresponding to rotation through the angle $\pi/2$ in a counterclockwise direction. That is, we identify a vector

$$\mathbf{x} = \begin{bmatrix} x \\ y \end{bmatrix}$$

with the point (x, y) in the plane, and then we rotate it to produce a new point (x', y') and finally put

$$f\left(\begin{bmatrix} x \\ y \end{bmatrix} \right) = \begin{bmatrix} x' \\ y' \end{bmatrix}.$$

The following diagram shows the three vectors

$$\mathbf{x} = \begin{bmatrix} -1/2 \\ 1/2 \end{bmatrix}, \quad \mathbf{y} = \begin{bmatrix} 3/2 \\ 1 \end{bmatrix}, \quad \text{and their sum,} \quad \mathbf{x} + \mathbf{y} = \begin{bmatrix} 1 \\ 3/2 \end{bmatrix},$$

together with the three vectors obtained by rotating all three counterclockwise through $\pi/2$ radians:

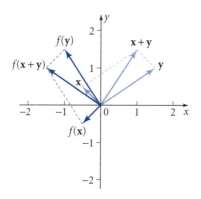

We have also drawn in the dashed lines completing the parallelograms used in the geometric construction of the vector sum. Now, since the whole parallelogram used in forming the vectors sum of $\mathbf{x}$ and $\mathbf{y}$ is rotated *as a unit*, you can see that

$$f(\mathbf{x} + \mathbf{y}) = f(\mathbf{x}) + f(\mathbf{y}).$$

That is, this transformation is additive. It is also geometrically clear, if you ponder a bit, that for any $\mathbf{x}$ in $\mathbb{R}^2$ and any number $a, f(a\mathbf{x}) = af(\mathbf{x})$. Hence this transformation is homogeneous. Since it is both additive and homogeneous, it is linear.

Notice two things: First, we proved that this transformation was linear just using its geometric definition; we did not use or need to use any algebraic formula for $f(\mathbf{x})$. We will soon use the knowledge that f is linear to write down an algebraic formula for it.

Second, it did not matter at all that the angle was $\pi/2$ or that the direction of rotation was counterclockwise. We just picked a specific angle and direction for the sake of drawing a picture. But think about it and you will see the following:

- *The linear transformation from $\mathbb{R}^2$ to $\mathbb{R}^2$ induced by rotation in the plane through any angle, clockwise or counterclockwise, is always linear.*

Here is another example of this type.

EXAMPLE 10   **(Linearity of reflections)** Consider the transformation from $\mathbb{R}^2$ to $\mathbb{R}^2$ given geometrically by reflection about the line $y = x$. Again, this reflection transforms the whole parallelogram representing the addition of $\mathbf{x}$ and $\mathbf{y}$ without any distortion so that reflection is clearly additive. Make sure you draw the picture; look back at the diagram in the previous example if you are not sure what to draw.

Also, the reflection of $a\mathbf{x}$ is the same as a times the reflection of $\mathbf{x}$, so reflection is homogeneous. Therefore, reflection about the line $y = x$ corresponds to a linear transformation of $\mathbb{R}^2$.

For the most part, in the rest of this book, we shall restrict ourselves to consideration of transformations from $\mathbb{R}^n$ to $\mathbb{R}^m$ with this very special property. You may be relieved that we are not being more ambitious. But then again, you may be concerned that we are giving up too much by focusing just on this special class of transformation. If so, please put your concern aside. First, there are a great many examples of linear transformation that come up everywhere in science and engineering. Second, through the method of "linearization," coming from differential calculus, the methods we study here for dealing with linear transformations can be applied in great generality to nonlinear transformations as well.

2.3 The matrix of a linear transformation

One really nice thing about linear transformations is that we can express a formula for any linear transformation f from $\mathbb{R}^n$ to $\mathbb{R}^m$ in terms of an $m \times n$ array of numbers, which is called the *matrix* of f.

It may seem strange to write down the formula for a function in terms of a rectangular array of numbers. But as we will see, this kind of formula is very convenient for computing the output $f(\mathbf{x})$ given the input $\mathbf{x}$, and that, after all, is what a formula for f should be good for.

The key to all of this is the nice way that linear transformations treat "linear combinations."

DEFINITION

(Linear combination) Given k vectors $\mathbf{x}_1, \mathbf{x}_2 \ldots, \mathbf{x}_k$ in $\mathbb{R}^n$, any vector of the form

$$a_1\mathbf{x}_1 + a_2\mathbf{x}_2 + \cdots + a_k\mathbf{x}_k$$

for any k numbers $a_1, a_2, \ldots, a_k$ is called a *linear combination* of $\mathbf{x}_1, \mathbf{x}_2 \ldots, \mathbf{x}_k$.*

---

*Notice the notation: $\mathbf{x}$ with the boldface $\mathbf{x}$ denotes the ith vector in a list of vectors, while x_i, with no boldface, is the ith entry of a vector $\mathbf{x}$. We would use $(\mathbf{x}_i)_j$ to denote the jth entry of the vector $\mathbf{x}_i$.

EXAMPLE 11 **(Linear combinations)** Let $x_1 = \begin{bmatrix} 1 \\ 0 \\ 2 \end{bmatrix}$ and $x_2 = \begin{bmatrix} 2 \\ 1 \\ 0 \end{bmatrix}$. Let $a_1 = 2$ and $a_2 = 1$. Then

$$a_1 x_1 + a_2 x_2 = 2 \begin{bmatrix} 1 \\ 0 \\ 2 \end{bmatrix} + \begin{bmatrix} 2 \\ 1 \\ 0 \end{bmatrix} = \begin{bmatrix} 4 \\ 1 \\ 4 \end{bmatrix}.$$

Hence the vector $\begin{bmatrix} 4 \\ 1 \\ 4 \end{bmatrix}$ is a linear combination of x_1 and x_2.

THEOREM 1 **(Before and after linear combinations)** *Consider any linear transformation f from $\mathbb{R}^n$ to $\mathbb{R}^m$. Then for any k vectors $x_1, x_2 \ldots, x_k$ in $\mathbb{R}^n$ and any k numbers $a_1, a_2, \ldots, a_k$,*

$$f(a_1 x_1 + a_2 x_2 + \cdots + a_k x_k) = a_1 f(x_1) + a_2 f(x_2) + \cdots + a_k f(x_k). \tag{2.4}$$

PROOF: When $k = 2$, (2.4) is just the defining property of a linear transformation. The general case is proved by induction. Suppose (2.4) has been established for any linear combination of $k - 1$ vectors. Then with w defined by

$$w = a_1 x_1 + a_2 x_2 + \cdots + a_{k-1} x_{k-1},$$

we have

$$a_1 x_1 + a_2 x_2 + \cdots + a_k x_k = w + a_k x_k.$$

By this proof, by (2.1), and by the inductive hypothesis,

$$f(a_1 x_1 + a_2 x_2 + \cdots + a_k x_k) = f(w + a_k x_k)$$
$$= f(w) + a_k f(x_k)$$
$$= (a_1 f(x_1) + a_2 f(x_2) + \cdots + a_{k-1} f(x_{k-1})) + a_k f(x_k).$$

But this equation is the same as (2.4). ∎

Theorem 1 enables us to represent any linear transformation from $\mathbb{R}^n$ to $\mathbb{R}^m$ by a *table*. Recall from the discussion of functions in Section 1 that if the domain of a function f is a finite set, say $\{1, 2, 3, \ldots, n\}$, one can specify the function in table form by listing the output values of the function in a "table" directly below the input values:

$$
\begin{array}{ccccc}
1 & 2 & 3 & \cdots & n \\
\downarrow & \downarrow & \downarrow & & \downarrow \\
f(1) & f(2) & f(3) & \cdots & f(n).
\end{array}
$$

Generally, when there are infinitely many input values, the table is infinitely long and ceases to be of use. However,

- *Despite the fact that there are infinitely many vectors in $\mathbb{R}^n$, we can still use a finite table to specify a linear transformation from $\mathbb{R}^n$ to $\mathbb{R}^m$.*

To explain this fundamental fact, we introduce the *standard basis vectors* in $\mathbb{R}^n$.

DEFINITION **(Standard basis vectors)** For any fixed positive integer n and any j with $j = 1, 2, 3, \ldots, n$, let e_j denote the vector in $\mathbb{R}^n$ that has 1 for its jth entry and 0 for all its other entries.

For example, the three standard basis vectors in $\mathbb{R}^3$ are

$$\mathbf{e}_1 = \begin{bmatrix} 1 \\ 0 \\ 0 \end{bmatrix}, \quad \mathbf{e}_2 = \begin{bmatrix} 0 \\ 1 \\ 0 \end{bmatrix}, \quad \text{and} \quad \mathbf{e}_3 = \begin{bmatrix} 0 \\ 0 \\ 1 \end{bmatrix}. \tag{2.5}$$

Here is the point of the definition: *Any* vector

$$\begin{bmatrix} x \\ y \\ z \end{bmatrix}$$

in $\mathbb{R}^3$ can be written as a linear combination of the standard basis vectors:

$$\begin{bmatrix} x \\ y \\ z \end{bmatrix} = x\mathbf{e}_1 + y\mathbf{e}_2 + z\mathbf{e}_3.$$

By Theorem 1, if f is a linear transformation from $\mathbb{R}^3$ to $\mathbb{R}^m$, any m, then

$$f\left(\begin{bmatrix} x \\ y \\ z \end{bmatrix}\right) = f(x\mathbf{e}_1 + y\mathbf{e}_2 + z\mathbf{e}_3) = xf(\mathbf{e}_1) + yf(\mathbf{e}_2) + zf(\mathbf{e}_3). \tag{2.6}$$

So if we have a table that lists the three output vectors $f(\mathbf{e}_1)$, $f(\mathbf{e}_2)$, and $f(\mathbf{e}_3)$, we can use it and (2.6) to compute

$$f\left(\begin{bmatrix} x \\ y \\ z \end{bmatrix}\right)$$

for arbitrary values of x, y, and z.

Suppose $\mathbf{v}_1$, $\mathbf{v}_2$, and $\mathbf{v}_3$ are three given vectors in $\mathbb{R}^m$ and

$$f(\mathbf{e}_1) = \mathbf{v}_1, \quad f(\mathbf{e}_2) = \mathbf{v}_2, \quad \text{and} \quad f(\mathbf{e}_3) = \mathbf{v}_3. \tag{2.7}$$

We can write (2.7) in table form as

$$\begin{array}{ccc} \mathbf{e}_1 & \mathbf{e}_2 & \mathbf{e}_3 \\ \downarrow & \downarrow & \downarrow \\ \mathbf{v}_1 & \mathbf{v}_2 & \mathbf{v}_3. \end{array} \tag{2.8}$$

Then the information provided in this table (2.8) suffices to compute $f(\mathbf{x})$ for *all the infinitely many* vectors $\mathbf{x}$ in $\mathbb{R}^3$ when f is linear, since, by (2.6) and (2.7),

$$f\left(\begin{bmatrix} x \\ y \\ z \end{bmatrix}\right) = x\mathbf{v}_1 + y\mathbf{v}_2 + z\mathbf{v}_3. \tag{2.9}$$

Actually, there are more symbols in (2.8) than we really need. We need only the list $[\mathbf{v}_1, \mathbf{v}_2, \mathbf{v}_3]$ if we remember that $\mathbf{v}_1$ comes from $\mathbf{e}_1$, $\mathbf{v}_2$ comes from $\mathbf{e}_2$, and so forth.

- *Thus the list of vectors $[\mathbf{v}_1, \mathbf{v}_2, \mathbf{v}_3]$ suffices to specify the linear transformation f.*

EXAMPLE 12   **(Computing $f(\mathbf{x})$ given $f(\mathbf{e}_1)$, $f(\mathbf{e}_2)$, and $f(\mathbf{e}_3)$)** Let

$$\mathbf{v}_1 = \begin{bmatrix} 1 \\ 2 \end{bmatrix}, \quad \mathbf{v}_2 = \begin{bmatrix} 2 \\ 3 \end{bmatrix}, \quad \text{and} \quad \mathbf{v}_3 = \begin{bmatrix} 1 \\ 1 \end{bmatrix},$$

and suppose f is a linear transformation from $\mathbb{R}^3$ to $\mathbb{R}^2$ such that $\mathbf{v}_1 = f(\mathbf{e}_1)$, $\mathbf{v}_2 = f(\mathbf{e}_2)$, and $\mathbf{v}_3 = f(\mathbf{e}_3)$, where $\mathbf{e}_1$, $\mathbf{e}_2$, and $\mathbf{e}_3$ are given by (2.5). We can now compute, for example,

$$f\left(\begin{bmatrix} 1 \\ 2 \\ 3 \end{bmatrix}\right)$$

using (2.9). Since

$$\begin{bmatrix} 1 \\ 2 \\ 3 \end{bmatrix} = 1\mathbf{e}_1 + 2\mathbf{e}_2 + 3\mathbf{e}_3,$$

the result is

$$f\left(\begin{bmatrix} 1 \\ 2 \\ 3 \end{bmatrix}\right) = \mathbf{v}_1 + 2\mathbf{v}_2 + 3\mathbf{v}_3 = \begin{bmatrix} 8 \\ 11 \end{bmatrix}.$$

Everything we needed to know to do the computation in Example 12 is given in the list

$$\left[\begin{bmatrix} 1 \\ 2 \end{bmatrix}, \begin{bmatrix} 2 \\ 3 \end{bmatrix}, \begin{bmatrix} 1 \\ 1 \end{bmatrix}\right]. \tag{2.10}$$

There are quite a few brackets in this expression, and they do not really tell us much. Let us suppress them and form the 2×3 rectangular array

$$\begin{bmatrix} 1 & 2 & 1 \\ 2 & 3 & 1 \end{bmatrix}. \tag{2.11}$$

(Whenever we refer to an $m \times n$ rectangular array, we *always* use the convention that m is the number of rows and n is the number of columns.) You can clearly recover (2.10) and hence f itself from (2.11). The jth column of this array is $\mathbf{v}_j$ for each $j = 1, 2, 3$, so it contains all you need to compute

$$f\left(\begin{bmatrix} x \\ y \\ z \end{bmatrix}\right),$$

through (2.9). We are at a crucial point; let us summarize:

- *The rectangular array of numbers in (2.11) is just an efficient way of recording the information required to compute $f(\mathbf{x})$ for any $\mathbf{x}$ in $\mathbb{R}^3$. Although we write it as a rectangular array of numbers, it is really a list of vectors $[\mathbf{v}_1, \mathbf{v}_2, \mathbf{v}_3]$ as in (2.10).*

DEFINITION

(Matrix) For positive integers m and n, an $m \times n$ *matrix* A is the $m \times n$ array of numbers $A_{i,j}$, $1 \le i \le m$, $1 \le j \le n$ with $A_{i,j}$ being the ith entry in the jth column. Given n vectors $\{\mathbf{v}_1, \mathbf{v}_2, \ldots, \mathbf{v}_n\}$ in $\mathbb{R}^m$, we let $[\mathbf{v}_1, \mathbf{v}_2, \ldots, \mathbf{v}_n]$ denote the $m \times n$ matrix whose jth column is $\mathbf{v}_j$ for each $j = 1, 2, \ldots, n$. We generally denote matrices by uppercase Roman letters.

DEFINITION

(Matrix corresponding to a linear transformation) If f is a linear transformation from $\mathbb{R}^n$ to $\mathbb{R}^m$, the matrix A_f *corresponding to* f is the $m \times n$ matrix

$$A_f = [f(\mathbf{e}_1), f(\mathbf{e}_2), \ldots, f(\mathbf{e}_n)]. \tag{2.12}$$

EXAMPLE 13   **(Planar rotation matrices)** Let f be the linear transformation of $\mathbb{R}^2$ induced by rotation through the angle $\pi/2$ radians in the counterclockwise direction. We saw in Example 9 that this is indeed a linear transformation. If we draw a picture representing $\mathbf{e}_1$ and $\mathbf{e}_2$ as arrows, and their rotations, we can see that

$$f(\mathbf{e}_1) = \mathbf{e}_2 \qquad \text{and} \qquad f(\mathbf{e}_2) = -\mathbf{e}_1.$$

Therefore, the matrix A_f corresponding to this linear transformation is

$$A_f = [f(\mathbf{e}_1), f(\mathbf{e}_2)] = [\mathbf{e}_2, -\mathbf{e}_1] = \begin{bmatrix} 0 & -1 \\ 1 & 0 \end{bmatrix}.$$

More generally, let f be the linear transformation of $\mathbb{R}^2$ induced by rotation through the angle θ radians in the counterclockwise direction. Then drawing the same sort of diagram (make sure you draw it!), we find

$$f(\mathbf{e}_1) = \begin{bmatrix} \cos(\theta) \\ \sin(\theta) \end{bmatrix} \quad \text{and} \quad f(\mathbf{e}_2) = \begin{bmatrix} -\sin(\theta) \\ \cos(\theta) \end{bmatrix}.$$

Therefore, the matrix A_f corresponding to this linear transformation is

$$A_f = [f(\mathbf{e}_1), f(\mathbf{e}_2)] = \begin{bmatrix} \cos(\theta) & -\sin(\theta) \\ \sin(\theta) & \cos(\theta) \end{bmatrix}. \tag{2.13}$$

The matrices in (2.13) are called 2×2 rotation matrices, and we will be seeing a lot of them.

We now define *matrix–vector* multiplication. We do so with a purpose in mind. The object is to set things up so that if f is a linear transformation from $\mathbb{R}^n$ to $\mathbb{R}^m$ and A_f is the corresponding matrix, then for every vector $\mathbf{x}$ in $\mathbb{R}^n$,

$$f(\mathbf{x}) = A_f \mathbf{x},$$

where the right-hand side is the matrix–vector product that we are about to define. Note that we indicate the product by writing the vector *just to the right of the matrix*, without any special symbol for the multiplication. Note also that the left-hand side is a vector in $\mathbb{R}^m$, so the product of an $m \times n$ matrix and a vector in $\mathbb{R}^n$ is another vector in $\mathbb{R}^m$. The definition is "cooked up" so that multiplication by A_f "does" the transformation f.

DEFINITION

(Matrix–vector multiplication) Let A be an $m \times n$ matrix whose jth column is the m-dimensional vector $\mathbf{v}_j$, and let

$$\mathbf{x} = \begin{bmatrix} x_1 \\ x_2 \\ \vdots \\ x_n \end{bmatrix}$$

be a vector in $\mathbb{R}^n$. Then the *matrix–vector product* of $A = [\mathbf{v}_1, \mathbf{v}_2, \ldots, \mathbf{v}_n]$ and $A\mathbf{x}$ is the vector in $\mathbb{R}^m$ given by

$$A\mathbf{x} = [\mathbf{v}_1, \mathbf{v}_2, \ldots, \mathbf{v}_n] \begin{bmatrix} x_1 \\ x_2 \\ \vdots \\ x_n \end{bmatrix} = x_1 \mathbf{v}_1 + x_2 \mathbf{v}_2 + \cdots + x_n \mathbf{v}_n. \tag{2.14}$$

EXAMPLE 14 **(Computing matrix–vector products)** Consider the matrix

$$\begin{bmatrix} 1 & 2 & 1 \\ 2 & 3 & 1 \end{bmatrix}$$

and the vector

$$\begin{bmatrix} 1 \\ 2 \\ 3 \end{bmatrix}.$$

Then

$$\begin{bmatrix} 1 & 2 & 1 \\ 2 & 3 & 1 \end{bmatrix} \begin{bmatrix} 1 \\ 2 \\ 3 \end{bmatrix} = 1 \begin{bmatrix} 1 \\ 2 \end{bmatrix} + 2 \begin{bmatrix} 2 \\ 3 \end{bmatrix} + 3 \begin{bmatrix} 1 \\ 1 \end{bmatrix} = \begin{bmatrix} 1 \\ 2 \end{bmatrix} + \begin{bmatrix} 4 \\ 6 \end{bmatrix} + \begin{bmatrix} 3 \\ 3 \end{bmatrix} = \begin{bmatrix} 8 \\ 11 \end{bmatrix}.$$

Likewise,

$$\begin{bmatrix} 1 & 2 \\ 2 & -3 \\ 1 & -2 \end{bmatrix} \begin{bmatrix} 3 \\ 2 \end{bmatrix} = 3 \begin{bmatrix} 1 \\ 2 \\ 1 \end{bmatrix} + 2 \begin{bmatrix} 2 \\ -3 \\ -2 \end{bmatrix} = \begin{bmatrix} 3 \\ 6 \\ 3 \end{bmatrix} + \begin{bmatrix} 4 \\ -6 \\ -4 \end{bmatrix} = \begin{bmatrix} 7 \\ 0 \\ -1 \end{bmatrix}.$$

However,

$$\begin{bmatrix} 1 & 2 \\ 2 & -3 \\ 1 & -2 \end{bmatrix} \begin{bmatrix} 1 \\ 2 \\ 3 \end{bmatrix}$$

is simply not defined. There is no definition for the product of a 3×2 matrix and a 3-dimensional vector. (One *could* make one up. But *useful* mathematical definitions are made for a reason. The purpose behind the definition we made was to get our hands on a convenient way to write down and work with linear transformations. The next theorem says we got it right.)

THEOREM 2   **(Correspondence between matrices and linear transformations)** *For any linear transformation f from $\mathbb{R}^n$ to $\mathbb{R}^m$, let $A_f = [f(\mathbf{e}_1), f(\mathbf{e}_2), \dots, f(\mathbf{e}_n)]$ be the corresponding $m \times n$ matrix. Then for all $\mathbf{x}$ in $\mathbb{R}^n$,*

$$f(\mathbf{x}) = A_f \mathbf{x}. \tag{2.15}$$

Conversely, given an $m \times n$ matrix A, define a transformation f_A from $\mathbb{R}^n$ to $\mathbb{R}^m$ by

$$f_A(\mathbf{x}) = A\mathbf{x}. \tag{2.16}$$

Then this transformation is linear, and A itself is the matrix corresponding to f_A. In particular, the jth column of A is $A\mathbf{e}_j$.

*   That is, we have a one-to-one correspondence between matrices and linear transformations, and the matrix–vector product gives the "action" of the transformation.*

PROOF: First, by definition,

$$f\left(\begin{bmatrix} x_1 \\ x_2 \\ \vdots \\ x_n \end{bmatrix} \right) = x_1 f(\mathbf{e}_1) + x_2 f(\mathbf{e}_2) + \cdots + x_n f(\mathbf{e}_n).$$

Also by definition, $A_f = [f(\mathbf{e}_1), f(\mathbf{e}_2), \ldots, f(\mathbf{e}_n)]$. Therefore, from (2.14),

$$
f\left(\begin{bmatrix} x_1 \\ x_2 \\ \vdots \\ x_n \end{bmatrix}\right) = [f(\mathbf{e}_1), f(\mathbf{e}_2), \ldots, f(\mathbf{e}_n)] \begin{bmatrix} x_1 \\ x_2 \\ \vdots \\ x_n \end{bmatrix},
$$

which proves (2.15).

For the second part, we have to show that (2.16) defines a linear transformation. Let us first show that it is additive. Let $\mathbf{x}$ and $\mathbf{y}$ be any two vectors in $\mathbb{R}^n$. Then writing $A = [\mathbf{v}_1, \mathbf{v}_2, \ldots, \mathbf{v}_n]$, from (2.14),

$$
f_A(\mathbf{x} + \mathbf{y}) = A(\mathbf{x} + \mathbf{y}) = [\mathbf{v}_1, \mathbf{v}_2, \ldots, \mathbf{v}_n] \begin{bmatrix} x_1 + y_1 \\ x_2 + y_2 \\ \vdots \\ x_n + y_n \end{bmatrix}
$$

$$
= (x_1 + y_1)\mathbf{v}_1 + (x_2 + y_2)\mathbf{v}_2 + \cdots + (x_n + y_n)\mathbf{v}_n
$$

$$
= (x_1\mathbf{v}_1 + x_2\mathbf{v}_2 + \cdots + x_n\mathbf{v}_n) + (y_1\mathbf{v}_1 + y_x\mathbf{v}_2 + \cdots + y_n\mathbf{v}_n)
$$

$$
= [\mathbf{v}_1, \mathbf{v}_2, \ldots, \mathbf{v}_n] \begin{bmatrix} x_1 \\ x_2 \\ \vdots \\ x_n \end{bmatrix} + [\mathbf{v}_1, \mathbf{v}_2, \ldots, \mathbf{v}_n] \begin{bmatrix} y_1 \\ y_2 \\ \vdots \\ y_n \end{bmatrix}
$$

$$
= A\mathbf{x} + A\mathbf{y} = f_A(\mathbf{x}) + f_A(\mathbf{y}).
$$

Hence f_A is additive. A simpler computation of the same type shows that f is also homogeneous; that is, for any a and any $\mathbf{x}$, $f_A(a\mathbf{x}) = af_A(\mathbf{x})$. Hence, f_A is linear.

Finally, note that by (2.16), $f_A(\mathbf{e}_j) = A\mathbf{e}_j$, so the matrix corresponding to f_A is, by definition, $[A\mathbf{e}_1, A\mathbf{e}_2, \ldots, A\mathbf{e}_n]$. But with $A = [\mathbf{v}_1, \mathbf{v}_2, \ldots, \mathbf{v}_n]$,

$$
A\mathbf{e}_j = [\mathbf{v}_1, \mathbf{v}_2, \ldots, \mathbf{v}_n] \begin{bmatrix} 0 \\ \vdots \\ 1 \\ \vdots \\ 0 \end{bmatrix} = \mathbf{v}_j,
$$

where the 1 is in the jth place by the definition of $\mathbf{e}_j$. Hence, $\mathbf{v}_j$, the jth column of A, is $A\mathbf{e}_j$, and $[A\mathbf{e}_1, A\mathbf{e}_2, \ldots, A\mathbf{e}_n] = A$. ∎

EXAMPLE 15   **(Rotating a vector)** Let $\mathbf{x}$ be the vector

$$
\mathbf{x} = \begin{bmatrix} 2 \\ 6 \end{bmatrix}.
$$

Let $\tilde{\mathbf{x}}$ be the vector obtained from $\mathbf{x}$ by rotating it counterclockwise through an angle $\pi/3$. What is $\tilde{\mathbf{x}}$?

Let f be the linear transformation of $\mathbb{R}^2$ given by rotation in a counterclockwise direction through an angle $\pi/3$ so that $\tilde{\mathbf{x}} = f(\mathbf{x})$. We know from Example 12 that the corresponding matrix is

$$
A_f = \begin{bmatrix} \cos(\pi/3) & -\sin(\pi/3) \\ \sin(\pi/3) & \cos(\pi/3) \end{bmatrix} = \frac{1}{2} \begin{bmatrix} 1 & -\sqrt{3} \\ \sqrt{3} & 1 \end{bmatrix}.
$$

Therefore, by Theorem 2,

$$\tilde{\mathbf{x}} = f(\mathbf{x}) = \frac{1}{2} \begin{bmatrix} 1 & -\sqrt{3} \\ \sqrt{3} & 1 \end{bmatrix} \begin{bmatrix} 2 \\ 6 \end{bmatrix} = \begin{bmatrix} 1 - 3\sqrt{3} \\ \sqrt{3} + 3 \end{bmatrix}.$$

There is another way to express the matrix–vector product of A and $\mathbf{x}$ simply in terms of the numerical entries of A and $\mathbf{x}$:

THEOREM 3   **(Entry form of matrix–vector multiplication)** *Let A be an $m \times n$ matrix, and let $\mathbf{x}$ be a vector in $\mathbb{R}^n$. Then the ith entry of the vector $A\mathbf{x}$ is*

$$\sum_{j=1}^{n} A_{i,j} x_j. \tag{2.17}$$

PROOF: By (2.14), the ith entry of $A\mathbf{x}$ is the ith entry of $x_1\mathbf{v}_1 + x_2\mathbf{v}_2 + \cdots + x_n\mathbf{v}_n = x_1(\mathbf{v}_1)_i + x_2(\mathbf{v}_2)_i + \cdots + x_n(\mathbf{v}_n)_i$. By definition $(\mathbf{v}_j)_i = A_{i,j}$, so $x_1(\mathbf{v}_1)_i + x_2(\mathbf{v}_2)_i + \cdots + x_n(\mathbf{v}_n)_i = \sum_{j=1}^{n} A_{i,j} x_j$. ∎

EXAMPLE 16   **(Computation of matrix–vector products entry by entry)** Consider the first product from Example 14, namely, the matrix

$$A = \begin{bmatrix} 1 & 2 & 1 \\ 2 & 3 & 1 \end{bmatrix}$$

and the vector

$$\mathbf{x} = \begin{bmatrix} 1 \\ 2 \\ 3 \end{bmatrix}.$$

The first entry of the product is

$$A_{1,1}x_1 + A_{1,2}x_2 + A_{1,3}x_3 = 1 + 4 + 3 = 8.$$

The second entry of the product is

$$A_{2,1}x_1 + A_{2,2}x_2 + A_{2,3}x_3 = 2 + 6 + 3 = 11.$$

Therefore, the product is $\begin{bmatrix} 8 \\ 11 \end{bmatrix}$, as we found before.

There is at least one more useful way to write the matrix–vector product using the "dot product of two vectors." (Recall that a matrix is just a list of vectors.) We will discuss that soon. In the meantime, try out Theorem 3 on the other matrix–vector products in Example 14. This way you can calculate the entries of the products one at a time instead of "in parallel." This method can be easier, and you may want to know only some of the entries of the product anyway. Most important of all, though, is to try computing a number of examples both ways.

Exercises

2.1 Let $\mathbf{a} = \begin{bmatrix} 1 \\ -1 \end{bmatrix}$ and let $\mathbf{b} = \begin{bmatrix} 2 \\ 1 \end{bmatrix}$.

(a) Compute $3\mathbf{a}$, $\mathbf{a} + \mathbf{b}$, and $\mathbf{a} - \mathbf{b}$.

(b) Draw a diagram in the plane showing $\mathbf{a}$ and $3\mathbf{a}$ as arrows, as in the examples of this section.

(c) Draw a diagram in the plane showing **a**, **b**, and **a** + **b** as arrows, as in the examples of this section.

(d) Draw a diagram in the plane showing **a**, **b**, and **a** − **b** as arrows, as in the examples of this section.

2.2 Let **a** = $\begin{bmatrix} 1 \\ 1 \end{bmatrix}$ and let **b** = $\begin{bmatrix} -2 \\ 3 \end{bmatrix}$.

(a) Compute 3**a**, **a** + **b**, and **a** − **b**.

(b) Draw a diagram in the plane showing **a** and −2**a** as arrows, as in the examples of this section.

(c) Draw a diagram in the plane showing **a**, **b**, and **a** + **b** as arrows, as in the examples of this section.

(d) Draw a diagram in the plane showing **a**, **b**, and **a** − **b** as arrows, as in the examples of this section.

2.3 Consider the following transformations from $\mathbb{R}^2$ to $\mathbb{R}^2$. Which ones are linear? Explain your answers, and for those that are linear, write down the corresponding matrix.

$$f\left(\begin{bmatrix} x \\ y \end{bmatrix}\right) = \begin{bmatrix} y + x \\ y - x \end{bmatrix} \qquad g\left(\begin{bmatrix} x \\ y \end{bmatrix}\right) = \begin{bmatrix} |x| \\ y - x \end{bmatrix} \qquad h\left(\begin{bmatrix} x \\ y \end{bmatrix}\right) = \begin{bmatrix} x^2 - y^2 \\ x^2 + y^2 \end{bmatrix}$$

2.4 Consider the following transformations from $\mathbb{R}^3$ to $\mathbb{R}^2$. Which ones are linear? Explain your answers, and for those that are linear, write down the corresponding matrix.

$$f\left(\begin{bmatrix} x \\ y \\ z \end{bmatrix}\right) = \begin{bmatrix} xz \\ xy \end{bmatrix} \qquad g\left(\begin{bmatrix} x \\ y \\ z \end{bmatrix}\right) = \begin{bmatrix} z \\ y \end{bmatrix} \qquad h\left(\begin{bmatrix} x \\ y \\ z \end{bmatrix}\right) = \begin{bmatrix} 1 \\ x \end{bmatrix}$$

2.5 Consider the following transformations from $\mathbb{R}^2$ to $\mathbb{R}^3$. Which ones are linear? Explain your answers, and for those that are linear, write down the corresponding matrix.

$$f\left(\begin{bmatrix} x \\ y \end{bmatrix}\right) = \begin{bmatrix} y \\ 1 \\ x \end{bmatrix} \qquad g\left(\begin{bmatrix} x \\ y \end{bmatrix}\right) = \begin{bmatrix} y \\ 0 \\ x \end{bmatrix} \qquad h\left(\begin{bmatrix} x \\ y \end{bmatrix}\right) = \begin{bmatrix} 0 \\ 0 \\ 0 \end{bmatrix}$$

2.6 Compute the single vector specified by the following linear combinations:

(a) $2\begin{bmatrix} 1 \\ 2 \end{bmatrix} - 3\begin{bmatrix} 2 \\ 1 \end{bmatrix}$ **(b)** $2\begin{bmatrix} 1 \\ 2 \end{bmatrix} - 3\begin{bmatrix} 2 \\ 1 \end{bmatrix} + \frac{1}{2}\begin{bmatrix} -3 \\ 1 \end{bmatrix} + \begin{bmatrix} 0 \\ 1 \end{bmatrix}$

(c) $2\begin{bmatrix} 1 \\ 2 \\ -1 \end{bmatrix} + 3\begin{bmatrix} 2 \\ -1 \\ -1 \end{bmatrix}$ **(d)** $\begin{bmatrix} 5 \\ 0 \\ -1 \end{bmatrix} - \begin{bmatrix} 3 \\ 3 \\ -3 \end{bmatrix}$

2.7 Consider the following list of 4 matrices and 4 vectors. There are 16 different pairs of matrices and vectors. Say for which pairs the matrix–vector product is defined and for which pairs it is not defined, and compute it when it is.

$$A = \begin{bmatrix} 1 & 1 & 1 & 2 \\ 1 & 2 & 0 & -1 \end{bmatrix} \qquad B = \begin{bmatrix} 1 & -1 & 1 \\ 2 & 1 & 0 \\ 4 & -1 & 2 \end{bmatrix} \qquad C = \begin{bmatrix} 2 & 1 \\ 0 & 2 \\ 1 & -2 \\ 5 & 1 \end{bmatrix} \qquad D = \begin{bmatrix} 0 & 1 \\ 1 & 0 \end{bmatrix}$$

$$\mathbf{v} = \begin{bmatrix} 1 \\ 1 \end{bmatrix} \qquad \mathbf{x} = \begin{bmatrix} 1 \\ -1 \end{bmatrix} \qquad \mathbf{y} = \begin{bmatrix} 2 \\ -1 \\ 2 \end{bmatrix} \qquad \mathbf{z} = \begin{bmatrix} -1 \\ -1 \\ 3 \\ 2 \end{bmatrix}.$$

2.8 Consider the following list of 4 matrices and 4 vectors. There are 16 different pairs of matrices and vectors. Say for which pairs the matrix–vector product is defined and for which pairs it is not defined, and compute it when it is.

$$A = \begin{bmatrix} 0 & 2 & 1 & 3 \\ 1 & 2 & 0 & -1 \\ 3 & -1 & 1 & 4 \end{bmatrix} \quad B = \begin{bmatrix} 0 & -2 & 1 \\ 2 & 1 & 0 \\ 3 & -1 & 2 \end{bmatrix} \quad C = \begin{bmatrix} 2 & 1 & 1 & 1 \\ 2 & 1 & 0 & -1 \\ 0 & -3 & 0 & 1 \\ 1 & 1 & 2 & -1 \end{bmatrix} \quad D = \begin{bmatrix} 1 & 1 \\ 2 & 0 \end{bmatrix}$$

$$\mathbf{v} = \begin{bmatrix} 1 \\ 2 \end{bmatrix} \quad \mathbf{x} = \begin{bmatrix} 2 \\ -1 \end{bmatrix} \quad \mathbf{y} = \begin{bmatrix} 1 \\ -3 \\ 1 \end{bmatrix} \quad \mathbf{z} = \begin{bmatrix} 2 \\ 1 \\ 3 \\ 4 \end{bmatrix}.$$

2.9 Let $A = \begin{bmatrix} 1 & 1 & 1 & 2 \\ 1 & 2 & 0 & -1 \\ 2 & 0 & 0 & -1 \\ 3 & -2 & 0 & 2 \end{bmatrix}$ and $\mathbf{x} = \begin{bmatrix} -1 \\ -2 \\ 2 \\ 1 \end{bmatrix}$. Compute the third entry of $A\mathbf{x}$ without computing the whole vector $A\mathbf{x}$.

2.10 Let f be a linear transformation from $\mathbb{R}^2$ to $\mathbb{R}^2$. Suppose that

$$f(\mathbf{e}_1) = \begin{bmatrix} -1 \\ 2 \end{bmatrix} \quad \text{and} \quad f(\mathbf{e}_2) = \begin{bmatrix} -2 \\ -1 \end{bmatrix}.$$

Find the matrix A_f corresponding to f.

2.11 Let f be a linear transformation from $\mathbb{R}^2$ to $\mathbb{R}^2$. Suppose that

$$f(\mathbf{e}_1) = \begin{bmatrix} 1 \\ 3 \end{bmatrix} \quad \text{and} \quad f(\mathbf{e}_2) = \begin{bmatrix} -3 \\ 1 \end{bmatrix}.$$

Find the matrix A_f corresponding to f.

2.12 Let f be the linear transformation from $\mathbb{R}^2$ to $\mathbb{R}^2$ given by rotation in the counterclockwise direction through an angle of $\pi/3$ radians. Let g be the linear transformation from $\mathbb{R}^2$ to $\mathbb{R}^2$ given by reflection about the line $y = x$. What is the matrix $A_{g \circ f}$ corresponding to the composition of g with f? What is the matrix $A_{f \circ g}$ corresponding to the composition of f with g?

2.13 Let f be the linear transformation from $\mathbb{R}^2$ to $\mathbb{R}^2$ given by first reflecting about the line $y = x$ and then reflecting about the line $x = 0$. What is the matrix A corresponding to this linear transformation?

2.14 Let f be the linear transformation from $\mathbb{R}^2$ to $\mathbb{R}^2$ given by first reflecting about the line $y = x$ and then reflecting about the line $y = -x$. What is the matrix A corresponding to this linear transformation?

2.15 Let f be the linear transformation from $\mathbb{R}^2$ to $\mathbb{R}^2$ given by reflection about the line through the origin with slope s. Let g be the linear transformation from $\mathbb{R}^2$ to $\mathbb{R}^2$ given by reflection about the x-axis. What is the matrix $A_{g \circ f}$ corresponding to the composition of g with f? (Your answer should be a matrix with entries depending on s.) How does your answer change if instead g reflects about the y-axis?

2.16 Let A and B be $m \times n$ matrices. If $A\mathbf{x} = B\mathbf{x}$ for all $\mathbf{x}$ in $\mathbb{R}^n$, does this mean that $A = B$? Explain why, or give a counterexample.

2.17 Consider the function from $\mathbb{R}^2$ to $\mathbb{R}^2$ defined by

$$f\left(\begin{bmatrix} 0 \\ 0 \end{bmatrix} \right) = \begin{bmatrix} 0 \\ 0 \end{bmatrix},$$

and for $\begin{bmatrix} x \\ y \end{bmatrix} \neq \begin{bmatrix} 0 \\ 0 \end{bmatrix}$, $f\left(\begin{bmatrix} x \\ y \end{bmatrix} \right) = \begin{bmatrix} x^3/(x^2 + y^2) \\ -y^3/(x^2 + y^2) \end{bmatrix}.$

(a) Show that for all **x** in $\mathbb{R}^2$ and all numbers a, $f(a\mathbf{x}) = af(\mathbf{x})$.

(b) Show that f is not linear.

2.18 Let f be any function from $\mathbb{R}$ to $\mathbb{R}$ such that for all numbers a and x, $f(ax) = af(x)$. Show that f is linear. Find a formula for f in case $f(1) = 3$.

2.19 Determine whether there is a linear transformation from $\mathbb{R}^2$ to $\mathbb{R}^2$ such that

$$f\left(\begin{bmatrix}1\\0\end{bmatrix}\right) = \begin{bmatrix}2\\3\end{bmatrix}, \quad f\left(\begin{bmatrix}0\\1\end{bmatrix}\right) = \begin{bmatrix}3\\2\end{bmatrix}, \quad \text{and} \quad f\left(\begin{bmatrix}1\\1\end{bmatrix}\right) = \begin{bmatrix}5\\5\end{bmatrix}.$$

If so, find the matrix of such a transformation. If not, explain why not.

2.20 Determine whether there is a linear transformation from $\mathbb{R}^2$ to $\mathbb{R}^2$ such that

$$f\left(\begin{bmatrix}1\\0\end{bmatrix}\right) = \begin{bmatrix}3\\3\end{bmatrix}, \quad f\left(\begin{bmatrix}1\\2\end{bmatrix}\right) = \begin{bmatrix}3\\1\end{bmatrix}, \quad \text{and} \quad f\left(\begin{bmatrix}1\\1\end{bmatrix}\right) = \begin{bmatrix}3\\2\end{bmatrix}.$$

If so, find the matrix of such a transformation. If not, explain why not.

2.21 Determine whether there is a linear transformation from $\mathbb{R}^2$ to $\mathbb{R}^2$ such that

$$f\left(\begin{bmatrix}1\\0\end{bmatrix}\right) = \begin{bmatrix}3\\3\end{bmatrix}, \quad f\left(\begin{bmatrix}1\\2\end{bmatrix}\right) = \begin{bmatrix}3\\1\end{bmatrix}, \quad \text{and} \quad f\left(\begin{bmatrix}1\\1\end{bmatrix}\right) = \begin{bmatrix}1\\1\end{bmatrix}.$$

If so, find the matrix of such a transformation. If not, explain why not.

2.22 Let f be a linear transformation from $\mathbb{R}^2$ to $\mathbb{R}^2$ such that

$$f\left(\begin{bmatrix}1\\0\end{bmatrix}\right) = \begin{bmatrix}a\\3\end{bmatrix} \quad \text{and} \quad f\left(\begin{bmatrix}0\\1\end{bmatrix}\right) = \begin{bmatrix}1\\1\end{bmatrix}.$$

Find all values of a, if any, for which

$$f\left(\begin{bmatrix}1\\2\end{bmatrix}\right) = \begin{bmatrix}5\\5\end{bmatrix}.$$

2.23 Let f be a linear transformation from $\mathbb{R}^2$ to $\mathbb{R}^2$ such that

$$f\left(\begin{bmatrix}1\\1\end{bmatrix}\right) = \begin{bmatrix}a\\3\end{bmatrix} \quad \text{and} \quad f\left(\begin{bmatrix}2\\1\end{bmatrix}\right) = \begin{bmatrix}1\\2a\end{bmatrix}.$$

Find all values of a, if any, for which $f\left(\begin{bmatrix}1\\2\end{bmatrix}\right) = \begin{bmatrix}5\\5\end{bmatrix}$.

2.24 Find all values of a, if any, for which there is a linear transformation from $\mathbb{R}^2$ to $\mathbb{R}^2$ such that

$$f\left(\begin{bmatrix}1\\0\end{bmatrix}\right) = \begin{bmatrix}3\\3\end{bmatrix}, \quad f\left(\begin{bmatrix}1\\2\end{bmatrix}\right) = \begin{bmatrix}3\\1\end{bmatrix}, \quad \text{and} \quad f\left(\begin{bmatrix}2\\1\end{bmatrix}\right) = \begin{bmatrix}a\\2\end{bmatrix}.$$

For each such a, find the corresponding matrix.

2.25 Find all values of a, if any, for which there is a linear transformation from $\mathbb{R}^2$ to $\mathbb{R}^2$ such that

$$f\left(\begin{bmatrix}0\\1\end{bmatrix}\right) = \begin{bmatrix}3\\3\end{bmatrix}, \quad f\left(\begin{bmatrix}2\\0\end{bmatrix}\right) = \begin{bmatrix}3\\2\end{bmatrix}, \quad \text{and} \quad f\left(\begin{bmatrix}1\\1\end{bmatrix}\right) = \begin{bmatrix}a\\3\end{bmatrix}.$$

For each such a, find the corresponding matrix.

2.26 Suppose that f is a linear transformation from $\mathbb{R}^n$ to $\mathbb{R}^n$, that g is a transformation from $\mathbb{R}^n$ to $\mathbb{R}^n$, but that g is not linear. Give an example in which $g \circ f$ is linear, and show that $g \circ f$ cannot be linear if f is invertible.

2.27 Suppose that f is a transformation from $\mathbb{R}^n$ to $\mathbb{R}^n$, that g is a transformation from $\mathbb{R}^n$ to $\mathbb{R}^n$, and that neither f nor g is linear but both are invertible. Does it follow that $g \circ f$ is not linear, or can it be that $g \circ f$ is a linear transformation from $\mathbb{R}^n$ to $\mathbb{R}^n$? Explain your answer.

Transformations of polynomials

The next exercises concern polynomials. As explained just before the final group of exercises in Section 1, polynomials can be specified by *listing* their coefficients, which allows us to think of polynomials as lists, that is, as vectors.

By means of this correspondence between polynomials and vectors, any transformation from polynomials of degree k to polynomials of degree ℓ induces a transformation from $\mathbb{R}^{k+1}$ to $\mathbb{R}^{\ell+1}$.

2.28 Define a transformation f from $\mathbb{R}^3$ to $\mathbb{R}^2$ as follows. Define

$$f\left(\begin{bmatrix} a \\ b \\ c \end{bmatrix}\right) = \begin{bmatrix} u \\ v \end{bmatrix},$$

where

$$u + vx = \frac{d}{dx}(a + bx + cx^2).$$

Is f linear? If so, find the corresponding matrix. If not, explain why not.

2.29 Define a transformation f from $\mathbb{R}^3$ to $\mathbb{R}^5$ as follows. Define

$$f\left(\begin{bmatrix} a \\ b \\ c \end{bmatrix}\right) = \begin{bmatrix} r \\ s \\ t \\ u \\ v \end{bmatrix},$$

where

$$r + sx + tx^2 + ux^3 + vx^4 = (a + bx + cx^2)^2.$$

Is f linear? If so, find the corresponding matrix. If not, explain why not.

2.30 Define a transformation f from $\mathbb{R}^3$ to $\mathbb{R}^2$ as follows. Define

$$f\left(\begin{bmatrix} a \\ b \\ c \end{bmatrix}\right) = \begin{bmatrix} u \\ v \\ t \end{bmatrix},$$

where

$$u + vx + tx^2 = \frac{d}{dx}(x(a + bx + cx^2)).$$

(The polynomial transformation is multiply by x, then differentiate.)

Is f linear? If so, find the corresponding matrix. If not, explain why not.

2.31 Define a transformation f from $\mathbb{R}^3$ to $\mathbb{R}^2$ as follows. Define

$$f\left(\begin{bmatrix} a \\ b \\ c \end{bmatrix}\right) = \begin{bmatrix} u \\ v \\ t \end{bmatrix},$$

where

$$u + vx + tx^2 = \frac{d^2}{dx^2}(x^2(a + bx + cx^2)).$$

(The polynomial transformation is multiply by x^2, then differentiate twice.)

Is f linear? If so, find the corresponding matrix. If not, explain why not.

2.32 Define a transformation f from $\mathbb{R}^3$ to $\mathbb{R}^4$ as follows. Define

$$f\left(\begin{bmatrix} a \\ b \\ c \end{bmatrix}\right) = \begin{bmatrix} s \\ t \\ u \\ v \end{bmatrix},$$

where $p(x) = a + bx + cx^2$ and

$$s = \int_0^1 p(x)\,dx \qquad t = \int_0^2 p(x)\,dx \qquad u = \int_0^3 p(x)\,dx \qquad v = \int_0^4 p(x)\,dx.$$

Is f linear? If so, find the corresponding matrix. If not, explain why not.

2.33 Define a transformation f from $\mathbb{R}^3$ to $\mathbb{R}^4$ as follows. Define

$$f\left(\begin{bmatrix} a \\ b \\ c \end{bmatrix}\right) = \begin{bmatrix} s \\ t \\ u \\ v \end{bmatrix},$$

where $p(x) = a + bx + cx^2$ and

$$s = p(1) \qquad t = p(2) \qquad u = p(3) \qquad v = p(4).$$

Is f linear? If so, find the corresponding matrix. If not, explain why not.

SECTION 3 | The Matrix Product

3.1 Matrix products and composition of linear transformations

Consider one linear transformation f from $\mathbb{R}^n$ to $\mathbb{R}^m$ and another linear transformation g from $\mathbb{R}^m$ to $\mathbb{R}^p$. Since the output of f is a vector in $\mathbb{R}^m$, which is what g takes as input, we can form the composition

$$h = g \circ f.$$

That is, for all $\mathbf{x}$ in $\mathbb{R}^n$, $h(\mathbf{x}) = g(f(\mathbf{x}))$. Now for any two vectors $\mathbf{x}_1$, $\mathbf{x}_2$ in $\mathbb{R}^n$ and any two numbers a_1, a_2,

$$
\begin{aligned}
h(a_1\mathbf{x}_1 + a_2\mathbf{x}_2) &= g(f(a_1\mathbf{x}_1 + a_2\mathbf{x}_2)) \\
&= g(a_1 f(\mathbf{x}_1) + a_2 f(\mathbf{x}_2)) \\
&= a_1 g(f(\mathbf{x}_1)) + a_2 g(f(\mathbf{x}_2)) \\
&= a_1 h(\mathbf{x}_1) + a_2 h(\mathbf{x}_2).
\end{aligned}
$$

The four equalities here are the definition of h, the linearity of f, the linearity of g, and then the definition of h again. This simple analysis yields the following fact.

THEOREM 4   **(The composition of linear transformations is linear)** *Suppose that f is a linear transformation from $\mathbb{R}^n$ to $\mathbb{R}^m$ and g is a linear transformation from $\mathbb{R}^m$ to $\mathbb{R}^p$. Then $g \circ f$ is a linear transformation from $\mathbb{R}^n$ to $\mathbb{R}^p$.*

There is a one-to-one correspondence between matrices and linear transformations. With f, g, and $g \circ f$ as in Theorem 4, let A_f, A_g, and $A_{g \circ f}$ be the corresponding matrices. What is the relation between these matrices? We define *matrix–matrix* multiplication so that

$$
A_{g \circ f} = A_g A_f.
$$

That is, the matrix product is defined so that it "does" composition of linear transformations. Note first of all that A_f is an $m \times n$ matrix, A_g is a $p \times m$ matrix, and $A_{g \circ f}$ is a $p \times n$ matrix.

DEFINITION   **(Matrix–matrix multiplication)** Let A be an $m \times n$ matrix and B be a $p \times m$ matrix. Let $\{\mathbf{v}_1, \mathbf{v}_2, \ldots, \mathbf{v}_n\}$ be the columns of A so that $A = [\mathbf{v}_1, \mathbf{v}_2, \ldots, \mathbf{v}_n]$. Then BA is the $p \times n$ matrix given by

$$
BA = [B\mathbf{v}_1, B\mathbf{v}_2, \ldots, B\mathbf{v}_n]. \tag{3.1}
$$

Notice that we are using matrix–vector multiplication to define matrix–matrix multiplication.

- *Matrix–matrix multiplication is just matrix–vector multiplication done in parallel.*

EXAMPLE 17   **(Computing the product of two matrices)** Let us compute the product BA where

$$
B = \begin{bmatrix} 1 & 2 \\ -2 & 1 \end{bmatrix} \quad \text{and} \quad A = \begin{bmatrix} 0 & -1 \\ 1 & 2 \end{bmatrix}.
$$

Since $A = [\mathbf{v}_1, \mathbf{v}_2]$, where $\mathbf{v}_1 = \begin{bmatrix} 0 \\ 1 \end{bmatrix}$ and $\mathbf{v}_2 = \begin{bmatrix} -1 \\ 2 \end{bmatrix}$, we just have to compute $B\mathbf{v}_1$ and $B\mathbf{v}_2$, since $BA = [B\mathbf{v}_1, B\mathbf{v}_2]$:

$$
B\mathbf{v}_1 = \begin{bmatrix} 1 & 2 \\ -2 & 1 \end{bmatrix}\begin{bmatrix} 0 \\ 1 \end{bmatrix} = 0\begin{bmatrix} 1 \\ -2 \end{bmatrix} + 1\begin{bmatrix} 2 \\ 1 \end{bmatrix} = \begin{bmatrix} 2 \\ 1 \end{bmatrix} \tag{3.2}
$$

and

$$
B\mathbf{v}_2 = \begin{bmatrix} 1 & 2 \\ -2 & 1 \end{bmatrix}\begin{bmatrix} -1 \\ 2 \end{bmatrix} = -1\begin{bmatrix} 1 \\ -2 \end{bmatrix} + 2\begin{bmatrix} 2 \\ 1 \end{bmatrix} = \begin{bmatrix} 3 \\ 4 \end{bmatrix}. \tag{3.3}
$$

Now put it all together:

$$
BA = [B\mathbf{v}_1, B\mathbf{v}_2] = \begin{bmatrix} 2 & 3 \\ 1 & 4 \end{bmatrix}.
$$

Notice that we can compute $B\mathbf{v}_1$ and $B\mathbf{v}_2$ "in parallel." It is helpful that the computations of the columns of the product are independent of one another. *There are a lot of numbers involved in a matrix product, but we need only concern ourselves with a few at a time.*

Now that we have seen how to compute matrix–matrix products, let us think back to where they came from. We said that our definition was cooked up so that the matrix–matrix product would "do" composition. The following theorem says that we got it right.

THEOREM 5   **(Matrix multiplication does composition)** *Let f be a linear transformation from $\mathbb{R}^n$ to $\mathbb{R}^m$, and g be a linear transformation from $\mathbb{R}^m$ to $\mathbb{R}^p$ so that the composition $h = g \circ f$ is defined. Let A_f, A_g, and $A_{g \circ f}$ be the corresponding matrices. Then*

$$A_{g \circ f} = A_g A_f. \tag{3.4}$$

PROOF: The jth column of $A_{g \circ f}$ is, by definition, $g(f(\mathbf{e}_j))$, so

$$A_{g \circ f} = [g(f(\mathbf{e}_1)), g(f(\mathbf{e}_2)), \ldots, g(f(\mathbf{e}_n))].$$

We defined matrix–vector multiplication so that for all $\mathbf{x}$, $g(\mathbf{x}) = A_g\mathbf{x}$. In particular, taking $\mathbf{x} = \mathbf{e}_j$, $g(f(\mathbf{e}_j)) = A_g f(\mathbf{e}_j)$, and therefore

$$A_{g \circ f} = [A_g f(\mathbf{e}_1), A_g f(\mathbf{e}_2), \ldots, A_g f(\mathbf{e}_n)]. \tag{3.5}$$

On the other hand, $A_f = [f(\mathbf{e}_1), f(\mathbf{e}_2), \ldots, f(\mathbf{e}_n)]$, so by (3.1),

$$A_g A_f = [A_g f(\mathbf{e}_1), A_g f(\mathbf{e}_2), \ldots, A_g f(\mathbf{e}_n)]. \tag{3.6}$$

Comparing (3.5) and (3.6), we see that indeed $A_{g \circ f} = A_g A_f$.   ■

It is useful to have a formula for matrix–matrix multiplication that is expressed directly in terms of the numerical entries of the matrices involved. Here is how to deduce this.

Let A be an $m \times n$ matrix, and let B be any $p \times m$ matrix. Then $(BA)_{i,j}$ is the ith entry of $(BA)\mathbf{e}_j$. Since the matrix product represents composition, $(BA)\mathbf{e}_j = B(A\mathbf{e}_j)$: We apply A, and then B. By Theorem 2, the jth column of A is $A\mathbf{e}_j$:

$$A\mathbf{e}_j = \begin{bmatrix} A_{1,j} \\ A_{2,j} \\ \vdots \\ A_{n,j} \end{bmatrix}.$$

Now apply B to $A\mathbf{e}_j$. By Theorem 3, the ith entry of

$$B(A\mathbf{e}_j) = B \begin{bmatrix} A_{1,j} \\ A_{2,j} \\ \vdots \\ A_{n,j} \end{bmatrix}$$

is $\sum_{k=1}^{m} B_{i,k} A_{k,j}$, which proves the following result.

THEOREM 6   **(Matrix–matrix multiplication in entry terms)** *Let A be an m × n matrix and B be a p × m matrix. Then BA is the p × n matrix whose i, jth entry is*

$$(BA)_{i,j} = \sum_{k=1}^{m} B_{i,k} A_{k,j}. \tag{3.7}$$

EXAMPLE 18   **(Entrywise computation of matrix products)** Let us compute the matrix product from Example 17 using Theorem 6. To get the 1, 1 entry in the product of 2×2 matrices A and B, we add up $B_{1,1}A_{1,1} + B_{1,2}A_{2,1}$, and so forth. In this example,

$$B_{1,1}A_{1,1} + B_{1,2}A_{2,1} = 1 \cdot 0 + 2 \cdot 1 = 2,$$

which gives us the upper left entry. Doing the same thing for the other three,

$$\begin{bmatrix} 1 & 2 \\ -2 & 1 \end{bmatrix} \begin{bmatrix} 0 & -1 \\ 1 & 2 \end{bmatrix} = \begin{bmatrix} 1 \cdot 0 + 2 \cdot 1 & 1 \cdot (-1) + 2 \cdot 2 \\ -2 \cdot 0 + 1 \cdot 1 & (-2) \cdot (-1) + 1 \cdot 2 \end{bmatrix} = \begin{bmatrix} 2 & 3 \\ 1 & 4 \end{bmatrix},$$

just as before. Notice the pattern here: $(BA)_{i,j}$ is the sum of the products of the corresponding entries of the ith row of B and the jth column of A. We will follow up on this pairing of the rows and columns in the sections to come.

3.2 Properties of matrix multiplication

Given any two specific matrices A and B where B is $m \times n$ and A is $n \times p$, we know how to compute the product BA. The first two examples in this section dealt with that. But what can we say about matrix multiplication *in general*?

- *Matrix multiplication is associative. That is, whenever C is an r × m matrix, B is an m × n matrix, and A is an n × p matrix,*

$$(CB)A = C(BA). \tag{3.8}$$

Why is this true? You could check it by using (3.7) and slogging through the sums. But Theorem 5 makes it easy. Let f_C, f_B, and f_A be the linear transformations corresponding to C, B, and A, respectively. Then by Theorem 5, $(CB)A$ corresponds to $(f_C \circ f_B) \circ f_A$ and $C(BA)$ corresponds to $f_C \circ (f_B \circ f_A)$. We know from the review of functions earlier in this chapter that $(f_C \circ f_B) \circ f_A = f_C \circ (f_B \circ f_A)$. Hence, (3.8) holds. In other words,

- *The matrix product is associative because the composition product is associative.*

This gives us another point of view on an identity we have already used, namely,

$$(CB)\mathbf{x} = C(B\mathbf{x}), \tag{3.9}$$

which holds for any $\mathbf{x}$ in $\mathbb{R}^n$. This follows directly from the fact that the matrix product represents composition. But we may also think of $\mathbf{x}$ as an $n \times 1$ matrix, and then (3.9) is just a special case of (3.8), with $\mathbf{x}$ in place of A.

Next, just as the composition product is not commutative in general, the matrix product is not commutative in general. Indeed, consider the matrices A and B from Example 17, where we computed

$$BA = \begin{bmatrix} 2 & 3 \\ 1 & 4 \end{bmatrix}.$$

Try computing AB now. You will find that

$$AB = \begin{bmatrix} 2 & -1 \\ -3 & 4 \end{bmatrix}.$$

Notice that $AB \neq BA$. In this case, at least both AB and BA were defined. But if B is $m \times n$ and A is $n \times p$ and $m \neq p$, then although BA is defined, AB is not.

What do we learn from this? If you are multiplying out a string of matrices, you do not need to worry about where any parentheses are; the result does not depend on that. You can just ignore any parentheses. But the product may well depend on the left-to-right order of the matrices in the product, so be careful about changing that.

3.3 The identity matrix and inverses

There is a very simple linear transformation from $\mathbb{R}^n$ to $\mathbb{R}^n$ that we have not discussed yet: the identity transformation $f(\mathbf{x}) = \mathbf{x}$ that simply returns the given input as output. (This transformation is linear; think about that.) We have a one-to-one correspondence between matrices and linear transformations. What matrix corresponds to the identity transformation $f_I(\mathbf{x}) = \mathbf{x}$? By Theorem 2, the matrix is just $[\mathbf{e}_1, \mathbf{e}_2, \ldots, \mathbf{e}_n]$.

DEFINITION

(Identity matrix) The $n \times n$ identity matrix I, or $I_{n \times n}$ when we need to indicate the size, is the matrix given by $I = [\mathbf{e}_1, \mathbf{e}_2, \ldots, \mathbf{e}_n]$. In terms of entries, the i, jth entry is

$$I_{i,j} = \begin{cases} 1 & \text{if } i = j \\ 0 & \text{if } i \neq j. \end{cases} \tag{3.10}$$

For example, the 3×3 identity matrix is

$$\begin{bmatrix} 1 & 0 & 0 \\ 0 & 1 & 0 \\ 0 & 0 & 1 \end{bmatrix}.$$

The pattern is the same in every dimension: Every diagonal entry is one, and every off-diagonal entry is zero.

If B is any $m \times n$ matrix, and I is the $n \times n$ identity matrix, by (3.10) and Theorem 6,

$$(BI)_{i,j} = \sum_{k=1}^{n} B_{i,k} I_{k,j} = B_{i,j}$$

since $I_{k,j} = 0$ for $k \neq j$, and $I_{j,j} = 1$. That is, $BI = B$. But this is also clear from the fact that I represents the identity transformation. If f_B is the linear transformation corresponding to B (as in Theorem 2), then BI corresponds to $f_B \circ f_I = f_B$. That is, $BI = B$.

In the same way, if A is any $n \times p$ matrix, $f_A = f_I \circ f_A$ and so by Theorem 2 again, $A = IA$. (We could also check this using (3.10) and Theorem 5.)

If A is an $n \times m$ matrix and B is an $m \times n$ matrix, then we can form the matrix product BA. Suppose that $BA = I_{m \times m}$. Then $BA\mathbf{x} = \mathbf{x}$ for all $\mathbf{x}$ in $\mathbb{R}^m$. Since matrix multiplication corresponds to the composition of linear functions, f_B, the linear transformation corresponding to B, *undoes* the effects of f_A, the linear transformation corresponding to A.

DEFINITION

(Left and right inverses) Let A be an $n \times m$ matrix. Then an $m \times n$ matrix B is a *left inverse* of A in case $BA = I_{m \times m}$, and an $m \times n$ matrix C is a *right inverse* of A in case $AC = I_{n \times n}$.

Notice that both left and right inverses of A, if either exist, have the same shape; that is, they both must be $m \times n$ for the multiplications to even make sense. But something even more is true when both a left and right inverse exist. Suppose that an $n \times m$ matrix A has both a left inverse B and a right inverse C. Then by (3.8), the associativity of matrix multiplication,

$$C = IC = (BA)C = B(AC) = BI = B.$$

Not only does $B = C$, but there is exactly one left inverse and exactly one right inverse, and they are the same. Indeed, suppose $\tilde{B}$ is any other left inverse. Then the same argument implies that $\tilde{B} = C$. Since $B = C$, $\tilde{B} = B$, and there is only one left inverse. In the same way, we see there is only one right inverse, which leads to the following definition.

DEFINITION

(Inverses and invertible matrices) A matrix A is *invertible* in the case where it has both a left and a right inverse. As we have seen, in this case it has just one left inverse and just one right inverse, and they are the same. This matrix is called the *inverse of A*, and it is usually denoted by A^{-1}.

EXAMPLE 19

(Inverse of a product) Let A and B be invertible $n \times n$ matrices. Is AB invertible? Yes, and in fact

$$(AB)^{-1} = B^{-1}A^{-1}. \tag{3.11}$$

To see this, note that by the associativity of matrix multiplication,

$$(B^{-1}A^{-1})(AB) = B^{-1}(A^{-1}A)B = B^{-1}IB = B^{-1}B = I.$$

The right inverse property follows the same way. The fact that the order of multiplication is switched is natural: To get $AB\mathbf{x}$, you first apply B to $\mathbf{x}$, and then you apply A to the result, $B\mathbf{x}$. To recover $\mathbf{x}$, you first undo A, then undo B.

Finding a right inverse of A means solving the matrix equation $AB = I$, where B is the unknown. This problem of solving for an unknown matrix can be reduced to solving for some unknown vectors, as we see in the next example.

EXAMPLE 20

(Inverse of a particular 2×2 matrix) Let $A = \begin{bmatrix} 0 & 2 \\ -1 & 3 \end{bmatrix}$. We are looking for a 2×2 matrix B such that $AB = I$. To reduce this unknown matrix problem to an unknown vector problem, write

$$B = [\mathbf{v}_1, \mathbf{v}_2].$$

Then by the definition of the matrix–matrix product,

$$AB = A[\mathbf{v}_1, \mathbf{v}_2] = [A\mathbf{v}_1, A\mathbf{v}_2].$$

Since $I_{2\times 2} = [\mathbf{e}_1, \mathbf{e}_2]$, $AB = I_{2\times 2}$ if and only if

$$[A\mathbf{v}_1, A\mathbf{v}_2] = [\mathbf{e}_1, \mathbf{e}_2].$$

That is, finding the matrix B we are looking for amounts to finding a pair of vectors $\mathbf{v}_1$ and $\mathbf{v}_2$ such that

$$A\mathbf{v}_1 = \mathbf{e}_1 \quad \text{and} \quad A\mathbf{v}_2 = \mathbf{e}_2.$$

First $\mathbf{v}_1$: Write

$$\mathbf{v}_1 = \begin{bmatrix} x \\ y \end{bmatrix}$$

so that $A\mathbf{v}_1 = \mathbf{e}_1$ becomes

$$\begin{bmatrix} 0 & 2 \\ -1 & 3 \end{bmatrix} \begin{bmatrix} x \\ y \end{bmatrix} = \begin{bmatrix} 1 \\ 0 \end{bmatrix}, \quad \text{or} \quad \begin{matrix} 2y = 1 \\ -x + 3y = 0. \end{matrix}$$

Evidently, $y = 1/2$, so $x = 3/2$. Therefore,

$$\mathbf{v}_1 = \begin{bmatrix} 3/2 \\ 1/2 \end{bmatrix}.$$

This gives us the first column of B.

Let

$$\mathbf{v}_2 = \begin{bmatrix} x \\ y \end{bmatrix}$$

so that $A\mathbf{v}_2 = \mathbf{e}_1$ becomes

$$\begin{bmatrix} 0 & 2 \\ -1 & 3 \end{bmatrix} \begin{bmatrix} x \\ y \end{bmatrix} = \begin{bmatrix} 0 \\ 1 \end{bmatrix}, \quad \text{or} \quad \begin{matrix} 2y = 0 \\ -x + 3y = 1. \end{matrix}$$

Evidently, $y = 0$, so $x = -1$. Therefore,

$$\mathbf{v}_2 = \begin{bmatrix} -1 \\ 0 \end{bmatrix}.$$

Putting it all together,

$$B = [\mathbf{v}_1, \mathbf{v}_2] = \frac{1}{2} \begin{bmatrix} 3 & -2 \\ 1 & 0 \end{bmatrix}.$$

You can check by direct multiplication that $AB = I$ and also that $BA = I$. Therefore, B is A^{-1}, the inverse of A.

The calculation of A^{-1} that we did in Example 20 can be adapted to find a general formula for the inverse in the 2×2 case. Indeed, let A be the general 2×2 matrix

$$A = \begin{bmatrix} a & b \\ c & d \end{bmatrix}.$$

Suppose that A has a right inverse B. As before, let $B = [\mathbf{v}_1, \mathbf{v}_2]$ so that

$$AB = A[\mathbf{v}_1, \mathbf{v}_2] = [A\mathbf{v}_1, A\mathbf{v}_2].$$

As before, since $I_{2\times2} = [\mathbf{e}_1, \mathbf{e}_2]$, $AB = I_{2\times2}$ if and only if

$$A\mathbf{v}_1 = \mathbf{e}_1 \quad \text{and} \quad A\mathbf{v}_2 = \mathbf{e}_2.$$

First $\mathbf{v}_1$: Write

$$\mathbf{v}_1 = \begin{bmatrix} x \\ y \end{bmatrix}.$$

Then

$$A\mathbf{v}_1 = \begin{bmatrix} ax + by \\ cx + dy \end{bmatrix}$$

so that $A\mathbf{v}_1 = \mathbf{e}_1$ is equivalent to the system of equations

$$\begin{matrix} ax + by = 1 \\ cx + dy = 0. \end{matrix} \tag{3.12}$$

To solve for x and y, multiply the top equation through by d and the bottom equation through by b:

$$\begin{matrix} dax + dby = d \\ bcx + bdy = 0. \end{matrix}$$

Subtracting, we find

$$(ad - bc)x = d. \tag{3.13}$$

In the same way, if we multiply the top equation in (3.12) through by c and the bottom equation through by a and subtract, we obtain

$$(ad - bc)y = -c. \tag{3.14}$$

Combining (3.13) and (3.14), we see that as long as $ad - bc \neq 0$, the unique solution is

$$\mathbf{v}_1 = \frac{1}{ad - bc} \begin{bmatrix} d \\ -c \end{bmatrix}.$$

We are halfway there, and we will leave the other half as an exercise. In the same way, you will see that as long as $ad - bc \neq 0$, $A\mathbf{x} = \mathbf{e}_2$ also has a unique solution:

$$\mathbf{v}_2 = \frac{1}{ad - bc} \begin{bmatrix} -b \\ a \end{bmatrix}.$$

Hence, when $A = \begin{bmatrix} a & b \\ c & d \end{bmatrix}$ satisfies $ad - bc \neq 0$, A has a unique right inverse B, given by the explicit formula

$$B = \frac{1}{ad - bc} \begin{bmatrix} d & -b \\ -c & a \end{bmatrix}. \tag{3.15}$$

At this point you should multiply out AB and check that indeed $AB = I$. You should also check that $BA = I$. That is, B is not just a right inverse, but it is the inverse of A.

In the case $ad - bc = 0$,

$$\begin{bmatrix} a & b \\ c & d \end{bmatrix} \begin{bmatrix} -b \\ a \end{bmatrix} = \begin{bmatrix} 0 \\ ad - bc \end{bmatrix} = \begin{bmatrix} 0 \\ 0 \end{bmatrix} \tag{3.16}$$

and

$$\begin{bmatrix} a & b \\ c & d \end{bmatrix} \begin{bmatrix} d \\ -c \end{bmatrix} = \begin{bmatrix} ad - bc \\ 0 \end{bmatrix} = \begin{bmatrix} 0 \\ 0 \end{bmatrix}. \tag{3.17}$$

As long as A is not the zero matrix, that is, as long as at least one of a, b, c, and d is nonzero, at least one of

$$\begin{bmatrix} -b \\ a \end{bmatrix} \begin{bmatrix} d \\ -c \end{bmatrix}$$

is not zero. Let $\mathbf{x}_0$ be the first one or, if that is zero, the second one. Then by (3.16) or (3.17),

$$A\mathbf{x}_0 = 0 \quad \text{while} \quad \mathbf{x}_0 \neq 0. \tag{3.18}$$

- *In (3.18), we are using the symbol* 0 *to denote the vector with every entry equal to zero, as on the right-hand side in (3.16) and (3.17). We will do this frequently in what follows. The context will make it clear whether* 0 *denotes a number or a vector.*

Whenever (3.18) holds, A is not invertible. Indeed, suppose that B is a left inverse of A and (3.18) holds. Then

$$\mathbf{x}_0 = I\mathbf{x}_0 = (BA)\mathbf{x}_0 = B(A\mathbf{x}_0) = B0 = 0,$$

which contradicts $\mathbf{x}_0 \neq 0$. Therefore, (3.18) is incompatible with the existence of a left inverse, let alone an inverse. We have proved the following result.

THEOREM 7 **(Inverting 2×2 matrices)** *Let A be the 2×2 matrix*

$$A = \begin{bmatrix} a & b \\ c & d \end{bmatrix}.$$

Then A is invertible if and only if $ad - bc \neq 0$, in which case

$$A^{-1} = \frac{1}{ad - bc} \begin{bmatrix} d & -b \\ -c & a \end{bmatrix}. \tag{3.19}$$

EXAMPLE 21 **(Using the 2×2 inverse formula)** Let

$$A = \begin{bmatrix} 0 & 2 \\ -1 & 3 \end{bmatrix},$$

as in Example 20. Then for this A, $ad - bc = 2$, and by (3.19),

$$A^{-1} = \frac{1}{2} \begin{bmatrix} 3 & -2 \\ 1 & 0 \end{bmatrix},$$

as we found before.

What are inverses good for? If a matrix A is invertible and you happen to know the inverse A^{-1}, then you can easily solve the vector equation $A\mathbf{x} = \mathbf{b}$ for any $\mathbf{b}$ in $\mathbb{R}^m$. Indeed,

$$A(A^{-1}\mathbf{b}) = (AA^{-1})\mathbf{b} = I\mathbf{b} = \mathbf{b},$$

so that $\mathbf{x}_0 = A^{-1}\mathbf{b}$ is a solution. *Here we used the fact that A has a right inverse, so whenever A has even just a right inverse B, $A\mathbf{x} = \mathbf{b}$ has at least one solution, namely, $B\mathbf{b}$.*

Are there any others? Suppose that $\mathbf{x}_1$ also satisfies $A\tilde{\mathbf{x}} = \mathbf{b}$. Then

$$A(\mathbf{x}_1 - \mathbf{x}_0) = A\mathbf{x}_1 - A\mathbf{x}_0 = \mathbf{b} - \mathbf{b} = 0.$$

Multiplying both sides of $A(\mathbf{x}_1 - \mathbf{x}_0) = 0$ on the left by A^{-1}, we get

$$\mathbf{x}_1 - \mathbf{x}_0 = A^{-1}A(\mathbf{x}_1 - \mathbf{x}_0) = A^{-1}0 = 0,$$

so $\mathbf{x}_1 = \mathbf{x}_0$ after all and there is no other solution.

Putting the two together, we see that *matrix inverses, when they exist, give us the unique solutions to vector equations such as $A\mathbf{x} = \mathbf{b}$ through $\mathbf{x} = A^{-1}\mathbf{b}$.*

EXAMPLE 22 **(Using the matrix inverse to solve equations)** Let us solve the equation $A\mathbf{x} = \mathbf{b}$ for

$$A = \begin{bmatrix} 0 & 2 \\ -1 & 3 \end{bmatrix} \quad \text{and} \quad \mathbf{b} = \begin{bmatrix} 1 \\ 2 \end{bmatrix}.$$

As we saw in Example 19, A is invertible, and by what we have learned, the unique solution is

$$A^{-1}\mathbf{b} = \frac{1}{2} \begin{bmatrix} 3 & -2 \\ 1 & 0 \end{bmatrix} \begin{bmatrix} 1 \\ 2 \end{bmatrix} = \begin{bmatrix} -1/2 \\ 1/2 \end{bmatrix}.$$

You might guess at this point that we will be busy in the rest of the book devising formulas for the inverses of larger matrices. This is not the case.

First, for larger matrices, computing an inverse takes many more steps than other methods of solving equations. Although the formula for inverses of 2×2 matrices is very useful to know, there are computational disadvantages to relying on matrix inversion formulas for solving matrix equations.

Second, only square matrices can possibly have inverses, as we shall show right after Example 23, and we do not want to limit ourselves to considering only square matrices.

To see that only square matrices can possibly have inverses, let us glean a general fact from the proof of Theorem 7:

- *If A is an $m \times n$ matrix and there is a nonzero vector $\mathbf{x}_0$ in $\mathbb{R}^n$ with $A\mathbf{x}_0 = 0$, then A does not have a left inverse (and therefore is not invertible).*

Just as we argued in the paragraph preceding the statement of Theorem 7, suppose that B is a left inverse of A and $A\mathbf{x}_0 = 0$. Then $\mathbf{x}_0 = I\mathbf{x}_0 = (BA)\mathbf{x}_0 = B(A\mathbf{x}_0) = B0 = 0$. Therefore, 0 is the only solution of $A\mathbf{x} = 0$ when A has a left inverse.

EXAMPLE 23   **(Checking for noninvertibility)** Consider

$$A = \begin{bmatrix} 1 & 2 & 3 \\ 2 & 4 & 2 \\ 3 & 6 & 0 \end{bmatrix}.$$

Since the second column is twice the first, $-2\mathbf{v}_1 + \mathbf{v}_2 = 0$, where $\mathbf{v}_j$ is the jth column of A. Therefore, let

$$\mathbf{x}_0 = \begin{bmatrix} -2 \\ 1 \\ 0 \end{bmatrix}$$

so that

$$A\mathbf{x}_0 = -2\mathbf{v}_1 + \mathbf{v}_2 = 0.$$

Since $\mathbf{x}_0$ is nonzero, A is not invertible.

This example may look artificial, but we found $\mathbf{x}_0$ by the strategy of looking for patterns in A, which is very often worth doing.

Now we can explain why only square matrices have a chance to be invertible if we "borrow" one simple result on vector equations. In the next chapter, where we systematically study equations of the type $A\mathbf{x} = \mathbf{b}$, we will see that

- *If A has more columns than rows, there is always a nonzero vector $\mathbf{x}_0$ that satisfies*

$$A\mathbf{x}_0 = 0. \tag{3.20}$$

Roughly speaking, this statement is true because if A is $m \times n$ with $n > m$, there are more input variables than there are output variables: $\mathbf{w}$ is in $\mathbb{R}^n$, and $\mathbf{b}$ is in $\mathbb{R}^m$. As we will see, this means that we can always find a nonzero solution of (3.20). Thus, according to the criterion in the bulleted point preceding Example 23, *matrices with left inverses cannot have more columns than rows*, since in this case there is a nonzero solution to (3.20).

If A is an invertible $m \times n$ matrix, then we must have $n \leq m$, since A has a left inverse. But A^{-1} is an $n \times m$ matrix, and it, too, has a left inverse, namely, A since $AA^{-1} = A^{-1}A = I$. But then A^{-1} cannot have more columns than rows, so $m \leq n$. Putting the two inequalities together, we have $m = n$, so A must be square to be invertible.*

There is one more thing we should say about matrix inverses; a word of caution: Multiplication by an inverse matrix A^{-1} undoes multiplication by A. That is,

$$A^{-1}(AB) = (A^{-1}A)B = B,$$

---

*We will state this as a theorem in the next chapter after we have proved that (3.20) always has nonzero solutions when A has more columns than rows. But it makes sense to discuss the facts here so that you do not wonder why we include only examples of inverses of square matrices in the exercises for this section!

as we have observed already. In this sense, multiplication by A^{-1} is a bit like "dividing by A." However, if C is an $n \times n$ matrix and A is an invertible $n \times n$ matrix, we have to be careful about the fact that often

$$A^{-1}C \neq CA^{-1}. \tag{3.21}$$

For this reason *we do not write*

$$\frac{C}{A}$$

to denote matrix division, as it is not clear which of the matrices in (3.21) this expression might mean.

3.4 Matrix algebra

We close this section with one more operation on matrices—one that is much simpler than the matrix product. Let A and B be two $m \times n$ matrices, and let f_A and f_B be the corresponding linear transformations from $\mathbb{R}^n$ to $\mathbb{R}^m$. We can define a new transformation g from $\mathbb{R}^n$ to $\mathbb{R}^m$ by adding the outputs of these two:

$$g(\mathbf{x}) = f_A(\mathbf{x}) + f_B(\mathbf{x}).$$

It is easy to check that g is additive and homogeneous since, after all, it is obtained by addition. (Still, check the details now.) Therefore $(f_A + f_B)$ corresponds to a matrix. How is this matrix related to A and B?

Let C be the matrix corresponding to g. Then the jth column of C is

$$g(\mathbf{e}_j) = f_A(\mathbf{e}_j) + f_B(\mathbf{e}_j) = A\mathbf{e}_j + B\mathbf{e}_j.$$

Therefore, $C_{i,j}$ is the ith entry of this vector, namely, $(A\mathbf{e}_j)_i + (B\mathbf{e}_j)_i = A_{i,j} + B_{i,j}$, and hence $C_{i,j} = A_{i,j} + B_{i,j}$. This brings us to the following definition.

DEFINITION

> **(Addition of matrices)** Let A and B be two $m \times n$ matrices. Then $A + B$ is the $m \times n$ matrix given by
>
> $$(A + B)_{i,j} = A_{i,j} + B_{i,j}. \tag{3.22}$$

EXAMPLE 24 **(Addition of matrices)** Let

$$A = \begin{bmatrix} 1 & 2 \\ 3 & 4 \end{bmatrix} \quad \text{and} \quad B = \begin{bmatrix} 3 & 1 \\ 1 & 3 \end{bmatrix}.$$

Then

$$A + B = \begin{bmatrix} 1 & 2 \\ 3 & 4 \end{bmatrix} + \begin{bmatrix} 3 & 1 \\ 1 & 3 \end{bmatrix} = \begin{bmatrix} 4 & 3 \\ 4 & 7 \end{bmatrix}.$$

Notice that $A + B$ is only defined if both A and B have the same size, $m \times n$. Also notice that, by definition, $A + B$ represents the linear transformation obtained by adding the outputs of the linear transformations represented by A and B. That is,

$$(A + B)\mathbf{x} = A\mathbf{x} + B\mathbf{x}, \tag{3.23}$$

the *distributive property* of matrix–vector multiplication.

EXAMPLE 25 **(The distributive property of matrix–vector multiplication)** Let

$$A = \begin{bmatrix} 1 & 2 \\ 3 & 4 \end{bmatrix} \quad \text{and} \quad B = \begin{bmatrix} 3 & 1 \\ 1 & 3 \end{bmatrix}$$

and let

$$\mathbf{x} = \begin{bmatrix} 1 \\ 1 \end{bmatrix}.$$

Then

$$A\mathbf{x} = \begin{bmatrix} 3 \\ 7 \end{bmatrix} \quad \text{and} \quad B\mathbf{x} = \begin{bmatrix} 4 \\ 4 \end{bmatrix}$$

so that

$$A\mathbf{x} + B\mathbf{x} = \begin{bmatrix} 7 \\ 11 \end{bmatrix}.$$

By Example 23,

$$(A + B)\mathbf{x} = \begin{bmatrix} 4 & 3 \\ 4 & 7 \end{bmatrix} \begin{bmatrix} 1 \\ 1 \end{bmatrix} = \begin{bmatrix} 7 \\ 11 \end{bmatrix}.$$

Finally, we define multiplication of numbers and matrices as follows: If A is any $m \times n$ matrix and a is any number, the product aA is defined by $(aA)_{i,j} = a(A_{i,j})$; that is, multiply each of the entries by a.

EXAMPLE 26 **(Multiplying a number and a matrix)** Let

$$A = \begin{bmatrix} 1 & 2 \\ 3 & 4 \end{bmatrix}.$$

Then

$$3A = 3\begin{bmatrix} 1 & 2 \\ 3 & 4 \end{bmatrix} = \begin{bmatrix} 3 & 6 \\ 9 & 12 \end{bmatrix}.$$

We now have introduced our full range of algebraic operations on matrices.

We close with an important remark: To evaluate a polynomial function $p(x)$ of a real variable x, the only operations on x that we need to perform are multiplying x with itself or some other numbers, and adding such multiples. Since we can do all of these operations for $n \times n$ matrices, we can plug square matrices into polynomials. We just interpret the constant term as a multiple of the identity, that is, we define $A^0 = I$. For example, if

$$p(x) = 1 - 2x + x^2 \quad \text{and} \quad A = \begin{bmatrix} 1 & 2 \\ 0 & 1 \end{bmatrix},$$

then

$$p(A) = \begin{bmatrix} 1 & 0 \\ 0 & 1 \end{bmatrix} - 2\begin{bmatrix} 1 & 2 \\ 0 & 1 \end{bmatrix} + \begin{bmatrix} 1 & 4 \\ 0 & 1 \end{bmatrix} = \begin{bmatrix} 0 & 0 \\ 0 & 0 \end{bmatrix}.$$

That is, A is a root of the polynomial p considered as a function on the class of 2×2 matrices.

Exercises

3.1 Consider the four matrices A, B, C, and D from Exercise 2.7. There are 16 ordered pairs of these matrices, allowing self-pairing, namely,

$$AA, \quad AB, \quad AC, \quad AD, \quad BA, \quad BB, \dots.$$

Which ones make sense as matrix products? Compute the product in each such case.

3.2 Consider the four matrices A, B, C, and D from Exercise 2.8. There are 16 ordered pairs of these matrices, allowing self-pairing, namely,

$$AA, \quad AB, \quad AC, \quad AD, \quad BA, \quad BB, \dots.$$

Which ones make sense as matrix products? Compute the product in each such case.

3.3 Let

$$A = \begin{bmatrix} 1 & 1 & 1 & 2 \\ 1 & 2 & 0 & -1 \\ 2 & 0 & 0 & -1 \\ 3 & -2 & 0 & 2 \end{bmatrix} \quad \text{and} \quad B = \begin{bmatrix} 0 & -1 & 1 & 1 \\ 1 & 1 & 1 & -1 \\ 2 & 3 & 1 & -3 \\ 0 & -3 & 0 & 0 \end{bmatrix}.$$

Compute the third column of AB by computing an appropriate matrix–vector product.

3.4 Let

$$A = \begin{bmatrix} 2 & 2 & 1 & 0 \\ 0 & 3 & 0 & -3 \\ 2 & 0 & 0 & -2 \\ 1 & -1 & 3 & 0 \end{bmatrix} \quad \text{and} \quad B = \begin{bmatrix} 0 & -6 & 6 & 2 \\ -1 & 1 & 1 & -2 \\ -2 & 3 & 1 & -1 \\ 0 & -3 & 0 & 1 \end{bmatrix}.$$

Compute the second column of AB by computing an appropriate matrix–vector product.

3.5 Let

$$A = \begin{bmatrix} 0 & 1 & 0 \\ 0 & 0 & 1 \\ 0 & 0 & 0 \end{bmatrix}.$$

Compute A^2 and A^3.

3.6 For any three numbers a, b, and c, let

$$A = \begin{bmatrix} 0 & a & b \\ 0 & 0 & c \\ 0 & 0 & 0 \end{bmatrix}.$$

Compute A^2 and A^3.

3.7 For any three numbers a, b, and c, let

$$A = \begin{bmatrix} a & b \\ 0 & c \end{bmatrix}.$$

(a) Compute A^2.

(b) Find all possible values of a, b, and c so that $A^2 = B$ where

$$B = \begin{bmatrix} 1 & 3 \\ 0 & 4 \end{bmatrix}.$$

How many different sets of values for a, b, and c are there for which $A^2 = B$? (The matrices A that you compute here are *square roots* of the matrix B.)

3.8 For any six numbers a, b, c, d, e, and f, let

$$A = \begin{bmatrix} a & b & c \\ 0 & d & e \\ 0 & 0 & f \end{bmatrix}.$$

(a) Compute A^2.

(b) Find all possible values of a, b, c, d, e, and f so that $A^2 = B$ where

$$B = \begin{bmatrix} 1 & 3 & 5 \\ 0 & 4 & 7 \\ 0 & 0 & 9 \end{bmatrix}.$$

How many different sets of values for a, b, c, d, e, and f are there for which $A^2 = B$? (The matrices A that you compute here are *square roots* of the matrix B.)

3.9 Given two $n \times n$ matrices A and B, the *commutator of A and B* is defined to be the $n \times n$ matrix $AB - BA$ and is denoted by $[A, B]$. The reason is that $AB = BA$ if and only if $[A, B] = 0$. More generally, but also directly from the definition, $AB = BA + [A, B]$.

(a) Compute $[A, B]$ for

$$A = \begin{bmatrix} 0 & 1 \\ 1 & 0 \end{bmatrix} \quad \text{and} \quad B = \begin{bmatrix} 1 & 0 \\ 0 & 2 \end{bmatrix}.$$

(b) For the same matrices A and B, compute $[A, [A, B]]$.

(c) Using only the associative property of matrix multiplication, show that for any three $n \times n$ matrices A, B, and C

$$[A, [B, C]] + [B, [C, A]] + [C, [A, B]] = 0.$$

This is known as *Jacobi's identity*. (We will not make use of it later. For our purposes, checking it is simply a good exercise in working with matrices.)

3.10 For any numbers a and b, let

$$A = \begin{bmatrix} 1 & a \\ 0 & 1 \end{bmatrix} \quad \text{and} \quad B = \begin{bmatrix} 1 & b \\ 0 & 1 \end{bmatrix}.$$

(a) Compute the product AB.

(b) Show that A is always invertible, and find the inverse. (Your answer will be a matrix that, like A, depends on a.)

3.11 For any three numbers a, b, and c, let

$$A = \begin{bmatrix} 1 & a & b \\ 0 & 1 & c \\ 0 & 0 & 1 \end{bmatrix}.$$

For any three numbers u, v, and w, let

$$B = \begin{bmatrix} 1 & u & v \\ 0 & 1 & w \\ 0 & 0 & 1 \end{bmatrix}.$$

(a) Compute the product AB.

(b) Show that A is always invertible, and find the inverse. (Your answer will be a matrix that, like A, depends on a, b, and c.)

3.12 Let $R(\theta)$ be the 2×2 rotation matrix

$$R(\theta) = \begin{bmatrix} \cos(\theta) & -\sin(\theta) \\ \sin(\theta) & \cos(\theta) \end{bmatrix}.$$

Using the formula for inversion of 2×2 matrices, check that $(R(\theta))^{-1} = R(-\theta)$.

3.13 Let

$$J = \begin{bmatrix} 0 & -1 \\ 1 & 0 \end{bmatrix},$$

and let I denote the 2×2 identity matrix

$$\begin{bmatrix} 1 & 0 \\ 0 & 1 \end{bmatrix}.$$

(a) Show that $J^2 = -I$ and that the transformation of $\mathbb{R}^2$ induced by J is simply the counterclockwise rotation about the origin through an angle of $\pi/2$. (Therefore $-I$ has a square root, and the square root is something quite natural. Indeed, two 90-degree turns do make a U-turn.)

(b) For any numbers x and y, form the matrix $xI + yJ$. Show that the product of any two such matrices is another matrix of the same form. More specifically,

$$(xI + yJ)(uI + vJ) = (xu - yv)I + (xv + yu)J.$$

(c) Find all values of x and y so that $(xI + yJ)^2 = 4I$.

(d) Find all values of x and y so that $(xI + yJ)^2 = -4I$. The point of this problem is that if we identify the real numbers with the real multiples of I, that is, the matrices of the form xI, then in the class of matrices of the form $uI + vJ$ there is a square root of x even if x is negative. We can identify the set of matrices of the form $xI + yJ$ with the complex numbers, and what we have just given amounts to a *construction* of the complex number system out of the real numbers, using matrices.

3.14 Show that for every choice of a and b with $b \neq 0$, there is a choice of c and d so that with

$$K = \begin{bmatrix} a & b \\ c & d \end{bmatrix},$$

we obtain

$$K^2 = -\begin{bmatrix} 1 & 0 \\ 0 & 1 \end{bmatrix}.$$

Find a formula for c and d in terms of a and b.

3.15 Find all 2×2 matrices A such that $A^2 = A$.

3.16 Find all 2×2 matrices A such that $A^2 = 0$.

3.17 Does there exist an invertible 2×2 matrix A such that $A^2 = 0$? If so, find an example. Otherwise, explain why not.

3.18 Give an example of *invertible* matrices A and B such that $A + B$ is not invertible.

3.19 Give an example of *noninvertible* matrices A and B such that $A + B$ is invertible.

3.20 Let A, B, and C be three invertible 2×2 matrices. Let $D = ABC$. Suppose that

$$D^{-1} = \begin{bmatrix} 5 & 2 \\ 2 & 1 \end{bmatrix}, \quad A = \begin{bmatrix} 3 & 2 \\ 4 & 3 \end{bmatrix}, \quad \text{and } B = \begin{bmatrix} 1 & 1 \\ 1 & 2 \end{bmatrix}.$$

Compute C.

3.21 Let A, B, and C be three invertible 2×2 matrices. Let $D = ABC$. Suppose that

$$D^{-1} = \begin{bmatrix} 4 & 5 \\ 5 & 6 \end{bmatrix}, \quad A = \begin{bmatrix} 5 & 7 \\ 7 & 10 \end{bmatrix}, \quad \text{and } C^{-1} = \begin{bmatrix} 1 & 1 \\ 1 & 2 \end{bmatrix}.$$

Compute B.

3.22 Let A be a 3×3 matrix. Show that if the second column is twice the first column, then A is not invertible.

3.23 Let A be a 3×3 matrix. Show that if the first column minus the second column equals the third column, then A is not invertible.

3.24 (a) Compute an explicit formula for the matrix power A^n, where

$$A = \begin{bmatrix} \cos(\pi/6) & -\sin(\pi/6) \\ \sin(\pi/6) & \cos(\pi/6) \end{bmatrix}.$$

(b) For any real number a, compute the limit

$$\lim_{n \to \infty} \begin{bmatrix} 1 & -a/n \\ a/n & 1 \end{bmatrix}^n.$$

Hint: See Exercise 3.13.

3.25 (a) Show that

$$\begin{bmatrix} 1 & 1/n \\ 0 & 1 \end{bmatrix}^n$$

is the same matrix for all values of n.

(b) Compute the limit

$$\lim_{n \to \infty} \begin{bmatrix} 1 & 1/n^3 \\ 0 & 1 \end{bmatrix}^n.$$

3.26 Find all matrices A, if any, such that

(a) $\begin{bmatrix} 1 & -1 \\ 1 & 4 \end{bmatrix} A = \begin{bmatrix} 2 & 0 \\ 1 & 3 \end{bmatrix}.$

(b) $\begin{bmatrix} -1 & 2 \\ 1 & 1 \end{bmatrix} A \begin{bmatrix} 1 & 1 \\ 2 & 0 \end{bmatrix} = \begin{bmatrix} 2 & 3 \\ 2 & 5 \end{bmatrix}.$

3.27 For any $n \times n$ matrix A, define the *trace* of A to be the number $\mathrm{tr}(A)$ given by

$$\mathrm{tr}(A) = \sum_{i=1}^{N} A_{i,i}.$$

Show that for any two $n \times n$ matrices B and C, $\mathrm{tr}(BC) = \mathrm{tr}(CB)$, even when $BC \neq CB$.

3.28 Using the result of the previous exercise, show that there do not exist two $n \times n$ matrices B and C such that $BC - CB = I_{n \times n}$.

Problems concerning polynomials of matrices

In the last paragraph of this section, we discussed the meaning of $p(A)$, where p is a polynomial function and A is an $n \times n$ matrix. The following exercises build on this discussion.

3.29 Consider the matrices

$$A = \begin{bmatrix} 1 & 2 \\ -1 & 3 \end{bmatrix} \quad \text{and} \quad B = \begin{bmatrix} 1 & -1 & 0 \\ 1 & 2 & 1 \\ 0 & 1 & 3 \end{bmatrix}.$$

Compute the polynomials $p(A)$ for the following polynomials:

(a) $p(x) = x^2 - 2x + 5.$

(b) $p(x) = x^2 - 2.$

3.30 Consider the matrices

$$A = \begin{bmatrix} 4 & 2 \\ -2 & 3 \end{bmatrix} \quad \text{and} \quad B = \begin{bmatrix} 2 & -2 & 0 \\ 0 & 2 & 1 \\ 0 & 1 & 1 \end{bmatrix}.$$

Compute the polynomials $p(A)$ and $p(B)$ for the following polynomials:

(a) $p(x) = x^2 - 7x + 16$.

(b) $p(x) = x^3 - 5x^2 + 7x - 2$.

3.31 Let A be an $n \times n$ matrix such that $A^{k+1} = 0$ for some k. Show that $I - A$ is invertible by checking that the matrix B defined by

$$B = I + A + A^2 + \cdots + A^k$$

is the inverse of $I - A$. (If $A^{k+1} = 0$, the inverse of $I - A$ is a polynomial in A of degree (at most) k.)

3.32 Using the result of the previous exercise, compute the inverse of

$$\begin{bmatrix} 1 & 2 & 4 \\ 0 & 1 & 2 \\ 0 & 0 & 1 \end{bmatrix}.$$

3.33 In fact, whenever an $n \times n$ matrix A is invertible, there is a polynomial p of degree $n - 1$ at most such that $A^{-1} = p(A)$. Why this is true and why this is useful will be explained later on, but we can see some examples now.

(a) Suppose that A is an $n \times n$ matrix and $q(x)$ is a polynomial of degree k such that $q(A) = 0$ but $q(0) \neq 0$. Define a function $p(x)$ by

$$p(x) = \frac{q(0) - q(x)}{xq(0)}.$$

Show that $p(x)$ is a polynomial of degree $k - 1$, that A is invertible, and that $A^{-1} = p(A)$.

(b) Let A and B be the matrices from Exercise 3.30. Find the inverses of A and B using part (a) of this exercise and the results of Exercise 3.30.

SECTION 4 | # The Dot Product and the Geometry of $\mathbb{R}^n$

4.1 Your laundry list, the price list, and your bill

Suppose that you and a friend take a load of laundry to the dry cleaners. Suppose that the load of laundry includes 4 shirts, 3 pairs of trousers, 2 dresses, 1 skirt, and 1 suit.

Suppose that the dry cleaners charges \$1.50 for cleaning a shirt, \$2.00 for cleaning a pair of trousers, \$3.00 for cleaning a dress, \$2.00 for cleaning a skirt, and \$4.50 for cleaning a suit. (Pressing included, of course!)

We have two lists of data here: a *laundry list* and a *price list*. We can organize them into two vectors **L** and **P**, where, in this example,

$$\mathbf{L} = \begin{bmatrix} \ell_1 \\ \ell_2 \\ \ell_3 \\ \ell_4 \\ \ell_5 \end{bmatrix} = \begin{bmatrix} 4 \\ 3 \\ 2 \\ 1 \\ 1 \end{bmatrix} \quad \text{and} \quad \mathbf{P} = \begin{bmatrix} p_1 \\ p_2 \\ p_3 \\ p_4 \\ p_5 \end{bmatrix} = \begin{bmatrix} 1.5 \\ 2.0 \\ 3.0 \\ 2.0 \\ 4.5 \end{bmatrix}.$$

When you go to pick up the laundry, the bill you have to pay is the number of shirts times the price of shirts plus the number of pairs of trousers times the price of a pair of trousers and so on:

$$\ell_1 p_1 + \ell_2 p_2 + \ell_3 p_3 + \ell_4 p_4 + \ell_5 p_5.$$

This quantity is the *dot product* of the two vectors

$$\mathbf{L} = \begin{bmatrix} \ell_1 \\ \ell_2 \\ \ell_3 \\ \ell_4 \\ \ell_5 \end{bmatrix} \quad \text{and} \quad \mathbf{P} = \begin{bmatrix} p_1 \\ p_2 \\ p_3 \\ p_4 \\ p_5 \end{bmatrix}.$$

It is denoted by $\mathbf{L} \cdot \mathbf{P}$,

$$\mathbf{L} \cdot \mathbf{P} = \ell_1 p_1 + \ell_2 p_2 + \ell_3 p_3 + \ell_4 p_4 + \ell_5 p_5.$$

With the numbers plugged in, it works out to 24.5, or $24.50.

You do not need linear algebra to compute your laundry bill—it is only a matter of simple arithmetic. But keep this example in mind as you read the rest of this section. It shows that dot products arise in the commonest of circumstances.

It turns out that the dot product is a kind of "bridge" between algebra and geometry. All sorts of algebraic problems, such as computing your dry cleaning bill, implicitly involve dot products. Making the dot products explicit brings a *geometric perspective* to a problem.

If the problem is just computing your dry cleaning bill, you do not need a geometric perspective—the problem is too simple. But in other problems, as we shall see, the geometric perspective can be advantageous.

4.2 Distance, the dot product, and its algebraic properties

So far we have emphasized vectors as lists of data. However, we can identify vectors in $\mathbb{R}^2$ or $\mathbb{R}^3$ with the corresponding points in Euclidean space, providing a useful geometric perspective on vectors and vector equations. Let us begin the discussion in two dimensions, where we can easily visualize the geometry.

Vectors were originally considered in $\mathbb{R}^2$ and $\mathbb{R}^3$ and were defined as quantities having both *magnitude* and *direction*. For example, consider a vector

$$\mathbf{x} = \begin{bmatrix} x_1 \\ x_2 \end{bmatrix}$$

in $\mathbb{R}^2$, and identify it with the corresponding point (x_1, x_2) in the plane. We are being a bit pedantic in using one notation for vectors in $\mathbb{R}^2$ and another for points in the plane. But here is the purpose of this distinction: We have already defined certain *algebraic operations,* namely, scalar multiplication and vector addition on $\mathbb{R}^2$, and we are already familiar with the *geometry* of the plane. Now we are going to *put the algebra and the geometry together,* through the identification of

$$\begin{bmatrix} x_1 \\ x_2 \end{bmatrix} \quad \text{with} \quad (x_1, x_2). \tag{4.1}$$

Through this identification, we might, for example, think of

$$\mathbf{x} = \begin{bmatrix} x_1 \\ x_2 \end{bmatrix}$$

as representing the *displacement* from the origin $(0,0)$ of a particle in two-dimensional Euclidean space.

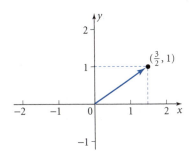

Regarding **x** as representing a displacement, we can call it a *displacement vector*. (This is not a mathematical definition of a new type of vector; it is rather convenient terminology that recalls what this particular list represents. The same is true of velocity vectors, acceleration vectors, and so on.) The magnitude of this *displacement vector* is the *Euclidean distance* from the origin $(0, 0)$ to the point (x_1, x_2). This distance is given by the Pythagorean formula

$$\sqrt{x_1^2 + x_2^2},$$

and it tells us the "magnitude," or "size," of the displacement. The plainest term for discussing the size of a displacement is "length," as in "length of travel." Therefore, for a vector

$$\mathbf{x} = \begin{bmatrix} x_1 \\ x_2 \end{bmatrix}$$

in $\mathbb{R}^2$, we define the length of **x**, denoted by $|\mathbf{x}|$, to be

$$|\mathbf{x}| = \sqrt{x_1^2 + x_2^2}. \qquad (4.2)$$

Let us put algebra and geometry together. Consider two vectors

$$\mathbf{x} = \begin{bmatrix} x_1 \\ x_2 \end{bmatrix} \qquad \text{and} \qquad \mathbf{y} = \begin{bmatrix} y_1 \\ y_2 \end{bmatrix}$$

in $\mathbb{R}^2$. What is $|\mathbf{x} - \mathbf{y}|$, the length of their difference? That is not hard to work out. By the definition,

$$\mathbf{x} - \mathbf{y} = \begin{bmatrix} x_1 - y_1 \\ x_2 - y_2 \end{bmatrix},$$

so by (4.2),

$$|\mathbf{x} - \mathbf{y}|^2 = (x_1 - y_1)^2 + (x_2 - y_2)^2.$$

The right-hand side is the square of the Euclidean distance from (x_1, x_2) to (y_1, y_2), so *the Euclidean distance from (x_1, x_2) to (y_1, y_2) is $|\mathbf{x} - \mathbf{y}|$, the length of the difference between* **x** *and* **y**.

Let us take the algebra a bit further:

$$\begin{aligned} |\mathbf{x} - \mathbf{y}|^2 &= (x_1 - y_1)^2 + (x_2 - y_2)^2 \\ &= x_1^2 - 2x_1y_1 + y_1^2 + x_2^2 - 2x_2y_2 + y_2^2 \\ &= (x_1^2 + x_2^2) + (y_1^2 + y_2^2) - 2(x_1y_1 + x_2y_2) \\ &= |\mathbf{x}|^2 + |\mathbf{y}|^2 - 2(x_1y_1 + x_2y_2), \end{aligned} \qquad (4.3)$$

where we used (4.2) in the first and last lines. If we *define* the "dot product" $\mathbf{x} \cdot \mathbf{y}$ by $\mathbf{x} \cdot \mathbf{y} = x_1 y_1 + x_2 y_2$, we can rewrite (4.3) as

$$|\mathbf{x} - \mathbf{y}|^2 = |\mathbf{x}|^2 + |\mathbf{y}|^2 - 2\mathbf{x} \cdot \mathbf{y}. \tag{4.4}$$

This may look like a simple notational convenience, but it goes much further than that. It will allow us to "leverage" our understanding of geometry in two and three dimensions to arbitrary dimension. To get started, we make the following definition.

DEFINITION

(Dot product in $\mathbb{R}^n$) Given two vectors $\mathbf{x}$ and $\mathbf{y}$ in $\mathbb{R}^n$, their *dot product* is the number $\mathbf{x} \cdot \mathbf{y}$ given by

$$\mathbf{x} \cdot \mathbf{y} = x_1 y_1 + x_2 y_2 + \cdots + x_n y_n. \tag{4.5}$$

EXAMPLE 27   **(Computing dot products)** Computing dot products is very simple. Here is an example:

$$\begin{bmatrix} 1 \\ 2 \end{bmatrix} \cdot \begin{bmatrix} 3 \\ 4 \end{bmatrix} = 1 \times 3 + 2 \times 4 = 11.$$

The same thing is perhaps even clearer if we do the products on the right before writing them down:

$$\begin{bmatrix} 1 \\ 2 \end{bmatrix} \cdot \begin{bmatrix} 3 \\ 4 \end{bmatrix} = 3 + 8 = 11.$$

In the same way,

$$\begin{bmatrix} 0 \\ 4 \\ 2 \end{bmatrix} \cdot \begin{bmatrix} 2 \\ -1 \\ 2 \end{bmatrix} = 0 - 4 + 4 = 0. \tag{4.6}$$

Notice that the dot product of two vectors is a number and that the dot product is defined only when the two vectors have the same dimension.

The next theorem summarizes some important algebraic properties of the dot product.

THEOREM 8   **(Algebraic properties of the dot product)** *Let $\mathbf{x}$ and $\mathbf{y}$ be any two vectors in $\mathbb{R}^n$. Then the dot product is commutative, which is to say*

$$\mathbf{x} \cdot \mathbf{y} = \mathbf{y} \cdot \mathbf{x}; \tag{4.7}$$

moreover,

$$\mathbf{x} \cdot \mathbf{x} = 0 \iff \mathbf{x} = 0. \tag{4.8}$$

Also, the dot product is distributive, which is to say that for any k vectors $\mathbf{x}_1, \mathbf{x}_2, \ldots, \mathbf{x}_k$ in $\mathbb{R}^n$, any other vector $\mathbf{y}$ in $\mathbb{R}^n$, and any k numbers $a_1, a_2, \ldots, a_k$,

$$\mathbf{y} \cdot (a_1 \mathbf{x}_1 + a_2 \mathbf{x}_2 + \cdots + a_k \mathbf{x}_k) = a_1 \mathbf{y} \cdot \mathbf{x}_1 + a_2 \mathbf{y} \cdot \mathbf{x}_2 + \cdots + a_k \mathbf{y} \cdot \mathbf{x}_k. \tag{4.9}$$

PROOF: To see the identities (4.7) and (4.9), write out both sides in terms of the entries. Similarly, writing $\mathbf{x} \cdot \mathbf{x}$ in terms of the entries of $\mathbf{x}$, $\mathbf{x} \cdot \mathbf{x} = x_1^2 + x_2^2 + \cdots + x_n^2$. Each term on the right is nonnegative, so $\mathbf{x} \cdot \mathbf{x} = 0$ if and only if $x_j = 0$ for each j, which is what $\mathbf{x} = 0$ means. ∎

We close this subsection with one more simple but useful result involving the dot product.

THEOREM 9   **(The dot product and matrix entries)** *Let A be an $m \times n$ matrix. Then for any i with $1 \leq i \leq m$ and any j with $1 \leq j \leq n$,*

$$A_{i,j} = \mathbf{e}_i \cdot (A\mathbf{e}_j), \tag{4.10}$$

where $\mathbf{e}_i$ is the ith standard basis vector in $\mathbb{R}^m$ and $\mathbf{e}_j$ is the jth standard basis vector in $\mathbb{R}^n$.

PROOF: By the last part of Theorem 2,

$$A\mathbf{e}_j = \begin{bmatrix} A_{1,j} \\ A_{2,j} \\ \vdots \\ A_{m,j} \end{bmatrix}.$$

By the definitions of $\mathbf{e}_i$ and the dot product, it follows that $\mathbf{e}_i \cdot (A\mathbf{e}_j)$ is just the ith entry of $A\mathbf{e}_j$, namely $A_{i,j}$.   ■

4.3 Geometry and the dot product

Let us return to formula (4.4) for the distance between two vectors in $\mathbb{R}^2$. This formula has a simple geometric interpretation in $\mathbb{R}^2$: Consider the triangle with vertices $(0, 0)$, (x_1, x_2), and (y_1, y_2). Let a, b, and c denote the lengths of the three sides of this triangle, with a being the length of the side opposite the vertex (x_1, x_2), b being the length of the side opposite the vertex (y_1, y_2), and c being the length of the side opposite the vertex $(0, 0)$.

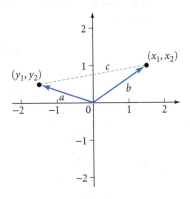

By the law of cosines,

$$c^2 = a^2 + b^2 - 2ab\cos(\theta), \tag{4.11}$$

where θ is the angle at the vertex $(0, 0)$. As you can see from the diagram, $a = |\mathbf{y}|$, $b = |\mathbf{x}|$, and $c = |\mathbf{x} - \mathbf{y}|$. Hence (4.11) is the same as

$$|\mathbf{x} - \mathbf{y}|^2 = |\mathbf{x}|^2 + |\mathbf{y}|^2 - 2|\mathbf{x}||\mathbf{y}|\cos(\theta). \tag{4.12}$$

Comparing (4.4) and (4.12),

$$|\mathbf{x}||\mathbf{y}|\cos(\theta) = \mathbf{x} \cdot \mathbf{y}. \tag{4.13}$$

Considering the diagram, we see that θ is the angle between the vectors $\mathbf{x}$ and $\mathbf{y}$. We see from (4.13) that we can express this angle in terms of the dot product through

$$\cos(\theta) = \frac{\mathbf{x} \cdot \mathbf{y}}{|\mathbf{x}||\mathbf{y}|},$$

and hence

$$\theta = \cos^{-1}\left(\frac{\mathbf{x} \cdot \mathbf{y}}{|\mathbf{x}||\mathbf{y}|}\right). \tag{4.14}$$

By the definition of the inverse cosine function, we always have $0 \le \theta \le \pi$, which is correct for an interior angle of a triangle.

Some special values of θ are especially interesting. Notice that $\cos(\theta) = 0$ if and only if $\mathbf{x} \cdot \mathbf{y} = 0$, and of course $\cos(\theta) = 0$ if and only if $\theta = \pi/2$. That is, *our triangle is a right triangle with $\mathbf{x}$ and $\mathbf{y}$ running along the orthogonal sides exactly when* $\mathbf{x} \cdot \mathbf{y} = 0$. Computing the dot product therefore gives us a test for when two vectors in $\mathbb{R}^2$ point in perpendicular directions: This is the case if and only if their dot product is zero.

The formula is also interesting when $\theta = 0$. Then $\cos(\theta) = 1$ and $\mathbf{x} \cdot \mathbf{y} = |\mathbf{x}||\mathbf{y}|$. Of course $\theta = 0$ when $\mathbf{y} = \mathbf{x}$, so

$$|\mathbf{x}|^2 = \mathbf{x} \cdot \mathbf{x}. \tag{4.15}$$

In fact, we do not need (4.14) to deduce this formula: From the definition of the length and from the definition of the dot product, we see that both sides of (4.15) equal $x_1^2 + x_2^2$. *Either way, we see that the dot product is useful for calculating lengths, angles, and distances.*

Let us generalize our formulas for computing angles and distance so that they can be applied in n dimensions. Since our formulas for distance and angle in the plane are expressed in terms of the dot product, which we have already defined in n dimensions, it is natural to proceed as follows.

DEFINITION

(Length, distance, and angle in $\mathbb{R}^n$) Let $\mathbf{x}$ be any vector in $\mathbb{R}^n$. The *length* of $\mathbf{x}$, denoted by $|\mathbf{x}|$, is defined by

$$|\mathbf{x}|^2 = \mathbf{x} \cdot \mathbf{x}. \tag{4.16}$$

The *distance* between two vectors $\mathbf{x}$ and $\mathbf{y}$ in $\mathbb{R}^n$ is the length of their difference, namely, $|\mathbf{x} - \mathbf{y}|$. Finally, the *angle* θ between any two nonzero vectors $\mathbf{x}$ and $\mathbf{y}$ in $\mathbb{R}^n$ is given by

$$\theta = \cos^{-1}\left(\frac{\mathbf{x} \cdot \mathbf{y}}{|\mathbf{x}||\mathbf{y}|}\right). \tag{4.17}$$

By the properties of the inverse cosine function, $0 \le \theta \le \pi$.

That was quick, but does it really make sense, and does it do anything for us computationally? The two questions are closely related, and the answer is "yes" in both cases. Let us start with one possible cause for concern: Since $|\cos(\theta)| \le 1$ no matter what the value of θ may be, we need to have

$$|\mathbf{x} \cdot \mathbf{y}| \le |\mathbf{x}||\mathbf{y}|,$$

otherwise (4.17) just does not define any angle at all. We know everything is consistent in $\mathbb{R}^2$, where we derived (4.17) using the law of cosines. Is it consistent in higher dimensions as well?

We are now in a position to address this question using algebra alone, since we have algebraic formulas for $|\mathbf{x} \cdot \mathbf{y}|$, $|\mathbf{x}|$ and $|\mathbf{y}|$. Here is how to do so. Consider any two nonzero vectors $\mathbf{x}$ and $\mathbf{y}$ in $\mathbb{R}^n$. Divide by their length to make them *unit vectors*, that is, vectors of unit length. In this way we define

$$\mathbf{u} = \frac{1}{|\mathbf{x}|}\mathbf{x} \quad \text{and} \quad \mathbf{v} = \frac{1}{|\mathbf{y}|}\mathbf{y}.$$

Notice that

$$\mathbf{u} \cdot \mathbf{v} = \frac{\mathbf{x} \cdot \mathbf{y}}{|\mathbf{x}||\mathbf{y}|}, \tag{4.18}$$

so to show that (4.17) makes sense we have to show only that $-1 \le \mathbf{u} \cdot \mathbf{v} \le 1$.

To get a handle on $\mathbf{u} \cdot \mathbf{v}$, let us compute $|\mathbf{u} - \mathbf{v}|^2$. (This is probably not an obvious thing to do at this stage, but you do not need to invent the idea—you only need to understand it.)* Using Theorem 8 several times, we find that

$$\begin{aligned}
|\mathbf{u} - \mathbf{v}|^2 &= (\mathbf{u} - \mathbf{v}) \cdot (\mathbf{u} - \mathbf{v}) \\
&= \mathbf{u} \cdot \mathbf{u} + \mathbf{v} \cdot \mathbf{v} - 2\mathbf{u} \cdot \mathbf{v} \\
&= 2 - 2\mathbf{u} \cdot \mathbf{v}.
\end{aligned}$$

This equation says that

$$\mathbf{u} \cdot \mathbf{v} = 1 - \frac{|\mathbf{u} - \mathbf{v}|^2}{2} \le 1.$$

In the exact same way, computing $|\mathbf{u} + \mathbf{v}|^2$, we find

$$\mathbf{u} \cdot \mathbf{v} = -1 + \frac{|\mathbf{u} + \mathbf{v}|^2}{2} \ge -1.$$

Putting it all together, $-1 \le \mathbf{u} \cdot \mathbf{v} \le 1$, and hence, by (4.18),

$$-1 \le \frac{\mathbf{x} \cdot \mathbf{y}}{|\mathbf{x}||\mathbf{y}|} \le 1. \tag{4.19}$$

Thus, (4.17) does provide a meaningful definition of θ.

The result we just deduced, (4.19), has many other uses besides showing that (4.17) makes sense. We can rewrite it in a more standard form if we multiply through by $|\mathbf{x}||\mathbf{y}|$:

$$|\mathbf{x} \cdot \mathbf{y}| \le |\mathbf{x}||\mathbf{y}|. \tag{4.20}$$

Better yet, we see from (4.18) that $|\mathbf{x} \cdot \mathbf{y}| = |\mathbf{x}||\mathbf{y}|$ if and only if $|\mathbf{u} \cdot \mathbf{v}| = 1$, which in turn happens only if either $|\mathbf{u} - \mathbf{v}| = 0$ or $|\mathbf{u} + \mathbf{v}| = 0$. This happens if and only if

$$\frac{1}{|\mathbf{x}|}\mathbf{x} = \pm \frac{1}{|\mathbf{y}|}\mathbf{y} \qquad \text{or equivalently} \qquad |\mathbf{y}|\mathbf{x} = \pm|\mathbf{x}|\mathbf{y}.$$

In geometric terms, there is equality in (4.20) if and only if $\mathbf{x}$ and $\mathbf{y}$ are proportional. This agrees with our definition of θ, since when equality holds in (4.20), we have $\cos(\theta) = \pm 1$, which means either $\theta = 0$ or $\theta = \pi$. If $\mathbf{x}$ and $\mathbf{y}$ are multiples of one another, then these vectors either point in the same direction, which corresponds to $\theta = 0$, or they point in opposite directions, which corresponds to $\theta = \pi$.

Inequality (4.20) is known as the *Schwarz inequality*. It is also the key to seeing that the length of $\mathbf{x} - \mathbf{y}$ is in fact a good measure of the distance between $\mathbf{x}$ and $\mathbf{y}$.

The point here is that anything that deserves to be called a measure of the distance between two vectors *had better act like a distance*. In mathematics, the notion of distance has a precise meaning, so it is only proper to call something a measure of distance between vectors when the following three properties hold:

1. For any two vectors $\mathbf{x}$ and $\mathbf{y}$, the *distance* from $\mathbf{x}$ to $\mathbf{y}$ is zero if and only if $\mathbf{x} = \mathbf{y}$.

2. For any two vectors $\mathbf{x}$ and $\mathbf{y}$, the *distance* from $\mathbf{x}$ to $\mathbf{y}$ is the same as the *distance* from $\mathbf{y}$ to $\mathbf{x}$.

3. For any three vectors $\mathbf{x}$, $\mathbf{y}$, and $\mathbf{z}$, the *distance* from $\mathbf{x}$ to $\mathbf{z}$ is no more than the sum of the distances from $\mathbf{x}$ to $\mathbf{y}$ and from $\mathbf{y}$ to $\mathbf{z}$.

*In fact, we are indebted to Michael Loss for this argument.

This list of three properties may seem somewhat arbitrary, but we will see that these properties are the ones that "come up all the time" in analysis of vector problems, so sometime down the road, it will become clear that this is the right list of properties to require.

Also, all these properties hold for the Euclidean distance in the plane. The first two are evident. The third one is called the *triangle inequality*. Consider three vectors **x**, **y**, and **z** in $\mathbb{R}^2$ as points in the plane. The distance from **x** to **y** is the length of one side of the triangle, the distance from **y** to **z** is another, and the distance from **x** to **z** is the third. Since the length of any side of a triangle in the plane is no greater than the sum of the lengths of the other two sides, the Euclidean distance in the plane satisfies the triangle inequality.

The intuitive meaning of the triangle inequality comes from thinking of the distance between two points as representing the "length of the shortest path between them." If you insist on going from **x** to **z** by passing though **y** on the way, that can only lengthen your path.

We have seen by a geometrical argument that the distance we have defined for vectors has this property in $\mathbb{R}^2$. We can use the Schwarz inequality to give an algebraic demonstration of the fact that the triangle inequality holds in every dimension.

Consider three vectors **x**, **y**, and **z** in $\mathbb{R}^n$. We want to show that

$$|\mathbf{x} - \mathbf{z}| \leq |\mathbf{x} - \mathbf{y}| + |\mathbf{y} - \mathbf{z}|, \tag{4.21}$$

which is to say that the distance from **x** to **z** is no more than the sum of the distances from **x** to **y** and then from **y** on to **z**. Let $\mathbf{v} = \mathbf{x} - \mathbf{y}$ and $\mathbf{w} = \mathbf{y} - \mathbf{z}$ so that

$$\mathbf{x} - \mathbf{z} = \mathbf{v} + \mathbf{w}.$$

Then

$$\begin{aligned}
|\mathbf{v} + \mathbf{w}|^2 &= (\mathbf{v} + \mathbf{w}) \cdot (\mathbf{v} + \mathbf{w}) \\
&= \mathbf{v} \cdot \mathbf{v} + \mathbf{w} \cdot \mathbf{w} + 2\mathbf{v} \cdot \mathbf{w} \\
&= |\mathbf{v}|^2 + |\mathbf{w}|^2 + 2\mathbf{v} \cdot \mathbf{w}.
\end{aligned}$$

Now apply the Schwarz inequality, (4.20), to conclude that

$$\begin{aligned}
|\mathbf{v} + \mathbf{w}|^2 &\leq |\mathbf{v}|^2 + |\mathbf{w}|^2 + 2\mathbf{w} \cdot \mathbf{v} \\
&\leq |\mathbf{v}|^2 + |\mathbf{w}|^2 + 2|\mathbf{w}||\mathbf{v}| \\
&= (|\mathbf{v}| + |\mathbf{w}|)^2,
\end{aligned}$$

and hence, taking the square root of both sides, that

$$|\mathbf{v} + \mathbf{w}| \leq |\mathbf{v}| + |\mathbf{w}|. \tag{4.22}$$

Recalling the definitions of **v** and **w**, we see that the triangle inequality, (4.21), does indeed hold. Thus we have seen that our *n*-dimensional distance function has the properties (1), (2), and (3) required of a measure of distance. Inequality (4.22), which we deduced along the way, is also interesting: It says that the length of a sum of vectors is no more than the sum of their lengths. Again, this inequality is easy to visualize in the plane. It is useful, too, so it has a name: the *Minkowski inequality*. We summarize these results in the following theorem.

THEOREM 10   **(The Schwarz, Minkowski, and triangle inequalities)** *For any two vectors* **x** *and* **y** *in* $\mathbb{R}^n$,

$$|\mathbf{x} \cdot \mathbf{y}| \leq |\mathbf{x}||\mathbf{y}|,$$

and there is equality if and only if **x** *is a multiple of* **y** *or if* **y** *is a multiple of* **x**. *For any two vectors* **v** *and* **w** *in* $\mathbb{R}^n$,

$$|\mathbf{v} + \mathbf{w}| \leq |\mathbf{v}| + |\mathbf{w}|,$$

and again, there is equality if and only if **v** *is a multiple of* **w** *or if* **w** *is a multiple of* **v**. *Finally, for any three vectors* **x**, **y**, *and* **z** *in* $\mathbb{R}^n$,

$$|\mathbf{x} - \mathbf{z}| \leq |\mathbf{x} - \mathbf{y}| + |\mathbf{y} - \mathbf{z}|.$$

4.4 Parallel and perpendicular components

We say that two nonzero vectors $\mathbf{x}$ and $\mathbf{y}$ in $\mathbb{R}^n$ are *parallel* if and only if one is a multiple of the other, for example, $\mathbf{x} = a\mathbf{y}$ for some number a. Then $|\mathbf{x}| = |a\mathbf{y}| = |a||\mathbf{y}|$, and

$$\mathbf{x} \cdot \mathbf{y} = a\mathbf{y} \cdot \mathbf{y} = a|\mathbf{y}|^2.$$

Hence

$$\frac{\mathbf{x} \cdot \mathbf{y}}{|\mathbf{x}||\mathbf{y}|} = \frac{a}{|a|} = \pm 1.$$

Therefore, the angle θ between $\mathbf{x}$ and $\mathbf{y}$ is either 0 or π, depending on whether a is positive or negative.

We say that two nonzero vectors $\mathbf{x}$ and $\mathbf{y}$ in $\mathbb{R}^n$ are *orthogonal*, which is just another word for "perpendicular," in the case where the angle between them is $\pi/2$. According to (4.17), this is the case if and only if $\mathbf{x} \cdot \mathbf{y} = 0$. We later make so much use of this fact that it deserves a separate definition:

DEFINITION

(Orthogonal vectors) Two vectors in $\mathbb{R}^n$, $\mathbf{x}$ and $\mathbf{y}$, are *orthogonal* to one another if and only if $\mathbf{x} \cdot \mathbf{y} = 0$. In the same way, we say that a set $\{\mathbf{x}_1, \mathbf{x}_2, \ldots, \mathbf{x}_k\}$ of k vectors in $\mathbb{R}^n$ is *orthogonal* if and only if

$$\mathbf{x}_i \cdot \mathbf{x}_j = 0 \tag{4.23}$$

for all $i \neq j$ in $1, 2, \ldots, k$.

EXAMPLE 28 **(Orthogonal vectors)** The simplest example is provided by the standard basis vectors $\mathbf{e}_1, \mathbf{e}_2, \ldots$. Here is another example in $\mathbb{R}^2$: $\mathbf{x}_1 = \begin{bmatrix} 1 \\ 1 \end{bmatrix}$ and $\mathbf{x}_2 = \begin{bmatrix} 3 \\ -3 \end{bmatrix}$.

In general, two nonzero vectors in $\mathbb{R}^n$ are neither parallel nor perpendicular. However, we can always take one of them apart into two components that are, respectively, parallel or perpendicular to the other vector. This fact will be very useful.

THEOREM 11 **(Parallel and perpendicular components)** *Let $\mathbf{x}$ and $\mathbf{y}$ be any two nonzero vectors in $\mathbb{R}^n$. Then there are vectors $\mathbf{x}_\parallel$ and $\mathbf{x}_\perp$ such that*

$$\mathbf{x} = \mathbf{x}_\parallel + \mathbf{x}_\perp \tag{4.24}$$

and such that $\mathbf{x}_\parallel$ is parallel to $\mathbf{y}$, while $\mathbf{x}_\perp$ is orthogonal to $\mathbf{y}$. Moreover, the vectors $\mathbf{x}_\parallel$ and $\mathbf{x}_\perp$ are uniquely determined and given by the formulas

$$\mathbf{x}_\parallel = (\mathbf{u} \cdot \mathbf{x})\mathbf{u} \quad \text{and} \quad \mathbf{x}_\perp = \mathbf{x} - (\mathbf{u} \cdot \mathbf{x})\mathbf{u}, \tag{4.25}$$

where $\mathbf{u}$ is the unit vector $\mathbf{u} = (1/|\mathbf{y}|)\mathbf{y}$ in the direction of $\mathbf{y}$.

The following diagram displays such a decomposition in $\mathbb{R}^2$.

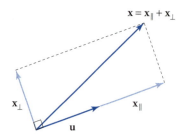

From the diagram, you can see why such a decomposition exists and even deduce the formulas in (4.25). Indeed, assume that there is such a decomposition $\mathbf{x} = \mathbf{x}_\| + \mathbf{x}_\perp$. Taking the dot product of both sides with $\mathbf{u}$,

$$\mathbf{x} \cdot \mathbf{u} = \mathbf{x}_\| \cdot \mathbf{u}, \tag{4.26}$$

since $\mathbf{u}$ is parallel to $\mathbf{y}$ and hence orthogonal to $\mathbf{x}_\perp$.

Since $\mathbf{x}_\|$ is parallel to $\mathbf{y}$ and hence to $\mathbf{u}$, there is some number a with $\mathbf{x}_\| = a\mathbf{u}$. But then

$$\mathbf{x}_\| \cdot \mathbf{u} = a\mathbf{u} \cdot \mathbf{u} = a. \tag{4.27}$$

Combining (4.26) and (4.27), $a = \mathbf{x} \cdot \mathbf{u}$, which gives the first formula in (4.25); then the second formula in (4.25) follows and $\mathbf{x} = \mathbf{x}_\| + \mathbf{x}_\perp$. This derivation of (4.25) proves the uniqueness of $\mathbf{x}_\|$ and $\mathbf{x}_\perp$, provided they exist: If there is any such decomposition, $\mathbf{x}_\|$ and $\mathbf{x}_\perp$ *must* be given by (4.25).

PROOF OF THEOREM 11: First, it is clear that the vectors $\mathbf{x}_\|$ and $\mathbf{x}_\perp$ in (4.25) satisfy $\mathbf{x} = \mathbf{x}_\| + \mathbf{x}_\perp$. Next, $\mathbf{x}_\|$ is defined to be a multiple of $\mathbf{u}$, which is in turn a multiple of $\mathbf{y}$, so $\mathbf{x}_\|$ is parallel to $\mathbf{y}$.

Checking that $\mathbf{x}_\perp$ is orthogonal to $\mathbf{y}$ amounts to checking that $\mathbf{x}_\perp$ is orthogonal to $\mathbf{u}$. Taking the dot product of these two vectors and using the distributive property from Theorem 8,

$$\mathbf{x}_\perp \cdot \mathbf{u} = (\mathbf{x} - (\mathbf{u} \cdot \mathbf{x})\mathbf{u}) \cdot \mathbf{u} = \mathbf{x} \cdot \mathbf{u} - (\mathbf{u} \cdot \mathbf{x})\mathbf{u} \cdot \mathbf{u} = 0,$$

since $\mathbf{u} \cdot \mathbf{u} = 1$. Therefore, the formulas (4.25) do give us an orthogonal decomposition as claimed. In our derivation of these formulas, we have already seen that there is no other way to do so. ∎

Exercises

4.1 Consider the following vectors:

$$\mathbf{v} = \begin{bmatrix} 1 \\ 1 \end{bmatrix} \qquad \mathbf{x} = \begin{bmatrix} 1 \\ -1 \end{bmatrix} \qquad \mathbf{y} = \begin{bmatrix} 2 \\ -1 \\ 2 \end{bmatrix} \qquad \mathbf{z} = \begin{bmatrix} -1 \\ -1 \\ 3 \\ 2 \end{bmatrix}.$$

Compute the length of each vector and the dot product of each pair of vectors for which it is defined.

4.2 Why does it make no sense to ask whether or not the dot product is associative?

4.3 Consider the vectors

$$\mathbf{v}_1 = \begin{bmatrix} 1 \\ 2 \end{bmatrix}, \qquad \mathbf{v}_2 = \begin{bmatrix} 2 \\ 1 \end{bmatrix}, \qquad \text{and} \qquad \mathbf{v}_3 = \begin{bmatrix} -2 \\ 1 \end{bmatrix}.$$

(a) Compute $\mathbf{v}_i \cdot \mathbf{v}_j$ for each $i, j = 1, 2, 3$.

(b) What is the length of each of the three vectors?

(c) What is the angle between each of the three pairs of vectors? Is any pair orthogonal?

(d) Draw a diagram showing each of the three vectors as arrows. Do the lengths and angles that you computed look right?

4.4 Consider the vectors

$$\mathbf{v}_1 = \begin{bmatrix} 3 \\ 1 \end{bmatrix}, \qquad \mathbf{v}_2 = \begin{bmatrix} 2 \\ 2 \end{bmatrix}, \qquad \text{and} \qquad \mathbf{v}_2 = \begin{bmatrix} -1 \\ 3 \end{bmatrix}.$$

(a) Compute $\mathbf{v}_i \cdot \mathbf{v}_j$ for each $i, j = 1, 2, 3$.

(b) What is the length of each of the three vectors?

(c) What is the angle between each of the three pairs of vectors? Is any pair orthogonal?

(d) Draw a diagram showing the three vectors as arrows. Do the lengths and angles that you computed look right?

4.5 Fix a number r with $-1 < r < 1$. For each n, let ℓ_n be the length of the vector

$$\begin{bmatrix} 1 \\ r \\ r^2 \\ \vdots \\ r^{n-1} \end{bmatrix}$$

in $\mathbb{R}^n$.

(a) Compute ℓ_n as a function of r.

(b) Now fix any other number s with $-1 < s < 1$. For each n, let α_n be the angle between the vectors

$$\begin{bmatrix} 1 \\ r \\ r^2 \\ \vdots \\ r^{n-1} \end{bmatrix} \qquad \text{and} \qquad \begin{bmatrix} 1 \\ s \\ s^2 \\ \vdots \\ s^{n-1} \end{bmatrix}$$

in $\mathbb{R}^n$. Compute α_n as a function of r and s.

(c) Compute $\lim_{n \to \infty} \ell_n$ and $\lim_{n \to \infty} \alpha_n$. Think about what the existence of these limits might say about the possibility of geometric considerations in infinitely many dimensions.

4.6 Consider the vectors

$$\mathbf{x} = \begin{bmatrix} 1 \\ 1 \\ 1 \end{bmatrix} \qquad \text{and} \qquad \mathbf{y} = \begin{bmatrix} 1 \\ -2 \\ 1 \end{bmatrix}.$$

(a) Compute $|\mathbf{x}|$, $|\mathbf{y}|$, and $\mathbf{x} \cdot \mathbf{y}$.

(b) Compute $|s\mathbf{x} + t\mathbf{y}|$ as a function of s and t.

(c) How does the angle between $\mathbf{x}$ and $\mathbf{x} + t\mathbf{y}$ depend on t?

4.7 Consider the vector

$$\mathbf{a} = \begin{bmatrix} a \\ b \end{bmatrix}.$$

Find all vectors

$$\begin{bmatrix} c \\ d \end{bmatrix}$$

that are orthogonal to **a**; that is, find conditions on c and d in terms of a and b that are necessary and sufficient for this orthogonality.

4.8 Let **u** be any unit vector in $\mathbb{R}^2$. Define a transformation f from $\mathbb{R}^2$ to $\mathbb{R}^2$ by

$$f(\mathbf{x}) = \mathbf{x} - 2(\mathbf{u} \cdot \mathbf{x})\mathbf{u}.$$

(a) Show that this transformation is linear and is *length preserving*; that is, for all **x**, the length of the output $|f(\mathbf{x})|$ equals the length of the input $|\mathbf{x}|$.

(b) Show that $f \circ f$ is the identity transformation.

(c) Specifically let

$$\mathbf{u} = \begin{bmatrix} \cos(\theta) \\ \sin(\theta) \end{bmatrix},$$

and find the matrix A_f.

4.9 Let **x** and **y** be two nonzero vectors in $\mathbb{R}^n$. Suppose that

$$|\mathbf{x} + \mathbf{y}|^2 = |\mathbf{x}|^2 + |\mathbf{y}|^2.$$

What is the angle between these two vectors?

4.10 Let **x** and **y** be any two vectors in $\mathbb{R}^n$. Show that the parallelogram identity, namely,

$$|\mathbf{x} + \mathbf{y}|^2 + |\mathbf{x} - \mathbf{y}|^2 = 2|\mathbf{x}|^2 + 2|\mathbf{y}|^2,$$

always holds.

4.11 The set of all vectors

$$\begin{bmatrix} x \\ y \\ z \end{bmatrix}$$

in $\mathbb{R}^3$ that are orthogonal to the fixed vector

$$\begin{bmatrix} 1 \\ 2 \\ -1 \end{bmatrix}$$

is a plane in $\mathbb{R}^3$. Find an equation specifying this plane.

4.12 Consider the vectors

$$\mathbf{x} = \begin{bmatrix} t \\ 1 \\ 1 \end{bmatrix} \quad \text{and} \quad \mathbf{y} = \begin{bmatrix} 1 \\ -2 \\ 1 \end{bmatrix}.$$

For which values of t, if any, are these vectors orthogonal? For which values of t, if any, are these vectors parallel?

4.13 Let $\{\mathbf{x}_1, \mathbf{x}_2, \ldots, \mathbf{x}_k\}$ be a collection of k orthogonal vectors in $\mathbb{R}^n$, and let $\mathbf{x} = \mathbf{x}_1 + \mathbf{x}_2 + \cdots + \mathbf{x}_k$ be their sum. Show that

$$|\mathbf{x}|^2 = |\mathbf{x}_1|^2 + |\mathbf{x}_2|^2 + \cdots + |\mathbf{x}_k|^2.$$

This equation generalizes the Pythagorean theorem to higher dimensions.

4.14 Let **u** be the unit vector $\mathbf{u} = \frac{1}{5}\begin{bmatrix} 3 \\ 4 \end{bmatrix}$ and let $\mathbf{x} = \begin{bmatrix} 1 \\ 2 \end{bmatrix}$. Compute $\mathbf{x}_{\parallel}$ and $\mathbf{x}_{\perp}$, where the direction is given by **u**.

4.15 Let **u** be the unit vector $\mathbf{u} = \frac{1}{5}\begin{bmatrix} -4 \\ 3 \end{bmatrix}$ and let $\mathbf{x} = \begin{bmatrix} 5 \\ 7 \end{bmatrix}$. Compute $\mathbf{x}_{\parallel}$ and $\mathbf{x}_{\perp}$, where the direction is given by **u**.

4.16 Let **u** be the unit vector $\mathbf{u} = \frac{1}{3}\begin{bmatrix} -2 \\ 1 \\ -2 \end{bmatrix}$ and let $\mathbf{x} = \begin{bmatrix} 5 \\ 1 \\ 2 \end{bmatrix}$. Compute $\mathbf{x}_{\parallel}$ and $\mathbf{x}_{\perp}$, where the direction is given by **u**.

4.17 Let **u** be the unit vector $\mathbf{u} = \frac{1}{\sqrt{6}}\begin{bmatrix} -2 \\ 1 \\ -1 \end{bmatrix}$ and let $\mathbf{x} = \begin{bmatrix} 1 \\ 1 \\ 1 \end{bmatrix}$. Compute $\mathbf{x}_{\parallel}$ and $\mathbf{x}_{\perp}$, where the direction is given by **u**.

SECTION 5 | Matrix Multiplication Revisited

5.1 Row and column forms

We have seen that it is natural to consider an $m \times n$ matrix A not only as a rectangular array of the mn numbers $A_{i,j}$ but as an array of vectors. Thinking of A as representing a linear transformation f_A from $\mathbb{R}^n$ to $\mathbb{R}^m$, the jth column of A, $\mathbf{c}_j$, is $\mathbf{c}_j = f_A(\mathbf{e}_j)$. Using this notation, we can write A as a horizontal array of n vectors in $\mathbb{R}^m$:

$$A = [\mathbf{c}_1, \mathbf{c}_2, \ldots, \mathbf{c}_n]. \tag{5.1}$$

This is the *column representation* of the matrix A.

Another way to write A as an array of vectors is to use the rows instead. Let $\mathbf{r}_i$ be the ith row of A. Consider A as a vertical array of m row vectors in $\mathbb{R}^n$:

$$A = \begin{bmatrix} \mathbf{r}_1 \\ \mathbf{r}_2 \\ \vdots \\ \mathbf{r}_m \end{bmatrix}. \tag{5.2}$$

This is the *row representation* of the matrix A.

These two representations are important in theory as well as in practice.* We have already had some practice in looking at matrix–vector and matrix–matrix multiplication in terms of the column representation. In this section, we bring in the row representation and the dot product.

Let us begin with matrix–vector multiplication.

---

*Those of you who have done programming with arrays know that programming languages generally do not *directly* support rectangular arrays. To work with rectangular arrays in a program, you have to store and operate on them as arrays of arrays, as in (5.1) and (5.2).

THEOREM 12   **(Matrix–vector multiplication in row and column terms)** *Let A be an m × n matrix with the column and row representations (5.1) and (5.2). Let*

$$\mathbf{v} = \begin{bmatrix} v_1 \\ v_2 \\ \vdots \\ v_n \end{bmatrix}$$

be any vector in $\mathbb{R}^n$. *Then A$\mathbf{v}$ is the linear combination of the columns of A given by*

$$A\mathbf{v} = v_1\mathbf{c}_1 + v_2\mathbf{c}_2 + \cdots + v_n\mathbf{c}_n, \tag{5.3}$$

and in terms of the rows of A,

$$A\mathbf{v} = \begin{bmatrix} \mathbf{r}_1 \cdot \mathbf{v} \\ \mathbf{r}_2 \cdot \mathbf{v} \\ \vdots \\ \mathbf{r}_m \cdot \mathbf{v} \end{bmatrix}. \tag{5.4}$$

PROOF: We do not need to prove (5.3); it is the definition of $A\mathbf{v}$. The new formula, (5.4), follows directly from (2.17), namely,

$$(A\mathbf{v})_i = \sum_{j=1}^{n} A_{i,j}v_j, \tag{5.5}$$

together with (5.2), which says that

$$A_{i,j} = (\mathbf{r}_i)_j. \tag{5.6}$$

Substituting (5.6) into (5.5) and using (4.5), the definition of the dot product, we deduce

$$(A\mathbf{v})_i = \sum_{j=1}^{n} (\mathbf{r}_i)_j v_j = \mathbf{r}_i \cdot \mathbf{v},$$

which is the same as (5.4).   ∎

EXAMPLE 29   **(Matrix-vector products using the row representation)** Let

$$A = \begin{bmatrix} 1 & 2 & 3 \\ 3 & 2 & 1 \end{bmatrix} \quad \text{and} \quad \mathbf{v} = \begin{bmatrix} 1 \\ -1 \\ 2 \end{bmatrix}.$$

The rows of A are

$$\mathbf{r}_1 = \begin{bmatrix} 1 \\ 2 \\ 3 \end{bmatrix} \quad \text{and} \quad \mathbf{r}_2 = \begin{bmatrix} 3 \\ 2 \\ 1 \end{bmatrix}.$$

Important convention: We write the rows vertically because that is our convention for writing vectors. We will always do this: *When we pull a row out of a matrix and think of it as a vector, we write it vertically with its left entry at the top and its right entry at the bottom.*

Continuing with the example, we compute that $\mathbf{r}_1 \cdot \mathbf{v} = 5$ and $\mathbf{r}_2 \cdot \mathbf{v} = 3$ with the result that

$$A\mathbf{v} = \begin{bmatrix} 5 \\ 3 \end{bmatrix}.$$

This is also what we get using (5.3):

$$A\mathbf{v} = \mathbf{c}_1 - \mathbf{c}_2 + 2\mathbf{c}_3 = \begin{bmatrix} 1 \\ 3 \end{bmatrix} - \begin{bmatrix} 2 \\ 2 \end{bmatrix} + 2\begin{bmatrix} 3 \\ 1 \end{bmatrix} = \begin{bmatrix} 5 \\ 3 \end{bmatrix}.$$

Why do we need two formulas for matrix–vector multiplication? Each has its particular advantages and uses. For example, the formula (5.4) brings *geometry* into the business of solving vector equations. Indeed, you see right away from (5.4) that $A\mathbf{v} = 0$ if and only if $\mathbf{v}$ is orthogonal to each row of A.*

Let us go on to the matrix–matrix product. Since this is just the matrix–vector product done in parallel, we can apply Theorem 12.

Let B be an $n \times p$ matrix so that the product AB, with A as before, is defined. Let $\mathbf{d}_j$ denote the jth column of B so that

$$B = [\mathbf{d}_1, \mathbf{d}_2, \ldots, \mathbf{d}_p] \tag{5.7}$$

and hence $AB = [A\mathbf{d}_1, A\mathbf{d}_2, \ldots, A\mathbf{d}_p]$. Theorem 12 gives us two ways to express each $A\mathbf{d}_j$, which in turn yields the following theorem:

THEOREM 13

(Matrix–matrix multiplication in row and column terms) *Let A be an $m \times n$ matrix with the column and row representations (5.1) and (5.2). Let B be an $n \times p$ matrix with the column representation (5.7). Then*

1. *The i,jth entry of AB is the dot product of the ith row of A with the jth column of B. Using the notation of (5.2) and (5.7),*

$$(AB)_{i,j} = \mathbf{r}_i \cdot \mathbf{d}_j. \tag{5.8}$$

2. *Each column of AB is a linear combination of the columns of A, and in particular, the jth column of AB is a linear combination of the columns of A with multiples taken from the jth column of B.*

PROOF: Since $(AB)_{i,j}$ is the ith component of the jth column of AB, $(AB)_{i,j} = (A\mathbf{d}_j)_i$ and (5.8) follows directly from (5.4), which proves part (1).

For the second part, we use (5.3). Since the jth column of AB is $A\mathbf{d}_j$ and since

$$\mathbf{d}_j = \begin{bmatrix} B_{1,j} \\ B_{2,j} \\ \vdots \\ B_{n,j} \end{bmatrix},$$

(5.3) with $\mathbf{d}_j$ in place of $\mathbf{v}$ gives

$$A\mathbf{d}_j = B_{1,j}\mathbf{c}_1 + B_{2,j}\mathbf{c}_2 + \cdots + B_{n,j}\mathbf{c}_n. \tag{5.9}$$

■

EXAMPLE 30

(Matrix products via row and column dot products) Let

$$A = \begin{bmatrix} 1 & 2 & 3 \\ 3 & 2 & 1 \end{bmatrix} \quad \text{and} \quad B = \begin{bmatrix} 1 & 0 & 3 \\ 0 & 1 & 4 \\ 2 & 0 & 1 \end{bmatrix}.$$

Let us use (5.9) to compute the third column of AB. This should clarify the meaning of (5.9), if nothing else. The columns of A are

$$\mathbf{c}_1 = \begin{bmatrix} 1 \\ 3 \end{bmatrix}, \quad \mathbf{c}_2 = \begin{bmatrix} 2 \\ 2 \end{bmatrix}, \quad \text{and} \quad \mathbf{c}_3 = \begin{bmatrix} 3 \\ 1 \end{bmatrix}.$$

*This may well seem like a mere curiosity right now, but it is our first hint of a very significant connection between the geometric ideas introduced in the last section and the problem of solving vector equations.

The third column of B is

$$\begin{bmatrix} 3 \\ 4 \\ 1 \end{bmatrix}.$$

Hence, according to (5.9), the third column of AB is

$$3\mathbf{c}_1 + 4\mathbf{c}_2 + \mathbf{c}_3 = 3\begin{bmatrix} 1 \\ 3 \end{bmatrix} + 4\begin{bmatrix} 2 \\ 2 \end{bmatrix} + \begin{bmatrix} 3 \\ 1 \end{bmatrix} = \begin{bmatrix} 14 \\ 18 \end{bmatrix}.$$

In the same way, we can easily compute the remaining columns in any order we like.

The next example uses the following definition.

DEFINITION

(Diagonal matrices) An $m \times n$ matrix is a *diagonal matrix* if and only if it is a square matrix (that is, $m = n$) such that $A_{i,j} = 0$ for all $i \neq j$. We write $\operatorname{diag}(a_1, a_2, \ldots, a_n)$ to denote the $n \times n$ diagonal matrix A with $A_{i,i} = a_i$ for $i = 1, 2 \ldots, n$.

EXAMPLE 31   **(Row and column representation of diagonal matrices)** Let $A = \operatorname{diag}(a_1, a_2, \ldots, a_n)$. Notice that A has the row and column representations

$$A = \begin{bmatrix} a_1\mathbf{e}_1 \\ a_2\mathbf{e}_2 \\ \vdots \\ a_n\mathbf{e}_n \end{bmatrix} \quad \text{and} \quad A = [a_1\mathbf{e}_1, a_2\mathbf{e}_2, \ldots, a_n\mathbf{e}_n].$$

Since similar formulas hold for any other $n \times n$ diagonal matrix $B = \operatorname{diag}(b_1, b_2, \ldots, b_n)$, from (5.8),

$$(AB)_{i,j} = (a_i\mathbf{e}_i) \cdot (b_j\mathbf{e}_j) = a_ib_j\mathbf{e}_i \cdot \mathbf{e}_j = \begin{cases} a_ib_i & \text{if } i = j \\ 0 & \text{if } i \neq j. \end{cases} \tag{5.10}$$

Therefore

$$AB = \operatorname{diag}(a_1b_1, a_2b_2, \ldots, a_nb_n) = \operatorname{diag}(b_1a_1, b_2a_2, \ldots, b_nb_n) = BA.$$

In other words, the product of any two diagonal matrices A and B is again diagonal, $AB = BA$, and the diagonal entries of the product are just products of the corresponding diagonal entries of A and B.

Here is another simple observation we can make using (5.8): Suppose that A is invertible, and that B is its inverse. Then by (3.10),

$$(AB)_{i,j} = \mathbf{r}_i \cdot \mathbf{d}_j = I_{i,j} = \begin{cases} 1 & \text{if } i = j \\ 0 & \text{if } i \neq j. \end{cases} \tag{5.11}$$

Therefore the jth column of the inverse of A is orthogonal to the ith row of A for all $j \neq i$. Again, this is a hint of the important role that geometry plays in solving vector equations.

5.2 The transpose of a matrix

There is another relation between the row and column representations of A and the dot product. To explain it, we need a definition.

DEFINITION

(Transpose of a matrix) Let A be an $m \times n$ matrix. Then the *transpose of A* is the $n \times m$ matrix A^t whose jth column is the jth row of A. In other words, the columns of A are the rows of A^t and vice versa. That is, the i, jth entry of A^t, $(A^t)_{i,j}$, is given by

$$(A^t)_{i,j} = A_{j,i}. \tag{5.12}$$

EXAMPLE 32   **(Matrix transposes)** Here are some examples of matrices and their transposes:

$$A = \begin{bmatrix} 1 & 3 & 2 \\ 0 & 2 & 1 \end{bmatrix} \quad \text{and} \quad A^t = \begin{bmatrix} 1 & 0 \\ 3 & 2 \\ 2 & 1 \end{bmatrix};$$

$$B = \begin{bmatrix} 1 & 0 \\ 3 & 2 \\ 2 & 1 \end{bmatrix} \quad \text{and} \quad B^t = \begin{bmatrix} 1 & 3 & 2 \\ 0 & 2 & 1 \end{bmatrix}.$$

Notice in the last example that $B = A^t$ and $B^t = A$, which means that $(A^t)^t = A$. That is, taking the transpose twice gets us back to the matrix we started from. This always happens: Applying (5.12) twice, we have $((A^t)^t)_{i,j} = (A^t)_{j,i} = A_{i,j}$, which means that we always have

$$(A^t)^t = A. \tag{5.13}$$

Here is the main theorem.

THEOREM 14

(The transpose and the dot product) *Let A be an $m \times n$ matrix. Then*

1. *For any $\mathbf{x}$ in $\mathbb{R}^n$ and any $\mathbf{y}$ in $\mathbb{R}^m$,*

$$\mathbf{y} \cdot A\mathbf{x} = (A^t\mathbf{y}) \cdot \mathbf{x}. \tag{5.14}$$

2. *If B is any $n \times p$ matrix, then*

$$(AB)^t = B^t A^t. \tag{5.15}$$

Before we examine the proof, note the resemblance of (5.15) to the formula (3.11) for the product of the inverse of a pair of invertible $n \times n$ matrices A and B, namely, $(AB)^{-1} = B^{-1}A^{-1}$.

PROOF: Write A in row form as

$$A = \begin{bmatrix} \mathbf{r}_1 \\ \vdots \\ \mathbf{r}_m \end{bmatrix}.$$

Then by Theorem 12,

$$A\mathbf{x} = \begin{bmatrix} \mathbf{r}_1 \cdot \mathbf{x} \\ \vdots \\ \mathbf{r}_m \cdot \mathbf{x} \end{bmatrix}.$$

Therefore,

$$\mathbf{y} \cdot A\mathbf{x} = \mathbf{y} \cdot \begin{bmatrix} \mathbf{r}_1 \cdot \mathbf{x} \\ \vdots \\ \mathbf{r}_m \cdot \mathbf{x} \end{bmatrix} = y_1 \mathbf{r}_1 \cdot \mathbf{x} + \cdots + y_n \mathbf{r}_n \cdot \mathbf{x}$$

$$= (y_1 \mathbf{r}_1 + \cdots + y_m \mathbf{r}_m) \cdot \mathbf{x}.$$

But by Theorem 12 and the definition of A^t,

$$y_1 \mathbf{r}_1 + \cdots + y_m \mathbf{r}_m = [\mathbf{r}_1, \ldots, \mathbf{r}_m]\mathbf{y} = A^t \mathbf{y}$$

so that $(y_1 \mathbf{r}_1 + \cdots + y_m \mathbf{r}_m) \cdot \mathbf{x} = (A^t \mathbf{y}) \cdot \mathbf{x}$.

For the second part, apply (5.14) in two steps. For all $\mathbf{x}$ in $\mathbb{R}^p$ and $\mathbf{y}$ in $\mathbb{R}^n$,

$$\mathbf{y} \cdot AB\mathbf{x} = (A^t \mathbf{y}) \cdot B\mathbf{x} = (B^t(A^t(\mathbf{y}))) \cdot \mathbf{x} = ((B^t A^t)\mathbf{y}) \cdot \mathbf{x}. \tag{5.16}$$

Applying (5.14) all at once,

$$\mathbf{y} \cdot AB\mathbf{x} = ((AB)^t \mathbf{y}) \cdot \mathbf{x}. \tag{5.17}$$

Combining (5.16) and (5.17),

$$((B^t A^t)\mathbf{y}) \cdot \mathbf{x} = ((AB)^t \mathbf{y}) \cdot \mathbf{x}$$

for all $\mathbf{x}$ and $\mathbf{y}$ in $\mathbb{R}^p$ and $\mathbb{R}^m$, respectively. Taking $\mathbf{x} = \mathbf{e}_i$ in $\mathbb{R}^n$ and taking $\mathbf{y} = \mathbf{e}_j$ in $\mathbb{R}^m$, we have from Theorem 9 that for all i and j

$$(B^t A^t)_{i,j} = \mathbf{e}_i \cdot (B^t A^t \mathbf{e}_j) = (B^t A^t \mathbf{e}_j) \cdot \mathbf{e}_i$$

$$= ((AB)^t \mathbf{e}_j) \cdot \mathbf{e}_i = \mathbf{e}_i \cdot ((AB)^t \mathbf{e}_j) = ((AB)^t)_{i,j},$$

which proves (5.15). ∎

We will apply Theorem 14 often. Here is one application.

5.3 Transposes and isometries

DEFINITION	**(Isometry)** An $m \times n$ matrix A is an *isometry* if and only if for all $\mathbf{x}$ in $\mathbb{R}^n$, $$\|A\mathbf{x}\| = \|\mathbf{x}\|. \tag{5.18}$$

In other words, A is an isometry if the length of the output is the same as the length of the input.

The term *isometry* comes from the classical Greek for "same measure." By definition, isometries preserve the lengths of vectors. However, they also preserve angles between vectors. To see why this is the case, let A be an $n \times n$ matrix that is an isometry, and let $\mathbf{x}$ and $\mathbf{y}$ be any two nonzero vectors in $\mathbb{R}^n$. Then $|A(\mathbf{x} - \mathbf{y})|^2 = |\mathbf{x} - \mathbf{y}|^2$ since A is an isometry. Since by linearity $A(\mathbf{x} - \mathbf{y}) = A\mathbf{x} - A\mathbf{y}$,

$$|A\mathbf{x} - A\mathbf{y}|^2 = |\mathbf{x} - \mathbf{y}|^2. \tag{5.19}$$

Writing the left side out in terms of the dot product,

$$|A\mathbf{x} - A\mathbf{y}|^2 = (A\mathbf{x} - A\mathbf{y}) \cdot (A\mathbf{x} - A\mathbf{y})$$

$$= A\mathbf{x} \cdot A\mathbf{x} + A\mathbf{y} \cdot A\mathbf{y} - 2A\mathbf{x} \cdot A\mathbf{y}$$

$$= |A\mathbf{x}|^2 + |A\mathbf{y}|^2 - 2A\mathbf{x} \cdot A\mathbf{y}$$

$$= |\mathbf{x}|^2 + |\mathbf{y}|^2 - 2A\mathbf{x} \cdot A\mathbf{y}.$$

Much more directly,

$$|\mathbf{x} - \mathbf{y}|^2 = |\mathbf{x}|^2 + |\mathbf{y}|^2 - 2\mathbf{x} \cdot \mathbf{y}.$$

Combining these computations with (5.19), we see that isometries also preserve dot products, that is,

$$Ax \cdot Ay = \mathbf{x} \cdot \mathbf{y}. \tag{5.20}$$

From here it is a short step to the preservation of angles. Let θ be the angle between $\mathbf{x}$ and $\mathbf{y}$, and let ϕ be the angle between $A\mathbf{x}$ and $A\mathbf{y}$. Then

$$\cos(\theta) = \frac{\mathbf{x} \cdot \mathbf{y}}{|\mathbf{x}||\mathbf{y}|} \quad \text{and} \quad \cos(\phi) = \frac{(A\mathbf{x}) \cdot (A\mathbf{y})}{|A\mathbf{x}||A\mathbf{y}|}.$$

But $(A\mathbf{x}) \cdot (A\mathbf{y}) = \mathbf{x} \cdot \mathbf{y}$ and $|A\mathbf{x}||A\mathbf{y}| = |\mathbf{x}||\mathbf{y}|$. Therefore,

$$\frac{\mathbf{x} \cdot \mathbf{y}}{|\mathbf{x}||\mathbf{y}|} = \frac{(A\mathbf{x}) \cdot (A\mathbf{y})}{|A\mathbf{x}||A\mathbf{y}|},$$

so $\theta = \phi$. This is what it means for A to be angle preserving.

In particular, if A is an isometry, then for any $\mathbf{x}$ and $\mathbf{y}$, $A\mathbf{x}$ and $A\mathbf{y}$ are orthogonal if and only if $\mathbf{x}$ and $\mathbf{y}$ are orthogonal.

- *A linear transformation that preserves lengths and hence distances automatically preserves angles, too.*

There is more to say about isometries: Suppose that $A = [\mathbf{v}_1, \mathbf{v}_2, \ldots, \mathbf{v}_n]$ is an $m \times n$ isometry. For each $j = 1, 2, \ldots, n$, $\mathbf{v}_j = A\mathbf{e}_j$. Then by (5.20), we have for each $i, j = 1, 2, \ldots, n$ that

$$\mathbf{v}_i \cdot \mathbf{v}_j = A\mathbf{e}_i \cdot A\mathbf{e}_j = \mathbf{e}_i \cdot \mathbf{e}_j = \begin{cases} 1 & \text{if} \quad i = j \\ 0 & \text{if} \quad i \neq j. \end{cases} \tag{5.21}$$

That is, each $\mathbf{v}_j$ is a unit vector, and $\mathbf{v}_i$ and $\mathbf{v}_j$ are orthogonal for $i \neq j$. In other words, $\{\mathbf{v}_1, \mathbf{v}_2, \ldots, \mathbf{v}_n\}$ is an *orthonormal* set of vectors in $\mathbb{R}^m$.

By definition, $\mathbf{v}_i$ is the ith row of A^t. Therefore, from (5.8) of Theorem 13,

$$(A^t A)_{i,j} = \mathbf{v}_i \cdot \mathbf{v}_j = \begin{cases} 1 & \text{if} \quad i = j \\ 0 & \text{if} \quad i \neq j. \end{cases} \tag{5.22}$$

Notice that (5.22) holds if and only if the columns of A are an orthonormal set of vectors in $\mathbb{R}^m$.

On the right-hand side we recognize the entries of the $n \times n$ identity matrix, so if A is an isometry, we must have that $A^t A = I$. That is, if A is an isometry, then the transpose of A is a left inverse of A.

This property is special. Indeed, if A is any $m \times n$ matrix such that $A^t A = I$, then for any $\mathbf{x}$ in $\mathbb{R}^n$,

$$|A\mathbf{x}|^2 = A\mathbf{x} \cdot A\mathbf{x} = \mathbf{x} \cdot A^t A\mathbf{x} = \mathbf{x} \cdot \mathbf{x} = |\mathbf{x}|^2,$$

which means that A is an isometry, proving the following theorem.

THEOREM 15   **(Isometries and the transpose)** *Let A be an $m \times n$ matrix. Then A is an isometry if and only if A^t is a left inverse of A, that is,*

$$A^t A = I, \tag{5.23}$$

which is the case if and only if the columns of A are an orthonormal set of vectors in $\mathbb{R}^m$.

EXAMPLE 33   **(Checking for isometry)** Let

$$A = \frac{1}{\sqrt{2}} \begin{bmatrix} 1 & 1 \\ -1 & 1 \end{bmatrix}$$

so that

$$A^t = \frac{1}{\sqrt{2}} \begin{bmatrix} 1 & -1 \\ 1 & 1 \end{bmatrix}.$$

Is A an isometry? Is A^t an isometry? Computing,

$$A^t A = \begin{bmatrix} 1 & 0 \\ 0 & 1 \end{bmatrix} \quad \text{and} \quad AA^t = \begin{bmatrix} 1 & 0 \\ 0 & 1 \end{bmatrix}.$$

The calculation on the left shows that A^t is a left inverse of A, and then by the theorem, A preserves lengths. The calculation on the right shows first that A^t is also a right inverse of A and hence that A is invertible, and second that A is a left inverse of A^t. Then since $(A^t)^t = A$, A^t is also length preserving.

5.4 Some very important formulas

The purpose of this section is to gather on one page some of the most useful formulas about vector and matrix products. These formulas will be used again and again and again. If later on in the book something seems mysterious, refer to the following formulas. Chances are that one of them will shed light on the matter.

SOME VERY IMPORTANT FORMULAS

(V.I.F.1): Let $A = [\mathbf{v}_1, \mathbf{v}_2, \ldots, \mathbf{v}_n]$ be an $m \times n$ matrix with columns $\mathbf{v}_1, \mathbf{v}_2, \ldots, \mathbf{v}_n$, and let $\mathbf{x} = \begin{bmatrix} x_1 \\ x_2 \\ \vdots \\ x_n \end{bmatrix}$

be any vector in $\mathbb{R}^n$. Then

$$A\mathbf{x} = [\mathbf{v}_1, \mathbf{v}_2, \ldots, \mathbf{v}_n] \begin{bmatrix} x_1 \\ x_2 \\ \vdots \\ x_n \end{bmatrix} = x_1\mathbf{v}_1 + x_2\mathbf{v}_2 + \cdots + x_n\mathbf{v}_n.$$

(V.I.F.2): Let A be any $m \times n$ matrix with rows $\mathbf{r}_1, \mathbf{r}_2, \ldots, \mathbf{r}_m$ so that

$$A = \begin{bmatrix} \mathbf{r}_1 \\ \mathbf{r}_2 \\ \vdots \\ \mathbf{r}_m \end{bmatrix}.$$

Let $\mathbf{x}$ be any vector in $\mathbb{R}^n$. Then

$$A\mathbf{x} = \begin{bmatrix} \mathbf{r}_1 \cdot \mathbf{x} \\ \mathbf{r}_2 \cdot \mathbf{x} \\ \vdots \\ \mathbf{r}_m \cdot \mathbf{x} \end{bmatrix}.$$

(V.I.F.3): Let $A = [\mathbf{v}_1, \mathbf{v}_2, \ldots, \mathbf{v}_n]$ be an $m \times n$ matrix with columns $\mathbf{v}_1, \mathbf{v}_2, \ldots, \mathbf{v}_n$. Let B be any $p \times m$ matrix. Then

$$BA = [B\mathbf{v}_1, B\mathbf{v}_2, \ldots, B\mathbf{v}_n].$$

(V.I.F.4): Let A be any $m \times n$ matrix, and let B be any $p \times m$ matrix. Then

$$(BA)_{i,j} = (\text{row } i \text{ of } B) \cdot (\text{column } j \text{ of } A).$$

(V.I.F.5): Let A be any $m \times n$ matrix, $\mathbf{x}$ any vector in $\mathbb{R}^m$, and $\mathbf{y}$ any vector in $\mathbb{R}^n$. Then

$$\mathbf{x} \cdot (A\mathbf{y}) = (A^t\mathbf{x}) \cdot \mathbf{y}.$$

Notice what is *not* on the list. The formula

$$(BA)_{i,j} = \sum_{k=1}^{m} B_{i,k} A_{k,j} \tag{5.24}$$

for multiplying a $p \times m$ matrix B and an $m \times n$ matrix A does not make the cut. This is because **V.I.F.4** says the same thing in a more useful geometric way. There is a principle here: As we shall see, it helps to make dot products explicit. There *is* a dot product in (5.24), but it is just a bit implicit, as are the vectors. Written in the form **V.I.F.4**, the geometry is explicit.

Exercises

5.1 Let

$$A = \begin{bmatrix} 1 & 2 & 3 \\ 2 & 0 & 2 \\ 0 & 1 & 1 \\ 1 & 2 & 3 \end{bmatrix} \quad \text{and} \quad \mathbf{v} = \begin{bmatrix} 1 \\ 2 \\ 1 \end{bmatrix}.$$

Use a single dot product to compute the second entry of $A\mathbf{v}$.

5.2 Let

$$A = \begin{bmatrix} 1 & 0 & 1 \\ 4 & 0 & 2 \\ 2 & 3 & 1 \\ 1 & 0 & 1 \end{bmatrix} \quad \text{and} \quad \mathbf{v} = \begin{bmatrix} 2 \\ 3 \\ 1 \end{bmatrix}.$$

Use a single dot product to compute the third entry of $A\mathbf{v}$.

5.3 Consider the matrices

$$A = \begin{bmatrix} 1 & 2 & 0 & 1 \\ 1 & 3 & 1 & 2 \\ 1 & 2 & 2 & 2 \end{bmatrix} \quad \text{and} \quad B = \begin{bmatrix} 1 & 1 & 1 \\ 2 & 0 & 2 \\ 0 & 0 & 1 \\ 3 & 2 & 1 \end{bmatrix}.$$

(a) Compute $(AB)_{2,3}$ without computing the whole matrix product AB.

(b) Write the second column of AB as a linear combination of the columns of A. That is, find numbers a, b, c, and d so that the third column of AB equals

$$a \begin{bmatrix} 1 \\ 1 \\ 1 \end{bmatrix} + b \begin{bmatrix} 2 \\ 3 \\ 2 \end{bmatrix} + c \begin{bmatrix} 0 \\ 1 \\ 2 \end{bmatrix} + d \begin{bmatrix} 1 \\ 2 \\ 2 \end{bmatrix}.$$

(c) Write the second row of AB as a linear combination of the rows of B.

5.4 Consider the matrices

$$A = \begin{bmatrix} 1 & 2 & 4 \\ 2 & 2 & 4 \\ 0 & 2 & -1 \end{bmatrix} \quad \text{and} \quad B = \begin{bmatrix} 1 & 2 & 3 \\ 2 & 1 & 3 \\ 1 & 1 & 2 \end{bmatrix}.$$

(a) Compute $(AB)_{2,2}$ and $(BA)_{2,2}$ without computing the whole matrix products AB and BA.

(b) Write the first column of AB as a linear combination of the columns of A. See Exercise 5.3(b).

(c) Write the first row of AB as a linear combination of the rows of B.

5.5 Let A be a 3×3 matrix whose column representation is $A = [\mathbf{v}_1, \mathbf{v}_2, \mathbf{v}_3]$. Find an explicit numerical 3×3 matrix B so that

$$AB = [\mathbf{v}_2 + \mathbf{v}_3, \mathbf{v}_1 + \mathbf{v}_2, \mathbf{v}_1 + \mathbf{v}_2].$$

(You are looking for a single B that works no matter what $\mathbf{v}_1$, $\mathbf{v}_2$, and $\mathbf{v}_3$ happen to be.)

5.6 Let B be a 3×3 matrix whose row representation is

$$B = \begin{bmatrix} \mathbf{v}_1 \\ \mathbf{v}_2 \\ \mathbf{v}_3 \end{bmatrix}.$$

Find an explicit numerical 3×3 matrix A so that

$$AB = \begin{bmatrix} \mathbf{v}_2 + \mathbf{v}_3 \\ \mathbf{v}_1 + \mathbf{v}_2 \\ \mathbf{v}_1 + \mathbf{v}_2 \end{bmatrix}.$$

(You are looking for a single A that works no matter what $\mathbf{v}_1$, $\mathbf{v}_2$, and $\mathbf{v}_3$ happen to be.)

5.7 (a) Let A be the $n \times n$ diagonal matrix $A = \mathrm{diag}(a_1, a_2, \ldots, a_j)$ in which, for some i with $1 \le i \le n$, $a_i = 1$, but $a_j = 0$ for $j \ne i$. In other words, $A_{i,i} = 1$, and every other entry is zero. Let B be any $n \times n$ matrix. Describe the rows of AB and the columns of BA.

(b) Is there any *nondiagonal* $n \times n$ matrix B that commutes with A (that is, $AB = BA$) for all $n \times n$ diagonal matrices A? (We already know that each diagonal $n \times n$ matrix B commutes with every diagonal matrix. The question is: Are there any others?) Give an example, or explain why not.

5.8 Let A be an $m \times n$ matrix and B be an $n \times p$ matrix. If A has no zero columns, does it then follow that AB has no zero columns? Explain why, or give a counterexample.

5.9 Let A be an $m \times n$ matrix and B be an $n \times p$ matrix. If B has at least one zero column, can AB ever have no zero columns? Explain why not, or give an example.

5.10 Find 2×2 matrices A and B such that $(A + B)^2 \ne A^2 + 2AB + B^2$. Can you do so if A and B are diagonal matrices?

5.11 Let A be the matrix

$$A = \begin{bmatrix} 1 & 0 & 0 & 0 \\ 0 & 1 & 0 & 0 \\ 0 & 0 & 0 & 0 \\ 0 & 0 & 0 & 0 \end{bmatrix}.$$

Let B be any other 4×4 matrix.

(a) Which rows of B, if any, can be freely modified without affecting the product AB?

(b) Which columns of B, if any, can be freely modified without affecting the product AB?

5.12 Let A be the matrix

$$A = \begin{bmatrix} 1 & 0 & 0 & 0 \\ 0 & 1 & 0 & 0 \\ 0 & 0 & 0 & 0 \\ 0 & 0 & 0 & 0 \end{bmatrix}.$$

Let B be any other 4×4 matrix.

(a) Which rows of B, if any, can be freely modified without affecting the product BA? (Note that B was on the right in the previous problem.)

(b) Which columns of B, if any, can be freely modified without affecting the product BA?

5.13 Let C be a 2×2 matrix such that

$$C \begin{bmatrix} 1 \\ 2 \end{bmatrix} = \begin{bmatrix} 2 \\ 1 \end{bmatrix} \quad \text{and} \quad C \begin{bmatrix} 2 \\ 1 \end{bmatrix} = \begin{bmatrix} -1 \\ 1 \end{bmatrix}.$$

Using the given information, find 2×2 matrices A and B so that $CA = B$, and then solve for C.

5.14 Let C be a 2×2 matrix such that

$$C\begin{bmatrix} 1 \\ 2 \end{bmatrix} = \begin{bmatrix} 2 \\ 1 \end{bmatrix} \quad \text{and} \quad C^2\begin{bmatrix} 1 \\ 2 \end{bmatrix} = \begin{bmatrix} -1 \\ 1 \end{bmatrix}.$$

Using the given information, find 2×2 matrices A and B so that $CA = B$, and then solve for C.

5.15 Are there any $m \times n$ isometries with $n > m$? Give an example, or explain why not.

5.16 Let A be an $n \times n$ matrix and m a positive integer. How is the mth power of A related to the mth power of A^t? Is it always true that $(A^m)^t = (A^m)^t$? Explain why, or give a counterexample.

5.17 Consider the matrices

$$A = \frac{1}{\sqrt{2}}\begin{bmatrix} 1 & 0 \\ 0 & \sqrt{2} \\ 1 & 0 \end{bmatrix}, \quad B = \begin{bmatrix} 0 & -1 & 1 \\ \sqrt{2} & 0 & 0 \\ 0 & 1 & 1 \end{bmatrix}, \quad \text{and} \quad C = \frac{1}{\sqrt{6}}\begin{bmatrix} \sqrt{3} & -1 \\ 0 & 2 \\ \sqrt{3} & 1 \end{bmatrix}.$$

Which, if any, are length preserving?

5.18 Let A be an $m \times n$ isometry, and let B be a $p \times m$ isometry. Is BA necessarily an isometry? Explain why, or give a counterexample.

5.19 Let A be a 3×3 matrix. Suppose the third row minus the first row equals the second row. Explain why A cannot be invertible.

5.20 An $m \times n$ matrix A is called *row stochastic* if each of its entries is nonnegative and if the sum of the entries in each row is one. Similarly, an $m \times n$ matrix A is called *column stochastic* if each of its entries is nonnegative and if the sum of the entries in each column is one.

(a) Let A be a $p \times m$ row stochastic matrix, and let B be an $m \times n$ row stochastic matrix. Show that AB is row stochastic.

(b) Let A be a $p \times m$ column stochastic matrix, and let B be an $m \times n$ column stochastic matrix. Show that AB is column stochastic.

5.21 Determine whether the following assertions are true or false. If true, explain why. If false, give a counterexample.

(a) If A is any $p \times m$ matrix, B is an $m \times n$ matrix, and the columns of B are all the same, then the columns of AB are all the same.

(b) If A is any $p \times m$ matrix, B is an $m \times n$ matrix, and the columns of B are all the same, then in each row of AB, all the entries are the same.

(c) If A is any $p \times m$ matrix, B is an $m \times n$ matrix, and the rows of A are all the same, then the rows of AB are all the same.

(d) If A is any $p \times m$ matrix, B is an $m \times n$ matrix, and the rows of B are all the same, then in each column of AB, all the entries are the same.

Problems about linear functionals

Linear transformations from $\mathbb{R}^n$ to $\mathbb{R}$ have a special name: They are called *linear functionals*. The explanation of the name is a long story that need not concern us here. But linear functionals *do matter*. The following exercises relate linear functionals to the dot product.

5.22 For any given vector $\mathbf{a}$ in $\mathbb{R}^n$, define a function f from $\mathbb{R}^n$ to $\mathbb{R}$ by $f(\mathbf{x}) = \mathbf{a} \cdot \mathbf{x}$. Show that f is a linear transformation from $\mathbb{R}^n$ to $\mathbb{R}$, that is, a linear functional on $\mathbb{R}^n$.

5.23 It turns out that *every* linear functional on $\mathbb{R}^n$ is of the type considered in the previous exercise. Use Theorem 2 and **V.I.F.2** to show that $f(\mathbf{x}) = A_f \mathbf{x} = \mathbf{a}_f \cdot \mathbf{x}$ where

$$\mathbf{a}_f = \begin{bmatrix} f(\mathbf{e}_1) \\ f(\mathbf{e}_2) \\ \vdots \\ f(\mathbf{e}_n) \end{bmatrix}.$$

5.24 Let f be a linear functional on $\mathbb{R}^n$. Suppose that for all $\mathbf{x}$ in $\mathbb{R}^n$, $f(\mathbf{x}) = \mathbf{a} \cdot \mathbf{x}$ and $f(\mathbf{x}) = \mathbf{b} \cdot \mathbf{x}$. Show that $\mathbf{b} = \mathbf{a}$. In other words, the vector $\mathbf{a}_f$ considered in Exercise 5.23 is the *only* vector $\mathbf{a}$ in $\mathbb{R}^n$ such that $f(\mathbf{x}) = \mathbf{a} \cdot \mathbf{x}$ for all $\mathbf{x}$ in $\mathbb{R}^n$.

The fact that any linear functional on $\mathbb{R}^n$ can be represented by the dot product with a uniquely determined vector $\mathbf{a}$ is sometimes called the *Riesz representation theorem*, though it is actually a special case of what Riesz proved.

5.25 For any vector

$$\begin{bmatrix} a \\ b \\ c \end{bmatrix},$$

define the polynomial $p(x) = a + bx + cx^2$. In turn, define the number $\int_0^1 p(x)\, dx$. Putting the pieces together, we get a function f from $\mathbb{R}^3$ to $\mathbb{R}$, defined by

$$f\left(\begin{bmatrix} a \\ b \\ c \end{bmatrix} \right) = \int_0^1 \left[a + bx + cx^2 \right] dx.$$

(a) Show that f is a linear functional on $\mathbb{R}^3$, that is, a linear transformation from $\mathbb{R}^3$ to $\mathbb{R}$.

(b) Find the vector $\mathbf{a}_f$ such that $f(\mathbf{x}) = \mathbf{a}_f \cdot \mathbf{x}$, which we know exists by Exercise 5.23 and is unique by Exercise 5.24.

5.26 For any vector

$$\begin{bmatrix} a \\ b \\ c \end{bmatrix},$$

define the polynomial $p(x) = a + bx + cx^2$. In turn, define the number $p(2)$. Putting the pieces together, we get a function f from $\mathbb{R}^3$ to $\mathbb{R}$, so we define

$$f\left(\begin{bmatrix} a \\ b \\ c \end{bmatrix} \right) = p(2) = a + b2 + c4.$$

(a) Show that f is a linear functional on $\mathbb{R}^3$, that is, a linear transformation from $\mathbb{R}^3$ to $\mathbb{R}$.

(b) Find the vector $\mathbf{a}_f$ such that $f(\mathbf{x}) = \mathbf{a}_f \cdot \mathbf{x}$, which we know exists by Exercise 5.23 and is unique by Exercise 5.24.

5.27 Let f and g be linear functionals on $\mathbb{R}^3$. Define a function h from $\mathbb{R}^3$ to $\mathbb{R}^2$ by

$$h(\mathbf{x}) = \begin{bmatrix} f(\mathbf{x}) \\ g(\mathbf{x}) \end{bmatrix}.$$

(a) Show that h is a linear transformation from $\mathbb{R}^3$ to $\mathbb{R}^2$.

(a) Let $\mathbf{a}_f$ be the vector corresponding to f, as in Exercise 5.23, and let $\mathbf{a}_g$ be the vector corresponding to g. If

$$\mathbf{a}_f = \begin{bmatrix} 1 \\ 2 \\ 3 \end{bmatrix} \quad \text{and} \quad \mathbf{a}_g = \begin{bmatrix} 3 \\ 2 \\ 1 \end{bmatrix},$$

find the matrix A_h corresponding to h, as in Theorem 2.

Visualizing Linear Transformations in $\mathbb{R}^2$

6.1 Graphing a linear transformation from $\mathbb{R}^2$ to $\mathbb{R}^2$

A picture really can be worth a thousand words. If f is a function of a single variable x, a lot of insight into the properties of f can be readily obtained from a graph of $y = f(x)$. For example, here is a graph of $y = f(x)$ for $f(x) = x^5 - x^2 + 1$ and $-1 \leq x \leq 1$, together with the horizontal line $y = 0.9$.

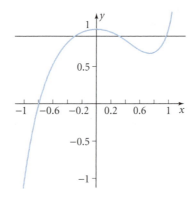

As you can see from the graph, there are three solutions to the equation $f(x) = 0.9$ with $-1 \leq x \leq 1$. In fact, you see that these solutions are approximately given by

$$x \approx -0.3, \qquad x \approx 0.3, \qquad \text{and} \qquad x \approx 0.95.$$

You also plainly see that the function has a local maximum near $x = 0$ and a local minimum near $x = 0.75$. (You can solve for these exactly—try it: The local maximum is at $x = 0$, and the local minimum is at $x = \sqrt[3]{2/5}$. The graph provides a good check on these analytic computations.)

The point is this: The graph gives a concise visual description of *what this function does*.

Now consider a function with vector input and vector output. To be specific, consider f_A, the linear transformation from $\mathbb{R}^2$ to $\mathbb{R}^2$ corresponding to the matrix

$$A = \begin{bmatrix} 1 & 1 \\ 0 & 1 \end{bmatrix}. \tag{6.1}$$

- *What sort of graph can we draw to represent the effect of this transformation and others like it?*

Here is one approach to this question. Let S be a set of points in the x, y plane. For example, S might be the unit circle or the x-axis or the y-axis. In principle, it could be any set, but for our purposes it should be a familiar, recognizable set. *We are going to draw a graph of the set that S is transformed into under the linear transformation f_A.*

To help us think of f_A as a transformation from $\mathbb{R}^2$ to $\mathbb{R}^2$, think of it as a transformation from the x, y plane to the u, v plane:

$$\begin{bmatrix} u \\ v \end{bmatrix} = f_A\left(\begin{bmatrix} x \\ y \end{bmatrix}\right) = A\begin{bmatrix} x \\ y \end{bmatrix} = \begin{bmatrix} x + y \\ y \end{bmatrix}. \tag{6.2}$$

So if x and y are the coordinates of an input point in the x, y plane, u and v are the coordinates of the transformed point in the u, v plane, and

$$u = x + y$$
$$v = y. \tag{6.3}$$

DEFINITION

(Image of a set) Let A be an $m \times n$ matrix, and let S be a subset of $\mathbb{R}^n$. Then the *image of S under A* is the set of all vectors in $\mathbb{R}^m$ of the form $A\mathbf{x}$ for some $\mathbf{x}$ in S.

The transformation f_A is invertible. Using the formula (3.19) of Theorem 7,

$$A^{-1} = \begin{bmatrix} 1 & -1 \\ 0 & 1 \end{bmatrix},$$

so from (6.2),

$$\begin{bmatrix} x \\ y \end{bmatrix} = A^{-1} \begin{bmatrix} u \\ v \end{bmatrix} = \begin{bmatrix} u - v \\ v \end{bmatrix}. \tag{6.4}$$

Hence, for any point (u, v) in the u, v plane, there is exactly one point (x, y) in the x, y plane whose image under f_A is (u, v), and its coordinates are

$$x = u - v$$
$$y = v. \tag{6.5}$$

The point (x, y) with x and y computed using (6.5) is the point that (u, v) "comes from" under the transformation f_A.

Together, (6.3) and (6.5) give us a lexicon that can be used to translate equations and formulas back and forth between x, y terms and u, v terms. It is very useful for finding images.

EXAMPLE 34

(Finding the equation for the image of the unit circle) Let S be the unit circle in the x, y plane. What is the image of S? Since S is given by the equation $x^2 + y^2 = 1$ in the x, y plane, (u, v) belongs to the image of S if and only if the point (x, y) from which it comes satisfies $x^2 + y^2 = 1$. By (6.5), $x = u - v$ and $y = v$ when (x, y) comes from (u, v). Substituting these expressions into the equation,

$$x^2 + y^2 = 1 \quad \rightarrow \quad (u - v)^2 + v^2 = 1,$$

which simplifies to

$$u^2 - 2uv + 2v^2 = 1. \tag{6.6}$$

This is the equation of the image, which is an ellipse in the u, v plane:

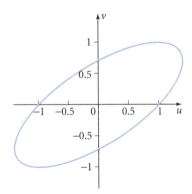

From this diagram you see how the transformation f_A stretches and pulls.

We can apply the same method to any region given by an equation and any invertible linear transformation. All we have to do is to work out the relations between the u, v coordinates and the x, y coordinates as in (6.3) and (6.5). As before, we can do so by inverting the matrix that defines the transformation. Then we substitute the formulas for x and y in terms of u and v into the equation. The result is the equation for the image.

There is another kind of set whose image is easy to find: any set bounded by a collection of straight line segments. The unit square is a perfect example. Before considering this example, let us make a key observation about linear transformations and straight line segments.

Let $\mathbf{a}$ and $\mathbf{b}$ be any two vectors in $\mathbb{R}^2$. These two vectors, considered as points in the plane, determine a line—the line through $\mathbf{a}$ and $\mathbf{b}$. The vectors on the line are of the form

$$\mathbf{a} + t(\mathbf{b} - \mathbf{a}) \tag{6.7}$$

for some number t. That is, you start from $\mathbf{a}$ and move along the direction of $\mathbf{b} - \mathbf{a}$. For $t = 1$, you get $\mathbf{a} + (\mathbf{b} - \mathbf{a}) = \mathbf{b}$, and of course for $t = 0$, you get $\mathbf{a}$. For $t = 1/2$, you get $(\mathbf{a} + \mathbf{b})/2$, the midpoint of the segment connecting $\mathbf{a}$ and $\mathbf{b}$. To get a representation of the segment joining $\mathbf{a}$ and $\mathbf{b}$, regroup terms in (6.7) to obtain $(1 - t)\mathbf{a} + t\mathbf{b}$. This gives a formula for the points on the line segment connecting $\mathbf{a}$ and $\mathbf{b}$: It is the set of points of the form

$$(1 - t)\mathbf{a} + t\mathbf{b} \qquad \text{for} \quad 0 \le t \le 1. \tag{6.8}$$

Notice how this formula "interpolates" between $\mathbf{a}$ and $\mathbf{b}$.

Here is the key observation: Under any *linear* transformation g from $\mathbb{R}^2$ to $\mathbb{R}^2$,

$$g((1 - t)\mathbf{a} + t\mathbf{b}) = (1 - t)g(\mathbf{a}) + tg(\mathbf{b}). \tag{6.9}$$

That is,

- *The image of the line segment connecting $\mathbf{a}$ and $\mathbf{b}$ is the line segment connecting their images.*

Let us apply this to find the image of the unit square under f_A.

EXAMPLE 35   **(Finding the image of the unit square)** The vertices of the unit square are the points corresponding to

$$\begin{bmatrix} 0 \\ 0 \end{bmatrix}, \qquad \begin{bmatrix} 1 \\ 0 \end{bmatrix}, \qquad \begin{bmatrix} 0 \\ 1 \end{bmatrix}, \qquad \text{and} \qquad \begin{bmatrix} 1 \\ 1 \end{bmatrix}.$$

The right side of the square is given by the line segment

$$(1 - t)\begin{bmatrix} 1 \\ 0 \end{bmatrix} + t\begin{bmatrix} 1 \\ 1 \end{bmatrix}.$$

By what we discussed earlier, its image is the line segment

$$(1 - t)f_A\left(\begin{bmatrix} 1 \\ 0 \end{bmatrix}\right) + tf_A\left(\begin{bmatrix} 1 \\ 1 \end{bmatrix}\right) = (1 - t)\begin{bmatrix} 1 \\ 0 \end{bmatrix} + t\begin{bmatrix} 2 \\ 1 \end{bmatrix}.$$

It is the same with the rest of the sides: All we need to do is to find the images of the vertices of the square and connect them:

$$f_A\left(\begin{bmatrix} 0 \\ 0 \end{bmatrix}\right) = \begin{bmatrix} 0 \\ 0 \end{bmatrix}, \qquad f_A\left(\begin{bmatrix} 1 \\ 0 \end{bmatrix}\right) = \begin{bmatrix} 1 \\ 0 \end{bmatrix}, \qquad f_A\left(\begin{bmatrix} 0 \\ 1 \end{bmatrix}\right) = \begin{bmatrix} 1 \\ 1 \end{bmatrix}, \qquad f_A\left(\begin{bmatrix} 1 \\ 1 \end{bmatrix}\right) = \begin{bmatrix} 2 \\ 1 \end{bmatrix}.$$

---

*We have just seen that linear transformations take straight line segments into straight line segments. The use of "linear" here is quite appropriate.

Drawing these points in and connecting the dots, we get the shaded parallelogram in the following graph:

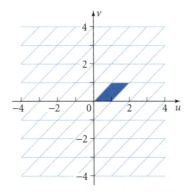

In this graph, there are a lot of other lines besides the ones that bound the image of the square. These lines are the lines on the image of the unit graph paper grid in the x, y plane. The next example combines the methods of the first two.

EXAMPLE 36 **(The image of a graph paper grid)** We can learn more about how the transformation f_A stretches and pulls by transforming the whole collection of lines on a graph paper grid. Consider the lines

$$x = i \quad \text{and} \quad y = j \quad \text{for} \quad i, j = \cdots - 3, -2, -1, 0, 1, 2, 3, \ldots. \tag{6.10}$$

Let S be the set of all of these lines. What is the image of S?

If we graph all these lines at once in the x, y plane, we get a graph paper grid with unit spacing:

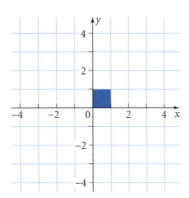

The shaded square in the positive quadrant is the unit square studied in Example 35.

We find the lines in the image by using (6.5) to translate the equations of the original lines into u, v terms. The line image of the line

$$x = i \quad \rightarrow \quad u - v = i \quad \text{and} \quad y = j \quad \rightarrow \quad v = j.$$

Graphing the lines $u - v = i$ and $v = j$ as i and j range of the integers gives us the distorted graph paper grid produced by the transformation f_A. These are the lines you see in the graph from Example 35, and the image of the shaded square here is the shaded parallelogram in the graph there.

Look back at the graph of Example 35. The whole u, v plane is covered by parallelogram tiles, each one of which is the image of a square in our graph paper grid in the x, y plane, as we saw in Example 35. The shaded tile in the second graph is the image of shaded tile in the first graph. Notice that all the tiles are of the exact same size and shape, and they are all lined up so that if you were shown where only one of them was, you could put the rest in the right places. Summarizing:

- *All the information contained in such a distorted graph paper picture is given in a graph of the image of the unit square in the positive quadrant of the x, y plane.*

  Let us examine this observation for another linear transformation.

EXAMPLE 37   **(Graph paper grid via unit tiles)** Consider the linear transformation f_B where

$$B = \begin{bmatrix} 1 & 3 \\ -3 & -1 \end{bmatrix}. \tag{6.11}$$

We find the image of the unit square under f_B and then "tile the plane" with it to find the image of the standard graph paper grid given by (6.10). Computing,

$$f_B(0) = 0, \qquad f_B(\mathbf{e}_1) = \begin{bmatrix} 1 \\ -3 \end{bmatrix}, \qquad f_B(\mathbf{e}_2) = \begin{bmatrix} 3 \\ -1 \end{bmatrix}, \qquad \text{and} \qquad f_B(\mathbf{e}_1 + \mathbf{e}_2) = \begin{bmatrix} 4 \\ -4 \end{bmatrix}.$$

Connecting as in Example 35 gives us the tile that is the image of the unit square:

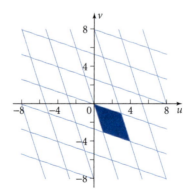

Tiling the rest of the plane with like tiles gives us the grid, exemplifying in a visual way the regularity and uniformity of linear transformations. What the transformation does to the unit square in the x, y plane is just stepped and repeated throughout the whole plane.

Think back to Example 34. The image of the unit circle under f_A was an ellipse. Is this true for f_B?

EXAMPLE 38   **(Another image of the unit circle)** For the matrix B given by (6.11), let us find the image of the unit circle. This time, we will not only draw it, but we will think harder about some of the "thousand words" it tells us.

Using Theorem 7,

$$B^{-1} = \frac{1}{8}\begin{bmatrix} -1 & -3 \\ 3 & 1 \end{bmatrix}.$$

Hence,

$$\begin{bmatrix} x \\ y \end{bmatrix} = B^{-1}\begin{bmatrix} u \\ v \end{bmatrix} = \frac{1}{8}\begin{bmatrix} -u - 3v \\ 3u + v \end{bmatrix},$$

which is the same as

$$x = -\frac{u + 3v}{8} \quad \text{and} \quad y = \frac{3u + v}{8}.$$

Using this lexicon to translate the equation for the unit circle, namely, $x^2 + y^2 = 1$, into u, v terms, we get

$$x^2 + y^2 = 1 \quad \rightarrow \quad \frac{1}{64}((u + 3v)^2 + (3u + v)^2) = 1,$$

which simplifies to

$$5u^2 + 6uv + 5v^2 = 32.$$

Again, this is the equation of an ellipse. Here is a graph of this ellipse, together with the centered circles of radius 2 and 4 in the u, v plane.

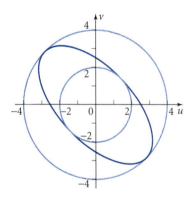

As you see, the larger circle circumscribes the ellipse, touching it along the line $v = -u$, while the smaller circle is inscribed within it, touching it along the line $v = u$. What does this tell us?

- *The radius of the inner circle is 2, so for all unit vectors* **x**, $|f_B(\mathbf{x})| \geq 2$.

The reason is that when **x** is a unit vector, it is on the unit circle, and its image is outside the circle of radius 2. Similarly, we conclude:

- *The radius of the outer circle is 4, so for all unit vectors* **x**, $|f_B(\mathbf{x})| \leq 4$.

What about vectors with other lengths? Let **y** be an arbitrary nonzero vector in $\mathbb{R}^2$, and define

$$\mathbf{x} = \frac{1}{|\mathbf{y}|}\mathbf{y} \quad \text{so that} \quad \mathbf{y} = |\mathbf{y}|\mathbf{x}$$

and $\mathbf{x}$ is a unit vector. Since f_B is linear and $|f_B(\mathbf{x})| \geq 2$,

$$|f_B(\mathbf{y})| = |f_B(|\mathbf{y}|\mathbf{x})| = |\mathbf{y}||f_B(\mathbf{x})| \geq 2|\mathbf{y}|.$$

That is, *the linear transformation f_B stretches every vector by a factor of at least 2.*

In the same way, we see that since the ellipse is inscribed within the circle of radius 4, $|f_B(\mathbf{x})| \leq 4$ for all unit vectors $\mathbf{x}$, so with the same notations,

$$|f_B(\mathbf{y})| = |f_B(|\mathbf{y}|\mathbf{x})| = |\mathbf{y}||f_B(\mathbf{x})| \leq 4|\mathbf{y}|.$$

That is, *the linear transformation f_B stretches every vector by a factor of at most 4.*

Putting it all together, for all $\mathbf{y}$ in $\mathbb{R}^2$,

$$2|\mathbf{y}| \leq |B\mathbf{y}| \leq 4|\mathbf{y}|. \tag{6.12}$$

These minimum and maximum stretching factors, and especially their ratio, are an important characteristic of a linear transformation from a computational standpoint, as we shall see.

6.2 The effect of linear transformations in $\mathbb{R}^2$ on area

In the remaining examples, we focus on the effect of a linear transformation on the area of a planar region.

EXAMPLE 39   **(How the area of the unit square changes)** Let B be given once more by (6.11) and S be the unit square. We have already seen that the image of S under f_B is the following parallelogram:

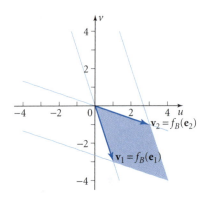

The two arrows drawn in the graph represent $f_B(\mathbf{e}_1)$ and $f_B(\mathbf{e}_2)$. By elementary planar geometry, the area of the image parallelogram is

$$|\mathbf{v}_1||\mathbf{v}_2||\sin(\theta)|, \tag{6.13}$$

where θ is the interior angle in the vertex at the origin, because $|\mathbf{v}_1|$ and $|\mathbf{v}_2|$ are the side lengths of the parallelogram. (In this example, the angle is acute, and the absolute value on $\sin(\theta)$ is superfluous, but with the absolute value, the formula works in general.)

Squaring (6.13) and using (4.17),

$$\begin{aligned}
(\text{Area})^2 &= |\mathbf{v}_1|^2|\mathbf{v}_2|^2 \sin^2(\theta) \\
&= |\mathbf{v}_1|^2|\mathbf{v}_2|^2(1 - \cos^2(\theta)) \\
&= |\mathbf{v}_1|^2|\mathbf{v}_2|^2 - |\mathbf{v}_1|^2|\mathbf{v}_2|^2 \cos^2(\theta) \\
&= |\mathbf{v}_1|^2|\mathbf{v}_2|^2 - (\mathbf{v}_1 \cdot \mathbf{v}_2)^2.
\end{aligned} \tag{6.14}$$

We could plug in numbers from B at this point, but to draw a general conclusion, write

$$\mathbf{v}_1 = \begin{bmatrix} a \\ c \end{bmatrix} \quad \text{and} \quad \mathbf{v}_2 = \begin{bmatrix} b \\ d \end{bmatrix}. \tag{6.15}$$

Then

$$
\begin{aligned}
|\mathbf{v}_1|^2 |\mathbf{v}_2|^2 - (\mathbf{v}_1 \cdot \mathbf{v}_2)^2 &= (a^2 + c^2)(b^2 + d^2) - (ab + cd)^2 \\
&= (a^2 b^2 + a^2 d^2 + c^2 b^2 + c^2 d^2) - (a^2 b^2 + c^2 d^2 + 2abcd) \\
&= a^2 d^2 + c^2 b^2 - 2abcd \\
&= (ad - bc)^2.
\end{aligned} \tag{6.16}
$$

Let us plug in numbers. For the matrix B given by (6.11), $|ad - bc| = |-1 - (-9)| = 8$. That is, this transformation magnifies the area of the unit square by a factor of 8.

Déja vu? We have seen the quantity $|ad - bc|$ before. Recall that the 2×2 matrix $[\mathbf{v}_1, \mathbf{v}_2] = \begin{bmatrix} a & b \\ c & d \end{bmatrix}$ is invertible if and only if $(ad - bc) \neq 0$.

We can understand, in geometric terms, the fact that

$$\begin{bmatrix} a & b \\ c & d \end{bmatrix}$$

is invertible if and only if $(ad - bc) \neq 0$: Combining (6.14) and (6.16), we see that the image of the unit square under the linear transformation corresponding to

$$\begin{bmatrix} a & b \\ c & d \end{bmatrix}$$

has area $|ad - bc|$. This quantity vanishes exactly when

$$\begin{bmatrix} a & b \\ c & d \end{bmatrix}$$

"squashes" the unit square down to a line segment. But in this case $\theta = 0$, and the images of the bounding lines, $x = 0$ and $y = 0$, land on top of each other. Evidently, in this case the transformation is not one-to-one and cannot be invertible. Otherwise, as long as the unit square is not "squashed" to a line segment, the transformation is invertible.

The transformation B magnified the area of the unit square by a factor of 8. It also magnifies the area of the unit circle by a factor of 8. We found in Example 38 that the image of the unit circle is an ellipse with major radius 4 and minor radius 2, which has area $\pi \times 4 \times 2 = 8\pi$, while the unit circle is π. There is a general fact here:

- *Let S be any subset of the plane for which the area is a well-defined number. Then the image of S under f_B has exactly 8 times as much area as S.*

The same is true for *any* linear transformation of the plane, except, of course, that you have to replace 8 by the value of $|ad - bc|$ for the transformation at hand.

The fine details of the explanation are a bit subtle, and in fact, although "area" seems to have a straightforward intuitive meaning, coming up with a precise mathematical definition for "general" sets would take us far afield from linear algebra. However, for nice sets, the meaning is quite clear, and we can learn something of general value by considering them. For example, consider the following region made up of 66 unit squares:

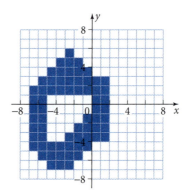

Evidently, its total area is 66 square units. What about its image under f_B? Then each one of these squares is transformed onto one of the tiles in the distorted graph paper of Example 37. *Each one of the 66 squares making up our region is transformed into a tile whose area is 8 square units, and none of these tiles overlap.* Therefore, the total area of the image region is 8×66 square units. This happens to be 528 square units, but the important thing is that it is *exactly* 8 *times as much.*

- *The key point in the argument is that all the transformed squares are congruent to one another, so we have to work out the "area magnification factor" for only one of them. This is a consequence of the "homogeneity and uniformity" of linear transformations, such as you see in the "distorted graph paper" graphs of Examples 35 and 37. It would not be true of a nonlinear transformation.*

This argument applies to *any* region made up of a finite number of little squares, and it does not matter what units we are using: inches, millimeters, angstroms, anything we want.

Not all regions are made up of a finite number of little squares—the circle is an example. But area is defined so that any region that has an area can be "approximated" arbitrarily closely by a region that is a finite collection of little squares. The remaining details belong to another subject (analysis), so we do not give them. But try to visualize a sequence of finer and finer approximations to a circle using finer and finer squares, and you can probably develop a good picture of how the argument would go. Let us summarize.

THEOREM 16   **(Area and linear transformations in $\mathbb{R}^2$)** *Let*

$$A = \begin{bmatrix} a & b \\ c & d \end{bmatrix}.$$

Then the area of the image of the unit square under the linear transformation corresponding to A is $|ad - bc|$ so that A is invertible if and only if this area is strictly positive.

Let us also make one final definition.

> **DEFINITION**
>
> **(Area-preserving linear transformation)** Let f be a linear transformation from $\mathbb{R}^2$ to $\mathbb{R}^2$, and let
> $$A_f = \begin{bmatrix} a & b \\ c & d \end{bmatrix}$$
> be the corresponding 2×2 matrix. Then f is an *area-preserving* linear transformation if and only if $|ad - bc| = 1$.

By Theorem 16, the condition $|ad - bc| = 1$ means that f preserves the area of the unit square and hence the area of any region made up of a finite number of squares. As we explained, this condition actually means that f preserves the area of any region that has an area, so the name is appropriate.

EXAMPLE 40   **(Area preserving or not?)** Let
$$A = \begin{bmatrix} 1 & 1 \\ 0 & 1 \end{bmatrix}$$
be the matrix (6.1). For this matrix, $|ad - bc| = |1 - 0| = 1$, so f_A is an area-preserving transformation.

Exercises

6.1 Let A be the matrix
$$A = \begin{bmatrix} 1 & 2 \\ 0 & 3 \end{bmatrix}.$$
Find the equation describing the image of S under the linear transformation corresponding to A and graph this image of S, where

(a) S is the y-axis.

(b) S is the line $x + y = 3$.

(c) S is the unit circle.

6.2 Let A be the matrix
$$A = \begin{bmatrix} 2 & -3 \\ 1 & 3 \end{bmatrix}.$$
Find the equation describing the image of S under the linear transformation corresponding to A and graph this image of S, where

(a) S is the x-axis.

(b) S is the line $x - y = 0$.

(c) S is the unit circle.

6.3 Let A be the matrix
$$A = \begin{bmatrix} 1 & 2 \\ 0 & 3 \end{bmatrix},$$
and let f_A be the corresponding linear transformation.

(a) Graph the image of the unit square under f_A and compute the area of this image.

(b) Let S be the triangle with vertices $(1, 1)$, $(1, 4)$, and $(3, -5)$. What is the area of S? Graph the image of S under f_a and compute the area of this image.

6.4 Let A be the matrix

$$A = \begin{bmatrix} 2 & -3 \\ 1 & 3 \end{bmatrix}.$$

(a) Graph the image of the unit square under the corresponding linear transformation and compute the area of this image.

(b) Let S be the triangle with vertices $(-1, 1)$, $(1, 1)$, and $(0, -1)$. What is the area of S? Graph the image of S under f_a and compute the area of this image.

6.5 Which, if either, of the following matrices corresponds to an area-preserving transformation?

$$A = \begin{bmatrix} 2 & -3 \\ 1 & 3 \end{bmatrix} \qquad B = \begin{bmatrix} 2 & 3 \\ 1 & 3 \end{bmatrix}.$$

6.6 Let

$$A = \begin{bmatrix} a & b \\ c & d \end{bmatrix}$$

be an invertible 2×2 matrix. We have asserted that the image of the unit circle under such a transformation is always an ellipse. The object here is to derive the equation of this ellipse in the u, v plane in terms of the entries of A.

(a) Explain why

$$\begin{bmatrix} u \\ v \end{bmatrix}$$

is in the image of the unit circle if and only if

$$\left| A^{-1} \begin{bmatrix} u \\ v \end{bmatrix} \right| = 1.$$

(b) Explain why

$$\begin{bmatrix} u \\ v \end{bmatrix}$$

is in the image of the unit circle if and only if

$$\begin{bmatrix} u \\ v \end{bmatrix} \cdot (A^{-1})^t A^{-1} \begin{bmatrix} u \\ v \end{bmatrix} = 1.$$

(c) Show that the equation for the image of S is

$$(d^2 + c^2)u^2 + (a^2 + b^2)v^2 - 2(bd + ac)uv = (ad - bc)^2.$$

6.7 Let

$$A = \begin{bmatrix} 1 & 2 \\ 3 & 4 \end{bmatrix}.$$

By the results of Exercise 6.6, the image of the unit circle under A is an ellipse. Find the equation of the ellipse and graph it.

6.8 Let

$$A = \begin{bmatrix} 2 & 1 \\ 2 & 4 \end{bmatrix}.$$

By the results of Exercise 6.6, the image of the unit circle under A is an ellipse. Find the equation of the ellipse and graph it.

The Solution Set of a Linear System

Overview of Chapter 2

This chapter is about solving equations of the form $A\mathbf{x} = \mathbf{b}$ where A is a given $m \times n$ matrix and $\mathbf{b}$ is a given vector in $\mathbb{R}^n$. The set of all vectors in $\mathbb{R}^m$, if any, that satisfy this equation is called its *solution set*.

- *Solving an equation means finding a clear description of its solution set.*

The first major issue arises right here: What do we mean by a "clear description" of the solution set? Can we just list the solutions? In general, no. It turns out, as we will soon see, that equations of the form $A\mathbf{x} = \mathbf{b}$ have either no solution, exactly one solution, or infinitely many solutions. If there are infinitely many solutions, we cannot list them. What do we do instead?

There is a good substitute for a list in this context: We can *parameterize* the solution set. This means finding vectors $\mathbf{x}_0, \mathbf{v}_1, \mathbf{v}_2, \ldots, \mathbf{v}_r$ so that for any given values of the parameters $t_1, t_2, \ldots, t_r$,

$$\mathbf{x}_0 + t_1\mathbf{v}_1 + t_1\mathbf{v}_2 + \cdots + t_r\mathbf{v}_r$$

is a solution of the equation, and each solution can be written in this way for *exactly one choice* of values for the parameters $t_1, t_2, \ldots, t_r$. A parameterization with this uniqueness property is a *one-to-one parameterization*. In many contexts, and as far as we are concerned here, they are the only useful kind.

This chapter explains how to parameterize the solution sets of systems of linear equations. There is an *algorithm* for doing this: the *row-reduction algorithm*. Analysis of this algorithm leads to a systematic way of determining the solution set of an arbitrary system of linear equations, no matter how many variables there are and no matter how many equations there are. So in this chapter, we will solve one of the main problems of linear algebra.

SECTION 1 | # Solving Equations and Parameterization

1.1 Planes and lines in $\mathbb{R}^3$

The *solution set* of an equation is the set consisting of all its solutions. The purpose of this section is to introduce the notion of *one-to-one parameterization*, which is used to describe the solution sets of equations with infinitely many solutions.

To grasp the relation between *systems of equations* and *parameterizations*, consider something familiar: planes and lines in $\mathbb{R}^3$.

A plane in $\mathbb{R}^3$ is the *solution set* of a single linear equation of the form

$$\mathbf{a} \cdot \mathbf{x} = d, \tag{1.1}$$

where

$$\mathbf{x} = \begin{bmatrix} x \\ y \\ z \end{bmatrix}.$$

That is, it is the set of all points $\mathbf{x}$ that satisfy the equation (1.1). If, for example,

$$\mathbf{a} = \begin{bmatrix} 1 \\ 2 \\ 1 \end{bmatrix} \quad \text{and} \quad d = 4,$$

(1.1) becomes

$$x + 2y + z = 4. \tag{1.2}$$

You are probably familiar with the fact that such an equation describes a plane in $\mathbb{R}^3$, but we will see it from a geometric point of view when we parameterize its solution set.

A line in $\mathbb{R}^3$ is the intersection of two planes:

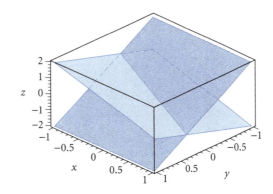

Therefore, we can specify a line as the solution set of a system of two equations specifying two planes:

$$\mathbf{a}_1 \cdot \mathbf{x} = d_1$$
$$\mathbf{a}_2 \cdot \mathbf{x} = d_2. \tag{1.3}$$

Introduce the 2×3 matrix A whose first row is $\mathbf{a}_1$ and whose second row is $\mathbf{a}_2$, that is,

$$A = \begin{bmatrix} \mathbf{a}_1 \\ \mathbf{a}_2 \end{bmatrix}.$$

Then by **V.I.F.2**,

$$A\mathbf{x} = \begin{bmatrix} \mathbf{a}_1 \cdot \mathbf{x} \\ \mathbf{a}_2 \cdot \mathbf{x} \end{bmatrix},$$

so if we introduce the vector

$$\mathbf{d} = \begin{bmatrix} d_1 \\ d_2 \end{bmatrix},$$

then (1.3) is equivalent to the matrix equation

$$A\mathbf{x} = \mathbf{d}. \tag{1.4}$$

For example, consider the case

$$A = \begin{bmatrix} 1 & 2 & 3 \\ 3 & 2 & 1 \end{bmatrix} \quad \text{and} \quad \mathbf{d} = \begin{bmatrix} 1 \\ -1 \end{bmatrix}. \tag{1.5}$$

Then the matrix equation $A\mathbf{x} = \mathbf{d}$ is equivalent to the system of equations

$$x + 2y + 3z = 1$$
$$3x + 2y + z = -1. \tag{1.6}$$

Its solution set is the line formed by the intersection of the two planes

$$x + 2y + 3z = 1 \quad \text{and} \quad 3x + 2y + z = -1. \tag{1.7}$$

That summarizes the description of lines and planes in terms of equations. What is the parametric description of planes and lines? The answer has two parts.

Let us begin with the equation (1.2). We can eliminate x by using the equation to write

$$x = 4 - 2y - z.$$

If

$$\mathbf{x} = \begin{bmatrix} x \\ y \\ z \end{bmatrix}$$

is any point in the solution set of (1.2), then

$$\mathbf{x} = \begin{bmatrix} x \\ y \\ z \end{bmatrix} = \begin{bmatrix} 4 - 2y - z \\ y \\ z \end{bmatrix}.$$

Expanding the right-hand side into a constant term, a constant multiple of y and a constant multiple of z,

$$\mathbf{x} = \begin{bmatrix} 4 \\ 0 \\ 0 \end{bmatrix} + y \begin{bmatrix} -2 \\ 1 \\ 0 \end{bmatrix} + z \begin{bmatrix} -1 \\ 0 \\ 1 \end{bmatrix}.$$

The leftover variables y and z are the parameters for the solution set. To emphasize their new status as parameters, we rename them s and t, putting $y = s$ and $z = t$. Then defining

$$\mathbf{x}_0 = \begin{bmatrix} 4 \\ 0 \\ 0 \end{bmatrix}, \qquad \mathbf{v}_1 = \begin{bmatrix} -2 \\ 1 \\ 0 \end{bmatrix}, \qquad \text{and} \qquad \mathbf{v}_2 = \begin{bmatrix} -1 \\ 0 \\ 1 \end{bmatrix},$$

we can write

$$\mathbf{x} = \mathbf{x}_0 + s\mathbf{v}_1 + t\mathbf{v}_2. \tag{1.8}$$

We have just seen that *every* solution of (1.2) can be written in the form (1.8). We now show two more things, first, that *every* vector of the form $\mathbf{x}_0 + s\mathbf{v}_1 + t\mathbf{v}_2$ is indeed a solution of (1.2) and second, that a solution of (1.2) can be written in the form (1.8) for *just one* pair of values for s and t.

To see the first point, compute that with $\mathbf{a} = \begin{bmatrix} 1 \\ 2 \\ 1 \end{bmatrix}$,

$$\mathbf{a} \cdot \mathbf{x}_0 = 4, \qquad \mathbf{a} \cdot \mathbf{v}_1 = 0, \qquad \text{and} \qquad \mathbf{a} \cdot \mathbf{v}_2 = 0$$

so that for every s and t, $\mathbf{a} \cdot (\mathbf{x}_0 + s\mathbf{v}_1 + t\mathbf{v}_2) = 4 + s0 + t0 = 4$. In other words, for every s and t, $\mathbf{x}_0 + s\mathbf{v}_1 + t\mathbf{v}_2$ is a solution of (1.2).

To see the second point, note that

$$\mathbf{x}_0 + s\mathbf{v}_1 + t\mathbf{v}_2 = \begin{bmatrix} 4 - 2s - t \\ s \\ t \end{bmatrix}.$$

Hence, if $\mathbf{x} = \mathbf{x}_0 + s\mathbf{v}_1 + t\mathbf{v}_2$, then $s = y$ and $t = z$. In this parameterization, the y and z values of $\mathbf{x}$ uniquely determine the values of the parameters s and t.

- *As the parameters s and t are varied, $\mathbf{x}_0 + s\mathbf{v}_1 + t\mathbf{v}_2$ sweeps through every point in the plane, stays in the plane, and sweeps through any given point in the plane for exactly one pair of values of s and t, setting up a one-to-one correspondence between pairs (s, t) and points in the plane. This is what we mean by a one-to-one parameterization of the plane.*

Next, consider lines. We shall parameterize the line in $\mathbb{R}^3$ given by the intersection of the planes described by the pair of equations in (1.6). As before, we proceed by eliminating variables. From the first equation,

$$x = 1 - 2y - 3z. \tag{1.9}$$

Substituting (1.9) into the second equation it becomes $3(1 - 2y - 3z) + 2y + z = -1$, or $-4y - 8z = -4$. We can use (1.9) to eliminate y: solving for y,

$$y = 1 - 2z. \tag{1.10}$$

Returning to (1.9) and using (1.10) to eliminate y from the right-hand side, we obtain $x = 1 - 2(1 - 2z) - 3z$, or

$$x = -1 + z. \tag{1.11}$$

Both x and y are now expressed in terms of the leftover variable z, which will be our parameter. If

$$\mathbf{x} = \begin{bmatrix} x \\ y \\ z \end{bmatrix}$$

is any point in the solution set of (1.6), then

$$\mathbf{x} = \begin{bmatrix} x \\ y \\ z \end{bmatrix} = \begin{bmatrix} -1 + z \\ 1 - 2z \\ z \end{bmatrix}.$$

Expanding the right-hand side into a constant term and a constant multiple of z,

$$\mathbf{x} = \begin{bmatrix} -1 \\ 1 \\ 0 \end{bmatrix} + z \begin{bmatrix} 1 \\ -2 \\ 1 \end{bmatrix}.$$

Defining

$$\mathbf{x}_0 = \begin{bmatrix} x \\ y \\ z \end{bmatrix} \quad \text{and} \quad \mathbf{v} = \begin{bmatrix} 1 \\ -2 \\ 1 \end{bmatrix}$$

and renaming z as t to emphasize its status as a parameter,

$$\mathbf{x} = \mathbf{x}_0 + t\mathbf{v}.$$

This is our parameterization of the line described by (1.6). We have already seen that every solution has the form $\mathbf{x}_0 + t\mathbf{v}$. To see that for every t, $\mathbf{x}_0 + t\mathbf{v}$ is a solution of (1.6), we work with the matrix form of the equation, (1.4), with A and $\mathbf{d}$ given by (1.5). Computing $A\mathbf{x}_0$ and $A\mathbf{v}$,

$$A\mathbf{x}_0 = \mathbf{d} \quad \text{and} \quad A\mathbf{v} = 0.$$

Therefore, $A(\mathbf{x}_0 + t\mathbf{v}) = A\mathbf{x}_0 + tA\mathbf{v} = \mathbf{d} + t0 = \mathbf{d}$, so $\mathbf{x}_0 + t\mathbf{v}$ is a solution for all values of t. Finally, we see that if $\mathbf{x}$ is any solution and $\mathbf{x}$ is written in the form $\mathbf{x}_0 + t\mathbf{v}$, the value of t is given by the z coordinate of $\mathbf{x}$ and is thus uniquely determined. Again, *by solving the equation, that is, eliminating variables, we have found a one-to-one parameterization.*

1.2 A geometric perspective

Let us think about how a parameterization of the form $\mathbf{x}_0 + s\mathbf{v}_1 + t\mathbf{v}_2$ "sweeps out" a plane in $\mathbb{R}^3$ as s and t are varied:

Call point $\mathbf{x}_0$ the *base point*. It could be any point in the plane. The plane is a two-dimensional "slice" of $\mathbb{R}^3$, so there are two "independent" directions one can move in while staying in the plane, specified by $\mathbf{v}_1$ and $\mathbf{v}_2$, which are called *direction vectors*. The idea is that if we start from $\mathbf{x}_0$, move in the direction of $\mathbf{v}_1$ to $\mathbf{x}_0 + s\mathbf{v}_1$, and from there in the direction of $\mathbf{v}_2$ to the point $\mathbf{x}_0 + s\mathbf{v}_1 + t\mathbf{v}_2$, we stay in the plane and can reach any point in the plane this way—for exactly one pair of values for s and t. The fact that we need two parameters is a reflection of the fact that a plane is "two dimensional."*

---

*The notion of dimension will be very important later on, and we will precisely define it. For now, we rely on your intuitive understanding of this concept in $\mathbb{R}^3$.

Any point in the plane can serve as a base point $\mathbf{x}_0$, and *any two* vectors $\mathbf{v}_1$ and $\mathbf{v}_2$ that are parallel to the plane can serve as the direction vectors as long as neither is a multiple of the other. Thus,

- *There are infinitely many ways to parameterize any given plane.*

Likewise, in a parameterization of a line of the form $\mathbf{x}_0 + t\mathbf{v}$, we again refer to $\mathbf{x}_0$ as the base point and $\mathbf{v}$ as the direction vector. Again, we can choose any point on the line as the base point and any (nonzero) vector parallel to the line as the direction vector. Thus again, there are infinitely many ways to parameterize any given line. Here is an illustration of the "base point, direction vector" scheme:

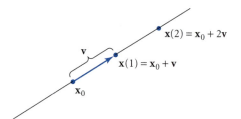

We have found parameterizations in the previous subsection by algebraic means. Here is another approach that is somewhat more geometric. Given the equation $\mathbf{a} \cdot \mathbf{x} = d$, we seek vectors $\mathbf{x}_0$, $\mathbf{v}_1$, and $\mathbf{v}_2$ such that

$$\mathbf{a} \cdot \mathbf{x}_0 = d, \qquad \mathbf{a} \cdot \mathbf{v}_1 = 0, \qquad \text{and} \qquad \mathbf{a} \cdot \mathbf{v}_2 = 0. \tag{1.12}$$

Then for any s and t, $\mathbf{a} \cdot (\mathbf{x}_0 + s\mathbf{v}_1 + t\mathbf{v}_2) = d + s0 + t0 = d$, so $\mathbf{x}_0 + s\mathbf{v}_1 + t\mathbf{v}_2$ is a solution of the equation and hence a point of the plane. Let us do this for

$$\mathbf{a} = \begin{bmatrix} 1 \\ 2 \\ 1 \end{bmatrix} \qquad \text{and} \qquad d = 4,$$

as before.

Finding a particular solution $\mathbf{x}_0$ is easy: We can arbitrarily set $y = z = 0$, and then $\mathbf{a} \cdot \mathbf{x} = 4$ reduces to $x = 4$:

$$\mathbf{x}_0 = \begin{bmatrix} 4 \\ 0 \\ 0 \end{bmatrix}.$$

This was not very geometric, but it was very easy. Since we just want some solution—any solution we happen to bump into—we can make the simplifying assumption $y = z = 0$.

We can use geometry to find vectors with $\mathbf{a} \cdot \mathbf{v}_1 = \mathbf{a} \cdot \mathbf{v}_2 = 0$. Consider a vector $\mathbf{v}$ of the form

$$\mathbf{v} = \begin{bmatrix} v \\ w \\ 0 \end{bmatrix}.$$

Then, because of the zero,

$$\mathbf{a} \cdot \mathbf{v} = \begin{bmatrix} 1 \\ 2 \\ 1 \end{bmatrix} \cdot \begin{bmatrix} v \\ w \\ 0 \end{bmatrix} = \begin{bmatrix} 1 \\ 2 \end{bmatrix} \cdot \begin{bmatrix} v \\ w \end{bmatrix}.$$

Hence $\mathbf{a} \cdot \mathbf{v} = 0$ if and only if

$$\begin{bmatrix} v \\ w \end{bmatrix} \quad \text{is orthogonal to} \quad \begin{bmatrix} 1 \\ 2 \end{bmatrix}.$$

To get a vector that is orthogonal to $\begin{bmatrix} 1 \\ 2 \end{bmatrix}$, rotate the vector $\begin{bmatrix} 1 \\ 2 \end{bmatrix}$ counterclockwise through the angle $\pi/2$. In Example 13 of Chapter 1, we computed the matrix of this transformation, and using the result we see that the transformed vector, denoted

$$\begin{bmatrix} 1 \\ 2 \end{bmatrix}^\perp,$$

is given by

$$\begin{bmatrix} 1 \\ 2 \end{bmatrix}^\perp = \begin{bmatrix} -2 \\ 1 \end{bmatrix},$$

leading to the choice

$$\mathbf{v}_1 = \begin{bmatrix} -2 \\ 1 \\ 0 \end{bmatrix}.$$

Now what about $\mathbf{v}_2$? If we simply choose $\mathbf{v}_2 = \mathbf{v}_1$, we *would* have all three equations in (1.12) satisfied. However, with this choice, $\mathbf{x}_0 + s\mathbf{v}_1 + t\mathbf{v}_2 = \mathbf{x}_0 + (s + t)\mathbf{v}_1$, so as s and t are varied, all that $\mathbf{x}_0 + s\mathbf{v}_1 + t\mathbf{v}_2$ sweeps out is a line, not a plane. Hence we *must* choose $\mathbf{v}_2$ so that it is not a multiple of $\mathbf{v}_1$.

To do so, suppose that

$$\mathbf{v} = \begin{bmatrix} 0 \\ v \\ w \end{bmatrix}.$$

The same reasoning that led us to $\mathbf{v}_1$ tells us that we will have $\mathbf{a} \cdot \mathbf{v} = 0$ if

$$\begin{bmatrix} v \\ w \end{bmatrix} = \begin{bmatrix} 2 \\ 1 \end{bmatrix}^\perp = \begin{bmatrix} -1 \\ 2 \end{bmatrix}.$$

Hence, we take

$$\mathbf{v}_2 = \begin{bmatrix} 0 \\ -1 \\ 2 \end{bmatrix},$$

which gives us another parameterization of our plane:

$$\begin{bmatrix} 4 \\ 0 \\ 0 \end{bmatrix} + s\begin{bmatrix} -2 \\ 1 \\ 0 \end{bmatrix} + t\begin{bmatrix} 0 \\ -1 \\ 2 \end{bmatrix}.$$

Comparing with what we found before, we see that the choices for $\mathbf{x}_0$ and $\mathbf{v}_1$ are the same, but the choices for $\mathbf{v}_2$ differ. This result should be no surprise: We know there are infinitely many parameterizations, and we used a different method to find this one. It is left as an exercise to show that this is in fact a one-to-one parameterization.

We can gain further familiarity with parameterization by considering the problem of parameterizing lines from a geometric starting point.

Recall that given two distinct points $\mathbf{x}_0$ and $\mathbf{x}_1$, a unique line runs through them. The points on this line are given by

$$\mathbf{x}_0 + t(\mathbf{x}_1 - \mathbf{x}_0) = (1 - t)\mathbf{x}_0 + t\mathbf{x}_1 \qquad (1.13)$$

as t varies over the real line. Here, $\mathbf{x}_0$ is the base point, and $\mathbf{v} = \mathbf{x}_1 - \mathbf{x}_0$ is the direction vector. In fact, in *any dimension*, there is a unique line through any two distinct points, so the formula (1.13) is valid in $\mathbb{R}^n$ for every n.

To see that the parameterization $\mathbf{x}_0 + t\mathbf{v}$ is one-to-one, notice that if for some t and $\tilde{t}$

$$\mathbf{x}_0 + t\mathbf{v} = \mathbf{x}_0 + \tilde{t}\mathbf{v},$$

then canceling $\mathbf{x}_0$ from both sides, we have $(t - \tilde{t})\mathbf{v} = 0$. Since $\mathbf{v} \neq 0$, $\tilde{t} = t$, so there is just one t value for each point on the line.

EXAMPLE 1   **(Parameterizing the line through two points in $\mathbb{R}^2$)** Consider the two points

$$\mathbf{x}_0 = \begin{bmatrix} 1 \\ 1 \end{bmatrix} \qquad \text{and} \qquad \mathbf{x}_1 = \begin{bmatrix} 2 \\ 3 \end{bmatrix}.$$

To parameterize the line passing through them, form

$$\mathbf{v} = \mathbf{x}_1 - \mathbf{x}_0 = \begin{bmatrix} 1 \\ 2 \end{bmatrix}.$$

The parameterization is

$$\mathbf{x}_0 + t\mathbf{v} = \begin{bmatrix} 1 \\ 1 \end{bmatrix} + t \begin{bmatrix} 1 \\ 2 \end{bmatrix} = \begin{bmatrix} 1 + t \\ 1 + 2t \end{bmatrix}.$$

As t varies over the real numbers, the point on the line with parameter value t, $\mathbf{x}(t)$, is given by

$$\mathbf{x}(t) = \begin{bmatrix} 1 + t \\ 1 + 2t \end{bmatrix}.$$

We can do something very similar for planes in $\mathbb{R}^3$. Recall that given any three points $\mathbf{p}_1$, $\mathbf{p}_2$, and $\mathbf{p}_3$ in $\mathbb{R}^3$ *that do not all lie on a single line*, there is a unique plane passing through them. To obtain a parameterization, pick one of them, say $\mathbf{p}_1$, as a base point $\mathbf{x}_0$. Then form the two direction vectors

$$\mathbf{v}_1 = \mathbf{p}_2 - \mathbf{p}_1 \qquad \text{and} \qquad \mathbf{v}_2 = \mathbf{p}_3 - \mathbf{p}_1.$$

Then a one-to-one parameterization of the plane is given by

$$\mathbf{x}_0 + s\mathbf{v}_1 + t\mathbf{v}_2, \qquad (1.14)$$

as s and t range over the real numbers. Here is why: If you start from a base point $\mathbf{x}_0$ and can move arbitrary distances—back and forth—in two nonparallel directions, then you will cover a plane. Therefore, (1.14) is a parameterization of the plane.

To see that it is also one-to-one, suppose that there are s, t, $\tilde{s}$, and $\tilde{t}$ so that

$$\mathbf{x}_0 + s\mathbf{v}_1 + t\mathbf{v}_2 = \mathbf{x}_0 + \tilde{s}\mathbf{v}_1 + \tilde{t}\mathbf{v}_2.$$

Canceling $\mathbf{x}_0$ from both sides and rearranging,

$$(s - \tilde{s})\mathbf{v}_1 = (\tilde{t} - t)\mathbf{v}_2.$$

Suppose that $s \neq \tilde{s}$. Then

$$\mathbf{v}_1 = \frac{\tilde{t} - t}{s - \tilde{s}}\mathbf{v}_2,$$

and then

$$\mathbf{p}_1 = \mathbf{x}_0 + \mathbf{v}_1 = \mathbf{x}_0 + \frac{\tilde{t} - t}{s - \tilde{s}} \mathbf{v}_2$$

would be on the line through $\mathbf{p}_1$ and $\mathbf{p}_3$, contrary to assumption. Therefore $s = \tilde{s}$. In the same way, one sees $t = \tilde{t}$, so the parameterization is one-to-one.

EXAMPLE 2   **(Parameterizing the plane through three points)**   Consider the three points

$$\mathbf{p}_1 = \begin{bmatrix} 1 \\ 0 \\ 2 \end{bmatrix}, \qquad \mathbf{p}_2 = \begin{bmatrix} 2 \\ 1 \\ 0 \end{bmatrix}, \qquad \text{and} \qquad \mathbf{p}_3 = \begin{bmatrix} -1 \\ -3 \\ 1 \end{bmatrix}.$$

We choose $\mathbf{x}_0 = \mathbf{p}_1$ and form

$$\mathbf{v}_1 = \mathbf{p}_2 - \mathbf{p}_1 = \begin{bmatrix} 1 \\ 1 \\ -2 \end{bmatrix} \qquad \text{and} \qquad \mathbf{v}_2 = \mathbf{p}_3 - \mathbf{p}_1 = \begin{bmatrix} -2 \\ -3 \\ -1 \end{bmatrix}.$$

The vectors $\mathbf{v}_1$ and $\mathbf{v}_2$ are not proportional (as you see from the pattern of minus signs), so the three points $\mathbf{p}_1, \mathbf{p}_2$, and $\mathbf{p}_3$ do not lie on the same line, and hence they do determine a plane; this plane is parameterized by

$$\mathbf{x}_0 + s\mathbf{v}_1 + t\mathbf{v}_2 = \begin{bmatrix} 1 \\ 0 \\ 2 \end{bmatrix} + s\begin{bmatrix} 1 \\ 1 \\ -2 \end{bmatrix} + t\begin{bmatrix} -2 \\ -3 \\ -1 \end{bmatrix} = \begin{bmatrix} 1 + s - 2t \\ s - 3t \\ 2 - 2s - t \end{bmatrix}.$$

As s and t vary, the vector

$$\mathbf{x}(s, t) = \begin{bmatrix} 1 + s - 2t \\ s - 3t \\ 2 - 2s - t \end{bmatrix}$$

ranges over the plane in a one-to-one way.

1.3   Recovering an equation from a parameterization

We have seen how to parameterize the solution set of an equation. It is often useful to go backwards. In particular, we can ask: Given a parameterization of a line in $\mathbb{R}^2$ or a plane in $\mathbb{R}^3$, how do we find an equation for the line or plane?

Consider, for example, the line passing through the points

$$\mathbf{x}_0 = \begin{bmatrix} 1 \\ 1 \end{bmatrix} \qquad \text{and} \qquad \mathbf{x}_1 = \begin{bmatrix} 2 \\ 3 \end{bmatrix}.$$

In Example 1, we found the parameterization

$$\mathbf{x}(t) = \begin{bmatrix} 1 + t \\ 1 + 2t \end{bmatrix}$$

for this line.

The general equation of a line in $\mathbb{R}^2$ has the form $ax + by = c$, where at least one of a and b is not zero. Plugging in $x = 1 + t$ and $y = 1 + 2t$ from the parameterization,

$$ax + by = a(1 + t) + b(1 + 2t) = (a + b) + t(a + 2b).$$

We want the right-hand side to be a constant, so choose $a = -2$ and $b = 1$. Then we have

$$-2x + y = -1.$$

The same principle can be applied to recover the equation of a plane from a parameterization.

EXAMPLE 3   **(The equation of a plane from a parameterization)** Consider the plane passing through

$$\mathbf{p}_1 = \begin{bmatrix} 1 \\ 0 \\ 2 \end{bmatrix}, \qquad \mathbf{p}_2 = \begin{bmatrix} 2 \\ 1 \\ 0 \end{bmatrix}, \qquad \text{and} \qquad \mathbf{p}_3 = \begin{bmatrix} -1 \\ -3 \\ 1 \end{bmatrix}.$$

In Example 2, we found the parameterization

$$\mathbf{x}(s, t) = \begin{bmatrix} 1 + s - 2t \\ s - 3t \\ 2 - 2s - t \end{bmatrix}$$

for this plane. The general equation for a plane in $\mathbb{R}^3$ is $ax + by + cz = d$, where at least one of a, b, or c is not zero. Plugging in $x = 1 + s - 2t$, $y = s - 3t$ and $z = 1 - 2s - t$ from the parameterization, we have

$$ax + by + cz = a(1 + s - 2t) + b(s - 3t) + c(2 - 2s - t) = (a + 2c) + s(a + b + 2c) + t(-2a - 3b - c). \quad (1.15)$$

Since the right-hand side must be a constant, namely d, the coefficients of s and t on the right must be zero:

$$a + b - 2c = 0$$
$$2a + 3b + c = 0.$$

This is a system of equations. In this chapter, we will see how to find all solutions of such systems. But now, we are looking for just one solution other than the obvious one $a = b = c = 0$.

To avoid this solution, we set $c = 1$. Then the system becomes

$$a + b = 2$$
$$2a + 3b = -1.$$

Subtracting the first equation from the second twice, we eliminate a from it, and get $b = -5$. Plugging $b = -5$ into the first equation, we get $a = 7$. Then, with $a = 7$, $b = -5$, and $c = 1$ in (1.15), we have the equation

$$7x - 5y + z = 9.$$

You can easily check that $\mathbf{p}_1$, $\mathbf{p}_2$, and $\mathbf{p}_3$ each solve this equation, as they must.

As you see from the examples, recovering a linear equation from a parameterization requires us to solve a *system of equations*. This chapter provides a systematic way to find all solutions, if any, of any number of equations in any number of variables. With only two or three variables it is easy enough to proceed "by hand," as we did here, especially since we seek just one nonzero solution and not all possible solutions.

Exercises

1.1 Find a one-to-one parameterization of the solution set in $\mathbb{R}^2$ of the equation

$$x - y = 4,$$

which is a line in $\mathbb{R}^2$. Determine two points in $\mathbb{R}^2$ that satisfy the preceding equation. Draw a picture of the solution set.

1.2 Find a one-to-one parameterization of the solution set in $\mathbb{R}^2$ of the equation

$$2x + y = 5,$$

which is a line in $\mathbb{R}^2$. Determine three points in $\mathbb{R}^2$ that satisfy the preceding equation. Draw a picture of the solution set.

1.3 Find a one-to-one parameterization of the solution set in $\mathbb{R}^2$ of the equation

$$x + 3y = 1,$$

which is a line. Graph this line, together with the unit circle. Are there any unit vectors $\begin{bmatrix} x \\ y \end{bmatrix}$ that satisfy the equation $x + 3y = 1$? If so, indicate them in your graph, and compute them.

1.4 Find a one-to-one parameterization of the solution set in $\mathbb{R}^2$ of the equation

$$x - 5y = 1,$$

which is a line. Graph this line, together with the unit circle. Are there any unit vectors $\begin{bmatrix} x \\ y \end{bmatrix}$ that satisfy the equation $x - 5y = 1$? If so, indicate them in your graph, and compute them.

1.5 Find an equation in $\mathbb{R}^2$ whose solution set is the line with the parametrization $\mathbf{x}_0 + t\mathbf{v}$ where

$$\mathbf{x}_0 = \begin{bmatrix} 1 \\ 2 \end{bmatrix} \quad \text{and} \quad \mathbf{v} = \begin{bmatrix} -2 \\ -3 \end{bmatrix}.$$

Graph the line. Indicate the points $\mathbf{x}_0 + \mathbf{v}$ and $\mathbf{x}_0 - 2\mathbf{v}$.

1.6 Find an equation in $\mathbb{R}^2$ whose solution set is the line with the parametrization $\mathbf{x}_0 + t\mathbf{v}$ where

$$\mathbf{x}_0 = \begin{bmatrix} 1 \\ -1 \end{bmatrix} \quad \text{and} \quad \mathbf{v} = \begin{bmatrix} 1 \\ 3 \end{bmatrix}.$$

Graph the line. Indicate the points $\mathbf{x}_0 + 2\mathbf{v}$ and $\mathbf{x}_0 - \mathbf{v}$.

1.7 Consider the line that passes through the points $\begin{bmatrix} 3 \\ 0 \end{bmatrix}$ and $\begin{bmatrix} 2 \\ 1 \end{bmatrix}$.

(a) Draw a picture of the line.

(b) Determine a one-to-one parametrization of the line.

(c) Determine an equation whose solution set is the line considered here.

1.8 Consider the line that passes through the points $\begin{bmatrix} 2 \\ 1 \end{bmatrix}$ and $\begin{bmatrix} -1 \\ 1 \end{bmatrix}$.

(a) Draw a picture of the line.

(b) Determine a one-to-one parameterization of the line.

(c) Determine an equation whose solution set is the line considered here.

1.9 Give two different parameterizations of the line in $\mathbb{R}^2$ that passes through the points $(1, 4)$ and $(-1, -2)$.

1.10 Give two different parameterizations of the line in $\mathbb{R}^3$ that passes through the points $(1, 0, 3)$ and $(-1, -3, 2)$.

1.11 Find a one-to-one parameterization of the solution set in $\mathbb{R}^3$ of the equation $2x_1 + 3x_2 - x_3 = 1$.

1.12 Find a one-to-one parameterization of the solution set in $\mathbb{R}^3$ of the equation $x_1 - 2x_2 + x_3 = 2$.

1.13 Find a one-to-one parameterization of the solution set in $\mathbb{R}^4$ of the equation $x_1 - 3x_2 + x_3 + x_4 = 0$.

1.14 Find a one-to-one parameterization of the solution set in $\mathbb{R}^4$ of the equation $3x_2 + x_3 = 0$.

1.15 Consider the equation $2x - y + z = 1$.

(a) Parametrize the solution set by eliminating x.

(b) Parametrize the solution set by eliminating y.

(c) Parametrize the solution set by eliminating z.

When are two parameterized lines the same?

As we have seen, there are infinitely many ways to parameterize any given line. Suppose $\mathbf{x}_0$ and $\mathbf{v}$ are some given pair of vectors and that $\tilde{\mathbf{x}}_0$ and $\tilde{\mathbf{v}}$ are another given pair of vectors. Suppose that both $\mathbf{v}$ and $\tilde{\mathbf{v}}$ are different from 0 so that as t varies, $\mathbf{x}_0 + t\mathbf{v}$ traces out a line, and as s varies, $\tilde{\mathbf{x}}_0 + s\tilde{\mathbf{v}}$ traces out a line. *When are these two lines the same?*

- *The lines traced out by $\mathbf{x}_0 + t\mathbf{v}$ as t varies and by $\tilde{\mathbf{x}}_0 + s\tilde{\mathbf{v}}$ as s varies are the same if and only if $\tilde{\mathbf{x}}_0 - \mathbf{x}_0$, $\tilde{\mathbf{v}}$, and $\mathbf{v}$ are all proportional—that is, multiples of a single vector.*

1.16 Show that the claim just made is true.

1.17 Consider the lines given by $\mathbf{x}_0 + t\mathbf{v}$ and $\tilde{\mathbf{x}}_0 + s\tilde{\mathbf{v}}$ where

$$\mathbf{x}_0 = \begin{bmatrix} 0 \\ 2 \end{bmatrix} \quad \text{and} \quad \mathbf{v} = \begin{bmatrix} 1 \\ -3/2 \end{bmatrix}$$

and where

$$\tilde{\mathbf{x}}_0 = \begin{bmatrix} 4/3 \\ 0 \end{bmatrix} \quad \text{and} \quad \tilde{\mathbf{v}} = \begin{bmatrix} -2/3 \\ 1 \end{bmatrix}.$$

Show that $\tilde{\mathbf{x}}_0 - \mathbf{x}_0$, $\tilde{\mathbf{v}}$, and $\mathbf{v}$ are all proportional and that the parameterized lines $\mathbf{x}_0 + t\mathbf{v}$ and $\tilde{\mathbf{x}}_0 + s\tilde{\mathbf{v}}$ coincide.

1.18 Consider the lines parameterized by $\begin{bmatrix} 1 \\ 0 \\ 1 \end{bmatrix} + s \begin{bmatrix} 2 \\ 3 \\ 1 \end{bmatrix}$ and $\begin{bmatrix} -1 \\ -3 \\ 0 \end{bmatrix} + t \begin{bmatrix} 4 \\ 6 \\ 2 \end{bmatrix}$ as s and t vary. Are these two different parameterizations of the same line, or are the lines different?

1.19 Consider the lines parameterized by $\begin{bmatrix} 1 \\ 2 \\ 2 \end{bmatrix} + s \begin{bmatrix} -1 \\ 1 \\ -3 \end{bmatrix}$ and $\begin{bmatrix} 0 \\ 3 \\ -1 \end{bmatrix} + t \begin{bmatrix} 2 \\ -2 \\ 6 \end{bmatrix}$ as s and t vary. Are these two different parameterizations of the same line, or are the lines different?

Lines intersecting planes

Which are more useful, parameterizations or equations? It depends on what we are trying to do. Consider the following type of problem: Suppose we are given a plane that is specified by giving three (noncolinear) points $\mathbf{p}_0$, $\mathbf{p}_1$, $\mathbf{p}_2$ that it contains and a line specified by giving two distinct points, $\mathbf{q}_0$, $\mathbf{q}_1$, that it contains.

- *Does the line intersect the plane, and if so, where?*

To answer this type of question, let $\mathbf{x}(t) = \mathbf{x}_0 + t\mathbf{v}$ be a parameterization of the line, and let $\mathbf{a} \cdot \mathbf{x} = d$ be an equation for the plane. Plug $\mathbf{x}(t)$ into the equation for the plane, getting $\mathbf{a} \cdot (\mathbf{x}_0 + t\mathbf{v}) = d$. As long as $\mathbf{a} \cdot \mathbf{v} \neq 0$, we can solve for t:

$$t = (d - \mathbf{a} \cdot \mathbf{x}_0)/(\mathbf{a} \cdot \mathbf{v}).$$

Then, with this value of t, $\mathbf{x}(t)$ is both on the line and on the plane. In this sort of problem, we use both a parameterization *and* an equation. Since we have seen in the examples how to find both of them, we can solve for the intersection.

1.20 The formula for t just given does not make sense if $\mathbf{a} \cdot \mathbf{v} = 0$. Show that in this case, either there is no intersection or else the entire line is contained in the plane.

1.21 Consider the plane in $\mathbb{R}^3$ that contains the three points $\mathbf{p}_0 = \begin{bmatrix} 1 \\ 2 \\ 1 \end{bmatrix}$, $\mathbf{p}_1 = \begin{bmatrix} 3 \\ 0 \\ 1 \end{bmatrix}$, and $\mathbf{p}_2 = \begin{bmatrix} 1 \\ 1 \\ 2 \end{bmatrix}$.

Consider also the line in $\mathbb{R}^3$ containing the two points $\mathbf{q}_0 = \begin{bmatrix} 1 \\ 1 \\ 1 \end{bmatrix}$ and $\mathbf{q}_1 = \begin{bmatrix} 2 \\ 2 \\ 2 \end{bmatrix}$.

(a) Find an equation for the plane.

(b) Parametrize the line.

(c) The line and plane intersect in a unique point. Compute it.

1.22 Consider the plane in $\mathbb{R}^3$ that contains the three points $\mathbf{p}_0 = \begin{bmatrix} 1 \\ 4 \\ 1 \end{bmatrix}$, $\mathbf{p}_1 = \begin{bmatrix} 1 \\ 2 \\ 0 \end{bmatrix}$, and $\mathbf{p}_2 = \begin{bmatrix} 0 \\ 2 \\ 1 \end{bmatrix}$.

Consider also the line in $\mathbb{R}^3$ containing the two points $\mathbf{q}_0 = \begin{bmatrix} 1 \\ 0 \\ 1 \end{bmatrix}$ and $\mathbf{q}_1 = \begin{bmatrix} 1 \\ 0 \\ 2 \end{bmatrix}$.

(a) Find an equation for the plane.

(b) Parametrize the line.

(c) The line and plane intersect in a unique point. Compute it.

1.23 Consider the plane in $\mathbb{R}^3$ that contains the three points $\mathbf{p}_0 = \begin{bmatrix} 1 \\ 1 \\ 1 \end{bmatrix}$, $\mathbf{p}_1 = \begin{bmatrix} 3 \\ 0 \\ 0 \end{bmatrix}$, and $\mathbf{p}_2 = \begin{bmatrix} 2 \\ 0 \\ 1 \end{bmatrix}$.

Consider also the line in $\mathbb{R}^3$ containing the two points $\mathbf{q}_0 = \begin{bmatrix} 1 \\ 1 \\ 1 \end{bmatrix}$ and $\mathbf{q}_1 = \begin{bmatrix} 2 \\ 0 \\ 0 \end{bmatrix}$.

(a) Find an equation for the plane

(b) Parametrize the line.

(c) The line and plane intersect in a unique point. Compute it.

The distance from a point to a line

Let ℓ be a line in $\mathbb{R}^3$, and let $\mathbf{y}$ be any point in $\mathbb{R}^3$. Which point on the line is closest to $\mathbf{y}$, and what is the distance between $\mathbf{y}$ and that point? This distance is, by definition, the distance between $\mathbf{y}$ and the line ℓ.

This sort of question can be readily answered using a parameterization.

1.24 Suppose that $\mathbf{x}(t) = \mathbf{x}_0 + t\mathbf{v}$ is a parameterization of ℓ. The question just raised amounts to the following: For which value of t is $|\mathbf{x}(t) - \mathbf{y}|^2$ the smallest?

(a) Show that

$$|\mathbf{x}(t) - \mathbf{y}|^2 = |\mathbf{v}|^2 t^2 - 2t\mathbf{v} \cdot (\mathbf{y} - \mathbf{x}_0) + |\mathbf{y} - \mathbf{x}_0|^2.$$

(b) By completing the square or by using calculus, show that the right-hand side is minimized at $t = \mathbf{v} \cdot (\mathbf{y} - \mathbf{x}_0)/|\mathbf{v}|^2$. Evaluating $\mathbf{x}(t)$ at this value of t, deduce that the point in ℓ that is closest to $\mathbf{y}$ is

$$\text{Closest point} = \mathbf{x}_0 + \frac{\mathbf{v} \cdot (\mathbf{y} - \mathbf{x}_0)}{|\mathbf{v}|^2} \mathbf{v}. \tag{1.16}$$

(c) Show that the distance from $\mathbf{y}$ to the closest point is

$$\left(|\mathbf{y} - \mathbf{x}_0|^2 - \frac{|\mathbf{v} \cdot (\mathbf{y} - \mathbf{x}_0)|^2}{|\mathbf{v}|^2} \right)^{1/2},$$

and that this quantity is equal to $|(\mathbf{y} - \mathbf{x}_0)_\perp|$, where $\mathbf{y} - \mathbf{x}_0 = (\mathbf{y} - \mathbf{x}_0)_\| + (\mathbf{y} - \mathbf{x}_0)_\perp$ is the decomposition of $\mathbf{y} - \mathbf{x}_0$ into its components that are parallel and perpendicular to $\mathbf{v}$. That is,

$$\text{Distance from } \mathbf{y} \text{ to } \ell = |(\mathbf{y} - \mathbf{x}_0)_\perp|. \tag{1.17}$$

(d) Draw a picture showing ℓ, the base point $\mathbf{x}_0$, the closest point in ℓ, and the decomposition $\mathbf{y} - \mathbf{x}_0 = (\mathbf{y} - \mathbf{x}_0)_\| + (\mathbf{y} - \mathbf{x}_0)_\perp$. In your picture, the validity of (1.16) and (1.17) should be visually clear.

1.25 Consider the line ℓ passing through $\mathbf{x}_0 = \begin{bmatrix} 1 \\ 2 \\ 3 \end{bmatrix}$ and $\mathbf{x}_1 = \begin{bmatrix} 3 \\ 1 \\ 1 \end{bmatrix}$. Let $\mathbf{y} = \begin{bmatrix} 1 \\ 1 \\ 1 \end{bmatrix}$. Find the point on ℓ that is closest to $\mathbf{y}$, and find the distance from $\mathbf{y}$ to ℓ.

1.26 Consider the line ℓ passing through $\mathbf{x}_0 = \begin{bmatrix} -1 \\ 2 \\ -2 \end{bmatrix}$ and $\mathbf{x}_1 = \begin{bmatrix} 1 \\ 3 \\ 1 \end{bmatrix}$. Let $\mathbf{y} = \begin{bmatrix} 1 \\ 1 \\ 2 \end{bmatrix}$. Find the point on ℓ that is closest to $\mathbf{y}$, and find the distance from $\mathbf{y}$ to ℓ.

The distance from a point to a plane

1.27 Consider the plane in $\mathbb{R}^3$ given by $ax + by + cz = d$.

(a) Let $\mathbf{a} = \begin{bmatrix} a \\ b \\ c \end{bmatrix}$. Let $\mathbf{x}_0$ be any point on the plane. Show that a point $\mathbf{w}$ is on the plane if and only if

$$\mathbf{a} \cdot (\mathbf{w} - \mathbf{x}_0) = 0,$$

that is, if and only if $\mathbf{w} - \mathbf{x}_0$ is perpendicular to $\mathbf{a}$.

(b) Let $\mathbf{y}$ be any point in $\mathbb{R}^3$, and decompose $\mathbf{y} - \mathbf{x}_0$ into its components that are parallel and perpendicular to $\mathbf{a}$: $\mathbf{y} - \mathbf{x}_0 = (\mathbf{y} - \mathbf{x}_0)_\| + (\mathbf{y} - \mathbf{x}_0)_\perp$. Define a point $\mathbf{z}_0$ by

$$\mathbf{z}_0 = \mathbf{x}_0 + (\mathbf{y} - \mathbf{x}_0)_\perp.$$

Show that $\mathbf{z}_0$ belongs to the plane and that $\mathbf{y} - \mathbf{z}_0 = (\mathbf{y} - \mathbf{x}_0)_\|$, which is, of course, parallel to $\mathbf{a}$.

(c) Using the Pythagorean theorem and the orthogonality properties established in parts (a) and (b), show that if $\mathbf{w}$ is any point on the plane,

$$|\mathbf{y} - \mathbf{w}|^2 = |\mathbf{y} - \mathbf{z}_0|^2 + |\mathbf{z}_0 - \mathbf{w}|^2.$$

In particular, $|\mathbf{y} - \mathbf{z}_0| < |\mathbf{y} - \mathbf{w}|$ for any other point on the plane, so $\mathbf{z}_0$ is the point on the plane that is closest to $\mathbf{y}$. By definition, the distance from $\mathbf{y}$ to the plane is the distance from $\mathbf{y}$ to $\mathbf{z}_0$.

(d) Using part (c), show that $\mathbf{a} \cdot (\mathbf{y} - \mathbf{x}_0) = \mathbf{a} \cdot (\mathbf{y} - \mathbf{z}_0)$ and that $\mathbf{a} \cdot (\mathbf{y} - \mathbf{z}_0) = \pm|\mathbf{a}||\mathbf{y} - \mathbf{z}_0|$. Therefore, the distance from $\mathbf{y}$ to the plane is

$$\text{Distance} = \frac{|\mathbf{a} \cdot (\mathbf{y} - \mathbf{x}_0)|}{|\mathbf{a}|}. \tag{1.18}$$

1.28 Consider the plane given by $x + y + z = 3$. Find a point $\mathbf{x}_0$ on the plane of the form $\mathbf{x}_0 = \begin{bmatrix} a \\ 0 \\ 0 \end{bmatrix}$,

and then compute the distance from $\mathbf{y} = \begin{bmatrix} 2 \\ 3 \\ 2 \end{bmatrix}$ to the plane. Also, find the point in the plane that is closest to $\mathbf{y}$.

1.29 Consider the plane in $\mathbb{R}^3$ that contains the three points $\mathbf{p}_0 = \begin{bmatrix} 1 \\ 1 \\ 1 \end{bmatrix}$, $\mathbf{p}_1 = \begin{bmatrix} 3 \\ 0 \\ 0 \end{bmatrix}$, and $\mathbf{p}_3 = \begin{bmatrix} 2 \\ 0 \\ 1 \end{bmatrix}$.

Let $\mathbf{y} = \begin{bmatrix} 2 \\ 0 \\ 0 \end{bmatrix}$, and compute the distance from $\mathbf{y}$ to the plane.

1.30 Consider the plane given by $x + 2y - z = 2$. Find the point in the plane that is closest to the origin $\mathbf{0}$ and the distance from $\mathbf{0}$ to the plane.

SECTION 2 | Systems of Linear Equations

2.1 Systems of linear equations

A system of m linear equations in $\mathbb{R}^n$ is simply a list of m such equations:

$$
\begin{aligned}
\mathbf{a}_1 \cdot \mathbf{x} &= b_1 \\
\mathbf{a}_2 \cdot \mathbf{x} &= b_2 \\
&\vdots \qquad \vdots \\
\mathbf{a}_m \cdot \mathbf{x} &= b_m.
\end{aligned} \tag{2.1}
$$

There are a number of ways of writing such a system, each with its own merits. One way is to introduce the vectors

$$\mathbf{x} = \begin{bmatrix} x_1 \\ x_2 \\ \vdots \\ x_n \end{bmatrix} \quad \text{and} \quad \mathbf{b} = \begin{bmatrix} b_1 \\ b_2 \\ \vdots \\ b_m \end{bmatrix}$$

and the matrix

$$A = \begin{bmatrix} \mathbf{a}_1 \\ \mathbf{a}_2 \\ \vdots \\ \mathbf{a}_m \end{bmatrix} = \begin{bmatrix} A_{1,1} & A_{1,2} & \cdots & A_{1,n} \\ A_{2,1} & A_{2,2} & \cdots & A_{2,n} \\ & & \vdots & \\ A_{m,1} & A_{m,2} & \cdots & A_{m,n} \end{bmatrix}.$$

Notice that $\mathbf{x}$ is in $\mathbb{R}^n$ and $\mathbf{b}$ is in $\mathbb{R}^m$ and that $A_{i,j}$ is the jth entry of $\mathbf{a}_i$. Then by **V.I.F.2**, (2.1) can be rewritten in the compact form

$$Ax = \mathbf{b}. \tag{2.2}$$

Going in the other direction, we can write (2.1) out in full detail as

$$
\begin{aligned}
A_{1,1}x_1 + A_{1,2}x_2 + \cdots + A_{1,n}x_n &= b_1 \\
A_{2,1}x_1 + A_{2,2}x_2 + \cdots + A_{2,n}x_n &= b_2 \\
\vdots \qquad\qquad\qquad \vdots \\
A_{m,1}x_1 + A_{m,2}x_2 + \cdots + A_{m,n}x_n &= b_m.
\end{aligned}
\tag{2.3}
$$

It should be clear that (2.1), (2.2), and (2.3) are just three ways of expressing the same thing and that you can pass back and forth between them easily.

DEFINITION

(Solution set of a linear system) The *solution set S* of a linear system such as (2.1) is simply the set of all vectors that satisfy each of the equations.

To solve a system of linear equations is to find an explicit description of its solution set. This means giving a parametric description of S in the form

$$\mathbf{x}_0 + t_1\mathbf{v}_1 + t_2\mathbf{v}_2 + \cdots + t_r\mathbf{v}_r \tag{2.4}$$

in terms of some number r of parameters.

2.2 Some examples in two and three variables

Suppose we have two linear equations in two variables: $\mathbf{a}_1 \cdot \mathbf{x} = d_1$ and $\mathbf{a}_2 \cdot \mathbf{x} = d_2$. As long as $\mathbf{a}_1 \neq 0$ and $\mathbf{a}_2 \neq 0$, each of the equations describes a line, and the solution set is the intersection of these two lines.

If these lines are not parallel, they will intersect in exactly one point $\mathbf{x}_0$, and in this case the solution set consists of a single vector $\mathbf{x}_0$, as in the first diagram on the next page. *In this case, no parameters are required.*

If, however, the lines are parallel, then either they do not intersect and so the solution set is empty, or else the two lines are the same and *every point* on this line is a solution.

Altogether, then, there are either no solutions, exactly one solution, or infinitely many solutions. (There are never exactly two solutions, for example—something quite common with quadratic equations in one variable.)

EXAMPLE 4 **(Two equations in two variables and intersecting lines)** Consider the system of equations

$$
\begin{aligned}
2x_1 + x_2 &= 3 \\
x_1 + 2x_2 &= 3.
\end{aligned}
\tag{2.5}
$$

Each of these equations is the equation of a line, the first one with slope -2, and the second one with slope $-1/2$. Since the slopes are different, the two lines will intersect.

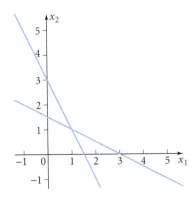

As you see in the graph and can easily check, the point of intersection is $\mathbf{x}_0 = \begin{bmatrix} 1 \\ 1 \end{bmatrix}$, which is the unique solution of this system.

On the other hand, consider the system

$$
\begin{aligned}
2x_1 + x_2 &= 3 \\
4x_1 + 2x_2 &= 3.
\end{aligned}
\tag{2.6}
$$

Each of these equations is the equation of a line, but this time both lines have the same slope, namely -2. Since they have different x_2 intercepts, they are distinct parallel lines, and there is no point of intersection and hence no solution.

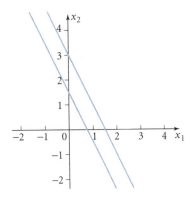

This is an example of a system for which the solution set is the empty set $\emptyset$.

The third possibility is exemplified by

$$
\begin{aligned}
2x_1 + x_2 &= 3 \\
4x_1 + 2x_2 &= 6.
\end{aligned}
\tag{2.7}
$$

There are not really two equations here: The second one is just a multiple of the first. Both equations describe the same line, so every point on this line is a solution of both equations. Here is the diagram:

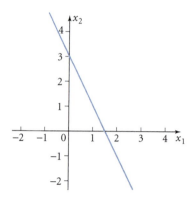

Consider three variables. Since a nontrivial linear equation in three variables describes a plane, the solution set of a system of two such equations is the intersection of two planes. If the planes are parallel, they either coincide, in which case every vector in the plane is a solution, or else they do not intersect at all, in which case there is no solution.

Usually, though, the planes are not parallel, and they intersect in a line. The vectors on this line are the ones satisfying both equations, so this line is the solution set.

EXAMPLE 5   **(Solutions in three variables and intersecting planes)** Consider the system of equations

$$2x_1 + x_2 + x_3 = 3$$
$$x_1 + 2x_2 - x_3 = 1.$$

(2.8)

Each of these equations is the equation of a plane. Here is a graph showing their intersection:

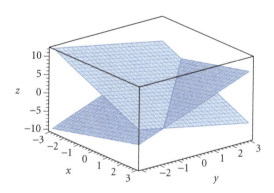

Using the methods explained in the previous section, we could parameterize this line.

When there are three equations in three variables, the solution set is the intersection of three planes. If no two of these planes are parallel, and the line produced by intersecting two of the planes is not parallel to the third, the intersection is a single point:

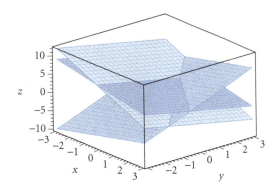

The three planes shown in this graph come from the system

$$2x_1 + x_2 + x_3 = 3$$
$$x_1 + 2x_2 - x_3 = 1 \tag{2.9}$$
$$x_1 + x_2 + x_3 = 3.$$

We know how to parameterize the line formed by the intersection of the first two planes. Let $\mathbf{x}(t)$ be such a parameterization of the intersection. Plugging it into the third equation, we can determine t and hence the point of intersection:

$$\frac{1}{3} \begin{bmatrix} 0 \\ 4 \\ 5 \end{bmatrix}.$$

You can check that this vector solves all three equations in (2.9). There is just one solution; no parameters are needed.

Having *more* equations leads to a *simpler* solution set—a single vector or possibly even the empty set. However, having many equations generally makes figuring out the solution set more work.

We now approach this problem in a systematic way. The key is the concept of an *equivalent system*.

2.3 Equivalent systems

DEFINITION

(Equivalent systems) Two systems of linear equations in the same n variables are *equivalent* if and only if they have the same solution set.

Both systems do not have to have the same number of equations, but they do have to involve the same set of variables.

If we are trying to solve a system, we are free to *change* the system as long as the result is an *equivalent* system. If we can change it into an equivalent system that is *easy to solve*, we have made progress.

EXAMPLE 6   **(Equivalent systems, one really easy)** Consider the system of equations

$$x_1 + x_2 = 2$$
$$2x_1 + 3x_2 = 3. \tag{2.10}$$

To simplify the system, multiply the top equation through by 2. We get the new system

$$2x_1 + 2x_2 = 4$$
$$2x_1 + 3x_2 = 3. \tag{2.11}$$

and we have not changed its solution set since $x + y = 2$ and $2x + 2y = 4$ are equivalent equations. (They describe the exact same line). So (2.10) and (2.11) are equivalent.

Now subtract the top equation in (2.11) from the bottom one:

$$2x_1 + 2x_2 = 4$$
$$x_2 = -1. \tag{2.12}$$

Subtracting equals from equals yields equals. So anything that was in the solution set of (2.11) is in the solution set of (2.12). But the passage from (2.11) to (2.12) is *reversible*: If you add the first equation in (2.12) to the second, you recover (2.11). Since *adding equals to equals yields equals*, anything in the solution set of (2.12) is in the solution set of (2.11). So these two solution sets contain each other, which means that they are the same. So (2.11) and (2.12) are equivalent.

Combining results, all three systems are equivalent. Moreover, (2.12) is *easy to solve*, since the second equation tells us right away that $x_2 = -1$. Plugging this expression into the first equation, we get $2x_1 = 6$, so $x_1 = 3$. Hence,

$$\begin{bmatrix} 3 \\ -1 \end{bmatrix}$$

is the solution of (2.12) and (2.10) as well.

In the next section, we develop a systematic approach to transforming any system of equations into an *easy-to-solve* system that is equivalent.

Exercises   **2.1** Let A be an $m \times n$ matrix, and let $\mathbf{b}$ be any vector in $\mathbb{R}^m$. Let C be an $m \times m$ matrix. Show that if C is invertible, $A\mathbf{x} = \mathbf{b}$ and $(CA)\mathbf{x} = (C\mathbf{b})$ are definitely equivalent systems. Give two examples showing that if C is not invertible, then $A\mathbf{x} = \mathbf{b}$ and $(CA)\mathbf{x} = (C\mathbf{b})$ may or may not be equivalent systems.

2.2 Every 2×2 matrix C of the form $C = \begin{bmatrix} a & 0 \\ c & b \end{bmatrix}$ is invertible as long as neither a nor b is zero. From our general formula for the inverse of a 2×2 matrix, $C^{-1} = \begin{bmatrix} 1/a & 0 \\ -c/(ab) & 1/b \end{bmatrix}$:

(a) Let A be a 2×3 matrix with rows $\mathbf{a}_1$ and $\mathbf{a}_2$, that is, $A = \begin{bmatrix} \mathbf{a}_1 \\ \mathbf{a}_2 \end{bmatrix}$.

Using the result of the previous exercise with $C = \begin{bmatrix} 1 & 0 \\ c & 1 \end{bmatrix}$, show that

$$
\begin{aligned}
\mathbf{a}_1 \cdot \mathbf{x} &= d_1 \\
\mathbf{a}_2 \cdot \mathbf{x} &= d_2
\end{aligned}
\tag{2.13}
$$

and

$$
\begin{aligned}
\mathbf{a}_1 \cdot \mathbf{x} &= d_1 \\
(\mathbf{a}_2 + c\mathbf{a}_1) \cdot \mathbf{x} &= (d_2 + cd_1)
\end{aligned}
\tag{2.14}
$$

are equivalent systems. This is the matrix perspective on the fact that adding a multiple of one equation to another results in an equivalent system.

(b) Suppose that neither $\mathbf{a}_1$ nor $\mathbf{a}_2$ is a multiple of the other in (2.13). Show that if one chooses $c = -(\mathbf{a}_1 \cdot \mathbf{a}_2)/|\mathbf{a}_1|^2$, then $\mathbf{a}_2 + c\mathbf{a}_1$ is orthogonal to $\mathbf{a}_1$.

(c) Two planes are said to be *orthogonal* in the case where their normal vectors are orthogonal. That is, the planes described by $\mathbf{a}_1 \cdot \mathbf{x} = d_1$ and $\mathbf{a}_2 \cdot \mathbf{x} = d_2$ are orthogonal in the case where $\mathbf{a}_1 \cdot \mathbf{a}_2 = 0$. Using the result of part (b), show that every line in $\mathbb{R}^3$ can be written as the intersection of two orthogonal planes.

2.3 In the previous exercise, we saw that when a line ℓ is given by the intersection of two planes, $\mathbf{a}_1 \cdot \mathbf{x} = d_1$ and $\mathbf{a}_2 \cdot \mathbf{x} = d_2$, it is possible to tilt the second plane about the line ℓ so that the two planes intersect in ℓ as before but now are orthogonal. Multiplying each equation through by a constant, we can arrange for the normal vectors to the planes to be unit vectors. Thus, given any system of equations for a line in $\mathbb{R}^3$, we can convert it into an equivalent system of the form

$$
\begin{aligned}
\mathbf{u}_1 \cdot \mathbf{x} &= b_1 \\
\mathbf{u}_2 \cdot \mathbf{x} &= b_2,
\end{aligned}
\tag{2.15}
$$

where $\mathbf{u}_1$ and $\mathbf{u}_2$ are orthogonal unit vectors.

Let ℓ be the line given by

$$
\begin{aligned}
2x - 2y + z &= 6 \\
-x + 4y + z &= -3.
\end{aligned}
$$

Find an equivalent system of the form (2.15). In particular, find a pair of equations for two orthogonal planes whose intersection is ℓ.

2.4 The system of equations for a line ℓ in $\mathbb{R}^3$ can always be written in the form $A\mathbf{x} = \mathbf{b}$ where the rows of A are orthogonal unit vectors. Here is a nice feature of this form: $|A\mathbf{y} - \mathbf{b}|$ is the distance from $\mathbf{y}$ to ℓ. Explain why this is true. Hint: With $A\mathbf{x} = \mathbf{b}$ written in the form (2.15), let $\mathbf{u}_3$ be a unit vector orthogonal to $\mathbf{u}_1$ and $\bar{n}_2$. Show that $(b_1\mathbf{u}_1 + b_2\mathbf{u}_2) + t\mathbf{u}_3$ is a parametrization of ℓ. Then use Exercise 1.24 to finish the exercise.

The point of this exercise is that although there are many equivalent systems of equations for the same line ℓ, some are more useful than others. How to go about "trading in" a given system on a "better" equivalent system is a main theme of this chapter.

SECTION 3 | Row Reduction

3.1 Augmented matrices

In this section we explain an algorithm for transforming any system of linear equations $A\mathbf{x} = \mathbf{b}$ into an *equivalent system* that is *really easy to solve*.

The matrix notation is particularly useful. First we simplify things by putting the variables in $\mathbf{x}$ out of sight.

<table>
<tr><td>DEFINITION</td><td>(Augmented matrix) Given any system of m equations in n variables, $A\mathbf{x} = \mathbf{b}$, form the $m \times (n + 1)$ matrix $[A|\mathbf{b}]$ whose first n columns are the corresponding columns of A and whose $(n+1)$st column is $\mathbf{b}$. This matrix is called the augmented matrix of the system $A\mathbf{x} = \mathbf{b}$. We refer to the solution set of the system corresponding to $[A|\mathbf{b}]$ simply as the solution set of $[A|\mathbf{b}]$. We say that two augmented matrices are equivalent if they have the same solution set in this sense.</td></tr>
</table>

EXAMPLE 7   **(Augmented matrices)** Let us write down the augmented matrix descriptions of each of the equivalent systems (2.10) and (2.12), respectively,

$$\begin{bmatrix} 1 & 1 & 2 \\ 2 & 3 & 3 \end{bmatrix} \tag{3.1}$$

and

$$\begin{bmatrix} 2 & 2 & 4 \\ 0 & 1 & -1 \end{bmatrix}. \tag{3.2}$$

By the definition and the fact that (2.10) and (2.12) are equivalent systems, they are equivalent augmented matrices.

Since the jth column corresponds to the jth variable x_j, we are not losing any information in passing to an augmented matrix description of a system. You can easily recover (2.10) and (2.12) from (3.1) and (3.2). We are just using "placeholding notation" for a clean expression of the system.

Now that we have introduced augmented matrices into the game, an $m \times n$ matrix with $n > 1$ has two possible interpretations. For example, the 2×3 matrix (3.1) could represent the system of equations (2.10) if we regard it as an augmented matrix, which is what we were doing in Example 7. But (3.1) could also represent a linear transformation from $\mathbb{R}^3$ to $\mathbb{R}^2$, as in Chapter 1. *However, whether a matrix represents a system of equations or a linear transformation is usually clear from context, so usually no special notation is needed to distinguish the two cases.* But sometimes, when we want to emphasize that, say, (3.1) represents a system of equations instead of a linear transformation, we write

$$\begin{bmatrix} 1 & 1 & | & 2 \\ 2 & 3 & | & 3 \end{bmatrix}. \tag{3.3}$$

The bar before the last column emphasizes that this expression is short for

$$\begin{bmatrix} 1 & 1 \\ 2 & 3 \end{bmatrix} \begin{bmatrix} x_1 \\ x_2 \end{bmatrix} = \begin{bmatrix} 2 \\ 3 \end{bmatrix}, \tag{3.4}$$

which in turn is short for (2.10). The shortest notation of all is just (3.1) without the bar, and we will use it when the context is clear.

The following theorem tells us when two augmented matrices correspond to equivalent systems.

THEOREM 1   **(Augmented matrices and equivalence)** *Two augmented matrices $[A|\mathbf{b}]$ and $[\tilde{A}|\tilde{\mathbf{b}}]$ are equivalent whenever each row of $[\tilde{A}|\tilde{\mathbf{b}}]$ is a linear combination of the rows of $[A|\mathbf{b}]$ and vice versa.*

Notice that for *any* row of $[\tilde{A}|\tilde{\mathbf{b}}]$ to be a linear combination of the rows of $[A|\mathbf{b}]$, both $[\tilde{A}|\tilde{\mathbf{b}}]$ and $[A|\mathbf{b}]$ must have the same number of columns or, in other words, refer to the same number of variables. However, the conditions of the theorem can easily hold when the two matrices have a different number of rows.

PROOF: We first express the statement $A\mathbf{x} = \mathbf{b}$ in terms of the rows of $[A|\mathbf{b}]$: Given $\mathbf{x}$ in $\mathbb{R}^n$, define the vector $\hat{\mathbf{x}}$ in $\mathbb{R}^{n+1}$ by

$$\hat{\mathbf{x}} = \begin{bmatrix} x_1 \\ x_2 \\ \vdots \\ x_n \\ -1 \end{bmatrix} \quad \text{where} \quad \mathbf{x} = \begin{bmatrix} x_1 \\ x_2 \\ \vdots \\ x_n \end{bmatrix}.$$

Let $\mathbf{a}_i$ denote the ith row of A, and let $\mathbf{r}_i$ denote the ith row of $[A|\mathbf{b}]$. From the definition of $\hat{\mathbf{x}}$, we see that

$$\mathbf{a}_i \cdot \mathbf{x} = b_i \iff \mathbf{r}_i \cdot \hat{\mathbf{x}} = 0.$$

Then from **V.I.F.2**,

$$A\mathbf{x} = \mathbf{b} \iff \mathbf{r}_i \cdot \hat{\mathbf{x}} = 0 \qquad \text{for each } i. \tag{3.5}$$

Now, suppose that each row $\tilde{\mathbf{r}}_i$ of $[\tilde{A}|\tilde{\mathbf{b}}]$ is a linear combination

$$\tilde{\mathbf{r}}_i = c_1 \mathbf{r}_1 + c_2 \mathbf{r}_2 + \cdots + c_m \mathbf{r}_m$$

of the rows of $[A|\mathbf{b}]$. Then if $A\mathbf{x} = \mathbf{b}$, by (3.5)

$$\begin{aligned} \tilde{\mathbf{r}}_i \cdot \hat{\mathbf{x}} &= (c_1 \mathbf{r}_1 + c_2 \mathbf{r}_2 + \cdots + c_m \mathbf{r}_m) \cdot \hat{\mathbf{x}} \\ &= c_1 \mathbf{r}_1 \cdot \hat{\mathbf{x}} + c_2 \mathbf{r}_2 \cdot \hat{\mathbf{x}} + \cdots + c_m \mathbf{r}_m \cdot \hat{\mathbf{x}} \\ &= c_1 0 + c_2 0 + \cdots + c_m 0 = 0, \end{aligned}$$

so that $\hat{\mathbf{x}}$ is orthogonal to this row and hence every row of $[\tilde{A}|\tilde{\mathbf{b}}]$.

Again by (3.5), applied this time with $\tilde{A}$ and $\tilde{\mathbf{b}}$, we conclude that $\tilde{A}\mathbf{x} = \tilde{\mathbf{b}}$. In other words, whenever each row of $[\tilde{A}|\tilde{\mathbf{b}}]$ is a linear combination of the rows of $[A|\mathbf{b}]$, every solution of $A\mathbf{x} = \mathbf{b}$ is a solution of $\tilde{A}\mathbf{x} = \tilde{\mathbf{b}}$.

By the same argument, if each row of $[A|\mathbf{b}]$ is a linear combination of the rows of $[\tilde{A}|\tilde{\mathbf{b}}]$, then each solution of $\tilde{A}\mathbf{x} = \tilde{\mathbf{b}}$ is also a solution of $A\mathbf{x} = \mathbf{b}$, and hence the systems are equivalent. ∎

This theorem tells us what we can do to the augmented matrix of a system without changing its solution set, leading to the following list.

DEFINITION

(Row operations) We use the term *row operation* to refer to any one of the following three operations on the rows of a matrix:

1. Exchanging the places of two rows

2. Multiplying one row through by a nonzero constant

3. Adding a multiple of one row to another

THEOREM 2   (**Row operations and equivalence**) *Let* [$A|\mathbf{b}$] *be the augmented matrix of some system, and let* [$\tilde{A}|\tilde{\mathbf{b}}$] *be the augmented matrix obtained from it by any row operation. Then* [$A|\mathbf{b}$] *and* [$\tilde{A}|\tilde{\mathbf{b}}$] *are equivalent.*

PROOF: It is quite clear that for a row operation of type (1), [$A|\mathbf{b}$] and [$\tilde{A}|\tilde{\mathbf{b}}$] actually have the *same* set of rows and so represent the same system of equations.

Consider a row operation of type (3): If we produce [$\tilde{A}|\tilde{\mathbf{b}}$] by adding a multiple of the ith row of [$A|\mathbf{b}$] to the jth row, $j \neq i$, then only the jth row has changed, and it is clearly a linear combination of the rows of [$A|\mathbf{b}$]. Therefore, each row of [$\tilde{A}|\tilde{\mathbf{b}}$] is a linear combination of the rows of [$A|\mathbf{b}$].

Next, we can recover the jth row of [$A|\mathbf{b}$] by subtracting the same multiple of the ith row of [$\tilde{A}|\tilde{\mathbf{b}}$], which equals the ith row of [$A|\mathbf{b}$], from the jth row of [$\tilde{A}|\tilde{\mathbf{b}}$]. We are subtracting off exactly what we added on, so we return to the original jth row. Hence, each row of [$A|\mathbf{b}$] is a linear combination of rows of [$\tilde{A}|\tilde{\mathbf{b}}$], and [$A|\mathbf{b}$] and [$\tilde{A}|\tilde{\mathbf{b}}$] are equivalent by Theorem 1.

Operation (2) is very much the same and is left as an exercise. ∎

EXAMPLE 8   (**Row operations in action**) Consider the augmented matrix

$$[A|\mathbf{b}] = \begin{bmatrix} 3 & 2 & | & 1 \\ 1 & 2 & | & 3 \end{bmatrix}. \tag{3.6}$$

By swapping the two rows, we obtain $\begin{bmatrix} 1 & 2 & | & 3 \\ 3 & 2 & | & 1 \end{bmatrix}$. Next, multiplying the first row through by -3 and adding it to the second row, we obtain

$$\begin{bmatrix} 1 & 2 & | & 3 \\ 0 & -4 & | & -8 \end{bmatrix}. \tag{3.7}$$

Since we used only operations from the list in Theorem 2, we know that at each step along the way, we had an augmented matrix equivalent to the original one. In particular, (3.7) is equivalent to (3.6).

This is good since the augmented matrix (3.7) represents a really easy system, namely,

$$\begin{array}{rcl} x_1 + 2x_2 & = & 3 \\ -4x_2 & = & -8. \end{array} \tag{3.8}$$

It is really easy since the second equation tells us directly that $x_2 = 2$, and using this we can solve the first equation to find $x_1 = -1$. Therefore,

$$\begin{bmatrix} -1 \\ 2 \end{bmatrix}$$

is the unique solution to the system represented by (3.6).

3.2 Row-reduced form

When does an augmented matrix [$A|\mathbf{b}$] represent a system that is really easy to solve? The answer, as we shall explain, is "*when* [$A|\mathbf{b}$] *is in row-reduced form.*"

Matrices that are in row-reduced form have a special pattern of zero and nonzero entries. Before we give a definition, here is the picture of a row-reduced matrix, which always has a sort of "staircase pattern" to it, such as

$$\begin{bmatrix} 0 & \bullet & * & * & * & * & * \\ 0 & 0 & \bullet & * & * & * & * \\ 0 & 0 & 0 & 0 & \bullet & * & * \\ 0 & 0 & 0 & 0 & 0 & \bullet & * \\ 0 & 0 & 0 & 0 & 0 & 0 & 0 \end{bmatrix}. \tag{3.9}$$

The bullets mark entries that are *definitely not zero*, and the asterisks mark entries that might or might not be zero. As we shall see, if $[A|\mathbf{b}]$ has this form, it is really easy to find the solution set for $A\mathbf{x} = \mathbf{b}$.

First, to capture this picture in formulas, we introduce some notation. If B is an $m \times n$ matrix whose ith row is not all zeros, let $j(i)$ be the number of the column that contains the first nonzero entry in the ith row of B. That is,

$$j(i) = \text{column number of the first nonzero entry in row } i. \tag{3.10}$$

If you prefer formulas to words, $j(i) = \min\{k = 1, 2, \ldots, n \mid B_{i,k} \neq 0\}$. If the ith row is all zeros, $j(i)$ is undefined.

- *In the matrix pictured in (3.9), $j(1) = 2$, since the bullet in the first row is in column 2. Likewise, $j(2) = 3$, $j(3) = 5$, $j(4) = 6$, and $j(7)$ is not defined.*

The "staircase pattern" in the picture is an important feature: The lower the row, the farther to the right the first nonzero entry is, which brings us to the definition.

DEFINITION

> **(Row-reduced form, pivots, and pivotal columns)** An $m \times p$ matrix B is in *row-reduced form* if and only if the first nonzero entry in each row comes strictly to the left of the first nonzero entry in the row below it. That is, if there are r rows that are not all zeros,
>
> $$j(1) < j(2) < \cdots < j(r). \tag{3.11}$$
>
> The r entries $B_{i,j(i)}$ are called the *pivots*. A column is called a *pivotal column* if and only if it contains a pivot so that columns $j(1), j(2), \ldots, j(r)$ are the pivotal columns.

- *Notice that there is exactly one pivot for each row that is not all zeros. In particular, the number of pivots cannot exceed the number of rows.*

EXAMPLE 9   **(Recognizing row-reduced form)** Consider the following matrices:

$$A = \begin{bmatrix} 1 & 6 & 5 & 1 \\ 2 & 1 & 0 & 3 \\ 0 & 0 & 1 & 0 \\ 0 & 0 & 0 & 1 \end{bmatrix} \quad B = \begin{bmatrix} 1 & 3 & 1 & 2 \\ 0 & 1 & 4 & 1 \\ 0 & 0 & 1 & 1 \\ 0 & 0 & 0 & 0 \end{bmatrix} \quad C = \begin{bmatrix} 1 & 0 & 2 & 0 \\ 0 & 1 & 1 & 1 \\ 0 & 0 & 0 & 1 \end{bmatrix}.$$

For matrix A, $j(1) = 1$, $j(2) = 1$, $j(3) = 3$, and $j(4) = 4$. Since it is not the case that $j(2) > j(1)$, A is not in row-reduced form.

For matrix B, $j(1) = 1$, $j(2) = 2$, $j(3) = 3$, and $j(4)$ is undefined since this row is all zeros. Since $j(1) < j(2) < j(3)$, B is in row-reduced form.

For matrix C, $j(1) = 1$, $j(2) = 2$, and $j(3) = 4$. Since $j(1) < j(2) < j(3)$, C is in row-reduced form.

From these examples, you can see that the decision about whether an augmented matrix is row-reduced or not is straightforward.

3.3 Back substitution

Suppose that $[A|\mathbf{b}]$ is an $m \times (n+1)$ matrix in row-reduced form and that there are r pivots. As long as $j(r) < n + 1$, the rth row of $[A|\mathbf{b}]$ represents the equation

$$0x_1 + \cdots + 0x_{j(r)-1} + A_{r,j(r)}x_{j(r)} + \cdots + A_{r,n}x_n = b_r.$$

By the definition of a pivot, $A_{r,j(r)} \neq 0$, so we can solve this equation for $x_{j(r)}$:

$$x_{j(r)} = \frac{1}{A_{r,j(r)}} \left(b_r - A_{r,j(r)+1}x_{j(r)+1} - \cdots - A_{r,n}x_n \right). \tag{3.12}$$

On the other hand, if $j(r) = n + 1$, the rth row of $[A|\mathbf{b}]$ represents the equation

$$0x_1 + \cdots + 0x_n = b_r,$$

with $b_r \neq 0$, and this equation has no solution. *Hence, if there is a pivot in the final column, the solution set is empty.*

Now, move on up to the next row above and solve for $x_{j(r-1)}$, the pivotal variable for that row. Do one more thing: Use (3.12) to substitute $x_{j(r)}$ out of the solution.

Keep doing this, working your way to the top of the matrix. At the end, you have expressed each of the pivotal variables (variables corresponding to pivotal columns) as a linear combination of the nonpivotal variables, which we also call *free variables.*

There are $m - r$ nonpivotal variables, and these leftover, or free, variables become the parameters in our parameterization of the solution set.

EXAMPLE 10   **(Back substitution)** Let

$$[A|\mathbf{b}] = \begin{bmatrix} 1 & 3 & 1 & 2 \\ 0 & 1 & 4 & 1 \\ 0 & 0 & 1 & 1 \\ 0 & 0 & 0 & 0 \end{bmatrix}.$$

To solve the corresponding system by back substitution, we note that there is no pivot in the final column, so there will be a solution. Clearly, $r = 3$ and $j(3) = 3$. The third row stands for the equation

$$x_3 = 1,$$

which basically solves itself.

Next, the second row corresponds to $x_2 + 4x_3 = 1$. Solving for $x_2 = x_{j(2)}$, we get $x_2 = 1 - 4x_3$. Substituting $x_3 = 1$,

$$x_2 = -3.$$

Finally, the first row corresponds to $x_1 + 3x_2 + x_3 = 2$. Solving for $x_1 = x_{j(1)}$, we get $x_1 = 2 - 3x_2 - x_3$. Substituting $x_3 = 1$ and $x_2 = -3$, the equation in the first row becomes

$$x_1 = 10.$$

There are no leftover variables to become parameters, and the unique solution is

$$\begin{bmatrix} x_1 \\ x_2 \\ x_3 \end{bmatrix} = \begin{bmatrix} 10 \\ -3 \\ 1 \end{bmatrix}.$$

Let us go on to another example in which not every variable is pivotal. This time, let

$$[A|\mathbf{b}] = \begin{bmatrix} 1 & 1 & 2 & 0 \\ 0 & 0 & 2 & 4 \\ 0 & 0 & 0 & 0 \end{bmatrix}.$$

Here, $r = 2, j(1) = 1$, and $j(2) = 3$, so x_2 is nonpivotal. The second row corresponds to $2x_3 = 4$, yielding $x_3 = 2$. The first row corresponds to $x_1 + x_2 + 2x_3 = 0$, yielding $x_1 = -x_2 - 2x_3$. Substituting x_3 with 2 from the previous equation yields $x_1 = -x_2 - 4$.

To parameterize the solution set, substitute

$$x_1 = -x_2 - 4 \qquad \text{and} \qquad x_3 = 2$$

into

$$\mathbf{x} = \begin{bmatrix} x_1 \\ x_2 \\ x_3 \end{bmatrix},$$

getting

$$\begin{bmatrix} -x_2 - 4 \\ x_2 \\ 2 \end{bmatrix} = \begin{bmatrix} -4 \\ 0 \\ 2 \end{bmatrix} + x_2 \begin{bmatrix} -1 \\ 1 \\ 0 \end{bmatrix}.$$

This is our parameterization of the solution set, which is a line in $\mathbb{R}^3$ with x_2 as the parameter.

THEOREM 3

(**Back substitution yields one-to-one parameterizations**) *Let* $[A|\mathbf{b}]$ *be an* $m \times (n+1)$ *augmented matrix, and suppose that* $[A|\mathbf{b}]$ *is in row-reduced form. If there is a pivot in the final column of* $[A|\mathbf{b}]$, *then* $A\mathbf{x} = \mathbf{b}$ *has no solution. Otherwise, the parameterization of the solution set provided by back substitution is one-to-one.*

PROOF: If there is a pivot in the final column, the last nonzero row corresponds to the impossible equation $0x_1 + \cdots + 0x_n = b \neq 0$, and there is no solution.

Otherwise, each nonzero row corresponds to an equation of the form (3.12),

$$ x_{j(i)} = \frac{1}{A_{i,j(i)}}(b_r - A_{i,j(i)+1}x_{i+1} - \cdots - A_{i,n}x_n) , $$

which specifies the ith pivotal variable as a linear combination of b_i and the variables to the right of $x_{j(i)}$, if any.

In the last of these equations, (3.12), all the remaining variables (if any) on the right are free variables since $x_{j(r)}$ is the last pivotal variable. Back substitution eliminates all the pivotal variables from the right-hand side of each of the other equations so that

- *After back substitution, the equations express the pivotal variables as linear combinations of the entries of* **b** *and the free variables, if any, as in Example 10.*

To achieve parameterization, promote the free variables to parameters. Why is this parameterization one-to-one? If **x** is any solution of $A\mathbf{x} = \mathbf{b}$, the free variable entries of **x** are the parameter values, so you can read off the parameter values looking at **x**. Also, the final system of equations specifies the values of the pivotal entries of any solution, including **x**, in terms of its nonpivotal entries—which are the parameters. Thus, there is a one-to-one correspondence between solutions and between sets of values for the parameters. ∎

3.4 The row-reduction algorithm

We now show how to start from *any* $m \times (n+1)$ augmented matrix and then, using fewer than $m(m+1)/2$ of the row operations described in Theorem 2, how to transform it into an equivalent augmented matrix that is in row-reduced form and therefore easy to deal with by back substitution.

The proof is given in careful detail so that we can count the number of steps the algorithm might use.*

THEOREM 4

(**Row reduction**) *Let* $[A|\mathbf{b}]$ *be an* $m \times (n+1)$ *augmented matrix. Then there is an equivalent augmented matrix* $[\tilde{A}|\tilde{\mathbf{b}}]$ *in row-reduced form that can be obtained from* $[A|\mathbf{b}]$ *through a sequence of at most* $m(m+1)/2$ *elementary row operations of the type listed in Theorem 2.*

PROOF: We suppose the matrix is not all zeros, or there is nothing to do. Go to the first column from the left that has a nonzero entry in it. Call this the *first pivotal column*. If the entry in this

---

*This is a crucial issue in the analysis of algorithms. If you are going to write a program implementing an algorithm, you certainly want to be sure that you do not write an infinite loop. A proof that the procedure always terminates in a finite number of steps provides that assurance. But if you have a formula for the number of steps it can take the algorithm to terminate in the worst case, then you do more; you can say something useful about the run time of the program. So the *proof* is an example of a type of *calculation* that is very important in applications.

column in the first row is zero, swap two rows to bring a nonzero entry in this column into the first row. *The swap is one row operation.* Now subtract an appropriate multiple of the first row from all the rows beneath it so that you "clean out" the rest of the first pivotal column; that is, make every entry below the first row in the first pivotal column equal to zero. *There are at most $m - 1$ rows to deal with, so those actions are at most $m - 1$ more row operations at this stage.* The total number of row operations used so far is at most m. At this stage, the matrix looks something like

$$\begin{bmatrix} 0 & \bullet & * & * & * & * \\ 0 & 0 & * & * & * & * \\ 0 & 0 & * & * & * & * \\ 0 & 0 & * & * & * & * \\ 0 & 0 & * & * & * & * \end{bmatrix}.$$

In this case, the second column is the first pivotal column. The bullet represents a nonzero entry, and the asterisks can be either zero or not.

If the matrix is all zeros below the first row, you are done. Otherwise, go to the first column that contains a nonzero entry *below the first row.* Call this column the *second pivotal column.* Observe that the second pivotal column is necessarily to the right of the first. If the entry of this column in the second row is zero, swap two rows to bring a nonzero entry in this column into the second row. *That is one more row operation.* Now subtract an appropriate multiple of the second row from all the rows beneath (but not above) it so that you clean out the rest of the second pivotal column as before. *There are $m - 2$ rows to deal with, at most $m - 2$ more row operations at this stage.* Altogether we used at most $m - 1$ row operations dealing with the second pivotal column, and the running total of row operations used so far is

$$m + (m - 1).$$

At this stage the matrix looks something like

$$\begin{bmatrix} 0 & \bullet & * & * & * & * \\ 0 & 0 & \bullet & * & * & * \\ 0 & 0 & 0 & * & * & * \\ 0 & 0 & 0 & * & * & * \\ 0 & 0 & 0 & * & * & * \end{bmatrix} \quad \text{or perhaps} \quad \begin{bmatrix} 0 & \bullet & * & * & * & * \\ 0 & 0 & 0 & \bullet & * & * \\ 0 & 0 & 0 & 0 & * & * \\ 0 & 0 & 0 & 0 & * & * \\ 0 & 0 & 0 & 0 & * & * \end{bmatrix},$$

among other possibilities, depending on how far to the right we have to go to find a nonzero entry below the first row.

If the matrix is all zeros below the second row, you are done. Otherwise, go to the first column that contains a nonzero entry below the second row. Call this the third pivotal column, and observe that it is necessarily to the right of the second pivotal column. If the entry of this column in the third row is zero, swap two rows to bring a nonzero entry in this column into the third row. *That is one more row operation.* Now subtract an appropriate multiple of the third row from all the rows beneath (but not above) it so that you clean out the rest of the third pivotal column as before. *There are at most $m - 3$ rows to deal with, at most $m - 2$ more row operations at this stage.* Altogether we used at most $m - 2$ row operations dealing with the third pivotal column, and the running total of row operations used so far is

$$m + (m - 1) + (m - 2).$$

Now repeat this procedure until you get to the bottom of the matrix or run out of nonzero rows. This procedure takes at most $m - 1$ stages, so the total row operation count is

$$m + (m - 1) + (m - 2) + \cdots + 2 = \frac{m(m + 1)}{2} - 1.$$

After you are done, you have produced a matrix that is in row-reduced form. And since you only used elementary row operations to effect the transformation, the resulting augmented matrix is equivalent to the one from which we started. ∎

The procedure described in the proof of Theorem 4 is often called *row reduction* or *the row-reduction algorithm*. Row reduction is an explicit procedure—an algorithm—for finding the solution set of any linear system of equations:

1. Write down the corresponding augmented matrix.

2. Apply the procedure described in the proof of the row-reduction theorem to reduce the augmented matrix to row-reduced form.

3. Convert back to a system, and solve it by back substitution.

EXAMPLE 11   **(Using the row-reduction algorithm)** Let us put everything together and find the solution set to the system

$$x_2 - x_3 + 2x_4 = 1$$
$$x_1 + 2x_2 + x_3 - x_4 = 3 \tag{3.13}$$
$$x_1 - x_2 + x_4 = -2.$$

The corresponding augmented matrix is

$$\left[\begin{array}{cccc|c} 0 & 1 & -1 & 2 & 1 \\ 1 & 2 & 1 & -1 & 3 \\ 1 & -1 & 0 & 1 & -2 \end{array} \right].$$

The first column contains nonzero entries, but not in the first row. So we swap rows one and two:

$$\left[\begin{array}{cccc|c} 1 & 2 & 1 & -1 & 3 \\ 0 & 1 & -1 & 2 & 1 \\ 1 & -1 & 0 & 1 & -2 \end{array} \right].$$

(We could just as well have swapped rows one and three, since that, too, would have brought a nonzero entry into the first row.) Now we subtract the first row from the third (or add -1 times the first row to the third, if you like):

$$\left[\begin{array}{cccc|c} 1 & 2 & 1 & -1 & 3 \\ 0 & 1 & -1 & 2 & 1 \\ 0 & -3 & -1 & 2 & -5 \end{array} \right].$$

We have just dealt with the first pivotal column; now for the second. The second column has nonzero entries below the first row, so it is the second pivotal column. It has a nonzero entry in the second row, so no swapping of rows is required this time. We add 3 times the second row to the third row:

$$\left[\begin{array}{cccc|c} 1 & 2 & 1 & -1 & 3 \\ 0 & 1 & -1 & 2 & 1 \\ 0 & 0 & -4 & 8 & -2 \end{array} \right].$$

This augmented matrix represents the system

$$x_1 + 2x_2 + x_3 - x_4 = 3$$
$$x_2 - x_3 + 2x_4 = 1 \tag{3.14}$$
$$2x_3 - 4x_4 = 1,$$

which we solve by back substitution, working our way up from the bottom.

The bottom equation contains the nonpivotal variable x_4, which we rename as the parameter t—that is, $x_4 = t$. We solve for x_3:

$$x_3 = (1 + 4t)/2. \tag{3.15}$$

We substitute $x_4 = t$ and (3.15) into the second equation and solve for x_2:

$$x_2 = 3/2. \tag{3.16}$$

We substitute $x_4 = t$, (3.15), and (3.16) into the first equation and solve for x_1:

$$x_1 = -t - 1/2. \tag{3.17}$$

Simplifying,

$$\begin{bmatrix} x_1 \\ x_2 \\ x_3 \\ x_4 \end{bmatrix} = \begin{bmatrix} -t - 1/2 \\ 3/2 \\ 2t + 1/2 \\ t \end{bmatrix} = \frac{1}{2}\begin{bmatrix} -1 \\ 3 \\ 1 \\ 0 \end{bmatrix} + t\begin{bmatrix} -1 \\ 0 \\ 2 \\ 1 \end{bmatrix}.$$

3.5 The kernel of a matrix

Let A be an $m \times n$ matrix. The equation $A\mathbf{x} = 0$ is always solvable since $\mathbf{x} = 0$ is a solution. The solution set of $A\mathbf{x} = 0$ has an important bearing on the number of solutions of $A\mathbf{x} = \mathbf{b}$ in general and in particular on whether there is a unique solution.

Suppose that the only solution of $A\mathbf{x} = 0$ is $\mathbf{x} = 0$. Then $A\mathbf{x} = \mathbf{b}$ cannot have more than one solution for any $\mathbf{b}$. Indeed, suppose that $A\mathbf{x}_1 = \mathbf{b}$ and $A\mathbf{x}_2 = \mathbf{b}$. Then

$$A(\mathbf{x}_1 - \mathbf{x}_2) = A\mathbf{x}_1 - A\mathbf{x}_2 = \mathbf{b} - \mathbf{b} = 0. \tag{3.18}$$

Since the only solution of $A\mathbf{x} = 0$ is $\mathbf{x} = 0$, by (3.18) $\mathbf{x}_1 - \mathbf{x}_2 = 0$. In other words, $\mathbf{x}_1 = \mathbf{x}_2$, and the two solutions are actually the same. Hence, *any* two solutions are the same, and there is just one solution.

On the other hand, suppose that $\mathbf{w}$ is a nonzero solution to $A\mathbf{x} = 0$. That is, $\mathbf{w} \neq 0$, and $A\mathbf{w} = 0$. Let $\mathbf{b}$ be any vector so that $A\mathbf{x} = \mathbf{b}$ has a solution $\mathbf{x}_0$. Then

$$A(\mathbf{x}_0 + \mathbf{w}) = A\mathbf{x}_0 + A\mathbf{w} = \mathbf{b} + 0 = \mathbf{b}.$$

Hence, both $\mathbf{x}_0$ and $\mathbf{x}_0 + \mathbf{w}$ solve $A\mathbf{x} = \mathbf{b}$, and since $\mathbf{w} \neq 0$, they are two different solutions. In short, when $A\mathbf{x} = 0$ has nonzero solutions, $A\mathbf{x} = \mathbf{b}$ cannot have a unique solution for any $\mathbf{b}$.

For the reasons just explained, the solution set of $A\mathbf{x} = 0$ deserves a name.

DEFINITION | **(Kernel)** Let A be an $m \times n$ matrix. The *kernel* of A is the solution set of the equation $A\mathbf{x} = 0$. It is denoted by Ker(A).

Since the kernel of a matrix is the solution set of a system of equations, namely, $A\mathbf{x} = 0$, we can specify which vectors are in it *exactly as we would for any other solution set*: by giving a one-to-one parameterization.

EXAMPLE 12   **(Parameterizing a kernel)** Let

$$A = \begin{bmatrix} 0 & 1 & -1 & 2 \\ 1 & 2 & 1 & -1 \\ 1 & -1 & 0 & 1 \end{bmatrix},$$

the matrix we would have if we wrote the system (3.13) in the form $A\mathbf{x} = \mathbf{b}$. As in Example 11, using the same sequence of row operations, $[A|0]$ row reduces to

$$\begin{bmatrix} 1 & 2 & 1 & -1 & | & 0 \\ 0 & 1 & -1 & 2 & | & 0 \\ 0 & 0 & -4 & 8 & | & 0 \end{bmatrix}.$$

The variable x_4 is nonpivotal. Turning it into a parameter through $x_4 = t$, we find by back substitution that the solution set of $A\mathbf{x} = 0$, or in other words Ker(A), is the line given by

$$t\begin{bmatrix} -1 \\ 0 \\ 2 \\ 1 \end{bmatrix}.$$

When 0 is the only vector belonging to Ker(A), we write Ker(A) = 0.* Since we can parameterize the solution set of $A\mathbf{x} = 0$ using the row-reduction algorithm, we can parameterize Ker(A) and in particular can determine whether or not Ker(A) = 0.

But parameterizing Ker(A) does much more than that: It *almost* gives a parameterization of the solution set of $A\mathbf{x} = \mathbf{b}$ for any $\mathbf{b}$. Here is what we mean: Of course, if $A\mathbf{x} = \mathbf{b}$ has no solutions, there is nothing to parameterize. So suppose that $A\mathbf{x} = \mathbf{b}$ does have at least one solution $\mathbf{x}_0$. Suppose that $\mathbf{x}_1$ is also a solution. Then

$$A(\mathbf{x}_1 - \mathbf{x}_0) = A\mathbf{x}_1 - A\mathbf{x}_0 = \mathbf{b} - \mathbf{b} = 0.$$

This means that $\mathbf{w} = \mathbf{x}_1 - \mathbf{x}_0$ belongs to the kernel of Ker(A), so $\mathbf{x}_1 = \mathbf{x}_0 + \mathbf{w}$, where $\mathbf{w}$ belongs to Ker(A).

Conversely, suppose that $\mathbf{w}$ is any vector that belongs to Ker(A), and define $\mathbf{x}_1 = \mathbf{x}_0 + \mathbf{w}$. Then

$$A(\mathbf{x}_0 + \mathbf{w}) = A\mathbf{x}_0 + A\mathbf{w} = \mathbf{b} + 0 = \mathbf{b}.$$

Thus, $\mathbf{x}_0 + \mathbf{w}$ solves $A\mathbf{x} = \mathbf{b}$. We conclude that the solution set of $A\mathbf{x} = \mathbf{b}$ is precisely the set of vectors of the form $\mathbf{x}_0 + \mathbf{w}$ where $\mathbf{x}_0$ is some particular solution of $A\mathbf{x} = \mathbf{b}$ and $\mathbf{w}$ belongs to Ker(A). Therefore, if

$$t_1\mathbf{v}_1 + t_2\mathbf{v}_2 + \cdots + t_k\mathbf{v}_k$$

is a one-to-one parameterization of Ker(A),

$$\mathbf{x}_0 + t_1\mathbf{v}_1 + t_2\mathbf{v}_2 + \cdots + t_k\mathbf{v}_k$$

is a one-to-one parameterization of the solution set of $A\mathbf{x} = \mathbf{b}$.

We summarize.

THEOREM 5 **(Uniqueness of solutions and Ker(A))** *Let A be an $m \times n$ matrix and let $\mathbf{b}$ be a vector in $\mathbb{R}^m$. Suppose that $A\mathbf{x} = \mathbf{b}$ is solvable. Then $A\mathbf{x} = \mathbf{b}$ has a unique solution if and only if Ker(A) = 0. Moreover, if Ker(A) $\neq 0$ and $\mathbf{x}_0$ is any particular solution of $A\mathbf{x} = \mathbf{b}$, then*

$$t_1\mathbf{v}_1 + t_2\mathbf{v}_2 + \cdots + t_k\mathbf{v}_k$$

is a one-to-one parameterization of Ker(A) if and only if

$$\mathbf{x}_0 + t_1\mathbf{v}_1 + t_2\mathbf{v}_2 + \cdots + t_k\mathbf{v}_k$$

is a one-to-one parameterization of the solution set of $A\mathbf{x} = \mathbf{b}$.

EXAMPLE 13 **(From Ker(A) to the solution set of Ax = b)** Let

$$A = \begin{bmatrix} 0 & 1 & -1 & 2 \\ 1 & 2 & 1 & -1 \\ 1 & -1 & 0 & 1 \end{bmatrix},$$

as in Example 12, and let

$$\mathbf{b} = \begin{bmatrix} 1 \\ 3 \\ -2 \end{bmatrix}.$$

Then $A\mathbf{x} = \mathbf{b}$ is the matrix form of (3.13), the system we solved in Example 11.

---

*Notice that Ker(A) is by definition a set of vectors, so "Ker(A) = 0" is a new use of the symbol 0, which has already been used for numbers, vectors, and matrices. This sounds risky, but the context will make things clear.

Suppose we know that

$$\mathbf{x}_0 = \frac{1}{2}\begin{bmatrix} -1 \\ 3 \\ 1 \\ 0 \end{bmatrix}$$

is one *particular* solution of $A\mathbf{x} = \mathbf{b}$, which is, in fact, the case. What then is the *general* solution? We found $\text{Ker}(A)$ in Example 12, so by Theorem 5, the general solution is

$$\frac{1}{2}\begin{bmatrix} -1 \\ 3 \\ 1 \\ 0 \end{bmatrix} + t\begin{bmatrix} -1 \\ 0 \\ 2 \\ 1 \end{bmatrix},$$

as we found in Example 11. As you can see, there is not much going on here that is really new. Still, the relation between the solution set of $A\mathbf{x} = \mathbf{b}$ and $A\mathbf{x} = 0$ is worth bearing in mind.

Now, given an $m \times n$ matrix A, how do you tell if $\text{Ker}(A) = 0$ or not? You row reduce, of course! More explicitly, you row reduce and count the number of pivots. If any columns of A are nonpivotal, the corresponding variables are "free," and become parameters in our method for parameterizing the solution set of $A\mathbf{x} = 0$. The kernel of A is the zero vector and nothing else, exactly when there are no free parameters, which is the case exactly when every column is pivotal:

- Let A be any $m \times n$ matrix. Then $\text{Ker}(A) = 0$ *if and only if when A is row reduced, there is a pivot in every column so that there are no free variables.*

We are now in a position to tie up a loose end left over from the last chapter. There, in our explanation of why only square matrices can possibly have inverses, we "borrowed" the fact that whenever A has more columns than rows, $A\mathbf{x} = 0$ has nonzero solutions.

To see why, recall the "bullet point" made before Example 9 that the number of pivots does not exceed the number of rows. Hence, if there are more columns than rows, there cannot be a pivot for every column.

Exercises

3.1 Find the intersection of the lines given by $x_1 + 2x_2 = 5$ and $2x_1 + x_2 = 5$.

3.2 Find the intersection of the lines given by $x_1 + 2x_2 = 5$ and $2x_1 + 2x_2 = 5$.

3.3 Consider the system of equations

$$x + 2y + z = b$$
$$2x + y + 2z = 2$$
$$3x + 3y + az = 3.$$

(a) For which values of a and b, if any, does this system have a unique solution? Give the solution for any such values of a and b.

(b) For which values of a and b, if any, does this system have no solution?

(c) For which values of a and b, if any, does this system have infinitely many solutions?

3.4 Consider the system of equations

$$x + z = 1$$
$$2x + ay + z = 1$$
$$2x - y = b.$$

(a) For which values of a and b, if any, does this system have a unique solution? Give the solution for any such values of a and b.

(b) For which values of a and b, if any, does this system have no solution?

(c) For which values of a and b, if any, does this system have infinitely many solutions?

3.5 Consider the system of equations

$$x - 2y + az = 2$$
$$x + y + z = 0$$
$$3y + z = 2.$$

(a) For which values of a, if any, does this system have a unique solution? Give the solution for any such values of a and b.

(b) For which values of a, if any, does this system have no solution?

(c) For which values of a, if any, does this system have infinitely many solutions?

3.6 Consider the system of equations

$$2x + ay - z = 1$$
$$x + by + -\frac{1}{2}z = 2$$
$$x + 2y + 3z = 0.$$

(a) For which values of a and b, if any, does this system have a unique solution? Give the solution for any such values of a and b.

(b) For which values of a and b, if any, does this system have no solution?

(c) For which values of a and b, if any, does this system have infinitely many solutions?

3.7 Consider the system of equations

$$x - 2y + z = 3$$
$$2x + y + 3z = a$$
$$3x + ay - z = 5.$$

(a) For which values of a, if any, does this system have a unique solution? Give the solution for any such values of a and b.

(b) For which values of a, if any, does this system have no solution?

(c) For which values of a, if any, does this system have infinitely many solutions?

3.8 Consider the following systems:

(a)
$$2x - 3y - w = 2$$
$$y + z + w = 2$$
$$-x + 2y + z + w = 1$$
$$3x - 2y + w = 0.$$

(b)
$$x + 2y - z = 2$$
$$2x - y + 2z = 1$$
$$x + 2y = 0.$$

Which of these systems has a unique solution—neither, both, just (a), or just (b)? In each case where there is a unique solution, find it.

3.9 Analyze the following systems by reducing them. Determine whether there is no solution, one solution, or are infinitely many solutions. If there is a unique solution, say what it is.

(a)
$$2x - y = 0$$
$$-x + 2y - z = 1$$
$$x - y + 3z - w = 0$$
$$y + 2w = 0.$$

(b)
$$2x + 2y + z = 0$$
$$x + z = 1$$
$$x + 2y = 5.$$

3.10 Let $A = \begin{bmatrix} 1 & 1 & 1 \\ 1 & 1 & 1 \\ 1 & 1 & 1 \end{bmatrix}$.

(a) Give a one-to-one parameterization of Ker(A).

(b) Let $\mathbf{b} = A\mathbf{e}_1$. Find a one-to-one parameterization of the solution set of $A\mathbf{x} = \mathbf{b}$.

3.11 Let $A = \begin{bmatrix} 1 & 2 & 3 \\ 4 & 4 & 4 \\ 3 & 2 & 1 \end{bmatrix}$.

(a) Give a one-to-one parameterization of Ker(A).

(b) Let $\mathbf{b} = A\mathbf{e}_3$. Find a one-to-one parameterization of the solution set of $A\mathbf{x} = \mathbf{b}$.

3.12 Let A and B be any two matrices such that the product AB is defined. Show that*

$$\text{Ker}(B) \subset \text{Ker}(AB).$$

3.13 Let A and B be any two matrices such that the product AB is defined. Suppose that Ker$(A) = 0$. Show that in this case

$$\text{Ker}(B) = \text{Ker}(AB).$$

3.14 There is exactly one 3×3 matrix A with Ker$(A) = \mathbb{R}^3$. What is A?

3.15 Let A and B be two matrices with the same number of columns.

(a) Show that if A and B have the same kernel, then A and B have the same sets of pivotal columns. (By "the same pivotal columns," we mean that the jth column of A is pivotal if and only if the jth column of B is pivotal, not that the columns themselves are the same.)

(b) If A and B have the same pivotal columns, then do A and B have the same kernel? If so, explain why; otherwise, give a counterexample.

SECTION 4 | # The Importance of Being Pivotal

4.1 The rank of a matrix

We now have a general method for finding the solution set S of any linear system $A\mathbf{x} = \mathbf{b}$: row reduction of $[A|\mathbf{b}]$, followed by back substitution.

As we have seen, once $[A|\mathbf{b}]$ is row reduced, an easy visual inspection makes it clear whether or not the solution set is empty: The solution set is empty exactly when there is a pivot in the final column. Recall the generic look of a row-reduced matrix, as explained in the

---

*Given any two sets X and Y, $X \subset Y$ says that X is a subset of Y, which means that every element of X also belongs to Y.

previous section:

$$\begin{bmatrix} 0 & \bullet & * & * & * & * & * \\ 0 & 0 & \bullet & * & * & * & * \\ 0 & 0 & 0 & 0 & \bullet & * & * \\ 0 & 0 & 0 & 0 & 0 & \bullet & * \\ 0 & 0 & 0 & 0 & 0 & 0 & 0 \end{bmatrix}. \tag{4.1}$$

The bullets represent pivots, and there is one for each row that is not all zeros. Here there is no pivot in the final column, so by Theorem 3, the solution set is not empty.

Something about this may seem a bit strange. *Row reduction involves arbitrary choices.* When it is necessary to swap rows, there are usually several ways to do it. You could swap rows even when it is not, strictly speaking, necessary, leading to different row-reduced forms of $[A|\mathbf{b}]$.

- *How different can two row-reduced forms of the same matrix be?*

EXAMPLE 14   **(Different row reductions)**   Consider the matrix

$$[A|\mathbf{b}] = \begin{bmatrix} 1 & 2 & 3 & \bigm| & 4 \\ 2 & 2 & 2 & \bigm| & 2 \end{bmatrix}.$$

If we subtract twice the first row from the second, we get

$$\begin{bmatrix} 1 & 2 & 3 & \bigm| & 4 \\ 0 & -2 & -4 & \bigm| & -6 \end{bmatrix}.$$

This matrix is in row-reduced form, and the first two columns are the pivotal ones.

We could also do the row reduction differently. Go back to $[A|\mathbf{b}]$ and divide the second row through by 2, then swap the rows, giving

$$\begin{bmatrix} 1 & 1 & 1 & \bigm| & 1 \\ 1 & 2 & 3 & \bigm| & 4 \end{bmatrix}.$$

Subtracting the first row from the second, the result is

$$\begin{bmatrix} 1 & 1 & 1 & \bigm| & 1 \\ 0 & 1 & 2 & \bigm| & 3 \end{bmatrix}.$$

This matrix is in row-reduced form, but it is different from the one we got before. Notice, though, that the pivots are in the same places in both.

The key observation about possible differences is this: Either $A\mathbf{x} = \mathbf{b}$ has solutions or it does not, and neither case depends on how we might decide to row reduce or whether we decide to do any row reduction at all. Either the solutions are there or they are not.

Therefore, if *one* way of row reducing $[A|\mathbf{b}]$ produces a pivot in the final column, then *all* ways must, since a pivot in the final column corresponds to $A\mathbf{x} = \mathbf{b}$ having no solutions.

- *If one way of row reducing $[A|\mathbf{b}]$ leads to a pivotal final column, they all do; if one does not, none of them does.*

Taking this line of reasoning a bit further leads to the following theorem.

THEOREM 6   **(Places of pivots always come out the same)**   *Let A be an $m \times n$ augmented matrix. Then any two ways of row reducing it produce the same number of pivots in the same places.*

PROOF: Write A in the column form $A = [\mathbf{v}_1, \mathbf{v}_2, \dots, \mathbf{v}_n]$, for $k = 2, 3, \dots, n$, and let

$$A_{k-1} = [\mathbf{v}_1, \mathbf{v}_2, \dots, \mathbf{v}_{k-1}].$$

The augmented matrix

$$[A_{k-1}|\mathbf{v}_k]$$

is the $m \times k$ matrix obtained from A by crossing out everything but the first k columns.

Observe: *Any sequence of row operations that row reduces A also row reduces $[A_{k-1}|\mathbf{v}_k]$ for each $k = 2, 3, \dots, n$.* Moreover, this row reduction of $[A_{k-1}|\mathbf{v}_k]$ is exactly what we would get if we row reduced A (with the same sequence of row operations) and then crossed out everything but the first k columns. The reason is that the row operations do not mix up the columns.

Hence we get a pivot in the kth column of $[A_{k-1}|\mathbf{v}_k]$—and hence A—when $A_{k-1}\mathbf{x} = \mathbf{v}_k$ has no solutions, and we do not get a pivot if $A\mathbf{x} = \mathbf{v}_k$ has solutions. Whether it has solutions or not is an objective fact, independent of any choices we make in going about the row reduction. ∎

Theorem 6 tells us that it makes sense to speak of "the pivotal columns of A" instead of just "pivotal columns of some row reduction of A" because whether or not a column ends up being pivotal does not depend on how the row reduction is done. It only depends on A itself. Therefore, Theorem 6 allows us to make the following definition.

DEFINITION | **(Rank of a matrix)** The *rank r* of an $m \times n$ matrix A is the number of pivotal columns in A.

Let A be any $m \times n$ matrix with rank r. Since there can be at most one pivot for each column, $r \le n$. Likewise, because of the staircase structure of a row-reduced matrix (see (4.1)), there can be at most one pivot for each row. Therefore $r \le m$, so

$$r \le \min\{\, m, n \,\}. \tag{4.2}$$

Nice things happen when either $r = m$ or when $r = n$. First, the good thing that happens when $r = m$ is

- *For any $m \times n$ matrix A, $A\mathbf{x} = \mathbf{b}$ is solvable for every $\mathbf{b}$ in $\mathbb{R}^m$ if and only if the rank of A is m.*

To see why, suppose, for example, that A is a 4×6 matrix and that after row reduction it looks like this:

$$\begin{bmatrix} \bullet & * & * & * & * & * \\ 0 & \bullet & * & * & * & * \\ 0 & 0 & 0 & \bullet & * & * \\ 0 & 0 & 0 & 0 & \bullet & * \end{bmatrix}. \tag{4.3}$$

There is a pivot in each of the rows, so $r = m$. Now choose *any* vector $\mathbf{b}$ in $\mathbb{R}^4$. Since we now reduce from left to right, row reducing $[A|\mathbf{b}]$ leads to

$$\left[\begin{array}{cccccc|c} \bullet & * & * & * & * & * & * \\ 0 & \bullet & * & * & * & * & * \\ 0 & 0 & 0 & \bullet & * & * & * \\ 0 & 0 & 0 & 0 & \bullet & * & * \end{array}\right]. \tag{4.4}$$

No matter how $\mathbf{b}$ is chosen, there is no pivot in the final column because there can be at most one pivot per row, and all rows have already been occupied by a pivot from a column of A.

Conversely, suppose that $r < m$. For instance, suppose in our 4×6 example, $r = 3$, and row reduction of A leads to the form

$$\begin{bmatrix} \bullet & * & * & * & * & * \\ 0 & \bullet & * & * & * & * \\ 0 & 0 & 0 & \bullet & * & * \\ 0 & 0 & 0 & 0 & 0 & 0 \end{bmatrix}. \tag{4.5}$$

Adjoining the vector $\mathbf{e}_4$ to this matrix gives us the augmented matrix of a system with no solutions:

$$\begin{bmatrix} \bullet & * & * & * & * & * & \bigm| & 0 \\ 0 & \bullet & * & * & * & * & \bigm| & 0 \\ 0 & 0 & 0 & \bullet & * & * & \bigm| & 0 \\ 0 & 0 & 0 & 0 & 0 & 0 & \bigm| & 1 \end{bmatrix}. \tag{4.6}$$

If we now apply to (4.6) the inverses of the row operations that transformed A into (4.5), in reverse order, we get the augmented matrix $[A|\mathbf{c}]$ where $\mathbf{c}$ is whatever the inverse row operations end up doing to $\mathbf{e}_4$. Since $[A|\mathbf{c}]$ is equivalent to (4.6) and the latter represents a system with no solutions, $A\mathbf{x} = \mathbf{c}$ has no solutions. The same reasoning applies in general and shows that if $r < m$, there exists a $\mathbf{c}$ so that $A\mathbf{x} = \mathbf{c}$ has no solutions, justifying the statement in the preceding bulleted point.

The good thing that happens when $r = n$ is that if $A\mathbf{x} = \mathbf{b}$ is solvable, then the solution is unique.

- *For any $m \times n$ matrix A and any $\mathbf{b}$ in $\mathbb{R}^m$ for which $A\mathbf{x} = \mathbf{b}$ is solvable, the solution is unique if and only if the rank of A is n.*

To see this, recall that the pivotal columns of A correspond to pivotal variables. There will be r pivotal variables and $n - r$ free variables, or parameters.

If $r = n$, there are no free variables, so if $A\mathbf{x} = \mathbf{b}$ is solvable, there is just one solution. On the other hand, if $r < n$, there will be at least one free variable. Therefore, if $A\mathbf{x} = \mathbf{b}$ is solvable at all, there is at least a one-parameter family of solutions and hence infinitely many solutions.

If A is a square matrix, then it is possible for *both* good things to happen at once: If A is an $n \times n$ matrix with rank n, then $A\mathbf{x} = \mathbf{b}$ has a unique solution for every $\mathbf{b}$ in $\mathbb{R}^n$. This means that A is invertible, and $A^{-1}\mathbf{b}$ is the unique solution. On the other hand, when A is invertible, $A\mathbf{x} = \mathbf{b}$ has a unique solution for every $\mathbf{b}$ in $\mathbb{R}^n$, namely, $A^{-1}\mathbf{b}$, so it must be that the rank of A equals both the number of rows and columns, which therefore must be the same. We summarize.

THEOREM 7 (**Existence, uniqueness, and rank**) *Let A be an $m \times n$ matrix with rank r. Then $r \le m$ and $r \le n$. Moreover,*

    1. The solutions of $A\mathbf{x} = \mathbf{b}$ are unique whenever they exist if and only if $r = n$.

    2. The equation $A\mathbf{x} = \mathbf{b}$ is solvable for every $\mathbf{b}$ in $\mathbb{R}^m$ if and only if $r = m$.

    3. If A is a square $n \times n$ matrix, then A is invertible if and only if $r = n$.

EXAMPLE 15 (**Using rank to check for existence and uniqueness**) Consider the following systems of equations, written as augmented matrices:

$$[A|\mathbf{a}] = \begin{bmatrix} 1 & 2 & 3 & \bigm| & 4 \\ 4 & 3 & 2 & \bigm| & 1 \end{bmatrix} \qquad [B|\mathbf{b}] = \begin{bmatrix} 1 & 2 & 3 & \bigm| & 4 \\ 4 & 3 & 2 & \bigm| & 1 \\ 5 & 5 & 5 & \bigm| & 6 \end{bmatrix} \qquad [C|\mathbf{c}] = \begin{bmatrix} 1 & 2 & 3 & \bigm| & 4 \\ 4 & 3 & 2 & \bigm| & 1 \\ 1 & 2 & 4 & \bigm| & 5 \end{bmatrix}.$$

Which, if any, has no solutions? Unique solutions? Infinitely many solutions?

To answer these questions, row reduce. Starting with $[A|\mathbf{a}]$, one row operation gives us

$$\begin{bmatrix} 1 & 2 & 3 & | & 4 \\ 0 & -5 & -10 & | & -15 \end{bmatrix}.$$

We see that the rank of A is 2 and A is 2×3. Therefore, $A\mathbf{x} = \mathbf{a}$ is solvable for *any* $\mathbf{a}$, but the solutions are never unique. (There is a one-parameter family of solutions.)

Next, $[B|\mathbf{b}]$ row reduces to

$$\begin{bmatrix} 1 & 2 & 3 & | & 4 \\ 0 & -5 & -10 & | & -15 \\ 0 & 0 & 0 & | & 1 \end{bmatrix}.$$

The rank of B is 2, and B is 3×3. Hence, $B\mathbf{x} = \tilde{\mathbf{b}}$ will be solvable for some $\tilde{\mathbf{b}}$ (like $\tilde{\mathbf{b}} = 0$) but not others and never uniquely. In this case, there is a pivot in the final column, so $B\mathbf{x} = \mathbf{b}$ has no solution.

Finally, $[C|\mathbf{c}]$ row reduces to

$$\begin{bmatrix} 1 & 2 & 3 & | & 4 \\ 0 & -5 & -10 & | & -15 \\ 0 & 0 & 1 & | & 1 \end{bmatrix}.$$

The rank of C is 3, and C is 3×3. So $C\mathbf{x} = \mathbf{c}$ has a unique solution for every $\mathbf{c}$, including this one in particular.

4.2 Kernels and inverses of square matrices

Theorem 7 has an important corollary.

COROLLARY TO THEOREM 7   **(Inverses of square matrices)** *A square matrix A is invertible if and only if $Ker(A) = 0$.*

PROOF: Suppose A is $n \times n$. If $Ker(A) = 0$, then by Theorem 5, the solutions of $A\mathbf{x} = \mathbf{b}$ are unique whenever they exist. Hence, by (1) of Theorem 7, the rank of A is n. Then by (3) of Theorem 7, A is invertible.

On the other hand, if A is invertible, then the solutions of $A\mathbf{x} = \mathbf{b}$ are unique whenever they exist, and $Ker(A) = 0$ by Theorem 5. ■

Here is an important class of matrices to which this theorem applies.

EXAMPLE 16   **(Square isometries)** Recall that an $m \times n$ matrix A is an *isometry* in the case where $|A\mathbf{x}| = |\mathbf{x}|$ for every $\mathbf{x}$. In particular, $A\mathbf{x} = 0$ if and only if $\mathbf{x} = 0$. Hence $Ker(A) = 0$.

Let A be an $n \times n$ isometry. By the corollary and the fact that $Ker(A) = 0$, A is invertible. Moreover, we already know from Theorem 15 of Chapter 1 that the transpose of A is the left inverse of A. That is, $A^t A = I$. Whenever a matrix is invertible, any left inverse is *the* inverse, and so A^t is the inverse of A when A is a square isometry.

The class of square isometries plays an important role in very many applications, as we shall see, so it has a name. It is called the class of *orthogonal matrices*.

Before making a formal definition, recall from Theorem 15 of Chapter 1 that an $m \times n$ matrix A is an isometry if and only if its columns are an orthonormal set of vectors in $\mathbb{R}^m$. Hence, an $n \times n$ matrix U is a square isometry if and only if its columns are an orthonormal set of vectors in $\mathbb{R}^n$. (The term "orthonormal set of vectors" is defined in Chapter 1 following (5.21); please review it now if it is unfamiliar.)

DEFINITION   **(Orthogonal matrices)** An $n \times n$ matrix U is *orthogonal* in the case where the columns of U are orthonormal.

If U is orthogonal, then U^t is the inverse of U, and both

$$U^t U = I \qquad \text{and} \qquad UU^t = I. \tag{4.7}$$

The first equality in (4.7) is a direct consequence of the definition and **V.I.F.4**. That is,

$$
\begin{aligned}
(U^t U)_{i,j} &= (\text{row } i \text{ of } U^t) \cdot (\text{column } j \text{ of } U) \\
&= (\text{column } i \text{ of } U) \cdot (\text{column } j \text{ of } U) \\
&= I_{i,j}
\end{aligned}
$$

since the columns of U are orthonormal, so the dot product is 1 if $i = j$ and is zero otherwise.

The second equality is more interesting and less direct. It says that the rows of U are also orthonormal. By **V.I.F.4**,

$$
\begin{aligned}
(UU^t)_{i,j} &= (\text{row } i \text{ of } U) \cdot (\text{column } j \text{ of } U^t) \\
&= (\text{row } i \text{ of } U) \cdot (\text{row } j \text{ of } U).
\end{aligned}
$$

Then by the second equality in (4.7),

$$(\text{row } i \text{ of } U) \cdot (\text{row } j \text{ of } U) = I_{i,j},$$

which says that the rows of U are orthonormal. This is a striking fact.

- *Any $n \times n$ matrix that has orthonormal columns automatically also has orthonormal rows.*

In particular, the transpose of an orthogonal matrix is an orthogonal matrix. Since the transpose is the inverse, the inverse of an orthogonal matrix is also an orthogonal matrix. In proving all this, we relied on the Corollary to Theorem 7, which told us that $UU^t = I$ while $U^t U = I$ holds just because U is an isometry.

The following theorem contains some of what we have learned.

THEOREM 8   **(Orthogonal matrices)** *Let U be an $n \times n$ orthogonal matrix. Then U is invertible, and $U^{-1} = U^t$, and U^{-1} is also an orthogonal matrix.*

EXAMPLE 17   **(A 3×3 orthogonal matrix)** Let

$$U = \frac{1}{3} \begin{bmatrix} 1 & 2 & 2 \\ 2 & 1 & -2 \\ -2 & 2 & -1 \end{bmatrix}.$$

As you can check, $U^t U = I = UU^t$. Hence the three columns of U are orthonormal, as are the three rows.

The next example contains another important application of the Corollary to Theorem 7; at the same time it leads to the definition of another important class of matrices.

EXAMPLE 18 **(The polynomial fitting problem)** Let $P(s)$ be a polynomial of degree n in the variable s so that

$$P(s) = a_0 + a_1 s + a_2 s^2 + \cdots + a_n s^n$$

for some coefficients $a_0, a_1, a_2, \ldots, a_n$. Suppose that we are given $n + 1$ points in the s, t plane

$$(s_1, t_1), (s_2, t_2), \ldots, (s_{n+1}, t_{n+1})$$

so that $s_i \neq s_j$ whenever $i \neq j$. That is, no two of our points can lie on the same vertical line.

The *polynomial fitting problem* is to choose the coefficients $a_0, a_1, a_2, \ldots, a_n$ so that the graph of $t = P(s)$ passes through each of these points. In other words, we want to choose the coefficients so that

$$P(s_i) = t_i \quad \text{for} \quad i = 1, 2, \ldots, n + 1. \tag{4.8}$$

The system of equations (4.8) may not look linear at first, but remember that the variables we are trying to determine are the coefficients $a_0, a_1, a_2, \ldots, a_n$, and it is a linear system in these variables, even if $P(s)$ is a nonlinear function of s. Indeed, let

$$\mathbf{a} = \begin{bmatrix} a_0 \\ a_1 \\ \vdots \\ a_n \end{bmatrix}, \quad \mathbf{b} = \begin{bmatrix} t_1 \\ t_2 \\ \vdots \\ t_{n+1} \end{bmatrix}, \quad \text{and } A = \begin{bmatrix} 1 & s_1 & s_1^2 & s_1^3 & \cdots & s_1^n \\ 1 & s_2 & s_2^2 & s_2^3 & \cdots & s_2^n \\ 1 & s_3 & s_3^2 & s_3^3 & \cdots & s_3^n \\ \vdots & & & & \vdots & \\ 1 & s_{n+1} & s_{n+1}^2 & s_{n+1}^3 & \cdots & s_{n+1}^n \end{bmatrix}. \tag{4.9}$$

Then (4.8) is equivalent to $A\mathbf{a} = \mathbf{b}$. If A is invertible, the vector of coefficients $\mathbf{a}$ is given by $\mathbf{a} = A^{-1}\mathbf{b}$.

According to the Corollary to Theorem 7, to show that A is invertible, we have to show that $\text{Ker}(A) = 0$. Hence, we focus on $A\mathbf{a} = 0$, which in this case corresponds to

$$P(s_i) = 0 \quad \text{for} \quad i = 1, 2, \ldots, n + 1. \tag{4.10}$$

This would mean that each of the $n + 1$ distinct numbers s_i is a root of $P(s)$. But a nonzero nth-degree polynomial can have at most n distinct roots. (Use the fundamental theorem of algebra to factor $P(s)$.) Thus, the only way to achieve (4.10) is with $a_0 = a_1 = \cdots = a_n = 0$, which means $\mathbf{a} = 0$. That is, $A\mathbf{x} = 0$ has only the zero solution $\mathbf{x} = 0$. That does it: $\text{Ker}(A) = 0$, so A is invertible. Hence, the required coefficients exist and are given by $\mathbf{a} = A^{-1}\mathbf{b}$.

The class of matrices that we encountered in Example 18 has a name.

DEFINITION

(Vandermonde matrices) An $(n + 1) \times (n + 1)$ matrix A of the form (4.9), in which each row consists of the zero through the nth power of some number, is called a *Vandermonde matrix*.

The name is Dutch for "from the moon," but you see that the Vandermonde matrices come from the polynomial fitting problem, since (4.8) is equivalent to $A\mathbf{a} = \mathbf{b}$. We have just learned two things as a consequence of the Corollary to Theorem 7:

- *Every Vandermonde matrix with distinct entries in its second column is invertible.*
- *Given any $n + 1$ distinct numbers $s_1, s_2, \ldots, s_{n+1}$ and any $n + 1$ numbers $t_1, t_2, \ldots, t_{n+1}$, there is one and only one way to choose the coefficients of an nth-degree polynomial $P(x)$ so that*

$$\begin{bmatrix} P(s_1) \\ P(s_2) \\ \vdots \\ P(s_{n+1}) \end{bmatrix} = \begin{bmatrix} t_1 \\ t_2 \\ \vdots \\ t_{n+1} \end{bmatrix}. \tag{4.11}$$

The vector of coefficients is given by

$$\begin{bmatrix} a_1 \\ a_2 \\ \vdots \\ a_{n+1} \end{bmatrix} = A^{-1} \begin{bmatrix} t_1 \\ t_2 \\ \vdots \\ t_{n+1} \end{bmatrix}, \tag{4.12}$$

where A is the Vandermonde matrix (4.9).

4.3 Computing inverses

Consider an $n \times n$ matrix A with rank $r = n$. Theorem 7 says that such an A is invertible. We know A^{-1} exists. But how do we *compute* A^{-1}?

Suppose we write A^{-1} in terms of its columns as

$$A^{-1} = [\mathbf{v}_1, \mathbf{v}_2, \ldots, \mathbf{v}_n].$$

If we can find each of the columns $\mathbf{v}_j$, we have found A^{-1}. To get an equation that $\mathbf{v}_j$ must satisfy, note that by **V.I.F.3**, $AA^{-1} = I$ is the same as

$$[A\mathbf{v}_1, A\mathbf{v}_2, \ldots, A\mathbf{v}_n] = [\mathbf{e}_1, \mathbf{e}_2, \ldots, \mathbf{e}_n].$$

That is, to find the jth column of A^{-1}, namely, $\mathbf{v}_j$, all we need to do is solve the equation

$$A\mathbf{v}_j = \mathbf{e}_j.$$

We can do this by row reduction.

EXAMPLE 19 **(Computing an inverse)** Consider the Vandermonde matrix

$$A = \begin{bmatrix} 1 & 1 & 1 \\ 1 & 2 & 4 \\ 1 & 3 & 9 \end{bmatrix}.$$

It has rank 3, as you can check and as we will soon see. Let us solve $A\mathbf{v}_1 = \mathbf{e}_1$. The augmented matrix is

$$[A|\mathbf{e}_1] = \left[\begin{array}{ccc|c} 1 & 1 & 1 & 1 \\ 1 & 2 & 4 & 0 \\ 1 & 3 & 9 & 0 \end{array}\right].$$

Row reducing with no row swaps leads to

$$\left[\begin{array}{ccc|c} 1 & 1 & 1 & 1 \\ 0 & 1 & 3 & -1 \\ 0 & 0 & 2 & 1 \end{array}\right].$$

At this stage, we *could* proceed to find the unique solution by back substitution, which is the next step according to our general strategy. But in this case, we can keep going and obtain the unique solution by *further row reduction and no back substitution*. Divide the last row through by 2 so that all the pivots are 1:

$$\left[\begin{array}{ccc|c} 1 & 1 & 1 & 1 \\ 0 & 1 & 3 & -1 \\ 0 & 0 & 1 & 1/2 \end{array}\right].$$

Now keep going with row operations and clean out the entries above the pivots. First subtract the second column from the first, giving

$$\left[\begin{array}{ccc|c} 1 & 0 & -2 & 2 \\ 0 & 1 & 3 & -1 \\ 0 & 0 & 1 & 1/2 \end{array}\right].$$

Now subtract the third row from the second twice, and add the third row to the first twice. Doing so cancels out everything above the last pivot and gives

$$\begin{bmatrix} 1 & 0 & 0 & | & 3 \\ 0 & 1 & 0 & | & -5/2 \\ 0 & 0 & 1 & | & 1/2 \end{bmatrix},$$

corresponding to the system

$$I\mathbf{x} = \begin{bmatrix} 3 \\ -5/2 \\ 1/2 \end{bmatrix}. \tag{4.13}$$

This equation solves itself. The unique solution is $\mathbf{v}_1$, so the first column of A^{-1} is the vector on the right side of (4.13).

Notice that in Example 19, we solved the equation specifying the first column of A^{-1} without back substitution; instead, we continued until we had row reduced A to the identity matrix. Let us go through this schematically. We will learn something interesting.

For *any* square matrix A, the augmented matrix corresponding to $A\mathbf{v}_j = \mathbf{e}_j$ is $[A|\mathbf{e}_j]$. If we row reduce it (and if A were 4×4, to be concrete), we get the row-reduced form

$$\begin{bmatrix} \bullet & * & * & * & | & * \\ 0 & \bullet & * & * & | & * \\ 0 & 0 & \bullet & * & | & * \\ 0 & 0 & 0 & \bullet & | & * \end{bmatrix},$$

where the bullets denote nonzero entries and the asterisks denote entries that may or may not be zero. The matrix looks like it does because the rank equals the number of columns and rows, so we can put the pivots only along the diagonal, filling it up.

At this stage, we *could* proceed to find the unique solution by back substitution. But as in Example 19, we can keep going and obtain the unique solution by *further row reduction and no back substitution.*

Multiply each row through by the inverse of its pivot. The pivot is nonzero, so you can do so. After this step, each pivot is 1. Now that we know the values of the pivots, we do not need to use bullet dots to denote them any more. We have

$$\begin{bmatrix} 1 & * & * & * & | & * \\ 0 & 1 & * & * & | & * \\ 0 & 0 & 1 & * & | & * \\ 0 & 0 & 0 & 1 & | & * \end{bmatrix}.$$

Now clean out the remaining entries in the pivotal columns. For example, multiply the second row through by whatever entry is in the first row, second column, and then subtract this product from the first row. The result is

$$\begin{bmatrix} 1 & 0 & * & * & | & * \\ 0 & 1 & * & * & | & * \\ 0 & 0 & 1 & * & | & * \\ 0 & 0 & 0 & 1 & | & * \end{bmatrix}.$$

Continuing in this way, we arrive at a matrix of the form

$$\begin{bmatrix} 1 & 0 & 0 & 0 & | & * \\ 0 & 1 & 0 & 0 & | & * \\ 0 & 0 & 1 & 0 & | & * \\ 0 & 0 & 0 & 1 & | & * \end{bmatrix}.$$

If we let $\mathbf{c}$ denote the final column of this new matrix, it represents a very simple equation: $I\mathbf{x} = \mathbf{c}$. This equation solves itself; $\mathbf{x} = \mathbf{c}$ is the unique solution. Since we used row reduction to arrive at $I\mathbf{x} = \mathbf{c}$ starting from $A\mathbf{x} = \mathbf{e}_j$, the two equations are equivalent, so $\mathbf{c}$ is the unique solution of $Ax = e_j$. That is, $c = v_j$. We now have a method for finding v_j using row reduction alone.

We could do the same operations one at a time for each of $v_1, v_2, \ldots, v_n$. But we can be more efficient if we use the fact that the operations of row reduction act separately on each column of a matrix. Let us adjoin all n vectors e_j at once, forming the $n \times 2n$ matrix

$$[A|e_1, e_2, \ldots, e_n],$$

which has the form

$$\begin{bmatrix} * & * & * & * & 1 & 0 & 0 & 0 \\ * & * & * & * & 0 & 1 & 0 & 0 \\ * & * & * & * & 0 & 0 & 1 & 0 \\ * & * & * & * & 0 & 0 & 0 & 1 \end{bmatrix}.$$

Since we are adjoining a copy of the identity matrix on the right, another way to describe this $n \times 2n$ matrix would be

$$[A|I].$$

Row reducing this matrix to standard row-reduced form yields

$$\begin{bmatrix} \bullet & * & * & * & * & * & * & * \\ 0 & \bullet & * & * & * & * & * & * \\ 0 & 0 & \bullet & * & * & * & * & * \\ 0 & 0 & 0 & \bullet & * & * & * & * \end{bmatrix},$$

provided A has full rank. (Otherwise, one or more of the entries marked by bullets would be zero.) In the case where A has full rank, we can proceed with more row operations as before to arrive at

$$\begin{bmatrix} 1 & 0 & 0 & 0 & * & * & * & * \\ 0 & 1 & 0 & 0 & * & * & * & * \\ 0 & 0 & 1 & 0 & * & * & * & * \\ 0 & 0 & 0 & 1 & * & * & * & * \end{bmatrix}, \tag{4.14}$$

The $(n + j)$th column of this matrix is the solution of $Ax = e_j$, which is the jth column of A^{-1}. Therefore, the $n \times 2n$ matrix (4.14) is $[I|A^{-1}]$, so the right half of this matrix is A^{-1}.

EXAMPLE 20

(An efficient matrix inverse computation) Let us apply this method to compute the inverse of the Vandermonde matrix

$$A = \begin{bmatrix} 1 & 1 & 1 \\ 1 & 2 & 4 \\ 1 & 3 & 9 \end{bmatrix}.$$

We augment this matrix with the identity, obtaining

$$\begin{bmatrix} 1 & 1 & 1 & 1 & 0 & 0 \\ 1 & 2 & 4 & 0 & 1 & 0 \\ 1 & 3 & 9 & 0 & 0 & 1 \end{bmatrix}.$$

Cleaning out the first column leads to

$$\begin{bmatrix} 1 & 1 & 1 & 1 & 0 & 0 \\ 0 & 1 & 3 & -1 & 1 & 0 \\ 0 & 2 & 8 & -1 & 0 & 1 \end{bmatrix}.$$

One more row operation takes us to the row-reduced form:

$$\begin{bmatrix} 1 & 1 & 1 & 1 & 0 & 0 \\ 0 & 1 & 3 & -1 & 1 & 0 \\ 0 & 0 & 2 & 1 & -2 & 1 \end{bmatrix}.$$

Dividing the final row through by 2, we get

$$\begin{bmatrix} 1 & 1 & 1 & 1 & 0 & 0 \\ 0 & 1 & 3 & -1 & 1 & 0 \\ 0 & 0 & 1 & 1/2 & -1 & 1/2 \end{bmatrix}.$$

Three more row operations clear out the upper right entries of the 3×3 part on the left, giving

$$\begin{bmatrix} 1 & 0 & 0 & \bigm| & 3 & -3 & 1 \\ 0 & 1 & 0 & \bigm| & -5/2 & 4 & -3/2 \\ 0 & 0 & 1 & \bigm| & 1/2 & -1 & 1/2 \end{bmatrix}.$$

This means that

$$A^{-1} = \frac{1}{2} \begin{bmatrix} 6 & -6 & 2 \\ -5 & 8 & -3 \\ 1 & -2 & 1 \end{bmatrix},$$

as you can easily check.

4.4 The reduced row echelon form of a matrix

Here is a schematic representation of a 4×6 augmented matrix $[A|\mathbf{b}]$ in row-reduced form:

$$\begin{bmatrix} \bullet & * & * & * & * & \bigm| & * \\ 0 & \bullet & * & * & * & \bigm| & * \\ 0 & 0 & 0 & \bullet & * & \bigm| & * \\ 0 & 0 & 0 & 0 & \bullet & \bigm| & * \end{bmatrix}. \tag{4.15}$$

We can go on and multiply each row through by the inverse of the pivotal entry to obtain

$$\begin{bmatrix} 1 & * & * & * & * & \bigm| & * \\ 0 & 1 & * & * & * & \bigm| & * \\ 0 & 0 & 0 & 1 & * & \bigm| & * \\ 0 & 0 & 0 & 0 & 1 & \bigm| & * \end{bmatrix}. \tag{4.16}$$

By further row operations, we can clean out the columns above the pivotal entries so that all entries above and below the pivots are zero. For instance, by adding an appropriate multiple of the second row to the first, we can cancel out the asterisk above the second pivot and so forth. The result is

$$\begin{bmatrix} 1 & 0 & * & 0 & 0 & \bigm| & * \\ 0 & 1 & * & 0 & 0 & \bigm| & * \\ 0 & 0 & 0 & 1 & 0 & \bigm| & * \\ 0 & 0 & 0 & 0 & 1 & \bigm| & * \end{bmatrix}. \tag{4.17}$$

This matrix is in what is known as *reduced row echelon form*. Since it was obtained from the row-reduced form (4.15) by further row operations, both (4.15) and (4.17) represent equivalent systems.

DEFINITION (**Reduced row echelon form**) A matrix is in reduced row echelon form if it is in row-reduced form, each pivotal entry is 1, all entries above and below a pivot are zero, and there are no all-zero rows.

EXAMPLE 21   (**Computing a reduced row echelon form**) Let

$$[A|\mathbf{b}] = \begin{bmatrix} 2 & 4 & 6 & 8 & 10 & \bigm| & 12 \\ 0 & 3 & 6 & 9 & 3 & \bigm| & 3 \\ 0 & 0 & 0 & 2 & 2 & \bigm| & 4 \\ 0 & 0 & 0 & 0 & 3 & \bigm| & 6 \end{bmatrix}.$$

Dividing each row through by its pivotal entry leads to

$$\begin{bmatrix} 1 & 2 & 3 & 4 & 5 & \bigm| & 6 \\ 0 & 1 & 2 & 3 & 1 & \bigm| & 1 \\ 0 & 0 & 0 & 1 & 1 & \bigm| & 2 \\ 0 & 0 & 0 & 0 & 1 & \bigm| & 2 \end{bmatrix}.$$

Cleaning out the entry above the second pivot leads to

$$\left[\begin{array}{ccccc|c} 1 & 0 & -1 & -2 & 3 & 4 \\ 0 & 1 & 2 & 3 & 1 & 1 \\ 0 & 0 & 0 & 1 & 1 & 2 \\ 0 & 0 & 0 & 0 & 1 & 2 \end{array}\right].$$

Doing the same for all of the other pivotal columns finally leads to

$$\left[\begin{array}{ccccc|c} 1 & 0 & -1 & 0 & 0 & -2 \\ 0 & 1 & 2 & 0 & 0 & -1 \\ 0 & 0 & 0 & 1 & 0 & 0 \\ 0 & 0 & 0 & 0 & 1 & 2 \end{array}\right].$$

As a practical matter, there is no point in computing a reduced row echelon form on paper if your goal is to solve a matrix equation $A\mathbf{x} = \mathbf{b}$. Once you have row reduced to find $[U|\mathbf{c}]$, you should stop and solve $U\mathbf{x} = \mathbf{c}$ by back substitution.

However, if you are using a computer to do the row reduction and your matrices are not extremely large, you will not notice the extra time it takes for your computer to go all the way to the row-reduced form. Then, because of all of the entries that are 0 or 1, it is particularly easy to parameterize the solution set.

Moreover, there are (noncomputational) reasons why you might want to know the reduced row echelon form of a matrix A. The first is implicit in the last statement, where we referred to "the" reduced row echelon form of a matrix A. It turns out that while matrices can have many row-reduced forms, they can have only one reduced row echelon form.

The key is the following:

- *Let A be an $m \times n$ matrix with rank r, and suppose that E is a reduced row echelon form matrix obtained by performing row operations on A. Then if the kth column of A is nonpivotal,*

$$\mathbf{v}_k = \sum_{i=1}^{r} E_{i,k}\mathbf{v}_{j(i)}, \tag{4.18}$$

where, as usual, $\mathbf{v}_{j(i)}$ denotes the ith pivotal column.

The formula (4.18) shows how to write a nonpivotal column as a linear combination of the pivotal columns: If $\mathbf{v}_k$ is not pivotal, the coefficients required for doing so can be found in the kth column of E. *This fact is what is special about the nonpivotal column of E.*

The point is this: If $\mathbf{v}_k$ is not pivotal, then there is exactly one way to write $\mathbf{v}_k$ as a linear combination of the pivotal columns of A. Indeed, $A\mathbf{e}_k = \mathbf{v}_k$, so there is at least one solution $\mathbf{x}$ to $A\mathbf{x} = \mathbf{v}_k$. When there is a solution, we can find another that has any values we choose for the free variables, and once the free variables are specified, the solution is unique. Therefore, there is exactly one $\mathbf{x}$ so that each nonpivotal entry in $\mathbf{x}$ is zero, and $A\mathbf{x} = \mathbf{v}_k$; in other words, $x_{j(1)}\mathbf{v}_{j(1)} + \cdots + x_{j(r)}\mathbf{v}_{j(r)} = \mathbf{v}_k$. After the next example, we will show that $E_{i,k} = x_{j(i)}$ for $i = 1, \ldots, r$ so that (4.18) is true.

Since the pivotal columns are linearly independent, there is only one set of coefficients $E_{1,k}, \ldots, E_{r,k}$ for which (4.18) is true, any two ways of row reducing A all the way to reduced row echelon form must yield the same kth column.

This reasoning applies to all the nonpivotal columns, and we already know that the ith pivotal column of E is by definition $\mathbf{e}_i$. Hence, there is just one reduced row echelon form of A.

- *Let A be an $m \times n$ matrix. Then any two ways of row reducing A all the way to reduced row echelon form result in the same matrix E, which we call the reduced row echelon form of A and denote by*

$$\text{rref}(A) = E. \tag{4.19}$$

Note that (4.19) defines a function on the set of $m \times n$ matrices with values in the set of $m \times n$ matrices. It is a very useful function, and a capability for computing $\mathrm{rref}(A)$ is built into every serious computer program for linear algebra.

In the next example, we check (4.18) in one particular case.

EXAMPLE 22   **(Linear combination of pivotal columns)** Consider $[A|\mathbf{b}]$ as in Example 14. As we saw there, $[A|\mathbf{b}]$ reduces to $[E|\mathbf{c}]$ where

$$\mathbf{c} = \begin{bmatrix} -2 \\ -1 \\ 0 \\ 2 \end{bmatrix}.$$

You can check (4.18) by

$$\begin{bmatrix} 12 \\ 3 \\ 4 \\ 6 \end{bmatrix} = -2 \begin{bmatrix} 2 \\ 0 \\ 0 \\ 0 \end{bmatrix} - 1 \begin{bmatrix} 4 \\ 3 \\ 0 \\ 0 \end{bmatrix} + 0 \begin{bmatrix} 8 \\ 9 \\ 2 \\ 0 \end{bmatrix} + 2 \begin{bmatrix} 10 \\ 3 \\ 2 \\ 3 \end{bmatrix}.$$

Since for A the pivotal columns are columns 1, 2, 4, and 5, so that $j(1) = 1, j(2) = 2, j(3) = 4$, and $j(4) = 5$, in agreement with (4.18).

To see why (4.18) is valid, let $\mathbf{w}_j$ denote the jth column of E and suppose that there are ℓ pivotal columns to the left of $\mathbf{v}_k$ in A. Form the matrix

$$[\mathbf{v}_{j(1)}, \dots, \mathbf{v}_{j(\ell)} | \mathbf{v}_k].$$

All we have done is to delete a number of columns of A. Since the row operations do not mix up columns, if we apply the same sequence of row operations that reduces $A \longrightarrow E$, we still reduce the columns we have kept in the same way, so

$$[\mathbf{v}_{j(1)}, \dots, \mathbf{v}_{j(\ell)} | \mathbf{v}_k] \longrightarrow [\mathbf{w}_{j(1)}, \dots, \mathbf{w}_{j(\ell)} | \mathbf{w}_k].$$

Since the ith pivotal column of a reduced row echelon form matrix is $\mathbf{e}_i$, $\mathbf{w}_{j(i)} = \mathbf{e}_i$, so

$$[\mathbf{w}_{j(1)}, \dots, \mathbf{w}_{j(\ell)} | \mathbf{w}_k] = [\mathbf{e}_1, \dots, \mathbf{e}_\ell | \mathbf{w}_k].$$

Delete the $m - \ell$ all-zero rows of the matrix on the right, and recall that $I_{\ell \times \ell} = [\mathbf{e}_1, \dots, \mathbf{e}_\ell]$. Then

$$[\mathbf{w}_{j(1)}, \dots, \mathbf{w}_{j(\ell)} | \mathbf{w}_k] \longrightarrow [I_{\ell \times \ell} | \hat{\mathbf{w}}_k],$$

where $\hat{\mathbf{w}}_k$ is the vector in $\mathbb{R}^\ell$ obtained by deleting the last $m - \ell$ entries in $\mathbf{w}_k$, which are all zero anyhow.

Since we obtained $[I_{\ell \times \ell} | \hat{\mathbf{w}}_k]$ from $[\mathbf{v}_{j(1)}, \dots, \mathbf{v}_{j(\ell)} | \mathbf{v}_k]$ by performing row operations and deleting the $m - \ell$ all-zero rows, these two augmented matrices represent equivalent systems.

Since the first one encodes the "self-solving" equation $I_{\ell \times \ell} \mathbf{y} = \hat{\mathbf{w}}_k$, the unique solution of $[\mathbf{v}_{j(1)}, \dots, \mathbf{v}_{j(\ell)}] \mathbf{y} = \mathbf{v}_k$ is also $\mathbf{y} = \hat{\mathbf{w}}_k$. That is,

$$[\mathbf{v}_{j(1)}, \dots, \mathbf{v}_{j(\ell)}] \hat{\mathbf{w}}_k = \mathbf{v}_k.$$

But since $\mathbf{w}_k$ is the kth column of E and $\hat{\mathbf{w}}_k$ consists of the first ℓ entries of $\mathbf{w}$,

$$\hat{\mathbf{w}}_k = \begin{bmatrix} E_{1,k} \\ \vdots \\ E_{\ell,k} \end{bmatrix},$$

so by V.I.F.1, $[\mathbf{v}_{j(1)}, \dots, \mathbf{v}_{j(\ell)}] \hat{\mathbf{w}}_k = \mathbf{v}_k$ is the same as

$$E_{1,k} \mathbf{v}_{j(1)} + \cdots + E_{\ell,k} \mathbf{v}_{j(\ell)} = \mathbf{v}_k.$$

We have almost arrived at (4.18). The last thing to observe is that since $E_{i,k} = 0$ for $i > \ell$, we do not change anything if we sum over all the pivotal rows, as in (4.18).

The identity (4.18) that we have just proved has many consequences in addition to the uniqueness of the reduced row echelon form. Here is another important one.

THEOREM 9 **(Checking for equivalence of systems)** *Let A be an $m \times n$ matrix, and let $\tilde{A}$ be an $\tilde{m} \times n$ matrix. Then*

$$\mathrm{rref}(A) = \mathrm{rref}(\tilde{A}) \Leftrightarrow \mathrm{Ker}(A) = \mathrm{Ker}(\tilde{A}).$$

Moreover, if $\mathbf{b}$ is any vector in $\mathbb{R}^m$ and $\tilde{\mathbf{b}}$ is any vector in $\mathbb{R}^{\tilde{m}}$, then $[A|\mathbf{b}]$ and $[\tilde{A}|\tilde{\mathbf{b}}]$ are nonempty and equivalent if and only if

$$\mathrm{rref}([A|\mathbf{b}]) = \mathrm{rref}([\tilde{A}|\tilde{\mathbf{b}}])$$

and there is no pivot in the final column.

PROOF: First, suppose that $\mathrm{rref}(A) = \mathrm{rref}(\tilde{A}) = E$. Then some sequence of row operations reduces $[A|0]$ to $[E|0]$, and another reduces $[\tilde{A}|0]$ to $[E|0]$. Thus both $[A|0]$ and $[\tilde{A}|0]$ are equivalent to $[E|0]$ and hence to one another. Then by Theorem 2, both $A\mathbf{x} = 0$ and $\tilde{A}\mathbf{x} = 0$ have the same solution sets, and hence $\mathrm{Ker}(A) = \mathrm{Ker}(\tilde{A})$.

Second, suppose that $\mathrm{Ker}(A) = \mathrm{Ker}(\tilde{A})$. Let us show that both A and $\tilde{A}$ have the same pivotal columns. Write $A = [\mathbf{v}_1, \ldots, \mathbf{v}_n]$, and $\tilde{A} = [\tilde{\mathbf{v}}_1, \ldots, \tilde{\mathbf{v}}_n]$. To be concrete, suppose that $n \geq 3$, and let us ask whether the third column of A is pivotal. As we have seen, this is the case if and only if $\mathbf{v}_3$ is a linear combination of the columns to its left:

$$\mathbf{v}_3 = x_1\mathbf{v}_1 + x_2\mathbf{v}_2 .$$

Then with $\mathbf{x}$ defined by $\mathbf{x} = \begin{bmatrix} x_1 \\ x_2 \\ -1 \\ 0 \\ \vdots \\ 0 \end{bmatrix}$, $A\mathbf{x} = 0$. Since $\mathrm{Ker}(A) = \mathrm{Ker}(\tilde{A})$, it follows that $\tilde{A}\mathbf{x} = 0$ also, and hence

$$\tilde{\mathbf{v}}_3 = x_1\tilde{\mathbf{v}}_1 + x_2\tilde{\mathbf{v}}_2 .$$

Therefore, if the third column of A is nonpivotal, the third column of $\tilde{A}$ is also nonpivotal.

The same argument works in reverse (starting from $\tilde{A}$) and with any other column, so we see that A and $\tilde{A}$ have the same pivotal columns whenever they have the same kernel. In particular, the ith pivotal columns of $\mathrm{rref}(A)$ and $\mathrm{rref}(\tilde{A})$ are both $\mathbf{e}_i$ and are both in the same place. Hence the pivotal columns of $\mathrm{rref}(A)$ and $\mathrm{rref}(\tilde{A})$ are the same.

Moreover, we have seen that if any column of A is a linear combination of the columns to its left, then the same is true for $\tilde{A}$, *with the exact same coefficients*. In particular, this statement applies if $\mathbf{v}_k$ is nonpivotal and (4.18) is the expression of $\mathbf{v}_k$ as a linear combination of the pivotal columns of A: The exact same linear combination of the same columns of $\tilde{A}$ yields $\tilde{\mathbf{v}}_k$. Since these coefficients give the kth column in $\mathrm{rref}(A)$, we see that the nonpivotal columns of $\mathrm{rref}(A)$ and $\mathrm{rref}(\tilde{A})$ are the same.

This argument shows that whenever $\text{Ker}(A) = \text{Ker}(\tilde{A})$, then $\text{rref}(A) = \text{rref}(\tilde{A})$, and it completes the proof that having the kernel is equivalent to having the same reduced row echelon form.

Next, suppose that $\text{rref}([A|\mathbf{b}]) = \text{rref}([\tilde{A}|\tilde{\mathbf{b}}]) = [E|\mathbf{c}]$. Then by Theorem 2, both systems are equivalent to $[E|\mathbf{c}]$ and hence to one another.

Finally, suppose that $A\mathbf{x} = \mathbf{b}$ and $\tilde{A}\mathbf{x} = \tilde{\mathbf{b}}$ are equivalent systems with nonempty solution sets. Let

$$\mathbf{x}_0 + t_1\mathbf{u}_1 + \cdots + t_k\mathbf{u}_k$$

be a one-to-one parameterization of their common solution set. Then by Theorem 5,

$$t_1\mathbf{u}_1 + \cdots + t_k\mathbf{u}_k$$

is a one-to-one parameterization of both $\text{Ker}(A)$ and $\text{Ker}(\tilde{A})$. Hence $\text{Ker}(A)$ and $\text{Ker}(\tilde{A})$. By the first part, $\text{rref}(A) = \text{rref}(\tilde{A})$. Consider the final columns of $\text{rref}([A|\mathbf{b}])$ and $\text{rref}([\tilde{A}|\tilde{\mathbf{b}}])$.

Let $\mathbf{c}$ denote the final column of $\text{rref}([A|\mathbf{b}])$. Define a vector $\mathbf{z}$ by setting $z_j = 0$ if the jth column is not pivotal, otherwise if the ith pivotal column is in the jth place; in other words, $j = j(i)$,

$$z_j = c_i .$$

Since the solution set is nonempty by hypothesis, the final column is nonpivotal. Then by (4.18) applied to the final column of $[A|\mathbf{b}]$,

$$A\mathbf{z} = c_1\mathbf{v}_{j(1)} + \cdots + c_r\mathbf{v}_{j(r)} = \mathbf{b} .$$

But since $\tilde{A}\mathbf{x} = \tilde{\mathbf{b}}$ is an equivalent system, $\mathbf{z}$ also solves $\tilde{A}\mathbf{x} = \tilde{\mathbf{b}}$:

$$\tilde{A}\mathbf{z} = c_1\mathbf{v}_{j(1)} + \cdots + c_r\mathbf{v}_{j(r)} = \tilde{\mathbf{b}} ,$$

which expresses $\tilde{\mathbf{b}}$ as a linear combination of the pivotal columns of $\tilde{A}$. By (4.18) again, $\mathbf{c}$ is also the final column of $\text{rref}([\tilde{A}|\tilde{\mathbf{b}}])$.   ∎

EXAMPLE 23   **(Checking for equivalence)**   Consider the systems of equations given by

$$[A|\mathbf{b}] = \begin{bmatrix} 1 & 2 & 3 & 2 & | & 9 \\ 2 & 4 & 0 & 3 & | & 6 \\ 2 & 3 & 1 & 3 & | & 9 \end{bmatrix} \quad \text{and} \quad [\tilde{A}|\tilde{\mathbf{b}}] = \begin{bmatrix} 1 & 4 & 1 & 2 & | & 3 \\ 4 & 7 & 1 & 6 & | & 15 \\ 6 & 10 & -4 & 8 & | & a \end{bmatrix} .$$

For which choices of the parameter value a in the second system, if any, are these two systems equivalent? To answer the question, we compute

$$\text{rref}([A|\mathbf{b}]) = \begin{bmatrix} 1 & 0 & 0 & 7/6 & | & 5 \\ 0 & 1 & 0 & 1/6 & | & -1 \\ 0 & 0 & 1 & 1/6 & | & 2 \end{bmatrix} ,$$

and

$$\text{rref}([A|\mathbf{b}]) = \begin{bmatrix} 1 & 0 & 0 & 7/6 & | & 23/4 - a/16 \\ 0 & 1 & 0 & 1/6 & | & a/16 - 7/4 \\ 0 & 0 & 1 & 1/6 & | & 17/4 - 3a/16 \end{bmatrix} .$$

You can easily check that the two reduced row echelon form matrices are the same if and only if $a = 12$. Hence, the systems are equivalent if and only if $a = -12$.

4.5 Lagrange polynomials and Vandermonde inverses

In Example 19, we computed the inverse of the Vandermonde matrix

$$A = \begin{bmatrix} 1 & 1 & 1 \\ 1 & 2 & 4 \\ 1 & 3 & 9 \end{bmatrix}$$

by row reduction, but there is a more profitable way to proceed.

Recall that a Vandermonde matrix relates the coefficients of a polynomial $P(x)$ to the values of $P(x)$ at the entries of the second column of the Vandermonde matrix. Hence, if we wish to find values of a_0, a_1, and a_2 so that with $P(x) = a_0 + a_1 x + a_2 x^2$

$$\begin{bmatrix} P(1) \\ P(2) \\ P(3) \end{bmatrix} = \begin{bmatrix} 1 \\ 3 \\ 1 \end{bmatrix}, \tag{4.20}$$

we need only to solve

$$\begin{bmatrix} 1 & 1 & 1 \\ 1 & 2 & 4 \\ 1 & 3 & 9 \end{bmatrix} \begin{bmatrix} a_0 \\ a_1 \\ a_2 \end{bmatrix} = \begin{bmatrix} 1 \\ 3 \\ 1 \end{bmatrix}.$$

Using the inverse of A computed in Example 20,

$$\begin{bmatrix} a_0 \\ a_1 \\ a_2 \end{bmatrix} = \frac{1}{2} \begin{bmatrix} 6 & -6 & 2 \\ -5 & 8 & -3 \\ 1 & -2 & 1 \end{bmatrix} \begin{bmatrix} 1 \\ 3 \\ 1 \end{bmatrix} = \begin{bmatrix} -5 \\ 8 \\ -2 \end{bmatrix}.$$

Hence, the desired polynomial is $P(x) = -5 + 8x - 2x^2$, which indeed satisfies (4.20).

We now reverse the line of argument. We will explain a *direct solution* of the polynomial fitting problem due to Lagrange and use it to invert the Vandermonde matrix.

Let t_1, t_2, t_3 be given. Let $P(x) = a_0 + a_1 x + a_2 x^2$ be the quadratic polynomial with

$$\begin{bmatrix} P(1) \\ P(2) \\ P(3) \end{bmatrix} = \begin{bmatrix} t_1 \\ t_2 \\ t_3 \end{bmatrix}.$$

Then from (4.11) and (4.12),

$$\begin{bmatrix} a_0 \\ a_1 \\ a_2 \end{bmatrix} = A^{-1} \begin{bmatrix} t_1 \\ t_2 \\ t_3 \end{bmatrix}. \tag{4.21}$$

Now if we choose $t_1 = 1$ and $t_2 = t_3 = 0$, then the right-hand side of (4.21) becomes $A^{-1}\mathbf{e}_1$, which by **V.I.F.1** is the first column of A^{-1}. Hence, the vector of coefficients of the quadratic polynomial $P_1(x)$ satisfying

$$\begin{bmatrix} P_1(1) \\ P_1(2) \\ P_1(3) \end{bmatrix} = \mathbf{e}_1 \tag{4.22}$$

is the first column of A^{-1}. Likewise, for $j = 2$ and $j = 3$, the vector of coefficients of the quadratic polynomial $P_j(x)$ satisfying

$$\begin{bmatrix} P_j(1) \\ P_j(2) \\ P_j(3) \end{bmatrix} = \mathbf{e}_j$$

is the jth column of A^{-1}. Once every column of A^{-1} is known, A^{-1} is known. Hence a direct method for writing down the polynomials P_1, P_2, and P_3 enables us to write down A^{-1}.

As Lagrange observed, it is actually quite easy to write down a quadratic polynomial that satisfies (4.22), and therefore it is easy to write down the first column of A^{-1}. Of course, what we can do for one column, we can do for the others. Hence Lagrange's observation makes it easy to write down A^{-1}.

Here is how this goes. To get a quadratic polynomial $P(x)$ such that $P(2) = P(3) = 0$, consider

$$(x - 2)(x - 3).$$

This certainly does the trick. If we divide by the value it takes at $x = 1$, we get a quadratic polynomial that satisfies (4.22):

$$P(x) = \frac{(x-2)}{(1-2)}\frac{(x-3)}{(1-3)} = \frac{x^2 - 5x + 6}{2}.$$

Notice that the right-hand side can be written

$$\frac{x^2 - 5x + 6}{2} = \begin{bmatrix} 3 \\ -5/2 \\ 1/2 \end{bmatrix} \cdot \begin{bmatrix} 1 \\ x \\ x^2 \end{bmatrix}$$

and that this equation corresponds to what we found in (4.13) in Example 19. Lagrange's observation gives us a very easy way to write down the columns of A^{-1}.

Lagrange's strategy for solving equations like (4.22) works for polynomials of any degree: Given $n + 1$ distinct numbers $\{s_1, s_2, \ldots, s_{n+1}\}$, define

$$P_1(x) = \frac{(x-s_2)}{(s_1-s_2)}\frac{(x-s_3)}{(s_1-s_3)} \cdots \frac{(x-s_{n+1})}{(s_1-s_{n+1})} = \prod_{j=2}^{n+1} \frac{(x-s_j)}{(s_1-s_j)}.$$

$P_1(x)$ is an nth-degree polynomial with

$$P_1(s_1) = 1 \qquad \text{and} \qquad P_1(s_j) = 0 \qquad \text{for } j \neq 1.$$

Likewise, for $i = 2, 3, \ldots, n + 1$, define

$$P_i(x) = \prod_{j=1, j \neq i}^{n+1} \frac{(x-s_j)}{(s_i-s_j)}.$$

$P_i(x)$ is an nth-degree polynomial with

$$P_1(s_i) = 1 \qquad \text{and} \qquad P_i(s_j) = 0 \qquad \text{for } j \neq i.$$

The polynomials $P_1(x), P_2(x), \ldots, P_{n+1}(x)$ are called the *Lagrange polynomials with nodes at* $s_1, s_2, \ldots, s_{n+1}$.

EXAMPLE 24   **(Computing Lagrange polynomials)** Let us take $n = 2$ and $\{s_1, s_2, s_3\} = \{1, 2, 3\}$. The Lagrange polynomials are

$$\begin{aligned} P_1(x) &= \frac{(x-2)}{(1-2)}\frac{(x-3)}{(1-3)} = \frac{x^2 - 5x + 6}{2} \\ P_2(x) &= \frac{(x-1)}{(2-1)}\frac{(x-3)}{(2-3)} = \frac{x^2 - 4x + 3}{-1} \\ P_3(x) &= \frac{(x-1)}{(3-1)}\frac{(x-2)}{(3-2)} = \frac{x^2 - 3x + 2}{2}. \end{aligned} \qquad (4.23)$$

Since, any linear combination of nth-degree polynomials is again an nth-degree polynomial, for any $n + 1$ numbers $\{t_1, t_2, \ldots, t_{n+1}\}$,

$$P(x) = t_1 P_1(x) + t_2 P_2(x) + \cdots + t_{n+1} P_{n+1}(x) \qquad (4.24)$$

is an nth degree polynomial. Moreover, if we evaluate it at s_j, only the jth term on the right is nonzero, and it is t_j. Thus, (4.24) defines a polynomial satisfying

$$P(s_j) = t_j \qquad \text{for } j = 1, 2, \ldots, n + 1.$$

EXAMPLE 25   **(Polynomial fitting using Lagrange polynomials)** To find the quadratic polynomial $P(x)$ satisfying (4.20) using Lagrange polynomials, we write

$$P(x) = 1P_1(x) + 3P_2(x) + 1P_3(x).$$

Using the explicit forms of P_1, P_2, and P_3 found in Example 21,

$$P(x) = \left(\frac{x^2 - 5x + 6}{2}\right) - 3(x^2 - 4x + 3) + \left(\frac{x^2 - 3x + 2}{2}\right)$$

$$= -2x^2 + 8x - 5.$$

This result is the same as the one we found before.

Taking linear combinations of Lagrange polynomials as in Example 22 is the practical way to solve a polynomial fitting problem; there is no need to explicitly invert a Vandermonde matrix. However, Vandermonde matrices do come up in many other contexts, and having an explicit solution to the polynomial fitting problem amounts to having an explicit formula for the inverse of any Vandermonde matrix A with distinct entries in the second column. Indeed, since

$$\begin{bmatrix} P_j(s_1) \\ P_j(s_2) \\ \vdots \\ P_j(s_{n+1}) \end{bmatrix} = \mathbf{e}_j,$$

(4.12) tells us that the vector of coefficients of $P_j(x)$ is the jth column of A^{-1}. In summary,

- If A is an $(n+1) \times (n+1)$ *Vandermonde matrix whose second column has distinct entries, and if the second column is*

$$\begin{bmatrix} s_1 \\ s_2 \\ \vdots \\ s_{n+1} \end{bmatrix},$$

  then the jth column of A^{-1} is the vector of coefficients of $P_j(x)$, where $P_j(x)$ is the jth Lagrange polynomial with nodes at $s_1, s_2, \ldots, s_{n+1}$.

Lagrange's direct approach to the polynomial fitting problem bypasses the application of the Corollary to Theorem 7 that we made in Example 17. However, it is often useful to know that the solution to a problem—such as the polynomial fitting problem—always exists. Knowing this, effort can confidently be invested in the search for a solution in any particular case. Another case in which the Corollary to Theorem 7 can be used to prove that solutions always exist is discussed in the exercises that follow. There are many others.

Exercises   **4.1** Let A and B be the following matrices:

$$A = \begin{bmatrix} 1 & 2 & 4 \\ 2 & 4 & 1 \\ 4 & 1 & 2 \end{bmatrix} \qquad B = \begin{bmatrix} 1 & 2 & 3 \\ 4 & 4 & 4 \\ 3 & 2 & 1 \end{bmatrix}.$$

(a) For each of A and B, compute the rank, and decide whether the matrix is invertible. If so, find the inverse.

(b) Find a vector $\mathbf{b}$ in $\mathbb{R}^3$ for which $A\mathbf{x} = \mathbf{b}$ has infinitely many solutions, or explain why the existence of such a vector is impossible.

(c) Find a vector $\mathbf{b}$ in $\mathbb{R}^3$ for which $B\mathbf{x} = \mathbf{b}$ has infinitely many solutions, or explain why the existence of such a vector is impossible.

4.2 Let A and B be the following matrices:

$$A = \begin{bmatrix} 1 & 1 & 3 \\ 1 & -1 & 1 \\ 1 & -2 & 0 \end{bmatrix} \qquad B = \begin{bmatrix} 1 & 1 & -1 \\ 1 & -1 & 2 \\ 1 & 2 & 0 \end{bmatrix}.$$

(a) For each of A and B, compute the rank, and decide whether the matrix is invertible. If so, find the inverse.

(b) Find a vector $\mathbf{b}$ in $\mathbb{R}^3$ for which $A\mathbf{x} = \mathbf{b}$ has infinitely many solutions, or explain why the existence of such a vector is impossible.

(c) Find a vector $\mathbf{b}$ in $\mathbb{R}^3$ for which $B\mathbf{x} = \mathbf{b}$ has infinitely many solutions, or explain why the existence of such a vector is impossible.

4.3 Compute the inverse of $B = \begin{bmatrix} 1 & 2 & 3 & 4 \\ 0 & 1 & 2 & 3 \\ 0 & 0 & 1 & 2 \\ 0 & 0 & 0 & 1 \end{bmatrix}.$

4.4 Compute the inverse of $B = \begin{bmatrix} 1 & 2 & -1 \\ 2 & -1 & 2 \\ 1 & 2 & 0 \end{bmatrix}.$

4.5 Let $A = \begin{bmatrix} 1 & 2 & 3 \\ 4 & 4 & 4 \\ 3 & 2 & 1 \end{bmatrix}$. Is there a vector $\mathbf{b}$ in $\mathbb{R}^3$ for which $A\mathbf{x} = \mathbf{b}$ has no solution? Either find such a vector $\mathbf{b}$, or explain why its existence is impossible.

4.6 Solving a system of linear equations, determine the second column of A^{-1} where $A = \begin{bmatrix} 1 & 2 & 0 \\ 0 & 4 & 1 \\ 0 & 3 & 5 \end{bmatrix}.$

4.7 Solving a system of linear equations, determine the third column of A^{-1} where $A = \begin{bmatrix} -1 & 0 & 1 \\ 0 & 1 & 1 \\ 2 & 1 & 5 \end{bmatrix}.$

4.8 Let C be a 2×2 matrix such that

$$C \begin{bmatrix} 1 \\ 2 \end{bmatrix} = \begin{bmatrix} 2 \\ 1 \end{bmatrix} \qquad \text{and} \qquad C^2 \begin{bmatrix} 1 \\ 2 \end{bmatrix} = \begin{bmatrix} -1 \\ 1 \end{bmatrix}.$$

What is C? (That is, give the entries.)

4.9 Let C be a 3×3 matrix such that

$$C \begin{bmatrix} 1 \\ 0 \\ 1 \end{bmatrix} = \begin{bmatrix} 2 \\ 1 \\ 1 \end{bmatrix}, \qquad C \begin{bmatrix} 2 \\ 1 \\ 3 \end{bmatrix} = \begin{bmatrix} -1 \\ 1 \\ -2 \end{bmatrix}, \qquad \text{and} \quad C \begin{bmatrix} 3 \\ 0 \\ 1 \end{bmatrix} = \begin{bmatrix} 0 \\ -1 \\ 1 \end{bmatrix}.$$

Find the matrix C.

4.10 Let $\{s_1, s_2, s_3, s_4\} = \{-1, 1, 0, 2\}$.

(a) Compute the Lagrange polynomials with nodes at s_1, s_2, s_3, s_4.

(b) Find the cubic polynomial $P(x)$ such that

$$P(-1) = 2, \qquad P(1) = -1, \qquad P(0) = 4, \quad \text{and} \quad P(2) = 3.$$

(c) Let A be the Vandermonde matrix $A = \begin{bmatrix} 1 & -1 & 1 & -1 \\ 1 & 1 & 1 & 1 \\ 1 & 0 & 0 & 0 \\ 1 & 2 & 4 & 8 \end{bmatrix}$ and compute A^{-1} using your answer to part (a).

4.11 Let $\{s_1, s_2, s_3, s_4\} = \{1, 2, 3, 4\}$.

(a) Compute the Lagrange polynomials with nodes at s_1, s_2, s_3, s_4.

(b) Find the cubic polynomial $P(x)$ such that

$$P(1) = 2, \qquad P(2) = -1, \qquad P(3) = 4, \quad \text{and} \quad P(4) = 3.$$

(c) Let A be the Vandermonde matrix $A = \begin{bmatrix} 1 & 1 & 1 & 1 \\ 1 & 2 & 4 & 8 \\ 1 & 3 & 9 & 27 \\ 1 & 4 & 16 & 64 \end{bmatrix}$ and compute A^{-1} using your answer to

part (a).

4.12 Let A and $\tilde{A}$ be the matrices

$$A = \begin{bmatrix} 1 & 2 & 3 & 1 \\ 2 & 3 & 5 & 1 \\ 3 & 1 & 4 & 3 \end{bmatrix} \quad \text{and} \quad \tilde{A} = \begin{bmatrix} 4 & 3 & 7 & 4 \\ -1 & 3 & 2 & -1 \\ 1 & 1 & 2 & 0 \end{bmatrix}.$$

(a) Compute $\text{rref}(A)$ and $\text{rref}(\tilde{A})$.

(b) Express the fourth column of A as a linear combination of the pivotal columns of A.

(c) Express the fourth column of $\tilde{A}$ as a linear combination of the pivotal columns of $\tilde{A}$.

(d) Compute a one-to-one parameterization of $\text{Ker}(A)$.

(e) Find a one-to-one parameterization of the solution set of $\tilde{A}\mathbf{x} = \mathbf{b}$ where $\mathbf{b} = \tilde{A}\mathbf{e}_3$.

4.13 Let A and $\tilde{A}$ be the matrices

$$A = \begin{bmatrix} 3 & 1 & 1 & 1 \\ 1 & 2 & -3 & 1 \\ 2 & 1 & 0 & 3 \end{bmatrix} \quad \text{and} \quad \tilde{A} = \begin{bmatrix} 5 & 2 & 1 & 4 \\ 4 & 1 & 2 & -1 \\ -2 & 1 & -4 & 0 \end{bmatrix}.$$

(a) Compute $\text{rref}(A)$ and $\text{rref}(\tilde{A})$.

(b) Express the fourth column of A as a linear combination of the pivotal columns of A.

(c) Express the fourth column of $\tilde{A}$ as a linear combination of the pivotal columns of $\tilde{A}$.

(d) Compute a one-to-one parameterization of $\text{Ker}(A)$.

(e) Find a one-to-one parameterization of the solution set of $\tilde{A}\mathbf{x} = \mathbf{b}$ where $\mathbf{b} = \tilde{A}\mathbf{e}_3$.

4.14 Consider the systems given by $[A|\mathbf{b}]$ and $[\tilde{A}|\tilde{\mathbf{b}}]$, where

$$[A|\mathbf{b}] = \begin{bmatrix} 1 & 2 & 3 & 1 & | & 3 \\ 2 & 3 & 5 & 1 & | & 2 \\ 3 & 1 & 4 & 3 & | & 4 \end{bmatrix} \quad \text{and} \quad [\tilde{A}|\tilde{\mathbf{b}}] = \begin{bmatrix} 4 & 3 & 7 & 4 & | & 7 \\ -1 & 3 & 2 & -1 & | & 2 \\ 1 & 1 & 2 & 0 & | & -1 \end{bmatrix}.$$

Are these equivalent systems?

4.15 Consider the systems given by $[A|\mathbf{b}]$ and $[\tilde{A}|\tilde{\mathbf{b}}]$, where

$$[A|\mathbf{b}] = \begin{bmatrix} 3 & 1 & 1 & 1 & | & 1 \\ 1 & 2 & -3 & 1 & | & 9 \\ 2 & 1 & 0 & 3 & | & 9 \end{bmatrix} \quad \text{and} \quad [\tilde{A}|\tilde{\mathbf{b}}] = \begin{bmatrix} 5 & 2 & 1 & 4 & | & 10 \\ 4 & 1 & 2 & -1 & | & -7 \\ -2 & 1 & -4 & 0 & | & 8 \end{bmatrix}.$$

Are these equivalent systems?

4.16 Consider the systems given by $[A|\mathbf{b}]$ and $[\tilde{A}|\tilde{\mathbf{b}}]$, where

$$[A|\mathbf{b}] = \begin{bmatrix} 1 & 2 & 3 & 1 & | & 3 \\ 2 & 3 & 5 & 1 & | & 2 \\ 3 & 1 & 4 & 3 & | & 4 \end{bmatrix} \quad \text{and} \quad [\tilde{A}|\tilde{\mathbf{b}}] = \begin{bmatrix} 4 & 3 & 7 & 4 & | & 5 \\ -1 & 3 & 2 & -1 & | & 2 \\ 1 & 1 & 2 & 0 & | & -3 \end{bmatrix}.$$

Are these equivalent systems?

4.17 Consider the systems given by $[A|\mathbf{b}]$ and $[\tilde{A}|\tilde{\mathbf{b}}]$, where

$$[A|\mathbf{b}] = \begin{bmatrix} 3 & 1 & 1 & 1 & | & 1 \\ 1 & 2 & -3 & 1 & | & 9 \\ 2 & 1 & 0 & 3 & | & 9 \end{bmatrix} \quad \text{and} \quad [\tilde{A}|\tilde{\mathbf{b}}] = \begin{bmatrix} 5 & 2 & 1 & 4 & | & 3 \\ 4 & 1 & 2 & -1 & | & -7 \\ -2 & 1 & -4 & 0 & | & 4 \end{bmatrix}.$$

Are these equivalent systems?

SECTION 5 | The *LU* Factorization

5.1 The virtue of keeping good records

Let A be an $m \times n$ matrix, and suppose that you need to solve $A\mathbf{x} = \mathbf{b}$ for many different vectors $\mathbf{b}$ in $\mathbb{R}^m$, a common situation in applications. If A happens to be an invertible square matrix, you could work out A^{-1} and then deal with each new $\mathbf{b}$ by multiplying out $A^{-1}\mathbf{b}$. But what if A is not invertible or not even square?

Applying the methods as they stand, we would have to row reduce $[A|\mathbf{b}]$, *starting from scratch*, for each different $\mathbf{b}$, which would quickly get repetitive. A good way forward lies in the following observation: *Since we row reduce from left to right, we can row reduce $[A|\mathbf{b}]$ using any sequence of row operations that would row reduce A itself, up until we get to the very last column.* So it behooves us to row reduce A first and keep a record of the operations we use to do it.

The key to doing so is to think of row operations as linear transformations. This makes sense because if we apply a row operation to a matrix A, it acts separately on each of the columns of A. For example, consider the operation of adding a times the ith row to the jth row. If $\mathbf{v}$ is any column of A, the operation results in

$$\mathbf{v} = \begin{bmatrix} v_1 \\ \vdots \\ v_i \\ \vdots \\ v_j \\ \vdots \\ v_m \end{bmatrix} \rightarrow \begin{bmatrix} v_1 \\ \vdots \\ v_i \\ \vdots \\ v_j + av_i \\ \vdots \\ v_m \end{bmatrix}.$$

Each entry of the output vector is a linear form in the entries of the input vector, so this transformation from $\mathbb{R}^m$ to $\mathbb{R}^m$ is a linear transformation. Let R be the corresponding matrix.

Since $I = [\mathbf{e}_1, \mathbf{e}_2, \ldots, \mathbf{e}_n]$ and $R = RI$, from **V.I.F.3**

$$R = RI = [R\mathbf{e}_1, R\mathbf{e}_2, \ldots, R\mathbf{e}_n].$$

That is, we can compute R by applying the corresponding operation to the columns of I.

For example, if the operation is to add twice the second row to the fourth row of a matrix with 5 rows, R would be given by

$$R = \begin{bmatrix} 1 & 0 & 0 & 0 & 0 \\ 0 & 1 & 0 & 0 & 0 \\ 0 & 0 & 1 & 0 & 0 \\ 0 & 2 & 0 & 1 & 0 \\ 0 & 0 & 0 & 0 & 1 \end{bmatrix} \tag{5.1}$$

since that is what we get when we apply this operation to $I_{5\times 5}$. Each of the row operations can be undone by another row operation of the same type, so all the matrices arising in this way are invertible. Indeed, to undo the operation just described, we would *subtract* the second row from the fourth twice. This corresponds to the matrix

$$\begin{bmatrix} 1 & 0 & 0 & 0 & 0 \\ 0 & 1 & 0 & 0 & 0 \\ 0 & 0 & 1 & 0 & 0 \\ 0 & -2 & 0 & 1 & 0 \\ 0 & 0 & 0 & 0 & 1 \end{bmatrix},$$

which we can confirm is the inverse of (5.1). We have focused here on row operations of the "adding a multiple of one row to another" type. But after a bit of thought, it will be clear that each row operation is represented by a matrix R in this way, R can be computed by applying the operation to the identity matrix, and R is always invertible since any row operation can be undone by a row operation of the same type.

Now let A be any $m \times n$ matrix, and consider the augmented matrix $[A|I]$, where I denotes the $m \times m$ identity matrix. We begin row reducing this matrix left to right. Let R_1 be the $m \times m$ matrix corresponding to the first row operation we use. Multiplying on the left by R_1 produces

$$R_1[A|I] = [R_1A|R_1I] = [R_1A|R_1]$$

since R_1 acts separately on each of the columns of $[A|I]$ and therefore separately on A and I. However, we do not *explicitly* multiply R_1 and $[A|I]$; we do the corresponding row-reduction operation on $[A|I]$. It is easier, and the result is the same.

Let R_2 denote the second row operation we use. Applying this operation to what we have so far, namely, $[R_1A|R_1]$, results in $[R_2R_1A|R_2R_1]$. If we keep going in this way, after some number of operations N we arrive at

$$[R_N \cdots R_2R_1A|R_N \cdots R_2R_1], \tag{5.2}$$

where $U = R_N \cdots R_2R_1A$ is in row-reduced form.* The matrix that records the row operations is

$$R = R_N \cdots R_2R_1. \tag{5.3}$$

With these definitions, $R[A|I] = [RA|RI] = [U|R]$ and

$$RA = U, \tag{5.4}$$

where U is in row-reduced form. Moreover, R is invertible. To see this, note that each R_j is invertible since any row operation can be undone by another row operation. A product of invertible matrices is invertible, and we have

$$R^{-1} = R_1^{-1}R_2^{-1} \cdots R_N^{-1}.$$

EXAMPLE 26   **(Finding *U* and *R*)** Let

$$A = \begin{bmatrix} 1 & 1 & 1 \\ 2 & 4 & 3 \\ 3 & 7 & 6 \end{bmatrix} \qquad \text{so that} \qquad [A|I] = \left[\begin{array}{ccc|ccc} 1 & 1 & 1 & 1 & 0 & 0 \\ 2 & 4 & 3 & 0 & 1 & 0 \\ 3 & 7 & 6 & 0 & 0 & 1 \end{array} \right].$$

Cleaning up the first column in two row operations leads to

$$\left[\begin{array}{ccc|ccc} 1 & 1 & 1 & 1 & 0 & 0 \\ 0 & 2 & 1 & -2 & 1 & 0 \\ 0 & 4 & 3 & -3 & 0 & 1 \end{array} \right].$$

---

*Since row-reduced matrices "live in the upper right corner," U is the traditional symbol to use in this context.

Cleaning up the second column in one more row operation finishes the job:

$$\left[\begin{array}{ccc|ccc} 1 & 1 & 1 & 1 & 0 & 0 \\ 0 & 2 & 1 & -2 & 1 & 0 \\ 0 & 0 & 1 & 1 & -2 & 1 \end{array}\right]. \tag{5.5}$$

Therefore,

$$U = \begin{bmatrix} 1 & 1 & 1 \\ 0 & 2 & 1 \\ 0 & 0 & 1 \end{bmatrix} \quad \text{and} \quad R = \begin{bmatrix} 1 & 0 & 0 \\ -2 & 1 & 0 \\ 1 & -2 & 1 \end{bmatrix}.$$

Here is the key to making use of the record matrix R: The fact that R is invertible means that with $\mathbf{c}$ defined by $\mathbf{c} = R\mathbf{b}$,

$$A\mathbf{x} = \mathbf{b} \quad \text{and} \quad U\mathbf{x} = \mathbf{c} \tag{5.6}$$

are equivalent systems of equations.

To see this, note that if $\mathbf{x}_0$ is any solution of $A\mathbf{x}_0 = \mathbf{b}$, then multiplying both sides by R, from (5.4) $U\mathbf{x}_0 = RA\mathbf{x}_0 = R\mathbf{b} = \mathbf{c}$, so $\mathbf{x}_0$ is also a solution of $U\mathbf{x} = \mathbf{c}$. On the other hand, if $\mathbf{x}_0$ is any solution of $U\mathbf{x}_0 = R\mathbf{b}$, then multiplying both sides by R^{-1}, $A\mathbf{x}_0 = R^{-1}U\mathbf{x}_0 = \mathbf{b}$, so $\mathbf{x}_0$ is also a solution of $A\mathbf{x} = \mathbf{b}$. Hence, the two equations in (5.6) have the same solution sets, leading to

- **The "A first and b later" procedure for solving $A\mathbf{x} = \mathbf{b}$.** *Let A be any $m \times n$ matrix, and write down $[A|I]$, where I is the $m \times m$ identity matrix. Apply row operations to transform this matrix into $[U|R]$, where U is in row-reduced form. Then for any $\mathbf{b}$ in $\mathbb{R}^m$, $A\mathbf{x} = \mathbf{b}$ is equivalent to $U\mathbf{x} = \mathbf{c}$, where $\mathbf{c} = R\mathbf{b}$. Since U is in row-reduced form, the latter equation is easily dealt with by back substitution.*

EXAMPLE 27   **(Using R and U to solve $A\mathbf{x} = \mathbf{b}$)** Let A be the 3×3 matrix considered in Example 26, for which we have already computed U and R. Let

$$\mathbf{b} = \begin{bmatrix} 1 \\ 1 \\ 1 \end{bmatrix}.$$

Then $A\mathbf{x} = \mathbf{b}$ is equivalent to $U\mathbf{x} = \mathbf{c}$, where

$$\mathbf{c} = R\mathbf{b} = \begin{bmatrix} 1 & 0 & 0 \\ -2 & 1 & 0 \\ 1 & -2 & 1 \end{bmatrix} \begin{bmatrix} 1 \\ 1 \\ 1 \end{bmatrix} = \begin{bmatrix} 1 \\ -1 \\ 0 \end{bmatrix}. \tag{5.7}$$

Next, $U\mathbf{x} = \mathbf{c}$ corresponds to the system of linear equations

$$\begin{aligned} x_1 + x_2 + x_3 &= 1 \\ 2x_2 + x_3 &= -1 \\ x_3 &= 0. \end{aligned} \tag{5.8}$$

Back substitution gives $x_3 = 0$, $x_2 = -1/2$, and $x_1 = 3/2$. The unique solution is

$$\mathbf{x} = \frac{1}{2}\begin{bmatrix} 3 \\ -1 \\ 0 \end{bmatrix}.$$

The procedure for finding R is much like the method for computing inverses by row reduction: We write the augmented matrix $[A|I]$ and apply row operations. The difference is that now A need not be square, and there is *even less work to do*, since we stop as soon as we get a row-reduced matrix U in the first n columns instead of continuing until we get the $n \times n$ identity matrix in the first n columns.

Just for comparison, let us compute A^{-1} for the matrix A in Examples 26 and 27. If you keep on going from (5.5), in three more row operations you will find that

$$A^{-1} = \frac{1}{2} \begin{bmatrix} 3 & 1 & -1 \\ -3 & 3 & -1 \\ 2 & -4 & 2 \end{bmatrix}.$$

Not only does it take more work to compute A^{-1} than to compute U and R, it takes more arithmetic operations to multiply out $A^{-1}\mathbf{b}$ than it does to *both* multiply out $R\mathbf{b}$ *and* solve $U\mathbf{x} = \mathbf{c}$.

5.2 What else the record tells us

- *The record R also provides us with a system of equations that $\mathbf{b}$ must satisfy for $A\mathbf{x} = \mathbf{b}$ to be solvable. In fact, $A\mathbf{x} = \mathbf{b}$ will have solutions if and only if these equations are satisfied.*

Let A be any $m \times n$ matrix with rank r. Suppose that $r < m$ so that $A\mathbf{x} = \mathbf{b}$ is not solvable for all choices of $\mathbf{b}$.

Suppose we have found an invertible $m \times m$ matrix R so that $U = RA$ is in row-reduced form. As we have seen, $A\mathbf{x} = \mathbf{b}$ is then equivalent to $U\mathbf{x} = R\mathbf{b}$, and $U\mathbf{x} = R\mathbf{b}$ is solvable exactly when there is no pivot in the final column of $[U|R\mathbf{b}]$.

Recall that the rank of A is the number of pivotal columns of A, which is also the number of nonzero rows in any row reduction of A. In particular, the bottom $m - r$ rows of U are all zeros. In this case, there is a pivot in the final column of $[U|R\mathbf{b}]$ exactly when any of the bottom $m - r$ entries of $R\mathbf{b}$ are nonzero. We now have a criterion for the solvability of $A\mathbf{x} = \mathbf{b}$:

- *Let A be any $m \times n$ matrix with rank $r < m$. Let R be an invertible $m \times m$ matrix so that $U = RA$ is in row-reduced form. Then, given $\mathbf{b}$ in $\mathbb{R}^m$, $A\mathbf{x} = \mathbf{b}$ is solvable if and only if the bottom $m - r$ entries of $R\mathbf{b}$ are zero.*

There is a more convenient way to express this. If we write R in terms of its rows,

$$R = \begin{bmatrix} \mathbf{r}_1 \\ \mathbf{r}_2 \\ \vdots \\ \mathbf{r}_m \end{bmatrix}; \qquad \text{then} \qquad R\mathbf{b} = \begin{bmatrix} \mathbf{r}_1 \cdot \mathbf{b} \\ \mathbf{r}_2 \cdot \mathbf{b} \\ \vdots \\ \mathbf{r}_m \cdot \mathbf{b} \end{bmatrix}.$$

Hence, the bottom $m - r$ entries of $R\mathbf{b}$ are all zero if and only if

$$\begin{bmatrix} \mathbf{r}_{r+1} \cdot \mathbf{b} \\ \mathbf{r}_{r+2} \cdot \mathbf{b} \\ \vdots \\ \mathbf{r}_m \cdot \mathbf{b} \end{bmatrix} = \begin{bmatrix} 0 \\ 0 \\ \vdots \\ 0 \end{bmatrix}.$$

If we let C denote the $(m - r) \times m$ matrix

$$C = \begin{bmatrix} \mathbf{r}_{r+1} \\ \mathbf{r}_{r+2} \\ \vdots \\ \mathbf{r}_m \end{bmatrix},$$

this last equation is the same as $C\mathbf{b} = 0$.

Notice that C consists exactly of the bottom $m - r$ rows of R, and we have just shown that $A\mathbf{x} = \mathbf{b}$ is solvable if and only if $C\mathbf{b} = 0$, that is, if $\mathbf{b}$ belongs to Ker(C).

EXAMPLE 28 **(Checking for solvability using the record matrix R)** Let

$$A = \begin{bmatrix} 1 & 2 \\ 2 & -1 \\ 0 & 1 \\ 1 & -3 \end{bmatrix}.$$

Row reducing $[A|I]$ to the form $[U|R]$, we find

$$U = \begin{bmatrix} 1 & 2 \\ 0 & -5 \\ 0 & 0 \\ 0 & 0 \end{bmatrix} \quad \text{and} \quad R = \frac{1}{5}\begin{bmatrix} 5 & 0 & 0 & 0 \\ -10 & 5 & 0 & 0 \\ -2 & 1 & 5 & 0 \\ 5 & -5 & 0 & 5 \end{bmatrix}.$$

We see that the rank of A is 2 since there are 2 pivots. Hence, since $4 - 2 = 2$, C consists of the bottom two rows of R so that

$$C = \begin{bmatrix} -2 & 1 & 5 & 0 \\ 5 & -5 & 0 & 5 \end{bmatrix}.$$

$Ax = b$ will be solvable if and only if $Cb = 0$.

Consider the vectors

$$\mathbf{b}_1 = \begin{bmatrix} 1 \\ 2 \\ 1 \\ 2 \end{bmatrix}, \quad \mathbf{b}_2 = \begin{bmatrix} 2 \\ -1 \\ 1 \\ -1 \end{bmatrix}, \quad \text{and } \mathbf{b}_3 = \begin{bmatrix} 1 \\ 1 \\ 1 \\ 1 \end{bmatrix}.$$

Let us apply our criterion for solvability. We find that

$$C\mathbf{b}_1 = \begin{bmatrix} 5 \\ 5 \end{bmatrix}, \quad C\mathbf{b}_2 = \begin{bmatrix} 0 \\ 0 \end{bmatrix}, \quad \text{and } C\mathbf{b}_3 = \begin{bmatrix} 4 \\ 5 \end{bmatrix}.$$

Hence $Ax = \mathbf{b}_2$ is solvable, while $Ax = \mathbf{b}_1$ and $Ax = \mathbf{b}_3$ are not.

In the following theorem we summarize what we have learned so far.

THEOREM 10 **(RA = U)** *Let A be any $m \times n$ matrix. Then there is an invertible $m \times m$ matrix R so that $U = RA$ is in row-reduced form. The matrices R and U can be found by row reducing $[A|I]$ to $[U|R]$. Furthermore,*

1. *For any vector $\mathbf{b}$ in $\mathbb{R}^m$, let $\mathbf{c} = R\mathbf{b}$. Then $Ax = \mathbf{b}$ is equivalent to $Ux = \mathbf{c}$.*

2. *Suppose that the rank of A is r, and $r < m$. Let C be the $(m - r) \times m$ matrix consisting of the bottom $m - r$ rows of R. Then $Ax = \mathbf{b}$ is solvable if and only if $C\mathbf{b} = 0$. In other words, $Ax = \mathbf{b}$ is solvable if and only if $\mathbf{b}$ belongs to $Ker(C)$.*

5.3 Factorizing A

So far in this section we have explained all the essential points. However, if you look into other texts or use standard computer programs for linear algebra, you will find that they employ a more intricate form of recordkeeping.

There is a reason they do. Up to this point, when we have solved systems of linear equations $Ax = \mathbf{b}$ by row reduction and back substitution, we have not made much use of the fact that A represents a linear transformation f_A. Writing $Ax = \mathbf{b}$ as $f_A(x) = \mathbf{b}$ suggests a different way of thinking about the system. It suggests the possibility of *factoring* f_A into the composition product of two or more simple transformations.

To see why doing so helps, consider the following equation built out of familiar functions of one variable:

$$e^{x^2} = 2. \tag{5.9}$$

We solve the equation by first taking the natural logarithm of both sides, getting $x^2 = \ln(2)$. We then take the square root of both sides, getting $x = \pm\sqrt{\ln(2)}$. The point is that the function $f(x) = e^{x^2}$ can be written

$$f = g \circ h, \qquad \text{where } g(y) = e^y \text{ and } h(x) = x^2.$$

The functions g and h are simple, familiar functions, and their inverses are included on any scientific calculator, so it is easy to deal with them one at a time. This "divide and conquer" strategy of factoring into easy pieces is so natural, you would probably do it to solve (5.9) without explicitly thinking about it as involving a factorization.

It pays to think about it in the context of matrix equations (and elsewhere) because it leads to divide-and-conquer strategies for solving them.

For example, we know that the record matrix R in $RA = U$ is invertible. Let $L = R^{-1}$. Then $A = LU$ is a factorization of A. This will be useful if there is something *simple* about both L and U. We already know that U is in row-reduced form, and there is definitely something simple about that: For any $\mathbf{c}$, $U\mathbf{x} = \mathbf{c}$ can be solved by back substitution.

What is simple about L? Very often, we do not need to do any row swaps at all to row reduce a matrix. In Example 26, for instance, all we did was subtract multiples of rows from other rows beneath them. For this reason, R has a very special form:

$$R = \begin{bmatrix} 1 & 0 & 0 \\ -2 & 1 & 0 \\ 1 & -2 & 1 \end{bmatrix}.$$

Every entry on the main diagonal is 1, and every entry above the main diagonal is 0. This matrix is an example of a *unit lower-triangular matrix*.

DEFINITION

(**Unit lower-triangular matrices**) An $n \times n$ matrix L is called *lower-triangular* if and only if $L_{i,j} = 0$ for $j > i$. L is called *unit lower-triangular* because it is lower triangular, and moreover, $L_{i,i} = 1$ for $i = 1, 2, \ldots, n$.

Unit lower-triangular matrices look like

$$\begin{bmatrix} 1 & 0 & 0 & 0 & 0 \\ * & 1 & 0 & 0 & 0 \\ * & * & 1 & 0 & 0 \\ * & * & * & 1 & 0 \\ * & * & * & * & 1 \end{bmatrix},$$

where, as usual, asterisks denote entries that could be nonzero.

The matrix R, which represents the operation of adding a multiple a of the ith row to the jth row, will always be unit lower-triangular for $j > i$: As explained at the beginning of this section, it is the identity except for an a in the j, ith place.

The next key fact is this:

- *The product of two unit lower-triangular matrices is also unit lower-triangular.*

Let us accept this for the moment, and turn to the consequences: Whenever A can be row reduced without swapping rows, the record matrix $R = R_N \ldots R_2 R_1$ is unit lower-triangular. The inverse of each R_j is also a matrix representing adding a multiple—with the opposite sign—of one row to another below it, so $L = R_1^{-1}R_2^{-1}\cdots R_N^{-1}$ is also unit lower-triangular. That is,

- *Whenever A can be row reduced without swapping rows, the result is a factorization $A = LU$, where L is unit lower-triangular and U is row reduced. Such a factorization of A is called an LU factorization.*

LU factorizations are useful because when L is unit lower-triangular, the equation $Ly = b$ can be solved very easily by *forward substitution*, as in the next example.

EXAMPLE 29   **(Solving $Ly = b$, where L is unit lower-triangular)**   Consider the matrices

$$L = \begin{bmatrix} 1 & 0 & 0 \\ 2 & 1 & 0 \\ 6 & 2 & 1 \end{bmatrix} \quad \text{and} \quad b = \begin{bmatrix} 1 \\ 2 \\ 3 \end{bmatrix}.$$

Then $Ly = b$ is equivalent to the system of equations

$$\begin{aligned} y_1 &= 1 \\ 2y_1 + y_2 &= 2 \\ 6y_1 + 2y_2 + y_3 &= 3. \end{aligned}$$

The first equation says that $y_1 = 1$. Substituting $y_1 = 1$ into the second equation, it says $2 + y_2 = 2$, or $y_2 = 0$. Substituting $y_1 = 1$ and $y_2 = 0$ into the third equation, it says $y_3 = -3$. Therefore, the unique solution is

$$y = \begin{bmatrix} 1 \\ 0 \\ -3 \end{bmatrix}.$$

In solving $Ly = c$ when L is unit lower-triangular, we can always determine the entries of the solution y by working our way down the rows: The top row tells us the value of y_1, then the second row tells us the value of y_2, and so on, as in the example.

Since we substitute the value for y_1 into the equation for y_2 and then the values of y_1 and y_2 into the equation for y_3 and so on, the process is called *forward substitution*.

This gives us a simple variant of the "A first, b later" approach: First, find an LU factorization of A. Then, to solve $LUx = b$, first solve $Ly = b$ by forward substitution, and then solve $Ux = y$, which is also easy since U is row reduced.

EXAMPLE 30   **(Using an LU factorization to solve $Ax = b$)**   Let

$$A = \begin{bmatrix} 1 & 1 & 1 \\ 2 & 4 & 3 \\ 3 & 7 & 6 \end{bmatrix}.$$

In Example 26 we found R and U. It is easy to row reduce $[R|I]$ to find $L = R^{-1}$; the result is that $A = LU$, where

$$L = \begin{bmatrix} 1 & 0 & 0 \\ 2 & 1 & 0 \\ 6 & 2 & 1 \end{bmatrix} \quad \text{and} \quad U = \begin{bmatrix} 1 & 1 & 1 \\ 0 & 2 & 1 \\ 0 & 0 & 1 \end{bmatrix}.$$

Suppose, say,

$$b = \begin{bmatrix} 1 \\ 2 \\ 3 \end{bmatrix}.$$

To find the solutions of $Ax = b$, we first solve $Ly = b$ and then $Ux = y$. As in Example 29, we use forward substitution to find that $Ly = b$ has the unique solution

$$y = \begin{bmatrix} 1 \\ 0 \\ -3 \end{bmatrix}.$$

Then $U\mathbf{x} = \mathbf{y}$ is equivalent to the system of equations

$$\begin{aligned}
x_1 + x_2 + x_3 &= 1 \\
2x_2 + x_3 &= 0 \\
x_3 &= -3.
\end{aligned}$$

By back substitution we learn that $x_3 = -3$, then $x_2 = 3/2$, and then $x_1 = 3/2$. Hence, the unique solution is

$$\mathbf{x} = \frac{1}{2}\begin{bmatrix} -6 \\ 3 \\ 3 \end{bmatrix}.$$

This version of the "*A* first, **b** later" method is particularly convenient, since computer programs for working with matrices will produce an *LU* factorization. It is not hard on a computer to arrange the recordkeeping so that the *L* instead of the *R* matrix is directly produced, though for pencil-and-paper work, it is easiest to find *R* first, as explained here.

Finally, here is the postponed demonstration that the product of lower-triangular matrices is lower triangular. Let *K* and *L* be two unit lower-triangular matrices. Let **r** be the *i*th row of *K* and **c** be the *j*th column of *L*. By **V.I.F.4**,

$$(KL)_{i,j} = \mathbf{r} \cdot \mathbf{c}.$$

However, each entry of **r** after the *i*th entry is zero, and each entry of **c** before the *j*th entry is zero. Hence, if $j > i$, there is no overlap in the nonzero entries of **r** and **c**, so $(KL)_{i,j} = \mathbf{r} \cdot \mathbf{c} = 0$. If $i = j$, the only overlap is in the *i*th place, where both entries are 1, so $(KL)_{i,i} = \mathbf{r} \cdot \mathbf{c} = 1$. Thus, *KL* is unit lower-triangular.

5.4 A more detailed record

R will only be unit lower-triangular when *A* can be row reduced without swapping rows. However, it is not always possible to avoid row swaps. Consider, for example,

$$A = \begin{bmatrix} 0 & 1 \\ 1 & 0 \end{bmatrix}. \tag{5.10}$$

As you can see, this matrix can be row reduced just by swapping the two rows; there is no way to proceed without swapping the two rows. What kind of record matrix do we get if we *only* do row swaps?

DEFINITION | **(Permutation matrices)** An $n \times n$ matrix *P* is a *permutation matrix* if and only if *P* can be obtained from $I_{n \times n}$ by simply rearranging the order of the rows.*

From what was said at the beginning of this chapter, *PA* is what you get by rearranging the rows of *A* in the same way that the rows of *I* are ordered in *P*. Moreover, since the rows of *P* are just a reordering of $\{\mathbf{e}_1, \mathbf{e}_2, \ldots, \mathbf{e}_n\}$, the rows of *P* and hence the columns of P^t are orthonormal. Hence, *P* is an orthogonal matrix, and $P^t = P^{-1}$.

EXAMPLE 31 | **(A permutation matrix)** The matrix

$$P = \begin{bmatrix} 0 & 1 & 0 \\ 0 & 0 & 1 \\ 1 & 0 & 0 \end{bmatrix}$$

---

*By convention, we include leaving the original order alone as a trivial sort of rearrangement, so that the identity matrix itself is a permutation matrix.

is a permutation matrix. If A is a 3×3 matrix with

$$A = \begin{bmatrix} \mathbf{a}_1 \\ \mathbf{a}_2 \\ \mathbf{a}_3 \end{bmatrix}, \quad \text{then} \quad PA = \begin{bmatrix} \mathbf{a}_2 \\ \mathbf{a}_3 \\ \mathbf{a}_1 \end{bmatrix}.$$

To see why, note that

$$P \begin{bmatrix} x_1 \\ x_2 \\ x_3 \end{bmatrix} = \begin{bmatrix} x_2 \\ x_3 \\ x_1 \end{bmatrix},$$

and then the formula for PA follows from **V.I.F.3**.

Like unit lower-triangular matrices, permutation matrices are nice and simple. For example, we have seen that every permutation matrix is invertible, and this inverse is just the transpose.

The general approach to recordkeeping during row reductions is to keep a separate record of the row swaps in a permutation matrix P and of the other row operations in a unit lower-triangular matrix L. Here is the simplest way to go about it. Suppose we are row reducing $[A|I]$ to find R and U, and at some stage we need to swap two rows. Let P be the permutation matrix that does this. We form the matrix PA, keep a record of P as well, and start over.[*] If we can now row reduce $[PA|I]$ to the form $[U|R]$ without swapping rows, R will be unit lower-triangular, and

$$R(PA) = U,$$

with R unit lower-triangular. Letting $L = R^{-1}$, as usual,

$$PA = LU,$$

where P is a permutation matrix, L is unit lower-triangular, and U is in row-reduced form.

If we are so unlucky as to need to swap rows again in the row reduction of $[PA|I]$, let P_1 be the first pair permutation we used—keep records!—and let P_2 be the second. Let P be their product, which does both row swaps. We start over with $[PA|I]$. Every time we swap rows, we get at least one more row in the right place, so the worst possible case is that we would have to do this m times for an $m \times n$ matrix A. We have arrived at the following conclusion.

THEOREM 11   **(LU factorization)** *Every $m \times n$ matrix A can be written in the form*

$$PA = LU,$$

where P is an $m \times m$ permutation matrix, L is a unit lower-triangular matrix, and U is in row-reduced form.

If A is written in this form, it is easy to solve $A\mathbf{x} = \mathbf{b}$: Multiply on the left by P to obtain $PA\mathbf{x} = P\mathbf{b}$, then let $\mathbf{c} = P\mathbf{b}$ so that $A\mathbf{x} = \mathbf{b}$ becomes $LU\mathbf{x} = \mathbf{c}$. Solve this result in two steps, as in Example 30. This is the ultimate form of the "A first, $\mathbf{b}$ later" method, and it is very useful in practice since computers work out the $PA = LU$ decomposition of any matrix A.

---

[*]By being clever about it, we can avoid starting over from scratch, but let us save some clever ideas for later.

In $PA = LU$, both P and L are "record matrices." P is a record of row-swap operations, and L is a record of row operations in which a multiple of one row is subtracted from a row beneath it. Keeping the records separate gives L and P their special form, though it makes the recordkeeping itself a bit more complicated. But since L is invertible, $PA = LU$ is the same as $(L^{-1}P)A = U$, and with $R = L^{-1}P$, $RA = U$. The matrix R is less pretty than L and P, but the combined record matrix of Theorem 10 is still quite effective for solving $Ax = \mathbf{b}$, as we have seen.

For our purposes in the rest of the book, Theorem 10 suffices. To work with computer programs for linear algebra, however, you need to at least understand Theorem 11 and how it relates to Theorem 10.

5.5 Positive definite matrices and Cholesky factorization

Perhaps you are familiar with the *second derivative test* in differential calculus: Consider a twice-differentiable function g of a single variable x. With primes denoting derivatives, if $g'(x_0) = 0$ and $g''(x_0) > 0$, then x_0 is a *local minimum* of g.

Minimization problems arise everywhere in science and engineering, sometimes in a single-variable setting but more often involving several variables. What is the analog of the second derivative test in several variables?

If f is a differentiable function on $\mathbb{R}^n$, then the analog of the second derivative of f is a symmetric matrix called the *Hessian of f*. (The indices of the matrix correspond to all possible pairs of variables.) It turns out that this matrix is always symmetric and that the analog of the second derivative test in several variables is based on the following notion of *positivity of an $n \times n$ symmetric matrix A*.

DEFINITION

(Positive definite) An $n \times n$ matrix A that is symmetric, i.e., $A = A^t$, is *positive definite* if

$$\mathbf{x} \neq 0 \quad \Rightarrow \quad \mathbf{x} \cdot A\mathbf{x} > 0 . \tag{5.11}$$

It is *positive semidefinite* if $\mathbf{x} \cdot A\mathbf{x} \geq 0$ for all $\mathbf{x}$.

The preamble about minimization problems in several variables has been intended only to show that the definition we have just made is well motivated and that there are good and practical reasons for trying to decide whether a given symmetric matrix is positive definite—or not. We will not be computing Hessian matrices because that is not a part of linear algebra. Instead, let us ask: *How can we decide whether a given symmetric matrix is positive definite?* Some cases are easy.

EXAMPLE 32　**(Some postive definite matrices)** First of all, the identity matrix is positive definite. Note that if $\mathbf{x} \neq 0$, then

$$\mathbf{x} \cdot I\mathbf{x} = \mathbf{x} \cdot \mathbf{x} = |\mathbf{x}|^2 > 0 .$$

For a more interesting example, consider $A = \begin{bmatrix} 1 & 2 \\ 2 & 5 \end{bmatrix}$. Is A positive definite? Let $\mathbf{x} = \begin{bmatrix} x \\ y \end{bmatrix}$. Then

$$\mathbf{x} \cdot A\mathbf{x} = \begin{bmatrix} x \\ y \end{bmatrix} \cdot \begin{bmatrix} x + 2y \\ 2x + 5y \end{bmatrix} = x^2 + 4xy + 5y^2 .$$

You might notice that

$$x^2 + 4xy + 5y^2 = (x + 2y)^2 + y^2 .$$

This sum of squares is never negative and is zero if and only if both $y = 0$ and $x = 0$. Therefore $\mathbf{x} \cdot A\mathbf{x} > 0$ whenever $\mathbf{x} \neq 0$, so A is positive definite.

For larger matrices, finding the right way to "complete the squares", as in the last example, is not so easy to do just by inspection. There is a better way to proceed.

EXAMPLE 33 $\quad$ **(Postive definite matrices and factorization)** Let $A = \begin{bmatrix} 1 & 2 \\ 2 & 5 \end{bmatrix}$ as before. Let $L = \begin{bmatrix} 1 & 0 \\ 2 & 1 \end{bmatrix}$. Although it is not clear at this point where L came from, we can easily check that

$$A = LL^t .$$

Once we know this, we can see that A is, in fact, positive definite. Here is how: Consider any nonzero vector $\mathbf{x}$ in $\mathbb{R}^2$. Then, by **V.I.F.5**,

$$\mathbf{x} \cdot A\mathbf{x} = \mathbf{x} \cdot LL^t\mathbf{x} = L^t\mathbf{x} \cdot L^t\mathbf{x} = |L^t\mathbf{x}|^2 \geq 0 .$$

Moreover, if $\mathbf{x} \cdot A\mathbf{x} = 0$, then $L^t\mathbf{x} = 0$, but since L^t is invertible, then $\mathbf{x} = 0$. Therefore, $\mathbf{x} \cdot A\mathbf{x} > 0$ unless $\mathbf{x} = 0$, which means that A is positive definite.

In Example 33, our demonstration that $A = \begin{bmatrix} 1 & 2 \\ 2 & 5 \end{bmatrix}$ is positive definite depended only on the fact that A had a factorization $A = LL^t$ with L invertible, and the argument applies to any matrix with such a factorization:

- *If an $n \times n$ matrix can be factored as $A = LL^t$ where L is invertible, then A is positive definite.*

This fact is useful because it turns out that whenever A is a positive definite matrix, then it has such a factorization, and better yet, we can find this factorization using the methods of this section—with one simple twist.

To see the main idea, consider a general 4×4 positive definite matrix A,

$$A = \begin{bmatrix} \bullet & * & * & * \\ * & \bullet & * & * \\ * & * & \bullet & * \\ * & * & * & \bullet \end{bmatrix} .$$

There are bullets in the diagonal entries because they are nonzero. In fact, by Theorem 9 of Chapter 1 and the fact that A is positive definite,

$$A_{i,i} = \mathbf{e}_i \cdot A\mathbf{e}_i > 0 . \tag{5.12}$$

This conclusion will be useful later on, so let us emphasize it:

- *If an $n \times n$ matrix is positive definite, then all its diagonal entries are strictly positive.*

Let us clean out the first column of A using row operations. Since the upper left entry is not zero, we do not need to do any swapping, and we simply subtract a multiple of the first row from each of the rows beneath it. Let R_1 be the record matrix so that

$$R_1A = \begin{bmatrix} \bullet & * & * & * \\ 0 & \bullet & * & * \\ 0 & * & \bullet & * \\ 0 & * & * & \bullet \end{bmatrix} .$$

Consider the matrix $(R_1A)^t$:

$$(R_1A)^t = \begin{bmatrix} \bullet & 0 & 0 & 0 \\ * & \bullet & * & * \\ * & * & \bullet & * \\ * & * & * & \bullet \end{bmatrix} .$$

The first column of $(R_1A)^t$ is the first row of R_1A. Since we did not change the first row of A in our row operations, the first row of R_1A is also the first row of A, which, since A is symmetric, is also the first column of A.

- *In conclusion, the first column of $(R_1A)^t$ equals the first column of A. Therefore, we can clean it out using exactly the same sequence of row operations. That is,*

$$R_1(AR_1)^t = \begin{bmatrix} \bullet & 0 & 0 & 0 \\ 0 & \bullet & * & * \\ 0 & * & \bullet & * \\ 0 & * & * & \bullet \end{bmatrix}.$$

Taking the transpose of this,

$$R_1AR_1^t = \begin{bmatrix} \bullet & 0 & 0 & 0 \\ 0 & \bullet & * & * \\ 0 & * & \bullet & * \\ 0 & * & * & \bullet \end{bmatrix}.$$

Now let R_2 be the record matrix of the sequence of row operations used to clean out the second column. Then

$$R_2(R_1AR_1^t) = \begin{bmatrix} \bullet & 0 & 0 & 0 \\ 0 & \bullet & * & * \\ 0 & 0 & \bullet & * \\ 0 & 0 & * & \bullet \end{bmatrix}.$$

The same sort of analysis we made just before shows that if we multiply by R_2^t on the right, we clean out the second row as well:

$$R_2(R_1AR_1^t)R_2^t = \begin{bmatrix} \bullet & 0 & 0 & 0 \\ 0 & \bullet & 0 & 0 \\ 0 & 0 & \bullet & * \\ 0 & 0 & * & \bullet \end{bmatrix}.$$

Continuing in this way,

$$R_3(R_2(R_1AR_1^t)R_2^t)R_3^t = \begin{bmatrix} \bullet & 0 & 0 & 0 \\ 0 & \bullet & 0 & 0 \\ 0 & 0 & \bullet & 0 \\ 0 & 0 & 0 & \bullet \end{bmatrix}. \tag{5.13}$$

Now define
$$R = R_3R_2R_1.$$

Note that $R^t = R_1^t R_2^t R_3^t$, so
$$R_3(R_2(R_1AR_1^t)R_2^t)R_3^t = RAR^t.$$

Since R is a product of invertible matrices, R is invertible.

Next we claim that each of the diagonal entries on the right in (5.13) is strictly positive. To see this, note that as in (5.12),

$$(RAR^t)_{i,i} = \mathbf{e}_i \cdot (RAR^t)\mathbf{e}_i = (R^t\mathbf{e}_i) \cdot A(R^t\mathbf{e}_i) > 0$$

since R is invertible, and A is positive definite.

Therefore, if we let D be the diagonal matrix whose entries are the square roots of the corresponding diagonal entries on the right in (5.13),

$$RAR^t = D^2. \tag{5.14}$$

Since R is invertible and D is diagonal, $(R^{-1}D)^t = D^t(R^{-1})^t = D(R^{-1})^t$, and we can define

$$L = R^{-1}D \tag{5.15}$$

and rewrite (5.14) as $A = R^{-1}D^2(R^t)^{-1}$, or

$$A = LL^t.$$

Notice that L is invertible since it is the product of two invertible matrices. Better yet, since it is the product of a unit lower-triangular matrix and a diagonal matrix, L is a lower-triangular matrix whose diagonal entries are the strictly positive diagonal entries of D.

What we have explained here in the general 4×4 case can be applied to any $n \times n$ positive definite matrix.

DEFINITION

(Cholesky factorization) A factorization of a positive definite matrix A of the form

$$A = LL^t,$$

where L is a lower-triangular matrix with strictly positive diagonal entries, is called a *Cholesky factorization*.

The following theorem summarizes our analysis so far.

THEOREM 12

(Cholesky factorization of positive definite matrices) *Let A be an $n \times n$ symmetric matrix. Then A is positive definite if and only if A has a Cholesky factorization. Moreover, the Cholesky factorization can be computed through the use of elementary row operations.*

Our strategy for deciding whether a given symmetric matrix is positive definite or not is to try and compute a Cholesky factorization for it. If we succeed, we know it is positive definite. If something goes wrong, we know it is not.

EXAMPLE 34   **(Computing a Cholesky factorization)** Consider the matrix

$$A = \begin{bmatrix} 4 & 2 & 4 \\ 2 & 2 & 3 \\ 4 & 3 & 9 \end{bmatrix}.$$

Is this matrix positive definite? To see, let us try to compute a Cholesky factorization. To clean out the first column, we take

$$R_1 = \begin{bmatrix} 1 & 0 & 0 \\ -1/2 & 1 & 0 \\ -1 & 0 & 1 \end{bmatrix}.$$

Then

$$R_1 A R_1^t = \begin{bmatrix} 4 & 0 & 0 \\ 0 & 1 & 1 \\ 0 & 1 & 5 \end{bmatrix}.$$

To clean out the second column, we take

$$R_2 = \begin{bmatrix} 1 & 0 & 0 \\ 0 & 1 & 0 \\ 0 & -1 & 1 \end{bmatrix}.$$

Then

$$R_2(R_1 A R_1^t)R_2^t = \begin{bmatrix} 4 & 0 & 0 \\ 0 & 1 & 0 \\ 0 & 0 & 4 \end{bmatrix}.$$

Therefore,

$$D = \begin{bmatrix} 2 & 0 & 0 \\ 0 & 1 & 0 \\ 0 & 0 & 2 \end{bmatrix} \quad \text{and} \quad L = (R_2 R_1)^{-1}D = \begin{bmatrix} 2 & 0 & 0 \\ 1 & 1 & 0 \\ 2 & 1 & 2 \end{bmatrix}.$$

We have found a Cholesky factorization, so A is positive definite.

EXAMPLE 35   **(Trying to compute a Cholesky factorization)** Consider the matrix

$$A = \begin{bmatrix} 4 & 2 & 8 \\ 2 & 2 & 3 \\ 8 & 3 & 9 \end{bmatrix}.$$

To clean out the first column, we take

$$R_1 = \begin{bmatrix} 1 & 0 & 0 \\ -1/2 & 1 & 0 \\ -2 & 0 & 1 \end{bmatrix}.$$

Then

$$R_1 A R_1^t = \begin{bmatrix} 4 & 0 & 0 \\ 0 & 1 & -1 \\ 0 & -1 & -7 \end{bmatrix}.$$

Notice the -7 on the diagonal. Therefore, $R_1 A R_1^t$ is not positive definite, so neither is A. In fact, let

$$\mathbf{x} = (R_1^t) \mathbf{e}_3 = \begin{bmatrix} -2 \\ 0 \\ 1 \end{bmatrix}.$$

Then

$$\mathbf{x} \cdot A\mathbf{x} = \mathbf{e}_3 \cdot (R_1 A R_1^t)\mathbf{e}_3 = \begin{bmatrix} 0 \\ 0 \\ 1 \end{bmatrix} \cdot \begin{bmatrix} 4 & 0 & 0 \\ 0 & 1 & -1 \\ 0 & -1 & -7 \end{bmatrix} \begin{bmatrix} 0 \\ 0 \\ 1 \end{bmatrix} = -7.$$

Notice, by the way, that all the entries of A are positive, but still it is not a positive definite matrix.

Exercises   **5.1** For the matrix $A = \begin{bmatrix} 1 & 2 & 4 \\ 2 & 5 & 1 \\ 1 & 1 & 1 \end{bmatrix}$, find an invertible matrix R and a row-reduced matrix U so

that $RA = U$. Also, compute $R\mathbf{b}$ where $\mathbf{b} = \begin{bmatrix} 1 \\ 1 \\ 1 \end{bmatrix}$, and solve the equation $U\mathbf{x} = R\mathbf{b}$. Finally, solve $A\mathbf{x} = \mathbf{b}$.

5.2 For the matrix $A = \begin{bmatrix} 1 & 2 & -1 \\ 2 & -1 & 2 \\ 1 & 2 & 0 \end{bmatrix}$, find an invertible matrix R and a row-reduced matrix U so that

$RA = U$. Also, compute $R\mathbf{b}$ where $\mathbf{b} = \begin{bmatrix} 1 \\ 0 \\ 2 \end{bmatrix}$, and solve the equation $U\mathbf{x} = R\mathbf{b}$. Finally, solve $A\mathbf{x} = \mathbf{b}$.

5.3 For the matrix $A = \begin{bmatrix} 1 & 2 & 4 \\ 2 & 5 & 1 \\ 1 & 1 & 1 \end{bmatrix}$, find a unit lower-triangular matrix L and a row-reduced matrix U

so that $A = LU$. Also, solve $L\mathbf{y} = \mathbf{b}$ where $\mathbf{b} = \begin{bmatrix} 1 \\ 1 \\ 1 \end{bmatrix}$, and solve the equation $U\mathbf{x} = \mathbf{y}$ for this $\mathbf{y}$. Finally,

solve $A\mathbf{x} = \mathbf{b}$.

5.4 For the matrix $A = \begin{bmatrix} 1 & 2 & -1 \\ 2 & -1 & 2 \\ 1 & 2 & 0 \end{bmatrix}$, find a unit lower-triangular matrix L and a row-reduced matrix

U so that $A = LU$. Also, solve $L\mathbf{y} = \mathbf{b}$ where $\mathbf{b} = \begin{bmatrix} 1 \\ 0 \\ 2 \end{bmatrix}$, and solve the equation $U\mathbf{x} = \mathbf{y}$ for this $\mathbf{y}$. Finally,

solve $A\mathbf{x} = \mathbf{b}$.

5.5 For the matrix $A = \begin{bmatrix} 1 & 2 & 4 \\ 2 & 4 & 1 \\ 4 & 1 & 2 \end{bmatrix}$, find a permutation matrix P, a unit lower-triangular matrix L, and

a row-reduced matrix U so that $PA = LU$. Also, compute $P\mathbf{b}$ where $\mathbf{b} = \begin{bmatrix} 1 \\ 1 \\ 1 \end{bmatrix}$, and solve the equation

$L\mathbf{y} = P\mathbf{b}$. Finally, solve $A\mathbf{x} = \mathbf{b}$.

5.6 For the matrix $A = \begin{bmatrix} 1 & 2 & 4 \\ -1 & -2 & 1 \\ -2 & -1 & 2 \end{bmatrix}$, find a permutation matrix P, a unit lower-triangular matrix L,

and a row-reduced matrix U so that $PA = LU$. Also, compute $P\mathbf{b}$ where $\mathbf{b} = \begin{bmatrix} 1 \\ 1 \\ 1 \end{bmatrix}$, and solve the equation

$L\mathbf{y} = P\mathbf{b}$. Finally, solve $A\mathbf{x} = \mathbf{b}$.

5.7 **(a)** Explain why whenever P is a permutation matrix, then P is length preserving; that is, $|P\mathbf{x}| = |\mathbf{x}|$ for all $\mathbf{x}$.

(b) Explain why whenever P is a permutation matrix, then P is invertible, and $P^{-1} = P^t$.

5.8 Let R be the matrix $R = \begin{bmatrix} 1 & 0 & 0 & 0 & 0 \\ 0 & 1 & 0 & 0 & 0 \\ 3 & 0 & 1 & 0 & 0 \\ 0 & 0 & 0 & 1 & 0 \\ 0 & 0 & 0 & 2 & 1 \end{bmatrix}$.

(a) Let A be any other matrix with 5 rows. Explain in words what multiplying A on the left by R does to the rows of A.

(b) Explain in words what you would do to undo the effect of multiplying A on the left by R.

(c) Without doing any computation, write down the inverse of R.

(d) Note that $(R^t A)^t = A^t((R^t)^t) = A^t R$. Let C be any matrix with 5 columns. Explain in words what multiplying C on the right by R does to the columns of C.

(e) How would your answers to parts (a), (b), and (c) change for $R = \begin{bmatrix} 1 & 0 & 0 & 0 & 0 \\ 0 & 1 & 0 & 0 & 0 \\ 3 & 0 & 1 & 0 & 0 \\ 0 & 0 & 2 & 1 & 0 \\ 0 & 0 & 0 & 0 & 1 \end{bmatrix}$? What feature

of this small change is responsible for the difference?

5.9 Let R be the matrix $R = \begin{bmatrix} 1 & 0 & 0 & 0 & 0 \\ 0 & 1 & 0 & 0 & 0 \\ 0 & 2 & 1 & 0 & 0 \\ 0 & 0 & 0 & 1 & 0 \\ 0 & 0 & 0 & 2 & 1 \end{bmatrix}$.

(a) Let A be any other matrix with 5 rows. Explain in words what multiplying A on the left by R does to the rows of A.

(b) Explain in words what you would do to undo the effect of multiplying A on the left by R.

(c) Without doing any computation, write down the inverse of R.

(d) Note that $(R^t A)^t = A^t((R^t)^t) = A^t R$. Let C be any matrix with 5 columns. Explain in words what multiplying C on the right by R does to the columns of C.

(e) How would your answers to parts (a), (b), and (c) change for $R = \begin{bmatrix} 1 & 0 & 0 & 0 & 0 \\ 2 & 1 & 0 & 0 & 0 \\ 0 & 2 & 1 & 0 & 0 \\ 0 & 0 & 2 & 1 & 0 \\ 0 & 0 & 0 & 2 & 1 \end{bmatrix}$? What feature

of this small change is responsible for the difference?

5.10 Let $A = \begin{bmatrix} 1 & 2 & 4 & 0 \\ 2 & 4 & 1 & 2 \\ 3 & 6 & 5 & 2 \\ 4 & 8 & 9 & 2 \end{bmatrix}$.

(a) Find a matrix C so that $A\mathbf{x} = \mathbf{b}$ is solvable if and only if $C\mathbf{b} = 0$.

(b) Does $A\mathbf{x} = \begin{bmatrix} 1 \\ 4 \\ 5 \\ 6 \end{bmatrix}$ have any solutions? If so, find them all. If not, explain how you know why not.

(c) Does $A\mathbf{x} = \begin{bmatrix} 1 \\ 2 \\ 1 \\ 2 \end{bmatrix}$ have any solutions? If so, find them all. If not, explain how you know why not.

5.11 Let $A = \begin{bmatrix} 1 & 1 & 2 & 1 \\ 2 & 1 & 3 & 2 \\ 1 & 0 & 1 & 1 \\ 3 & 2 & 5 & 3 \end{bmatrix}$.

(a) Find a matrix C so that $A\mathbf{x} = \mathbf{b}$ is solvable if and only if $C\mathbf{b} = 0$.

(b) Does $A\mathbf{x} = \begin{bmatrix} 2 \\ 3 \\ 4 \\ 5 \end{bmatrix}$ have any solutions? If so, find them all. If not, explain how you know why not.

(c) Does $A\mathbf{x} = \begin{bmatrix} 2 \\ 3 \\ 1 \\ 5 \end{bmatrix}$ have any solutions? If so, find them all. If not, explain how you know why not.

5.12 Let A and $\tilde{A}$ be $m \times n$ matrices. By considering the record matrices R and $\tilde{R}$ such that $\text{rref}(A) = RA$ and $\text{rref}(\tilde{A}) = \tilde{R}\tilde{A}$, show that $\text{Ker}(A) = \text{Ker}(\tilde{A})$ if and only if there is an invertible $m \times m$ matrix B such that $\tilde{A} = BA$.

5.13 Let $A = \begin{bmatrix} 9 & -3 & 6 \\ -3 & 2 & -5 \\ 6 & -5 & 17 \end{bmatrix}$. Is A positive definite or not? If so, compute a Cholesky factorization of A. If not, find a vector $\mathbf{x} \neq 0$ for which $\mathbf{x} \cdot A\mathbf{x} \leq 0$.

5.14 Let $A = \begin{bmatrix} 1 & -1 & 4 \\ -1 & 2 & -4 \\ 4 & -4 & 20 \end{bmatrix}$. Is A positive definite or not? If so, compute a Cholesky factorization of A. If not, find a vector $\mathbf{x} \neq 0$ for which $\mathbf{x} \cdot A\mathbf{x} \leq 0$.

5.15 Let $A = \begin{bmatrix} 9 & -3 & 6 \\ -3 & 2 & -6 \\ 6 & -6 & 17 \end{bmatrix}$. Is A positive definite or not? If so, compute a Cholesky factorization of A. If not, find a vector $\mathbf{x} \neq 0$ for which $\mathbf{x} \cdot A\mathbf{x} \leq 0$.

5.16 Let $A = \begin{bmatrix} 1 & -1 & 5 \\ -1 & 2 & -4 \\ 5 & -4 & 20 \end{bmatrix}$. Is A positive definite or not? If so, compute a Cholesky factorization of A. If not, find a vector $\mathbf{x} \neq 0$ for which $\mathbf{x} \cdot A\mathbf{x} \leq 0$.

5.17 Show that if A is any positive definite matrix and B is any invertible matrix, then $C = BAB^t$ is positive definite. Show that if B is not invertible, C is positive semidefinite

5.18 Find all values of c for which $\begin{bmatrix} c & -1 & -1 \\ -1 & c & -1 \\ -1 & -1 & c \end{bmatrix}$ is positive definite.

5.19 Let A and B be two $n \times n$ matrices.
(a) Suppose that A is positive definite and B is positive semidefinite. Show that $A + B$ is positive definite.

(b) Suppose that A is positive definite. What is the rank of A?

5.20 **(a)** Let A be an $n \times n$ matrix. Using the Schwarz inequality, show that

$$|\mathbf{x} \cdot A\mathbf{x}| \leq \left(\sum_{i,j=1}^{n} A_{i,j}^2 \right)^{1/2} |\mathbf{x}|^2 .$$

(b) Using the result of part (a), show that if A is any symmetric matrix, and a is any number with $a > \left(\sum_{i,j=1}^{n} A_{i,j}^2 \right)^{1/2}$, then $A + aI$ is positive definite.

(c) Using the result of part (b), show that while the diagonal entries of a positive definite matrix must be strictly positive, any (symmetric) pattern of signs is possible in the off diagonal entries.

5.21 Let A be an $n \times n$ positive definite matrix. Show that there is a number $c > 0$ so that $A - cI$ is positive definite.

5.22 Consider the matrix $A = \begin{bmatrix} 1 & 1 & 1 \\ 1 & 1 & 1 \\ 1 & 1 & 1 \end{bmatrix}$. Show that A has a factorization $A = LL^t$, where L is lower-triangular but not invertible.

5.23 Let A be an $n \times n$ positive definite matrix. Show that A^n is positive definite for all positive integers.

Solving equations with the Cholesky factorization

Suppose that you want to solve $A\mathbf{x} = \mathbf{b}$, where you know that A is positive definite. As we have explained, positive definite matrices arise in many applications, so this is not unreasonable to suppose. Notice also that by Theorem 12, positive definite matrices are invertible, so there will always be exactly one solution.

If you compute a Cholesky factorization $A = LL^t$, then you can solve $A\mathbf{x} = \mathbf{b}$ by using back substitution to solve $L\mathbf{y} = \mathbf{b}$ and then use forward substitution to solve $L^t\mathbf{x} = \mathbf{y}$. This is an efficient and effective way to solve the equation.

5.24 Use the Cholesky factorization computed in Example 34 to solve $A\mathbf{x} = \mathbf{b}$, where $\mathbf{b} = \begin{bmatrix} 3 \\ 4 \\ 1 \end{bmatrix}$, using the method just explained.

5.25 Compute a Cholesky factorization of $A = \begin{bmatrix} 1 & 1 & 1 \\ 1 & 2 & 2 \\ 1 & 2 & 3 \end{bmatrix}$ to solve $A\mathbf{x} = \mathbf{b}$, where $\mathbf{b} = \begin{bmatrix} 2 \\ 4 \\ 1 \end{bmatrix}$, using the method just explained.

5.26 Compute a Cholesky factorization of $A = \begin{bmatrix} 1 & 1 & 2 \\ 1 & 5 & 4 \\ 2 & 4 & 6 \end{bmatrix}$ to solve $A\mathbf{x} = \mathbf{b}$, where $\mathbf{b} = \begin{bmatrix} 2 \\ 4 \\ 1 \end{bmatrix}$, using the method just explained.

Quadratic forms and Cholesky factorization

The general *quadratic form* in three variables x, y, and z, is the general linear combination of all purely quadratic products of these variables:

$$q(x, y, z) = \alpha x^2 + \beta y^2 + \gamma z^2 + 2\delta xy + 2\varepsilon xz + 2\zeta yz \,,$$

where α, β, γ, δ, ε, and ζ are any six numbers. As you can easily compute,

$$\alpha x^2 + \beta y^2 + \gamma z^2 + 2\delta xy + 2\varepsilon xz + 2\zeta yz = \begin{bmatrix} x \\ y \\ z \end{bmatrix} \cdot \begin{bmatrix} \alpha & \delta & \varepsilon \\ \delta & \beta & \zeta \\ \varepsilon & \zeta & \gamma \end{bmatrix} \begin{bmatrix} x \\ y \\ z \end{bmatrix} \,.$$

Hence, there is a one-to-one correspondence between quadratic forms in three variables and symmetric 3×3 matrices.

Let q be a quadratic form in x, y, and z, and let A be the corresponding matrix so that

$$q(\mathbf{x}) = \mathbf{x} \cdot A\mathbf{x} \,.$$

A quadratic form is said to be *positive definite* or *positive semidefinite* if and only if the corresponding matrix is such.

Suppose that q is a positive definite quadratic form and its matrix A has a Cholesky factorization $A = LL^t$ where

$$L^t = \begin{bmatrix} a & d & e \\ 0 & b & f \\ 0 & 0 & c \end{bmatrix}$$

with a, b, and c strictly positive. As you can easily compute,

$$q(\mathbf{x}) = |L^t\mathbf{x}|^2 = \left| \begin{bmatrix} ax + dy + ez \\ by + fz \\ cz \end{bmatrix} \right|^2 \,.$$

Therefore,

$$q(x, y, z) = (ax + dy + ez)^2 + (by + fz)^2 + (cz)^2 \,.$$

That is, finding the Cholesky factorization of A allows us to write $q(x, y, z)$ as a sum of squares. It is possible only if q is at least positive semidefinite. Otherwise, there is a vector $\mathbf{x}$ with $q(\mathbf{x}) = \mathbf{x} \cdot A\mathbf{x} < 0$, which is not possible if $q(x, y, z)$ can be written as a sum of squares. Thus the Cholesky factorization allows us to decide when a quadratic form can be written as a sum of squares, and when this is possible, it gives us one way of doing so.

What we have said here about three variables generalizes to any number of variables.

5.27 Consider the quadratic form

$$q(x, y, z) = x^2 + 5y^2 + 6z^2 + 2xy + 4xz + 8yz .$$

Determine whether $q(x, y, z) \geq 0$ for all x, y, and z. If so, write $q(x, y, z)$ as a sum of squares. If not, find values of x, y, and z for which $q(x, y, z) < 0$.

5.28 Consider the quadratic form

$$q(x, y, z) = x^2 + 5y^2 + 14z^2 + 2xy + 4xz + 16yz .$$

Determine whether $q(x, y, z) \geq 0$ for all x, y, and z. If so, write $q(x, y, z)$ as a sum of squares. If not, find values of x, y, and z for which $q(x, y, z) < 0$.

5.29 Consider the quadratic form

$$q(x, y, z) = x^2 + y^2 + z^2 + 2xy + 4xz + 8yz .$$

Determine whether $q(x, y, z) \geq 0$ for all x, y, and z. If so, write $q(x, y, z)$ as a sum of squares. If not, find values of x, y, and z for which $q(x, y, z) < 0$.

5.30 Consider the quadratic form

$$q(x, y, z) = x^2 + 5y^2 + z^2 + 2xy + 4xz + 16yz .$$

Determine whether $q(x, y, z) \geq 0$ for all x, y, and z. If so, write $q(x, y, z)$ as a sum of squares. If not, find values of x, y, and z for which $q(x, y, z) < 0$.

5.31 Let A be an $n \times n$ symmetric matrix. Show that a necessary condition for A to be positive definite is that

$$A_{i,i} A_{j,j} > (A_{i,j})^2$$

for all i and j. Deduce a corresponding statement about the coefficients of quadratic forms. Show that this condition is also sufficient when $n = 2$. Is it also sufficient for $n = 3$?

5.32 Let q be a positive definite quadratic form on $\mathbb{R}^n$. Define the *q-length* of a vector $\mathbf{x}$, $|\mathbf{x}|_q$ by

$$|\mathbf{x}|_q = \sqrt{q(\mathbf{x})} .$$

Show that the *q-length* satisfies the the *Minkowski inequality* and the *triangle inequality*. (It may help to review the discussion of these inequalities in Section 4 of Chapter 1.) That is, show that for all $\mathbf{x}$, $\mathbf{y}$, and $\mathbf{z}$ in $\mathbb{R}^n$,

$$|\mathbf{x} + \mathbf{y}|_q \leq |\mathbf{x}|_q + |\mathbf{y}|_q ,$$

and

$$|\mathbf{x} - \mathbf{z}|_q \leq |\mathbf{x} - \mathbf{y}|_q + |\mathbf{y} - \mathbf{z}|_q .$$

The Image of a Linear Transformation

Overview of Chapter 3

Chapter 2 presented a general method for solving a linear system $A\mathbf{x} = \mathbf{b}$. Now suppose that you are asked to solve such a linear system, and you really, really need to know the solution—your job depends on producing this list of numbers. You go to work, row reduce, and there it is: a pivot in the final column.

Now what do you do? There is a Woody Allen routine about the "new math" in which previously unsolvable equations are dealt with by means of threats of reprisal. But that is comedy, and this is your job. What are you going to do?

As long as we are only working with equations in an algebraic setting, things are either equal or they are not. This black-and-white point of view is very limiting. Let us try to distinguish the many shades of gray in between. Geometry provides the way forward. Geometry allows us to think about solving equations in terms of *distance*. Distinguishing shades of gray then means distinguishing between how *close* or *far* a vector **x** may be from solving an equation.

Here is how this goes: When we are trying to solve $A\mathbf{x} = \mathbf{b}$, we are looking for a vector **x** that makes $|A\mathbf{x} - \mathbf{b}| = 0$. That is, we are looking for a vector **x** such that the distance from $A\mathbf{x}$ to **b** is zero. It is usually easier to work with the squared distance since it gets rid of some square roots, so we can restate the main problem that occupied us in Chapter 2 as follows:

- *How do we find all the vectors* **x** *that make* $|A\mathbf{x} - \mathbf{b}|^2 = 0$?

    Our new goal, if the solution set of $A\mathbf{x} = \mathbf{b}$ is empty, will be to answer the question

- *How do we find all the vectors* **x** *that make* $|A\mathbf{x} - \mathbf{b}|^2$ *as small as possible?*

Stating the two questions this way emphasizes their similarity. A vector **x** that makes $|A\mathbf{x} - \mathbf{b}|^2$ as small as possible is called a *least squares solution* of $A\mathbf{x} = \mathbf{b}$.

We will see that a least squares solution of $A\mathbf{x} = \mathbf{b}$ is an ordinary solution of $A\mathbf{x} = \mathbf{c}$ for some vector **c**, and that **c** can often be interpreted as a "corrected" version of **b**, with the effects of experimental error "projected out." This projection procedure is very useful in many other situations, and so we will study it in some detail.

Some amount of mathematical abstraction is needed to build a bridge between algebra and geometry, but the way has been prepared, and you are already familiar with concrete examples of the few abstract notions introduced here. None of the abstraction is pointless; the point is to develop and understand an effective machinery for computing least squares solutions. As we will see, *every* linear system $A\mathbf{x} = \mathbf{b}$ has least squares solutions. We will not only learn how to find them, but will also learn why they may well be a perfectly good substitute for an actual solution.

SECTION 1   Images of Linear Transformations

1.1 Images and why they matter

Suppose we are given an $m \times n$ matrix A. Most of our work in the previous chapter went into developing a general method for finding the solution set of $A\mathbf{x} = \mathbf{b}$ for any **b** in $\mathbb{R}^m$. As we have seen, though, sometimes there are no solutions.

Let us look at a concrete example and try to understand *why* there is no solution. That will help us to decide what we can do if we really, really need a solution, despite that pivot in the final column.

In Example 3 of Chapter 1, we considered a simple electric circuit with two voltages V_1 and V_2 and three currents I_1, I_2, and I_3. These voltages and currents were related by*

$$\frac{1}{3R} \begin{bmatrix} -2 & -1 \\ 1 & 2 \\ 1 & -1 \end{bmatrix} \begin{bmatrix} V_1 \\ V_2 \end{bmatrix} = \begin{bmatrix} I_1 \\ I_2 \\ I_3 \end{bmatrix}. \tag{1.1}$$

---

*We did not use the matrix notation in the first section of Chapter 1, but now that you know it, you can easily translate the formulas from Example 3 of Chapter 1.

Suppose you have measured the three currents I_1, I_2, and I_3 in a laboratory experiment and have obtained the data

$$I_1 = 1.02 \text{ amps}, \qquad I_2 = -2.04 \text{ amps}, \qquad \text{and} \qquad I_3 = 0.99 \text{ amps}. \tag{1.2}$$

If you then wanted to know what the voltages V_1 and V_2 that produced these currents were, you might try to solve

$$\frac{1}{3R} \begin{bmatrix} -2 & -1 \\ 1 & 2 \\ 1 & -1 \end{bmatrix} \begin{bmatrix} V_1 \\ V_2 \end{bmatrix} = \begin{bmatrix} 1.02 \\ -2.04 \\ 0.99 \end{bmatrix}. \tag{1.3}$$

Unfortunately, this equation has no solution, so you cannot possibly solve it to find V_1 and V_2. *What went wrong?*

There is nothing wrong with (1.1), which is a statement of physical law. In particular, as explained in Example 3 of Chapter 1, (1.1) incorporates

$$I_1 + I_2 + I_3 = 0, \tag{1.4}$$

one of Kirchhoff's rules. It expresses the conservation of electrical charge, an exact and absolute law of nature.

Now we can see why (1.3) has no solution and also what to do about it. The data (1.2) do not satisfy (1.4), which is an *exact* physical law. But laboratory data are *essentially never* exact. There is almost always some experimental error, some "noise" in the data. The vector

$$\begin{bmatrix} 1.02 \\ -2.04 \\ 0.99 \end{bmatrix}$$

is not the "right" right-hand side of (1.3); it is only an approximate measurement of it.

We could go back and measure the currents more carefully, but there will always be some noise, so we had better figure out how to deal with it. There is a natural thing to do:

- *Replace the vector*

$$\begin{bmatrix} 1.02 \\ -2.04 \\ 0.99 \end{bmatrix}$$

with the closest vector to it that does satisfy (1.4).

Now (1.4) is the equation of a plane in $\mathbb{R}^3$ considered as the set of "current vectors"

$$\mathbf{x} = \begin{bmatrix} I_1 \\ I_2 \\ I_3 \end{bmatrix}.$$

If we introduce

$$\mathbf{a} = \begin{bmatrix} 1 \\ 1 \\ 1 \end{bmatrix},$$

we can write (1.4) in the form $\mathbf{a} \cdot \mathbf{x} = 0$. As we know, such an equation describes a plane through the origin in $\mathbb{R}^3$.

To get the vector in the plane that is closest to $\mathbf{b}$, decompose $\mathbf{b}$ into the sum $\mathbf{b}_\parallel + \mathbf{b}_\perp$ of its parts parallel and perpendicular to $\mathbf{a}$. The perpendicular part lies in the plane and is the closest vector: Draw a picture to confirm it.

Since $|\mathbf{a}|^2 = 3$, the closest vector is

$$\mathbf{b}_\perp = \begin{bmatrix} 1.02 \\ -2.04 \\ 0.99 \end{bmatrix} - \frac{1}{3}\left(\begin{bmatrix} 1.02 \\ -2.04 \\ 0.99 \end{bmatrix} \cdot \mathbf{a}\right)\mathbf{a} = \begin{bmatrix} 1.02 \\ -2.04 \\ 0.99 \end{bmatrix} + \begin{bmatrix} 0.01 \\ 0.01 \\ 0.01 \end{bmatrix} = \begin{bmatrix} 1.03 \\ -2.03 \\ 1.00 \end{bmatrix}$$

(see Section 4 of Chapter 1).

To find the voltages, use the "corrected" experimental data as the right-hand side in (1.1) and solve for V_1 and V_2. This time, there is no problem, and we find the unique solution

$$V_1 = -0.01 \qquad \text{and} \qquad V_2 = -1.01.$$

Notice what went on here. We were trying to solve an equation $A\mathbf{x} = \mathbf{b}$ that was not solvable because the right-hand side $\mathbf{b}$ consisted of "noisy data." We corrected the data by changing the right-hand side from the given vector $\mathbf{b}$ into the closest vector $\mathbf{c}$ for which $A\mathbf{x} = \mathbf{c}$ does have a solution, in this case $\mathbf{c} = \mathbf{b}_\perp$, thus "washing out" the noise in the laboratory data and allowing us to proceed with the computation.

In this particular case, the set of "right-hand sides" satisfying the condition (1.4) constituted a plane, and since we know how to find the closest vector in a plane, we had no problem computing the "corrected right-hand side."

In general, though, if we are going to compute the closest vector in the set of vectors $\mathbf{b}$ for which $A\mathbf{x} = \mathbf{b}$ is solvable, we need to know something about what this set is. First, let us give it a name.

D E F I N I T I O N	**(Image of a matrix A)** Let A be an $m \times n$ matrix. The *image* of A is the set of vectors $\mathbf{b}$ in $\mathbb{R}^m$ for which $A\mathbf{x} = \mathbf{b}$ has a solution. It is denoted by $\text{Img}(A)$.

Now suppose that we are given a vector $\mathbf{b}$ in $\mathbb{R}^m$ that does not belong to $\text{Img}(A)$. Suppose that we can find a vector $\mathbf{c}$ in $\text{Img}(A)$ that is closer to $\mathbf{b}$ than any other vector in $\text{Img}(A)$. By the definition of $\text{Img}(A)$, since $\mathbf{c}$ belongs to $\text{Img}(A)$, there is at least one vector $\mathbf{x}_0$ in $\mathbb{R}^n$ with $A\mathbf{x}_0 = \mathbf{c}$. Also, by the definition of $\text{Img}(A)$, for any $\mathbf{x}$ in $\mathbb{R}^n$, $A\mathbf{x}$ belongs to $\text{Img}(A)$; $\mathbf{x}$ itself is the required solution.

Since $\mathbf{c}$ is closer to $\mathbf{b}$ than is any other vector in $\text{Img}(A)$,

$$|\mathbf{c} - \mathbf{b}|^2 \leq |A\mathbf{x} - \mathbf{b}|^2.$$

Since $\mathbf{c} = A\mathbf{x}_0$, we then have

$$|A\mathbf{x}_0 - \mathbf{b}|^2 \leq |A\mathbf{x} - \mathbf{b}|^2.$$

Since $\mathbf{x}$ can be any vector in $\mathbb{R}^n$, this means that choosing $\mathbf{x} = \mathbf{x}_0$ makes $|A\mathbf{x} - \mathbf{b}|^2$ as small as possible. In the terminology of the chapter overview, $\mathbf{x}_0$ is a least squares solution to $A\mathbf{x} = \mathbf{b}$.

- *Given an $m \times n$ matrix A and a vector $\mathbf{b}$ in $\mathbb{R}^m$, if there is a vector $\mathbf{c}$ in the $\text{Img}(A)$ that is closest to $\mathbf{b}$, then any solution $\mathbf{x}_0$ of $A\mathbf{x} = \mathbf{c}$ is a least squares solution of the original equation $A\mathbf{x} = \mathbf{b}$.*

Once we find $\mathbf{c}$, if it exists, we are back on familiar ground. By the definition of $\text{Img}(A)$, $A\mathbf{x} = \mathbf{c}$ is solvable, and we can use row reduction to solve it. But is there a closest vector $\mathbf{c}$, and if so, how do we find it?

To get going on this question, we need to have some way to decide whether or not a vector belongs to $\text{Img}(A)$. How about an equation for $\text{Img}(A)$? That should help!

1.2 Finding an equation for Img(A)

We have already learned how to find an equation for Img(A): The second part of Theorem 10 from Chapter 2 states that for any $m \times n$ matrix A, if we row reduce $[A|I]$ to find $[U|R]$ and let C consist of the bottom $m - r$ rows of R, where r is the rank of A, then $A\mathbf{x} = \mathbf{b}$ is solvable if and only if $C\mathbf{b} = 0$. In our new terminology, we can express the same thing by saying that $\mathbf{b}$ belongs to Img(A) if and only if $\mathbf{b}$ belongs to Ker(C). This is good news:

- *The image of every matrix A is the kernel of another matrix C.*

We know all about parameterizing kernels, so we already know all about parameterizing images.

EXAMPLE 1 **(Finding an equation for Img(A))** Consider

$$A = \begin{bmatrix} -2 & -1 \\ 1 & 2 \\ 1 & -1 \end{bmatrix}.$$

Apart from a scalar multiple, this is the matrix from (1.1) in our electrical circuit example. If A and B are two matrices and B is a nonzero scalar multiple of A, then Img(A) = Img(B); the simple proof is left as an exercise. Hence, in computing an equation for Img(A), we are computing an equation for the image of the matrix in our circuit example.

Row reducing $[A|I]$,

$$[U|R] = \begin{bmatrix} -2 & -1 & \bigg| & 1 & 0 & 0 \\ 0 & 3 & \bigg| & 1 & 2 & 0 \\ 0 & 0 & \bigg| & 2 & 2 & 2 \end{bmatrix}.$$

We see that the rank of A is 2, so C is the 1×3 matrix

$$C = [2 \quad 2 \quad 2].$$

Therefore,

$$\mathbf{b} = \begin{bmatrix} x \\ y \\ z \end{bmatrix}$$

belongs to Img(A) if and only if $C\mathbf{b} = 0$. Since $C\mathbf{b} = 2x + 2y + 2z$, $C\mathbf{b} = 0$ is the same as

$$x + y + z = 0. \tag{1.5}$$

Notice that if we change the names of the variables in (1.5) from x, y, and z to I_1, I_2, and I_3, (1.5) becomes (1.4). Thus, (1.4) is actually a necessary and sufficient condition for the solvability of (1.1).

Example 1 highlights an important point. When the image of A is not all of $\mathbb{R}^m$ in real applications, there is often a reason *why* it is not. In this case, the reason is (1.4), Kirchhoff's rule expressing the physical law of charge conservation. Although A itself may be subject to experimental error—perhaps the resistances were not measured carefully enough—we can be sure about the equation for the image. It expresses an exact physical law and is immune to experimental error. This is one reason we focus on "correcting" $\mathbf{b}$ and not A when $A\mathbf{x} = \mathbf{b}$ is not solvable.

There is another way to get an equation for Img(A) that you might find simpler if you are working things out on paper. Suppose, for example, that A has three rows. Let

$$\mathbf{b} = \begin{bmatrix} x \\ y \\ z \end{bmatrix},$$

and form the augmented matrix $[A|\mathbf{b}]$.* Row reduce this matrix and see for which values of x, y, and z there is no pivot in the final column: Whether there is a pivot or not tells you if $A\mathbf{x} = \mathbf{b}$ is solvable.

EXAMPLE 2   **(Finding an equation for Img(A) another way)**   Consider

$$A = \begin{bmatrix} 1 & -1 & 0 \\ 2 & 0 & 2 \\ 0 & 1 & 1 \end{bmatrix}.$$

Let

$$\mathbf{b} = \begin{bmatrix} x \\ y \\ z \end{bmatrix}$$

denote an arbitrary element of $\mathbb{R}^3$. To determine Img(A), form the augmented matrix corresponding to the equation $A\mathbf{x} = \mathbf{b}$:

$$[A|\mathbf{b}] = \begin{bmatrix} 1 & -1 & 0 & \bigm| & x \\ 2 & 0 & 2 & \bigm| & y \\ 0 & 1 & 1 & \bigm| & z \end{bmatrix}$$

Row reducing it to upper-triangular form, we find

$$\begin{bmatrix} 1 & -1 & 0 & \bigm| & x \\ 0 & 2 & 2 & \bigm| & y - 2x \\ 0 & 0 & 0 & \bigm| & z - (y/2) + x \end{bmatrix}.$$

Evidently there will be a solution if and only if the final entry in the last row is zero, that is, if

$$2x - y + 2z = 0.$$

In other words, $\mathbf{x}$ belongs to Img(A) if and only if $\mathbf{x}$ satisfies this equation.

Once we have an equation for the image of A, we can test to see whether or not any given vector belongs to it, giving us a criterion for solvability of the equation $A\mathbf{x} = \mathbf{b}$. The equation will be solvable if and only if $\mathbf{b}$ satisfies the equation for Img(A).

EXAMPLE 3   **(Testing for membership in Img(A))**   Let A be the matrix of Example 2, and consider the following vectors:

$$\mathbf{b}_1 = \begin{bmatrix} 0 \\ 4 \\ 2 \end{bmatrix}, \qquad \mathbf{b}_2 = \begin{bmatrix} 1 \\ 2 \\ 3 \end{bmatrix}, \qquad \text{and} \qquad \mathbf{b}_3 = \begin{bmatrix} 1 \\ 4 \\ 1 \end{bmatrix}.$$

Which, if any, belong to Img(A)? All we have to do to check is to plug these vectors into the equation $2x - y + 2z = 0$ for Img(A), which we can also write in the form $\mathbf{a} \cdot \mathbf{x} = 0$, where

$$\mathbf{a} = \begin{bmatrix} 2 \\ -1 \\ 2 \end{bmatrix}.$$

Then $\mathbf{a} \cdot \mathbf{b}_1 = \mathbf{a} \cdot \mathbf{b}_3 = 0$, while $\mathbf{a} \cdot \mathbf{b}_2 = 6$. Hence $\mathbf{b}_1$ and $\mathbf{b}_3$ belong to Img(A) but not $\mathbf{b}_2$. So $A\mathbf{x} = \mathbf{b}_1$ and $A\mathbf{x} = \mathbf{b}_3$ are solvable, but $A\mathbf{x} = \mathbf{b}_2$ is not.

---

*Since we do not work explicitly with $\mathbf{x}$ but rather with $\mathbf{b}$, we let x, y, and z have their usual starring role and use them in $\mathbf{b}$.

We can get further insight into Img(A), with A still being the matrix of Example 2, by recognizing that the equation we found for it, namely, $2x - y + 2z = 0$, is the equation of a plane through the origin in $\mathbb{R}^3$:

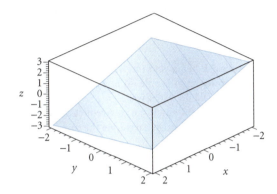

The lines drawn on the plane are lines of constant height z. Notice the different scale used for the z axes to adjust the proportions of the graph. This plane (imagine the patch extended indefinitely) is the set of points $\mathbf{b}$ in $\mathbb{R}^3$ for which the equation $A\mathbf{x} = \mathbf{b}$ has a solution. As you see, this is only a "slice" of $\mathbb{R}^3$.

In Examples 1 and 2, Img(A) turned out to be the solution set of a single linear equation. More generally, it will turn out to be the solution set of a system of linear equations. Let us do another example in more variables.

EXAMPLE 4   **(Finding an equation for Img(A), more variables)** Consider

$$A = \begin{bmatrix} 1 & 2 & 1 & 2 \\ 2 & 3 & 2 & 3 \\ 1 & 1 & 1 & 1 \\ 0 & 1 & 0 & 1 \end{bmatrix}.$$

Let

$$\mathbf{b} = \begin{bmatrix} x \\ y \\ z \\ w \end{bmatrix}$$

denote an arbitrary element of $\mathbb{R}^4$. To determine Img(A), form the augmented matrix corresponding to the equation $A\mathbf{x} = \mathbf{b}$:

$$[A|\mathbf{b}] = \begin{bmatrix} 1 & 2 & 1 & 2 & | & x \\ 2 & 3 & 2 & 3 & | & y \\ 1 & 1 & 1 & 1 & | & z \\ 0 & 1 & 0 & 1 & | & w \end{bmatrix}.$$

Row reducing $[A/\mathbf{b}]$ with no row swaps,

$$\begin{bmatrix} 1 & 2 & 1 & 2 & | & x \\ 2 & 3 & 2 & 3 & | & y \\ 1 & 1 & 1 & 1 & | & z \\ 0 & 1 & 0 & 1 & | & w \end{bmatrix} \rightarrow \begin{bmatrix} 1 & 2 & 1 & 2 & | & x \\ 0 & -1 & 0 & -1 & | & y - 2x \\ 0 & -1 & 0 & -1 & | & z - x \\ 0 & 1 & 0 & 1 & | & w \end{bmatrix}$$

$$\rightarrow \begin{bmatrix} 1 & 2 & 1 & 2 & | & x \\ 0 & -1 & 0 & -1 & | & y - 2x \\ 0 & 0 & 0 & 0 & | & z - y + x \\ 0 & 0 & 0 & 0 & | & w + y - 2x \end{bmatrix}.$$

Hence there is a solution of $A\mathbf{x} = \mathbf{b}$ if and only if the system

$$x - y + z = 0$$
$$-2x + y + w = 0$$

is satisfied. This system can be written in matrix form as $C\mathbf{b} = 0$, where

$$C = \begin{bmatrix} 1 & -1 & 1 & 0 \\ -2 & 1 & 0 & 1 \end{bmatrix}.$$

Therefore, $\mathbf{b}$ belongs to $\text{Img}(A)$ if and only if $\mathbf{b}$ belongs to the solution set of $C\mathbf{y} = 0$, so $C\mathbf{y} = 0$ is an equation for $\text{Img}(A)$. We do not call it *the* equation for $\text{Img}(A)$, since any other equivalent system of equations will do just as well.

If you went about the row reduction in a different way, you would get a different but equivalent system. You could also have done the row reduction with one row swap,

$$\begin{bmatrix} 1 & 2 & 1 & 2 & x \\ 2 & 3 & 2 & 3 & y \\ 1 & 1 & 1 & 1 & z \\ 0 & 1 & 0 & 1 & w \end{bmatrix} \rightarrow \begin{bmatrix} 1 & 2 & 1 & 2 & x \\ 0 & 1 & 0 & 1 & w \\ 0 & -1 & 0 & -1 & y - 2x \\ 0 & -1 & 0 & -1 & z - x \end{bmatrix} \rightarrow \begin{bmatrix} 1 & 2 & 1 & 2 & x \\ 0 & 1 & 0 & 1 & w \\ 0 & 0 & 0 & 0 & w + y - 2x \\ 0 & 0 & 0 & 0 & w + z - x \end{bmatrix},$$

which gives the system of equations

$$-2x + y + w = 0$$
$$-x + z + w = 0.$$

This system is equivalent to the one just obtained. Check it.

In Example 4, we got a system of two linear equations for the image of A, which we could write as a single matrix equation. Sometimes it is convenient to refer to a "system of equations for an image," sometimes to just an "equation for an image." In the latter case, we have in mind a matrix equation like $C\mathbf{y} = 0$, which of course represents a system.

If you ponder the matter a bit, you will see from these examples that

- *For any $m \times n$ matrix A, $\text{Img}(A)$ will be the solution set of a homogeneous equation $C\mathbf{y} = 0$ for some matrix C that can always be found by row reducing $[A|\mathbf{x}]$.*

We now have two closely related methods for finding an equation for $\text{Img}(A)$. The approach in which we row reduce $[A|I]$ may seem more cumbersome, but it has the advantage that it requires us only to row reduce matrices with numerical entries. This is useful since many computer programs have trouble row reducing a matrix like $[A|\mathbf{x}]$ that has variable entries.

EXAMPLE 5   **(An equation for Img(A) via row reduction of [A|I])** Let

$$A = \begin{bmatrix} 1 & 2 \\ 2 & -1 \\ 0 & 1 \\ 1 & -3 \end{bmatrix}.$$

Row reducing $[A|I]$ to the form $[U|R]$ requires no swapping of rows. Hence $P = I$, which simplifies things a bit. (It happens quite often.) The U and R that we find are

$$U = \begin{bmatrix} 1 & 2 \\ 0 & -5 \\ 0 & 0 \\ 0 & 0 \end{bmatrix} \quad \text{and} \quad R = \frac{1}{5}\begin{bmatrix} 5 & 0 & 0 & 0 \\ -10 & 5 & 0 & 0 \\ -2 & 1 & 5 & 0 \\ 5 & -5 & 0 & 5 \end{bmatrix}.$$

We see that the rank of A is 2 since there are 2 pivots. Hence, the bottom $4 - 2 = 2$ rows of R give us the matrix

$$C = \begin{bmatrix} -2 & 1 & 5 & 0 \\ 5 & -5 & 0 & 5 \end{bmatrix},$$

and $C\mathbf{x} = 0$ is an equation for $\text{Img}(A)$.

1.3 Parameterizing images

We know all about finding one-to-one parameterizations of the solution sets of equations $Cy = 0$. Since we now know how to find such an equation for the image of a matrix A, we know how to find one-to-one parameterizations of images.

EXAMPLE 6

(Parameterizing Img(A)) Let C and A be as in Example 4. To parameterize the solution set of $Cy = 0$, which is $\text{Img}(A)$, we row reduce $[C|0]$. In one step we obtain

$$\left[\begin{array}{cccc|c} 1 & -1 & 1 & 0 & 0 \\ 0 & -1 & 2 & 1 & 0 \end{array}\right].$$

The pivotal variables are y_1 and y_2. Putting $t_1 = y_3$ and $t_2 = y_4$, $y_2 = 2t_1 + t_2$ and $y_1 = (2t_1 + t_2) - t_1 = t_1 + t_2$. Hence, $\mathbf{y}$ is in the solution set when

$$\begin{bmatrix} y_1 \\ y_2 \\ y_3 \\ y_4 \end{bmatrix} = \begin{bmatrix} t_1 + t_2 \\ 2t_1 + t_2 \\ t_1 \\ t_2 \end{bmatrix} = t_1 \begin{bmatrix} 1 \\ 2 \\ 1 \\ 0 \end{bmatrix} + t_2 \begin{bmatrix} 1 \\ 1 \\ 0 \\ 1 \end{bmatrix}.$$

This parameterizes $\text{Img}(A)$.

There is a way to get a one-to-one parameterization of $\text{Img}(A)$ without first finding an equation for $\text{Img}(A)$:

- *If A is a matrix of rank r, and $\mathbf{v}_{j(1)}, \mathbf{v}_{j(2)}, \ldots, \mathbf{v}_{j(r)}$ are the pivotal columns of A, then*

$$\mathbf{b} = t_1 \mathbf{v}_{j(1)} + t_2 \mathbf{v}_{j(2)} + \cdots + t_r \mathbf{v}_{j(r)}$$

  is a one-to-one parameterization of $\text{Img}(A)$.

Here is why: By definition, if $\mathbf{b}$ belongs to $\text{Img}(A)$, there is a vector $\mathbf{y}$ in $\mathbb{R}^n$ so that $A\mathbf{y} = \mathbf{b}$. Writing

$$\mathbf{y} = \begin{bmatrix} y_1 \\ y_2 \\ \vdots \\ y_n \end{bmatrix}$$

and $A = [\mathbf{v}_1, \mathbf{v}_2, \ldots, \mathbf{v}_n]$,

$$\mathbf{b} = A\mathbf{y} = y_1\mathbf{v}_1 + y_2\mathbf{v}_2 + \cdots + y_n\mathbf{v}_n. \tag{1.6}$$

We know that as long as there is a solution, there is one solution with any values we like for the nonpivotal variables. Therefore, let us set all the nonpivotal variables to zero. This means that all the nonpivotal columns in (1.6) are zeroed out, and only the pivotal columns remain. So $\mathbf{b}$ is a linear combination of the pivotal columns.

Next, once the values of the nonpivotal variables are fixed (we have fixed them to be zero), the values of the pivotal variables are uniquely determined. Therefore, this parameterization is one-to-one.

EXAMPLE 7

(Parameterizing Img(A) using the pivotal columns) Let A be the same 4×4 matrix discussed in Example 4. Let $\mathbf{v}_1, \mathbf{v}_2, \mathbf{v}_3$, and $\mathbf{v}_4$ be its columns so that $A = [\mathbf{v}_1, \mathbf{v}_2, \mathbf{v}_3, \mathbf{v}_4]$. In the row reduction we did in Example 4, we saw that $\mathbf{v}_1$ and $\mathbf{v}_2$ are the pivotal columns of A. This means that

$$\mathbf{b} = t_1\mathbf{v}_1 + t_2\mathbf{v}_2 \tag{1.7}$$

is a one-to-one parameterization of $\text{Img}(A)$.

We summarize.

THEOREM 1   **(Parameterization of images)** *Let A be an m × n matrix. We can obtain a one-to-one parameterization of* Img(A) *by either*

1. *finding an equation* $C\mathbf{x} = 0$ *for* Img(A) *and then solving that equation for a parameterization of its solution set*
   or
2. *row reducing A to find the pivotal columns of A. Then we obtain a one-to-one parameterization by*

$$t_1 \mathbf{v}_{j(1)} + t_2 \mathbf{v}_{j(2)} + \cdots + t_r \mathbf{v}_{j(r)}$$

where j(k) is the index of the kth pivotal column, and $\mathbf{v}_{j(k)}$ *is the corresponding pivotal column of A.*

EXAMPLE 8   **(Parameterizing Img(A) two ways)** Let

$$A = \begin{bmatrix} 1 & 0 & 4 & 2 \\ 3 & 1 & 7 & 1 \\ 1 & 2 & -6 & -8 \\ 4 & 1 & 11 & 3 \end{bmatrix}.$$

Let us parameterize Img(A) using both methods. Row reducing A, we find $RA = U$, where

$$R = \begin{bmatrix} 1 & 0 & 0 & 0 \\ -3 & 1 & 0 & 0 \\ 5 & -2 & 1 & 0 \\ -1 & -1 & 0 & 1 \end{bmatrix} \quad \text{and} \quad U = \begin{bmatrix} 1 & 0 & 4 & 2 \\ 0 & 1 & -5 & 5 \\ 0 & 0 & 0 & 0 \\ 0 & 0 & 0 & 0 \end{bmatrix}.$$

Hence, an equation for Img(A) is $C\mathbf{x} = 0$, where

$$C = \begin{bmatrix} 5 & -2 & 1 & 0 \\ -1 & -1 & 0 & 1 \end{bmatrix}.$$

Parameterizing the solution set of $C\mathbf{x} = 0$ using the method of Chapter 2,

$$(1/7) \begin{bmatrix} -s + 2t \\ s + 5t \\ 7s \\ 7t \end{bmatrix} = (s/7) \begin{bmatrix} -1 \\ 1 \\ 7 \\ 0 \end{bmatrix} + (t/7) \begin{bmatrix} 2 \\ 5 \\ 0 \\ 7 \end{bmatrix}.$$

We can simplify by absorbing the factors of $1/7$ into the parameters. Define $s_1 = s/7$ and $s_2 = t/7$ so that our parameterization becomes

$$s_1 \begin{bmatrix} -1 \\ 1 \\ 7 \\ 0 \end{bmatrix} + s_2 \begin{bmatrix} 2 \\ 5 \\ 0 \\ 7 \end{bmatrix}.$$

This is one way to get a parameterization of Img(A). Here is the other. As we have just seen, the pivotal columns of A are the first two. Hence,

$$t_1 \begin{bmatrix} 1 \\ 3 \\ 1 \\ 4 \end{bmatrix} + t_2 \begin{bmatrix} 0 \\ 1 \\ 2 \\ 1 \end{bmatrix}$$

is another one-to-one parameterization.

The first method used in Example 8 required two row reductions and the second method required only one. There is an advantage to the first method, though: It usually gives a simpler parameterization—notice there are more zeros in the vectors from the first parameterization. This will happen in general.

Exercises

1.1 Find an equation for Img(A) where $A = \begin{bmatrix} 1 & 2 & 3 \\ 2 & 1 & 3 \\ 1 & 1 & 2 \end{bmatrix}$.

1.2 Find an equation for Img(A) where $A = \begin{bmatrix} 1 & 2 & 1 \\ 2 & 0 & 1 \\ 3 & 2 & 2 \end{bmatrix}$.

1.3 Let $A = \begin{bmatrix} 1 & 0 & 4 & 2 \\ 3 & 1 & 7 & 1 \\ 1 & 2 & -6 & -8 \\ 4 & 1 & 11 & 3 \end{bmatrix}$. Find an LU decomposition for A, and use it to find an equation for Img(A). Also, find an equation for Img(A) by row reducing [A|$\mathbf{x}$]. How much overlap is there in the computations?

1.4 Let $A = \begin{bmatrix} 1 & 3 & 1 & 4 \\ 0 & 1 & 2 & 1 \\ 4 & 7 & -6 & 11 \\ 2 & 1 & -8 & 3 \end{bmatrix}$. Find an LU decomposition for A, and use it to find an equation for Img(A). Also, find an equation for Img(A) by row reducing [A|$\mathbf{x}$]. How much overlap is there in the computations?

1.5 Let A be a 3×3 invertible matrix. Find a matrix C so that $C\mathbf{x} = 0$ is an equation for Img(A).

1.6 Let $A = \begin{bmatrix} 1 & 2 & 3 \\ 2 & 1 & 3 \\ 1 & 1 & 2 \end{bmatrix}$. Use both methods of Theorem 2 to parameterize Img(A).

1.7 Let $A = \begin{bmatrix} 1 & 2 & 1 \\ 2 & 0 & 1 \\ 3 & 2 & 2 \end{bmatrix}$. Use both methods of Theorem 2 to parameterize Img(A).

1.8 Let $A = \begin{bmatrix} 1 & 1 & 0 & 1 \\ 4 & 2 & 6 & 0 \\ 4 & 3 & 1 & 2 \\ 1 & 2 & 0 & 3 \end{bmatrix}$. Use both methods of Theorem 2 to parameterize Img(A).

1.9 Let $A = \begin{bmatrix} 1 & 3 & 1 & 4 \\ 0 & 1 & 2 & 1 \\ 4 & 7 & -6 & 11 \\ 2 & 1 & -8 & 3 \end{bmatrix}$. Use both methods of Theorem 2 to parameterize Img(A).

1.10 Let A and B be two matrices, and suppose that A is a nonzero scalar multiple of B; that is, $A = cB$ for some $c \neq 0$. Show that Img(A) = Img(B).

1.11 Let A be an $m \times n$ matrix, and let B be an invertible $n \times n$ matrix. Must it be true that Img(A) = Img(AB)? Explain why, or give a counterexample.

1.12 Are there any matrices A such that Img(A) = Ker(A)? Explain why not, or give an example of such a matrix.

| ## The Closest Vector Problem and the Normal Equations

2.1 The normal equations and the closest vector in Img(A)

Suppose that $A\mathbf{x} = \mathbf{b}$ has no solution so that no choice of $\mathbf{x}$ makes $|A\mathbf{x} - \mathbf{b}|^2 = 0$. Then let us do the best we can: We choose $\mathbf{x}$ to make $|A\mathbf{x} - \mathbf{b}|^2$ as small as possible.

DEFINITION	**(Least squares solution)** Given an $m \times n$ matrix A and a vector $\mathbf{b}$ in $\mathbb{R}^m$, a vector $\mathbf{x}_0$ in $\mathbb{R}^n$ is called a *least squares solution* of $A\mathbf{x} = \mathbf{b}$ if and only if for all $\mathbf{x}$ in $\mathbb{R}^n$, $$|A\mathbf{x}_0 - \mathbf{b}|^2 \leq |A\mathbf{x} - \mathbf{b}|^2. \qquad (2.1)$$

Notice that if $\mathbf{x}_0$ solves $A\mathbf{x} = \mathbf{b}$ in the usual sense, then $|A\mathbf{x}_0 - \mathbf{b}|^2 = 0$ so that (2.1) is certainly true. Thus, every solution is a least squares solution.

Least squares problems come up in many contexts. Here is a simple example. Suppose that some experiment has given you three data points in the u, v plane,

$$(u_1, v_1) = (1, 2), \qquad (u_2, v_2) = (2, 5), \qquad \text{and} \qquad (u_3, v_3) = (3, 7). \qquad (2.2)$$

Suppose that on theoretical grounds, you expect a relation between v and u of the form

$$v = a + bu$$

for some unknown a and b. To find a and b from your data, just draw a line through your three points. Then b is the slope and a is the v intercept.

The problem is that there is no line through these three points, as you see in the following graph:

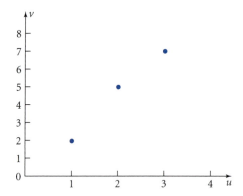

Clearly, the three points do not lie on any one line, though we can draw some lines that come close to passing through all three. Which line does best?

To answer this question, we first rephrase it as a linear algebra problem. Since we expect $au + b = v$, we are looking for values of a and b so that

$$a + bu_1 = v_1$$
$$a + bu_2 = v_2$$
$$a + bu_3 = v_3.$$

For the data (2.2),

$$a + b = 2$$
$$a + 2b = 5 \qquad\qquad (2.3)$$
$$a + 3b = 7,$$

a system of linear equations for the unknowns a and b. We know what to do next—write it in matrix form. Let

$$A = \begin{bmatrix} 1 & 1 \\ 1 & 2 \\ 1 & 3 \end{bmatrix} \quad \text{and} \quad \mathbf{b} = \begin{bmatrix} 2 \\ 5 \\ 7 \end{bmatrix}. \qquad\qquad (2.4)$$

Then with

$$\mathbf{x} = \begin{bmatrix} a \\ b \end{bmatrix},$$

(2.3) is equivalent to

$$A\mathbf{x} = \mathbf{b}. \qquad\qquad (2.5)$$

As we see from the graph, there is no solution. So we seek a least squares solution.

How do we do this? The definition of least squares solutions is based on an *inequality*, (2.1). We know a lot about solving *equations*, but this is something different. What do we do? Here is the answer:

- *Multiply both sides of (2.5) by A^t.*

This is not an obvious thing to do. We will explain later why it is the right thing to do. For now, note that it gives the equation

$$A^t A\mathbf{x} = A^t \mathbf{b}. \qquad\qquad (2.6)$$

Computing,

$$A^t A = \begin{bmatrix} 3 & 6 \\ 6 & 14 \end{bmatrix} \quad \text{and} \quad A^t \mathbf{b} = \begin{bmatrix} 14 \\ 33 \end{bmatrix}.$$

The 2×2 matrix $A^t A$ is invertible, and from the formula for 2×2 inverses, we have

$$(A^t A)^{-1} = \frac{1}{6} \begin{bmatrix} 14 & -6 \\ -6 & 3 \end{bmatrix}.$$

This is good news. While (2.5) has no solutions, (2.6) has a unique solution, namely,

$$\begin{bmatrix} a \\ b \end{bmatrix} = (A^t A)^{-1} \mathbf{b} = \frac{1}{6} \begin{bmatrix} 14 & -6 \\ -6 & 3 \end{bmatrix} \begin{bmatrix} 14 \\ 33 \end{bmatrix} = \begin{bmatrix} -1/3 \\ 5/2 \end{bmatrix},$$

and using these values of a and b in $v = a + bu$, we get the line

$$v = -\frac{1}{3} + \frac{5}{2} u.$$

Here is a diagram of the best-fit line plotted with the data points:

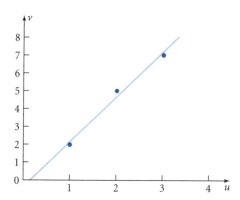

As you can see, the line fits pretty well. We now claim that it is the *best fit* in the least squares sense. That is, these values of a and b minimize

$$|A\mathbf{x} - \mathbf{b}|^2 = ((a + bu_1) - v_1)^2 + ((a + bu_2) - v_2)^2 + ((a + bu_3) - v_3)^2.$$

It is not immediately clear why this is true, but notice that the procedure for finding the best-fit line is very simple. We just multiplied the unsolvable equation (2.5) through by A^t, and got a new equation (2.6), which was not only solvable, it had a unique solution, and better yet, it involved a smaller matrix: 2×2 instead of 3×2.

Things are not always quite so simple, but the points we want to make with this introduction follow:

- *Least squares problems are practical and not artificial; curve fitting is one application of many we shall discuss.*
- *We will find least squares solutions by transforming unsolvable systems into solvable systems whose solutions are exactly the least squares solutions of the original system.*

We now explain why the transformation of (2.5) into (2.6) was the right transformation for solving the line-fitting problem. We want a general method, so we give the explanation in general terms.

Let A be a given $m \times n$ matrix and $\mathbf{b}$ a given vector in $\mathbb{R}^m$. To get going, make the optimistic assumption that a least squares solution $\mathbf{x}_0$ does in fact exist.* We will now turn the *inequality* (2.1) into an *equation* for $\mathbf{x}_0$.

Turning an inequality into an equation may sound like alchemy. The key is that (2.1) holds for every $\mathbf{x}$ in $\mathbb{R}^n$, so it is really the statement of infinitely many inequalities.

Although two wrongs do not make a right, enough inequalities can make an equality. For example, if you are told

$$x \geq 0 \quad \text{and} \quad x \leq 0,$$

you know that $x = 0$. Here is a useful vector version of this: Suppose $\mathbf{w}$ is some vector in $\mathbb{R}^n$ with the property

$$0 \leq \mathbf{w} \cdot \mathbf{v} \qquad \text{for all vectors } \mathbf{v} \text{ in } \mathbb{R}^n. \tag{2.7}$$

---

*Right now, this is an assumption, and making implicit assumptions always leads to trouble, so we make this assumption explicit.

Taking the choice $\mathbf{v} = -\mathbf{w}$, we see that $0 \le -|\mathbf{w}|^2$, which means that $|\mathbf{w}| = 0$, so

$$\mathbf{w} = 0. \tag{2.8}$$

Conversely, if $\mathbf{w} = 0$, then certainly (2.7) is true. Hence the inequality (2.7) is just another way of expressing the equation (2.8). We now use this reasoning to extract an equation from (2.1).

Let $\mathbf{v}$ be any vector in $\mathbb{R}^n$ and s any strictly positive number. Then with $\mathbf{x} = \mathbf{x}_0 + s\mathbf{v}$, we have from (2.1) and **V.I.F.5**, which brings in the transpose

$$
\begin{aligned}
|A\mathbf{x}_0 - \mathbf{b}|^2 &\le |A\mathbf{x} - \mathbf{b}|^2 \\
&= |A(\mathbf{x}_0 + s\mathbf{v}) - \mathbf{b}|^2 \\
&= |(A\mathbf{x}_0 - \mathbf{b}) + sA\mathbf{v}|^2 \\
&= |A\mathbf{x}_0 - \mathbf{b}|^2 + 2s(A\mathbf{x}_0 - \mathbf{b}) \cdot A\mathbf{v} + s^2|A\mathbf{v}|^2 \\
&= |A\mathbf{x}_0 - \mathbf{b}|^2 + 2s(A^tA\mathbf{x}_0 - A^t\mathbf{b}) \cdot \mathbf{v} + s^2|A\mathbf{v}|^2.
\end{aligned}
\tag{2.9}
$$

Canceling $|A\mathbf{x}_0 - \mathbf{b}|^2$ off both sides and then dividing by s,

$$0 \le 2(A^tA\mathbf{x}_0 - A^t\mathbf{b}) \cdot \mathbf{v} + s|A\mathbf{v}|^2$$

for all $\mathbf{v}$ in $\mathbb{R}^n$ and all $s > 0$. But since s can be arbitrarily small, we can get no significant help from the term $s|A\mathbf{v}|^2$, and it must be the case that

$$0 \le (A^tA\mathbf{x}_0 - A^t\mathbf{b}) \cdot \mathbf{v}$$

for all $\mathbf{v}$ in $\mathbb{R}^n$, which means that

$$A^tA\mathbf{x}_0 = A^t\mathbf{b}. \tag{2.10}$$

We have just proved that if there is any least squares solution $\mathbf{x}_0$ of $A\mathbf{x} = \mathbf{b}$, then $\mathbf{x}_0$ must be a solution in the usual sense of (2.10). It is standard practice to refer to the matrix equation (2.10), considered as a system of linear equations, as the *normal equations* for the least squares solutions of $A\mathbf{x} = \mathbf{b}$. The normal equations $A^tA\mathbf{x} = A^t\mathbf{b}$ have several nice features.

First, A^tA is a square matrix, and hence it is invertible if and only if $\mathrm{Ker}(A^tA) = 0$. Hence whenever $\mathrm{Ker}(A^tA) = 0$, (2.10) has a unique solution—*no matter what* $\mathbf{b}$ *is*.

Second,

$$\mathrm{Ker}(A^tA) = \mathrm{Ker}(A). \tag{2.11}$$

To see this, suppose that $\mathbf{x}$ belongs to $\mathrm{Ker}(A)$. This means that $A\mathbf{x} = 0$, and hence $A^tA\mathbf{x} = A^t(A\mathbf{x}) = 0$, so $\mathbf{x}$ belongs to $\mathrm{Ker}(A^tA)$. On the other hand, suppose that $\mathbf{x}$ belongs to $\mathrm{Ker}(A^tA)$. Then $0 = \mathbf{x} \cdot 0 = \mathbf{x} \cdot (A^tA\mathbf{x}) = A\mathbf{x} \cdot A\mathbf{x} = |A\mathbf{x}|^2$, so $\mathbf{x}$ belongs to $\mathrm{Ker}(A)$. In summary, $\mathbf{x}$ belongs to the kernel of A if and only if $\mathbf{x}$ belongs to the kernel of A^tA, which proves (2.11).

Recall that for any $m \times n$ matrix A, $\mathrm{Ker}(A) = 0$ if and only if the rank of A is n. But according to (2.11), the kernel of the square matrix A^tA is then also zero, and hence A^tA is invertible, guaranteeing a unique solution of the normal equations!

- *When the rank of A is n, the number columns in A, then (2.10) has a unique solution no matter what* $\mathbf{b}$ *is.*

One more point before some examples: We showed earlier that any least squares solution $A\mathbf{x} = \mathbf{b}$ has to be a solution of the normal equations $A^tA\mathbf{x} = A^t\mathbf{b}$. We now show that any solution of the normal equations $A^tA\mathbf{x} = A^t\mathbf{b}$ is in fact a least squares solution of $A\mathbf{x} = \mathbf{b}$.

Let $\mathbf{x}_0$ satisfy $A^t A \mathbf{x} = A^t \mathbf{b}$, and for any vector $\mathbf{x}$ in $\mathbb{R}^n$, let $\mathbf{v} = \mathbf{x} - \mathbf{x}_0$. Then, as in (2.9),

$$
\begin{aligned}
|A(\mathbf{x} - \mathbf{b})|^2 &= |A(\mathbf{x}_0 + \mathbf{v}) - \mathbf{b}|^2 \\
&= |(A\mathbf{x}_0 - \mathbf{b}) + A\mathbf{v}|^2 \\
&= |A\mathbf{x}_0 - \mathbf{b}|^2 + 2(A\mathbf{x}_0 - \mathbf{b}) \cdot A\mathbf{v} + |A\mathbf{v}|^2 \\
&= |A\mathbf{x}_0 - \mathbf{b}|^2 + 2(A^t A\mathbf{x}_0 - A^t\mathbf{b}) \cdot \mathbf{v} + |A\mathbf{v}|^2 \\
&= |A\mathbf{x}_0 - \mathbf{b}|^2 + |A\mathbf{v}|^2 \\
&\geq |(A\mathbf{x}_0 - \mathbf{b})|^2 ,
\end{aligned}
$$

since the cross terms vanish by the normal equations. Thus, (2.1) holds for all $\mathbf{x}$ in $\mathbb{R}^n$, so every solution of the normal equations is in fact a least squares solution of $A\mathbf{x} = \mathbf{b}$.

Let us summarize what we have learned so far:

- *Let A be an $m \times n$ matrix and $\mathbf{b}$ be any vector in $\mathbb{R}^n$. Then a vector $\mathbf{x}_0$ in $\mathbb{R}^n$ is a least squares solution of $A\mathbf{x} = \mathbf{b}$ if and only if $\mathbf{x}_0$ is a solution of the normal equations $A^t A\mathbf{x} = A^t\mathbf{b}$. Moreover, when $\mathrm{rank}(A) = n$, the normal equations have a unique solution for every $\mathbf{b}$ in $\mathbb{R}^m$.*

This conclusion has many important applications. Our next example is another line-fitting problem in which the data points are not all integers.

EXAMPLE 9   **(Fitting a line to five points)**   Suppose the data points (u_i, v_i), $i = 1, 2, \ldots, 5$ are

$$(1, 1.9), \qquad (2, 3.7), \qquad (3, 6.2), \qquad (4, 7.7), \qquad \text{and } (5, 10.5). \tag{2.12}$$

We seek numbers a and b so that

$$v_i = a + bu_i \qquad \text{for } i = 1, 2, \ldots, 5. \tag{2.13}$$

To write this in matrix form, define

$$
A = \begin{bmatrix} 1 & 1 \\ 1 & 2 \\ 1 & 3 \\ 1 & 4 \\ 1 & 5 \end{bmatrix} \qquad \text{and} \qquad \mathbf{b} = \begin{bmatrix} 1.9 \\ 3.7 \\ 6.2 \\ 7.7 \\ 10.5 \end{bmatrix}
$$

so that

$$\mathbf{x} = \begin{bmatrix} a \\ b \end{bmatrix}$$

satisfies $A\mathbf{x} = \mathbf{b}$ if and only if a and b satisfy (2.13) with the values of u_i and v_i from (2.12).

Here is a plot of the data:

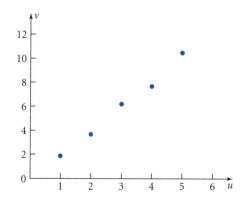

The five points do not lie on any one line, so we seek a least squares solution to $Ax = \mathbf{b}$. First, it is clear from the nature of the columns of A that the rank is 2. Therefore, the normal equations will have a unique solution, which will be the least squares solution.

Computing,

$$A^t A = \begin{bmatrix} 5 & 15 \\ 15 & 55 \end{bmatrix} \quad \text{and} \quad A^t \mathbf{b} = \begin{bmatrix} 30.0 \\ 111.2 \end{bmatrix}.$$

To solve the normal equations, row reduce the augmented matrix

$$\begin{bmatrix} 5 & 15 & | & 30.0 \\ 15 & 55 & | & 111.2 \end{bmatrix}$$

to find the solution

$$a = \frac{-9}{25} \quad \text{and} \quad b = \frac{53}{25}.$$

Here is a graph showing the best-fit line:

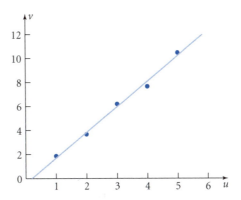

Now that we have computed a few least squares solutions, let us return to the "closest vector" question. By definition of a least squares solution of $Ax = \mathbf{b}$, $\mathbf{c} = Ax_0$ comes as close to $\mathbf{b}$ as any vector in $\text{Img}(A)$ can. Therefore, $|\mathbf{c} - \mathbf{b}| = |Ax_0 - \mathbf{b}|$ is a measure of the distance from $\mathbf{b}$ to $\text{Img}(A)$.

If $\mathbf{b}$ is obtained through laboratory measurements, and we believe on theoretical grounds that if measured exactly, it must lie in $\text{Img}(A)$, then $|Ax_0 - \mathbf{b}|$ provides a measure of the quality of our measurements. If $|Ax_0 - \mathbf{b}|$ is large, then there is a lot of "noise" in our measurements. For this reason, it is often useful to compute the closest vector $\mathbf{c} = Ax_0$ and $|\mathbf{c} - \mathbf{b}|^2$.

EXAMPLE 10 **(Computing the distance to Img(A))** Let A and $\mathbf{b}$ be as given in (2.4). At the beginning of this subsection, we found that the unique least squares solution to $Ax = \mathbf{b}$ is

$$x_0 = \begin{bmatrix} -1/3 \\ 5/2 \end{bmatrix}.$$

Therefore the vector $\mathbf{c}$ in $\text{Img}(A)$ that comes closest to $\mathbf{b}$ is

$$\mathbf{c} = Ax_0 = \begin{bmatrix} 1 & 1 \\ 1 & 2 \\ 1 & 3 \end{bmatrix} \begin{bmatrix} -1/3 \\ 5/2 \end{bmatrix} = \frac{1}{6} \begin{bmatrix} 1 \\ 2 \\ -1 \end{bmatrix},$$

so

$$\mathbf{b} - \mathbf{c} = \begin{bmatrix} 2 \\ 5 \\ 7 \end{bmatrix} \begin{bmatrix} -1/3 \\ 5/2 \end{bmatrix} - \frac{1}{6} \begin{bmatrix} 13 \\ 28 \\ 43 \end{bmatrix} = \frac{1}{6} \begin{bmatrix} 1 \\ 2 \\ -1 \end{bmatrix}.$$

Hence

$$|\mathbf{b} - \mathbf{c}| = \frac{1}{\sqrt{6}}.$$

The result is less than 5 percent of the length of $\mathbf{b}$, which is $\sqrt{78}$, so it takes only a 5 percent change in the measured data $\mathbf{b}$ to render the system solvable. If we are happy with 5 percent accuracy, then we are done. If not, we have to go back to the lab and measure whatever we were measuring a bit more carefully.

2.2 Exponential rates of growth

Another common application of the least squares method concerns determining *exponential rates of growth* from laboratory data, which occurs in almost every branch of science. Here we take an example from biology.

If a population of bacteria is grown with an ample supply of nutrients and space, the population will grow exponentially, meaning that if the population is measured at equal time intervals and p_n denotes the population at the nth time interval, there will be a formula

$$p_n = Ae^{bn} \tag{2.14}$$

for some numbers A and b. Here A is the initial population, and b is the exponential rate of growth.

The formula is not an exact formula for any finite value of A. In fact, the right-hand side usually will not even be an integer. Rather, it is a statistical law based on the law of large numbers, and it becomes exact only as the initial population tends to infinity. However, b is a characteristic of the type of bacteria and does not depend on the initial population. The problem of calculating this exponential rate of growth b from laboratory data often arises. The computation involves finding a least squares solution that "washes out" the effects of random fluctuations associated with a small laboratory population. Here is how it goes.

Any sequence of positive numbers $\{a_n\}$ given by the rule

$$a_n = Ae^{bn}$$

is called a *geometric progression*, and b is called its *exponential rate of growth*. Taking the natural logarithm of both sides,

$$\ln(a_n) = \ln(A) + bn.$$

The sequence $\{\ln(a_n)\}$ is an *arithmetic progression*: The points $(n, \ln(a_n))$ lie on a line. Moreover, the slope of this line is b, the exponential rate of growth of the original geometric progression $\{a_n\}$. That is,

- If $\{a_n\}$ is a geometric progression, the points $(n, \ln(a_n))$, $n = 1, 2, \ldots$, lie on a line, and b, the exponential rate of growth of $\{a_n\}$, is the slope of the line through the points $(n, \ln(a_n))$.

To find the exponential rate of growth b, we compute a best linear fit to the points $(n, \ln(a_n))$ for some range of n.

EXAMPLE 11   **(The exponential rate of growth of a bacteria colony)** Suppose that a bacteria colony is observed in a laboratory at hourly intervals. Let p_0 be the initial population and let p_n be the population at the nth hour. Suppose that the data are collected for 8 hours, with the following results for p_0 through p_8:

$$200, \quad 260, \quad 312, \quad 401, \quad 502, \quad 643, \quad 802, \quad 990, \quad 1253.$$

If the population is growing exponentially, then there would be a formula

$$p_n = Ae^{bn}$$

for some numbers A and b. In this case, taking the natural logarithm of both sides,

$$\ln(p_n) = a + bn \qquad \text{for } n = 0, 2, \ldots, 7, \tag{2.15}$$

where $a = \ln(A)$.

To write this system of equations for a and b in matrix form, introduce

$$A = \begin{bmatrix} 1 & 0 \\ 1 & 1 \\ 1 & 2 \\ 1 & 3 \\ 1 & 4 \\ 1 & 5 \\ 1 & 6 \\ 1 & 7 \\ 1 & 8 \end{bmatrix} \quad \text{and} \quad \mathbf{b} = \begin{bmatrix} \ln(200) \\ \ln(260) \\ \ln(312) \\ \ln(401) \\ \ln(502) \\ \ln(643) \\ \ln(802) \\ \ln(990) \\ \ln(1253) \end{bmatrix}.$$

Then a and b satisfy (2.15) for $n = 0, 1, \ldots, 7$ if and only if $A \begin{bmatrix} a \\ b \end{bmatrix} = \mathbf{b}$.

Before trying to solve this, let us plot the data point $(n \ln(p_n))$:

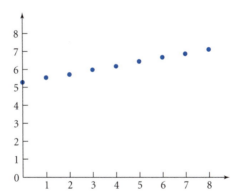

This plot is pretty encouraging; the data points do seem to lie more or less on a line, but not exactly. To find the line that fits best, we solve the normal equations

$$A^t A \begin{bmatrix} a \\ b \end{bmatrix} = A^t \mathbf{b}.$$

Computing $A^t A$ and $A^t \mathbf{b}$,

$$\begin{bmatrix} 9 & 36 \\ 36 & 204 \end{bmatrix} \begin{bmatrix} a \\ b \end{bmatrix} = \begin{bmatrix} 56.00 \\ 237.71 \end{bmatrix},$$

keeping three digits. Solving,

$$\begin{bmatrix} a \\ b \end{bmatrix} = \begin{bmatrix} 5.31 \\ 0.23 \end{bmatrix}.$$

Therefore, the best linear fit is

$$\ln(p_n) = 5.31 + (0.23)n.$$

Here is a graph showing this line fit to the data:

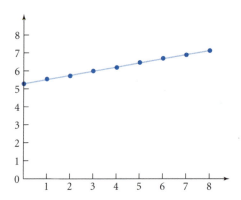

We can recover a formula for the populations p_n themselves by exponentiating:

$$p_n = e^{5.31 + (0.23)n}$$
$$= 202 \times (1.26)^n.$$

Plotting this against the actual data points, we can compare them to the best-fit exponential growth curve:

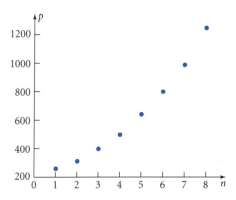

2.3 An approximate formula for the Fibonacci numbers

The sequence $\{f_n\}$ of Fibonacci numbers is generated by the recursive rule $f_{n+1} = f_n + f_{n-1}$ for $n \geq 2$, together with the starting values $f_1 = 1$ and $f_2 = 1$. Following are the first dozen terms:

$$1, \quad 1, \quad 2, \quad 3, \quad 5, \quad 8, \quad 13, \quad 21, \quad 34, \quad 55, \quad 89, \quad 144.$$

Certainly, this sequence is growing too fast to be a linear progression. Could it be a geometric progression?

To answer this question, we plot the points $(n, \ln(f_n))$ for $n = 1, 2, \ldots, 12$:

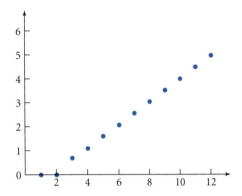

This graph looks pretty encouraging. Except for the very beginning, the points do look as if they would fit pretty well on a line. To find out which line, we seek numbers a and b so that

$$\ln(f_n) \approx a + bn.$$

To compute values for a and b, let us use the last half dozen points, where the linear relation looks quite good, and define

$$A = \begin{bmatrix} 1 & 7 \\ 1 & 8 \\ 1 & 9 \\ 1 & 10 \\ 1 & 11 \\ 1 & 12 \end{bmatrix} \quad \text{and} \quad \mathbf{b} = \begin{bmatrix} \ln(13) \\ \ln(21) \\ \ln(34) \\ \ln(55) \\ \ln(89) \\ \ln(144) \end{bmatrix}.$$

Then

$$A^t A = \begin{bmatrix} 6 & 57 \\ 57 & 559 \end{bmatrix} \quad \text{and} \quad A^t \mathbf{b} = \begin{bmatrix} 22.6016151\ldots \\ 223.134161\ldots \end{bmatrix}.$$

Solving

$$A^t A \begin{bmatrix} a \\ b \end{bmatrix} = A^t \mathbf{b},$$

we find

$$\begin{bmatrix} a \\ b \end{bmatrix} = \begin{bmatrix} -0.803279123\ldots \\ 0.481075262\ldots \end{bmatrix}. \tag{2.16}$$

Here is the graph of the data and the line:

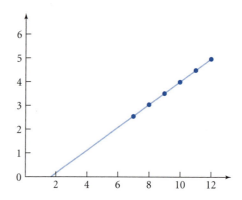

The graph shows a very good fit. Keeping only the digits just shown, the least squares approximation to $\ln(f_n)$ is given by

$$\ln(f_n) \approx -0.803279123 + (0.481075262)n,$$

which means that

$$f_n \approx e^{-0.803279123 + (0.481075262)n}$$
$$\approx (0.447857972) \times (1.61781304)^n. \tag{2.17}$$

Not only does the graph fit the data points well, we can extrapolate the line to estimate f_n for higher n, and it still does pretty well.

If we use this formula to extrapolate a value for f_{18}, we find

$$f_{18} \approx 2581.37013,$$

and since f_{18} must be an integer, we expect f_{18} to be an integer close to 2581. The actual value is 2584. Not bad!

Going out farther, the formula (2.17) predicts $f_{34} = 5684648$, while in fact $f_{34} = 5702887$. Percentagewise, pretty good.

We will use linear algebra methods to derive an exact formula for f_n in Chapter 5 and will see that

$$f_n = \frac{1}{\sqrt{5}} \left(\left(\frac{1 + \sqrt{5}}{2} \right)^n - \left(\frac{1 - \sqrt{5}}{2} \right)^n \right).$$

This is an amazing formula—at first glance, it may even seem preposterous to claim that it always defines an integer. But try it out: It does.

We will see how to derive the formula later on; for now we mention that the number

$$\frac{1 + \sqrt{5}}{2}$$

is called the *golden mean*. Its decimal value is $1.618033989\ldots$, while $(1 - \sqrt{5})/2 = -0.6180339890\ldots$. Therefore, the absolute value of

$$\frac{1}{\sqrt{5}} \left(\frac{1 - \sqrt{5}}{2} \right)^n$$

is less than $1/2$ for all n, so f_n is the integer closest to

$$\frac{1}{\sqrt{5}}\left(\frac{1+\sqrt{5}}{2}\right)^n.$$

This fact explains why we found a good exponential progression for the Fibonacci numbers. The natural logarithm of the golden mean has the decimal value $0.4812118252\ldots$, which we recognize in (2.16).

2.4 The distance between two lines

Least squares problems are everywhere. Here is another one: Consider two lines in $\mathbb{R}^3$ given parametrically by $\mathbf{x}_1(s) = \mathbf{x}_1 + s\mathbf{v}_1$ and $\mathbf{x}_2(t) = \mathbf{x}_2 + t\mathbf{v}_2$. The distance between the points $\mathbf{x}_1(s)$ and $\mathbf{x}_2(t)$ is $|\mathbf{x}_1(s) - \mathbf{x}_2(t)|$. As s and t vary, so does the distance between the points. The *distance between the two lines* is, by definition, the minimum value of $|\mathbf{x}_1(s) - \mathbf{x}_2(t)|$ as s and t range independently over the real numbers. Computing this distance is a least squares problem, as we now explain.

First of all, we seek the values of s and t that minimize the squared distance $|\mathbf{x}_1(s) - \mathbf{x}_2(t)|^2$. If we can find s_0 and t_0 so that

$$|\mathbf{x}_1(s_0) - \mathbf{x}_2(t_0)|^2 \le |\mathbf{x}_1(s) - \mathbf{x}_2(t)|^2 \qquad \text{for all } s, t, \tag{2.18}$$

then taking square roots of both sides, we have

$$|\mathbf{x}_1(s_0) - \mathbf{x}_2(t_0)| \le |\mathbf{x}_1(s) - \mathbf{x}_2(t)| \qquad \text{for all } s, t, \tag{2.19}$$

since the square root function is monotone increasing. Hence, $|\mathbf{x}_1(s_0) - \mathbf{x}_2(t_0)|$ is the distance between the lines.

Note that

$$\mathbf{x}_1(s) - \mathbf{x}_2(t) = s\mathbf{v}_1 - t\mathbf{v}_2 - (\mathbf{x}_2 - \mathbf{x}_1).$$

If we define A to be the matrix $A = [\mathbf{v}_1, -\mathbf{v}_2]$, and $\mathbf{b} = \mathbf{x}_2 - \mathbf{x}_1$ and finally $\mathbf{s} = \begin{bmatrix} s \\ t \end{bmatrix}$, we can rewrite the equation for $\mathbf{x}_1(s) - \mathbf{x}_2(t)$ as

$$\mathbf{x}_1(s) - \mathbf{x}_2(t) = A\mathbf{s} - \mathbf{b}.$$

Therefore,

$$|\mathbf{x}_1(s) - \mathbf{x}_2(t)|^2 = |A\mathbf{s} - \mathbf{b}|^2.$$

So if

$$\mathbf{s}_0 = \begin{bmatrix} s_0 \\ t_0 \end{bmatrix}$$

is the least squares solution of $A\mathbf{s} = \mathbf{b}$, then (2.18) holds, and the distance between the two lines is $|A\mathbf{s}_0 - \mathbf{b}|$.

EXAMPLE 12   **(The distance between two lines)** Let $\mathbf{x}_1, \mathbf{x}_2, \mathbf{v}_1$, and $\mathbf{v}_2$ be given by

$$\mathbf{x}_1 = \begin{bmatrix} 1 \\ 1 \\ 1 \end{bmatrix}, \qquad \mathbf{x}_2 = \begin{bmatrix} 3 \\ 1 \\ 0 \end{bmatrix}, \qquad \mathbf{v}_1 = \begin{bmatrix} 2 \\ 2 \\ 1 \end{bmatrix}, \qquad \text{and} \qquad \mathbf{v}_2 = \begin{bmatrix} -1 \\ 0 \\ -1 \end{bmatrix}.$$

Then

$$A = \begin{bmatrix} 2 & 1 \\ 2 & 0 \\ 1 & 1 \end{bmatrix} \quad \text{and} \quad \mathbf{b} = \begin{bmatrix} 2 \\ 0 \\ -1 \end{bmatrix}.$$

The normal equations are

$$A^t A \mathbf{s} = A^t \mathbf{b}.$$

We compute

$$A^t A = \begin{bmatrix} 9 & 3 \\ 3 & 2 \end{bmatrix},$$

so

$$(A^t A)^{-1} = \frac{1}{9} \begin{bmatrix} 2 & -3 \\ -3 & 9 \end{bmatrix}.$$

We then compute

$$A^t \mathbf{b} = \begin{bmatrix} 3 \\ 1 \end{bmatrix},$$

so

$$\begin{bmatrix} s_0 \\ t_0 \end{bmatrix} = \frac{1}{9} \begin{bmatrix} 2 & -3 \\ -3 & 9 \end{bmatrix} \begin{bmatrix} 3 \\ 1 \end{bmatrix} = \frac{1}{3} \begin{bmatrix} 1 \\ 0 \end{bmatrix}.$$

Hence, $s_0 = 1/3$ and $t_0 = 0$.

We now compute

$$\mathbf{x}_1(1/3) = \frac{1}{3} \begin{bmatrix} 5 \\ 5 \\ 4 \end{bmatrix} \quad \text{and} \quad \mathbf{x}_2(0) = \begin{bmatrix} 3 \\ 1 \\ 0 \end{bmatrix}.$$

Hence,

$$\mathbf{x}_1(1/3) - \mathbf{x}_2(0) = \frac{1}{3} \begin{bmatrix} -4 \\ 2 \\ 4 \end{bmatrix},$$

so $|\mathbf{x}_1(1/3) - \mathbf{x}_2(0)| = \frac{\sqrt{34}}{3}$, the distance between the two lines; $|\mathbf{x}_1(s) - \mathbf{x}_2(t)|$ is no smaller for any values of s and t, the value at $s = 1/3$, and $t = 0$.

2.5 What if not all of the columns of A are pivotal?

In all the problems we have considered so far, the rank of A was equal to the number of its columns, so every column was pivotal. We saw that in this case the normal equations $A^t A \mathbf{x} = A^t \mathbf{b}$ have a unique solution. What do we do if not all the columns are pivotal? Easy: *We just discard the nonpivotal columns.*

In a bit more detail, let A be an $m \times n$ matrix with rank r, and suppose that $r < n$ so that there are $n - r$ nonpivotal columns. Let B denote the $m \times r$ matrix obtained by deleting the nonpivotal columns of A and keeping the others in the same order. First we note that

$$\text{Img}(B) = \text{Img}(A).$$

Indeed, B and A have the same pivotal columns, and the pivotal columns of a matrix provide a one-to-one parameterization of its image by Theorem 1.

Therefore, for any vector $\mathbf{b}$ in $\mathbb{R}^m$, there is a unique least squares solution $\mathbf{x}_0$ to $B\mathbf{x} = \mathbf{b}$:

$$\mathbf{x}_0 = (B^t B)^{-1} B^t \mathbf{b}.$$

But then

$$\mathbf{c} = B\mathbf{x}_0 = B(B^t B)^{-1} B^t \mathbf{b}$$

is the vector in $\text{Img}(B)$ that is closest to $\mathbf{b}$. But $\text{Img}(B) = \text{Img}(A)$, *so the same vector $\mathbf{c}$ is the vector in* $\text{Img}(A)$ *that is closest to* $\mathbf{b}$. Moreover, since there is just one least squares solution, the closest vector $\mathbf{c}$ is uniquely determined; there are no ties in this competition.

EXAMPLE 13 **(Least squares with deficient rank)** Let

$$A = \begin{bmatrix} 1 & 2 & 1 & 2 \\ 2 & 3 & 2 & 3 \\ 1 & 1 & 1 & 1 \\ 0 & 1 & 0 & 1 \end{bmatrix},$$

as in Example 3. We first find the vector $\mathbf{c}$ in $\text{Img}(A)$ that is closest to

$$\mathbf{b} = \begin{bmatrix} 1 \\ 1 \\ 1 \\ 1 \end{bmatrix}.$$

In Example 5, we found that the first two columns of A were pivotal. Hence,

$$B = \begin{bmatrix} 1 & 2 \\ 2 & 3 \\ 1 & 1 \\ 0 & 1 \end{bmatrix}.$$

We now compute

$$B^t B = \begin{bmatrix} 6 & 9 \\ 9 & 15 \end{bmatrix} \quad \text{and} \quad B^t \mathbf{b} = \begin{bmatrix} 4 \\ 7 \end{bmatrix}.$$

The equation

$$\begin{bmatrix} 6 & 9 \\ 9 & 15 \end{bmatrix} \mathbf{x} = \begin{bmatrix} 4 \\ 7 \end{bmatrix}$$

has the unique solution $\mathbf{x}_0$ where

$$\mathbf{x}_0 = \frac{1}{3} \begin{bmatrix} -1 \\ 2 \end{bmatrix}.$$

The vector $\mathbf{c}$ in $\text{Img}(A) = \text{Img}(B)$ that is closest to $\mathbf{b}$ is therefore

$$\mathbf{c} = B\mathbf{x}_0 = \frac{1}{3} \begin{bmatrix} 3 \\ 4 \\ 1 \\ 2 \end{bmatrix}.$$

To get the set of all least squares solutions of $A\mathbf{x} = \mathbf{b}$, we could solve $A\mathbf{x} = \mathbf{c}$ by row reduction. Row reducing $[A|\mathbf{c}]$ leads to

$$\begin{bmatrix} 1 & 2 & 1 & 2 & | & 1 \\ 0 & 3 & 0 & 3 & | & 2 \\ 0 & 0 & 0 & 0 & | & 0 \\ 0 & 0 & 0 & 0 & | & 0 \end{bmatrix},$$

which gives the system of equations

$$x + 2y + z + 2w = 1$$
$$3y + 3w = 2.$$

Setting $z = t_1$ and $w = t_2$, $y = 2/3 - t_2$ and $x = 1 - (4/3 - 2t_2) - t_1 - 2t_2 = -1/3 - t_1$. Hence the solutions are given by

$$\begin{bmatrix} x \\ y \\ z \\ w \end{bmatrix} = \begin{bmatrix} -1/3 - t_1 \\ 2/3 - t_2 \\ t_1 \\ t_2 \end{bmatrix} = \frac{1}{3} \begin{bmatrix} -1 \\ 2 \\ 0 \\ 0 \end{bmatrix} + t_1 \begin{bmatrix} -1 \\ 0 \\ 1 \\ 0 \end{bmatrix} + t_2 \begin{bmatrix} 0 \\ -1 \\ 0 \\ 1 \end{bmatrix}.$$

This is a one-to-one parameterization of the set of least squares solutions of $A\mathbf{x} = \mathbf{b}$.

If we want just *some* solution, there is an easier way to finish the analysis. Let $A = [\mathbf{v}_1, \mathbf{v}_2, \mathbf{v}_3, \mathbf{v}_4]$ so that $B = [\mathbf{v}_1, \mathbf{v}_2]$. We have just seen that with $x_1 = -1/3$ and $x_2 = 2/3$,

$$
\begin{aligned}
\mathbf{c} = B \begin{bmatrix} x_1 \\ x_2 \end{bmatrix} &= [\mathbf{v}_1, \mathbf{v}_2] \begin{bmatrix} x_1 \\ x_2 \end{bmatrix} \\
&= x_1\mathbf{v}_1 + x_2\mathbf{v}_2 \\
&= [\mathbf{v}_1, \mathbf{v}_2, \mathbf{v}_3, \mathbf{v}_4] \begin{bmatrix} x_1 \\ x_2 \\ 0 \\ 0 \end{bmatrix} = A \begin{bmatrix} x_1 \\ x_2 \\ 0 \\ 0 \end{bmatrix}.
\end{aligned}
$$

That is,

$$
\frac{1}{3} \begin{bmatrix} -1 \\ 2 \\ 0 \\ 0 \end{bmatrix}
$$

is a least squares solution of $A\mathbf{x} = \mathbf{b}$. To go from the unique least squares solution of $B\mathbf{x} = \mathbf{b}$ to one particular least squares solution of $A\mathbf{x} = \mathbf{b}$, we insert zeros in the places corresponding to free variables.

We summarize what we have learned so far about finding least squares solutions as follows (the proof lies in the discussion just before Example 13).

THEOREM 2   **(Normal equations and least squares solutions)** *Let A be an $m \times n$ matrix, and $\mathbf{b}$ be any vector in $\mathbb{R}^m$. Then there always exists a unique vector $\mathbf{c}$ in $\mathrm{Img}(A)$ that is closer to $\mathbf{b}$ than any other vector in $\mathrm{Img}(A)$, and thus $A\mathbf{x} = \mathbf{b}$ always possesses at least one least squares solution.*

If $\mathrm{rank}(A) = n$, $A^t A$ is invertible and there is exactly one least squares solution $\mathbf{x}_0$ of $A\mathbf{x} = \mathbf{b}$: the unique solution of the normal equations $A^t A\mathbf{x} = A^t\mathbf{b}$. The closest vector $\mathbf{c}$ is given by

$$
\mathbf{c} = A\mathbf{x}_0 = A(A^t A)^{-1} A^t \mathbf{b}.
$$

In case $\mathrm{rank}(A) < n$, let B be the $m \times r$ matrix obtained from A by deleting nonpivotal columns. Then $B^t B$ is invertible, and the closest vector $\mathbf{c}$ is given by

$$
\mathbf{c} = B\mathbf{x}_0 = B(B^t B)^{-1} B^t \mathbf{b},
$$

and the least squares solutions of $A\mathbf{x} = \mathbf{b}$ are the exact solutions of $A\mathbf{x} = \mathbf{c}$.
In both cases, the normal equations $A^t A\mathbf{x} = A^t\mathbf{b}$ are solvable for all $\mathbf{b}$ in $\mathbb{R}^m$.

The theorem has a simple corollary that turns out to be very useful.

COROLLARY   **(Kernel and image of $A^t A$)** *Let A be any $m \times n$ matrix. Then*

$$
\mathrm{Img}(A^t A) = \mathrm{Img}(A^t) \quad \text{and} \quad \mathrm{Ker}(A^t A) = \mathrm{Ker}(A). \tag{2.20}
$$

PROOF: We have already shown the second half of (2.20) in (2.11). To prove the first half, let $\mathbf{y}$ be any vector in $\mathrm{Img}(A^t A)$. Then by definition, there is a vector $\mathbf{x}$ in $\mathbb{R}^n$ so that $\mathbf{y} = A^t A\mathbf{x} = A^t(A\mathbf{x})$. This means that whenever $\mathbf{y}$ belongs to $\mathrm{Img}(A^t A)$, $\mathbf{y}$ also belongs to $\mathrm{Img}(A^t)$.

On the other hand, suppose that $\mathbf{y}$ belongs to $\mathrm{Img}(A^t)$. Then by definition, there is a vector $\mathbf{b}$ in $\mathbb{R}^m$ so that $\mathbf{y} = A^t\mathbf{b}$. By Theorem 2, $A^t A\mathbf{x} = \mathbf{y} = A^t\mathbf{b}$ has a solution, so whenever $\mathbf{y}$ belongs to $\mathrm{Img}(A^t)$, $\mathbf{y}$ also belongs to $\mathrm{Img}(A^t A)$, thus proving that $\mathrm{Img}(A^t A) = \mathrm{Img}(A^t)$. ∎

As we can see from the proof, the fact that $\text{Img}(A^tA) = \text{Img}(A^t)$ is essentially equivalent to the fact that the normal equations are always solvable.

2.6 Dangerous cases

Consider the matrix

$$A = \begin{bmatrix} 1 & 1 \\ 1 & 1+a \\ 1 & 1-a \end{bmatrix}.$$

As long as $a \neq 0$, $\text{rank}(A) = 2$, but when $a = 0$, $\text{rank}(A) = 1$, and we would use a different procedure in trying to solve

$$A\mathbf{x} = \mathbf{b}$$

where, say,

$$\mathbf{b} = \begin{bmatrix} 1 \\ 2 \\ 1 \end{bmatrix}.$$

What happens when a is small so that the rank is "almost" 1 and not 2? Very nasty things, if you are working on a computer doing a floating-point calculation so that numbers in your calculations are rounded off.

Here is why. Computing A^tA, we find

$$A^tA = \begin{bmatrix} 3 & 3 \\ 3 & 3+2a^2 \end{bmatrix} \quad \text{and} \quad A^t\mathbf{b} = \begin{bmatrix} 4 \\ 4+a \end{bmatrix}.$$

Now if you are working with, say, 16 significant digits, and $a = 10^{-9}$, then your computer will round off the $3 + 2a^2$ in A^tA to 3, but it will not round off the $4 + a$ in $A^t\mathbf{b}$. Hence your computer will try to solve

$$\begin{bmatrix} 3 & 3 \\ 3 & 3 \end{bmatrix} \begin{bmatrix} x \\ y \end{bmatrix} = \begin{bmatrix} 4 \\ 4+a \end{bmatrix}.$$

Obviously this equation has no solution. Indeed, when $3 + 2a^2$ was rounded off to 3, the invertibility of A^tA was lost.

This problem is serious, and when matrices are large, it is not always easy to tell by looking if the rank is "almost deficient." A good way to proceed is a variant on what we would do if not all the columns were pivotal: We would replace A by another matrix Q, where

$$\text{Img}(Q) = \text{Img}(A),$$

but where Q is well behaved. We are going to have to figure out what "well behaved" really means, but by way of a preview, if the columns of Q are orthonormal, then Q is very well behaved.

Better yet, if the columns of Q are orthonormal so that Q is an isometry, the formulas in Theorem 2 simplify considerably. If Q is an $m \times r$ isometry and $\mathbf{b}$ is any vector in $\mathbb{R}^m$, Theorem 2 states that the vector $\mathbf{c}$ in $\text{Img}(Q)$ closest to $\mathbf{b}$ is given by

$$\mathbf{c} = Q(Q^tQ)Q^t\mathbf{b}. \tag{2.21}$$

But since Q is an isometry, $Q^tQ = I$, which reduces to a much simpler formula:

$$\mathbf{c} = QQ^t\mathbf{b}.$$

More important than the simplicity is the fact that the relation $Q^tQ = I$ is "robust" in a way that will make our computations relatively immune to round-off errors. Therefore, we end this section with an important question:

- *Given an $m \times n$ matrix A, can we always find an isometry Q so that* $\text{Img}(Q) = \text{Img}(A)$, *and if so, how?*

The answer to the first part of the question is "Yes, always." Explaining how takes some preparation. We begin with that in the next section.

Exercises

2.1 Find the least squares solution to $Ax = b$, where $A = \begin{bmatrix} 1 & 1 & 2 \\ 1 & 0 & 1 \\ 0 & 1 & 1 \end{bmatrix}$ and $b = \begin{bmatrix} 1 \\ 2 \\ 3 \end{bmatrix}$. Also, find the

vector in $\text{Img}(A)$ that is closest to b and the distance from b to $\text{Img}(A)$.

2.2 Find the least squares solution to $Ax = b$, where $A = \begin{bmatrix} 1 & 1 & 2 \\ 1 & -1 & 0 \\ 0 & 1 & 1 \end{bmatrix}$ and $b = \begin{bmatrix} 1 \\ 2 \\ 3 \end{bmatrix}$. Also, find the

vector in $\text{Img}(A)$ that is closest to b and the distance from b to $\text{Img}(A)$.

2.3 Find the least squares solution to $Ax = b$, where $A = \begin{bmatrix} 1 & 3 & 3 \\ 0 & 1 & 2 \\ 2 & 1 & -4 \\ 1 & 2 & 3 \end{bmatrix}$ and $b = \begin{bmatrix} 1 \\ 2 \\ 3 \\ 4 \end{bmatrix}$. Also, find the

vector in $\text{Img}(A)$ that is closest to b and the distance from b to $\text{Img}(A)$.

2.4 Find the least squares solution to $Ax = b$, where $A = \begin{bmatrix} 1 & 0 & 2 \\ 2 & 1 & 3 \\ 1 & 2 & 4 \\ 3 & 1 & 5 \end{bmatrix}$ and $b = \begin{bmatrix} 1 \\ 2 \\ 3 \\ 4 \end{bmatrix}$. Also, find the

vector in $\text{Img}(A)$ that is closest to b and the distance from b to $\text{Img}(A)$.

2.5 Consider the points (x_i, y_i), $i = 1, \ldots, 5$ given by

$$(1, 0), \qquad (2, 3), \qquad (3, 7), \qquad (4, 14), \qquad \text{and} \qquad (5, 22).$$

(a) Compute a best linear fit $y = a + bx$ for these data. Using your values for a and b, compute

$$\sum_{i=1}^{5} |y_i - (a + bx_i)|^2.$$

(b) Compute a best *quadratic* fit $y = a + bx + cx^2$ for these data. Using your values for a, b, and c, compute

$$\sum_{i=1}^{5} |y_i - (a + bx_i + cx_i^2)|^2.$$

(c) Based on your results from parts (a) and (b), would you say it is likely that these data points come from measuring quantities that are linearly related?

2.6 Consider two lines in $\mathbb{R}^3$ given parametrically by $x_1(s) = x_1 + sv_1$ and $x_2(t) = x_2 + tv_2$, where

$$x_1 = \begin{bmatrix} 1 \\ 2 \\ 1 \end{bmatrix}, \qquad x_2 = \begin{bmatrix} 1 \\ -1 \\ 0 \end{bmatrix}, \qquad v_1 = \begin{bmatrix} 1 \\ 0 \\ -1 \end{bmatrix}, \qquad \text{and} \qquad v_2 = \begin{bmatrix} 2 \\ 1 \\ 1 \end{bmatrix}.$$

Compute the distance between these two lines.

2.7 Consider two lines in $\mathbb{R}^3$ given parametrically by $\mathbf{x}_1(s) = \mathbf{x}_1 + s\mathbf{v}_1$ and $\mathbf{x}_2(t) = \mathbf{x}_2 + t\mathbf{v}_2$, where

$$\mathbf{x}_1 = \begin{bmatrix} 1 \\ 2 \\ 3 \end{bmatrix}, \qquad \mathbf{x}_2 = \begin{bmatrix} 2 \\ 0 \\ 2 \end{bmatrix}, \qquad \mathbf{v}_1 = \begin{bmatrix} 1 \\ 2 \\ 2 \end{bmatrix}, \qquad \text{and} \qquad \mathbf{v}_2 = \begin{bmatrix} -2 \\ 1 \\ 1 \end{bmatrix}.$$

Compute the distance between these two lines.

2.8 Consider two lines in $\mathbb{R}^3$ given parametrically by $\mathbf{x}_1(s) = \mathbf{x}_1 + s\mathbf{v}_1$ and $\mathbf{x}_2(t) = \mathbf{x}_2 + t\mathbf{v}_2$, where

$$\mathbf{x}_1 = \begin{bmatrix} 1 \\ 2 \\ -1 \end{bmatrix}, \qquad \mathbf{x}_2 = \begin{bmatrix} 2 \\ 1 \\ -5 \end{bmatrix}, \qquad \mathbf{v}_1 = \begin{bmatrix} 1 \\ -4 \\ -2 \end{bmatrix}, \qquad \text{and} \qquad \mathbf{v}_2 = \begin{bmatrix} 1 \\ 1 \\ -2 \end{bmatrix}.$$

Compute the distance between these two lines.

2.9 Consider two lines in $\mathbb{R}^3$ given parametrically by $\mathbf{x}_1(s) = \mathbf{x}_1 + s\mathbf{v}_1$ and $\mathbf{x}_2(t) = \mathbf{x}_2 + t\mathbf{v}_2$, where

$$\mathbf{x}_1 = \begin{bmatrix} 3 \\ 2 \\ 1 \end{bmatrix}, \qquad \mathbf{x}_2 = \begin{bmatrix} 1 \\ 1 \\ -1 \end{bmatrix}, \qquad \mathbf{v}_1 = \begin{bmatrix} 3 \\ -5 \\ -1 \end{bmatrix}, \qquad \text{and} \qquad \mathbf{v}_2 = \begin{bmatrix} -1 \\ 3 \\ 3 \end{bmatrix}.$$

Compute the distance between these two lines.

SECTION 3 | Subspaces

3.1 Subspaces of $\mathbb{R}^n$

Images and kernels are closely related. In the first section of this chapter, we saw that the image of every matrix A is the kernel of another matrix C.

The converse is also true: Suppose that $t_1\mathbf{v}_1 + t_2\mathbf{v}_2 + \cdots + t_k\mathbf{v}_k$ is a one-to-one parameterization of $\text{Ker}(A)$. Form the matrix $B = [\mathbf{v}_1, \mathbf{v}_2, \ldots, \mathbf{v}_k]$, and let

$$\mathbf{x} = \begin{bmatrix} t_1 \\ t_2 \\ \vdots \\ t_n \end{bmatrix}.$$

Then by **V.I.F.1**,

$$B\mathbf{x} = [\mathbf{v}_1, \mathbf{v}_2, \ldots, \mathbf{v}_k] \begin{bmatrix} t_1 \\ t_2 \\ \vdots \\ t_n \end{bmatrix} = t_1\mathbf{v}_1 + t_2\mathbf{v}_2 + \cdots + t_k\mathbf{v}_k.$$

As $\mathbf{x}$ ranges over $\mathbb{R}^k$, $B\mathbf{x}$ ranges over $\text{Ker}(A)$, and hence, $\text{Ker}(A) = \text{Img}(B)$. We conclude that the kernel of A is the image of another matrix B. As sets of vectors, images and kernels are essentially the same thing.

Since evidently kernels and images are closely related and since both are important in the theory of linear equations, it is useful to have a definition of a single class of subsets of $\mathbb{R}^n$ that includes both images and kernels at once.

DEFINITION

(Subspace) A nonempty subset S of $\mathbb{R}^n$ is called a *subspace* if and only if it is closed under vector addition and multiplication by numbers. That is,

1. $\mathbf{x} + \mathbf{y}$ belongs to S for all $\mathbf{x}$ and $\mathbf{y}$ in S.

2. $a\mathbf{x}$ belongs to S for all $\mathbf{x}$ in S and all numbers a.

The first two examples show that this definition encompasses both images and kernels.

EXAMPLE 14   **(Ker(A) is a subspace)** Let A be a matrix and let $S = \text{Ker}(A)$. If $\mathbf{x}$ and $\mathbf{y}$ belong to S,

$$A(\mathbf{x} + \mathbf{y}) = A\mathbf{x} + A\mathbf{y} = 0 + 0 = 0,$$

so requirement (1) is satisfied. Next, for any number a and any $\mathbf{x}$ in S,

$$A(a\mathbf{x}) = a(A\mathbf{x}) = a0 = 0.$$

Therefore, requirement (2) is also satisfied, and $S = \text{Ker}(A)$ is a subspace.

EXAMPLE 15   **(Img(A) is a subspace)** Let A be a matrix, and let $S = \text{Img}(A)$. If $\mathbf{x}$ and $\mathbf{y}$ belong to S, then by definition there are vectors $\mathbf{u}$ and $\mathbf{v}$ so that $\mathbf{x} = A\mathbf{u}$ and $\mathbf{y} = A\mathbf{v}$. Then

$$\mathbf{x} + \mathbf{y} = A\mathbf{u} + A\mathbf{v} = A(\mathbf{u} + \mathbf{v}),$$

so requirement (1) is satisfied. Next, for any number a and any $\mathbf{x}$ in S, there is a vector $\mathbf{u}$ with $\mathbf{x} = A\mathbf{u}$ so that

$$a\mathbf{x} = a(A\mathbf{u}) = A(a\mathbf{u}).$$

Therefore, $a\mathbf{x}$ is in the image of A, and requirement (2) is also satisfied. Hence $S = \text{Img}(A)$ is a subspace.

- *Because of Examples 14 and 15, any theorem we prove about subspaces applies to images and kernels.*

EXAMPLE 16   **(The zero subspace)** The subset of $\mathbb{R}^n$ consisting of only the zero vector 0 is a subspace since for any number a, $a0 = 0$ and since $0 + 0 = 0$.

Important note about notation: We use 0 to denote the zero subspace, that is, the subspace consisting of just the zero vector 0. The symbol 0 has three meanings: the *number* 0, the *zero vector* in $\mathbb{R}^n$, any n, and the *zero subspace* of $\mathbb{R}^n$, any n. This practice may seem dangerous, but it is standard, and the context will always make the meaning of the symbol 0 evident.

Notice that if S is a subspace of $\mathbb{R}^n$, the zero vector always belongs to S. The reason is that for any $\mathbf{x}$ in $\mathbb{R}^n$, $0 = 0\mathbf{x}$, and since S is closed under scalar multiplication and nonempty, it contains 0.

In $\mathbb{R}^3$, lines and planes through the origin are examples of subspaces. This conclusion follows from Example 14, since lines and planes are the solution sets of equations of the form $C\mathbf{x} = 0$.

Lines or planes that *do not* pass through the origin *are not* subspaces by what we have observed earlier about subspaces always containing the zero vector.

THEOREM 3

(Subspaces and linear combinations) *A subset S of $\mathbb{R}^n$ is a subspace if and only if any linear combination of vectors chosen from S also belongs to S.*

PROOF: Suppose S is a subspace. Let $t_1\mathbf{v}_1 + t_2\mathbf{v}_2 + \cdots + t_k\mathbf{v}_k$ be any combination of vectors $\mathbf{v}_j$ in S. We have to show that this linear combination belongs to S.

Define $\mathbf{w}_j = t_j\mathbf{v}_j$. Since S is closed under multiplication by numbers, each $\mathbf{w}_j$ belongs to S. Since

$$t_1\mathbf{v}_1 + t_2\mathbf{v}_2 + \cdots + t_k\mathbf{v}_k = \mathbf{w}_1 + \mathbf{w}_2 + \cdots + \mathbf{w}_k,$$

it suffices to show that the sum on the right belongs to S. But since S is closed under addition, $\mathbf{w}_1 + \mathbf{w}_2$ is in S. Again since S is closed under addition,

$$(\mathbf{w}_1 + \mathbf{w}_2) + \mathbf{w}_3 = \mathbf{w}_1 + \mathbf{w}_2 + \mathbf{w}_3$$

is in S. Continuing, we eventually conclude that $\mathbf{w}_1 + \mathbf{w}_2 + \cdots + \mathbf{w}_k$ is in S. The other implication is clear. ∎

EXAMPLE 17

(The "integer lattice" is not a subspace) Let S be the subset of $\mathbb{R}^n$ consisting of vectors with integer entries. This is sometimes called the *integer lattice* in $\mathbb{R}^n$. Is it a subspace? By Theorem 3, we only need to check whether it is closed under taking linear combinations. If $\mathbf{x}$ and $\mathbf{y}$ have integer entries, then so does $\mathbf{x} + \mathbf{y}$. However, taking scalar multiples can take us outside of S, for instance, when $t = 1/2$ and $\mathbf{x}$ has an odd entry. So S is not a subspace.

3.2 Spanning sets

With these examples in hand, we turn to the parameterization of subspaces. Suppose we are trying to find a one-to-one parameterization of a subspace S of $\mathbb{R}^n$. Suppose we have a set $\{\mathbf{v}_1, \mathbf{v}_2, \ldots, \mathbf{v}_k\}$ of vectors in S, and every vector $\mathbf{w}$ in S is a linear combination of them:

$$\mathbf{w} = x_1\mathbf{v}_1 + x_2\mathbf{v}_2 + \cdots + x_k\mathbf{v}_k$$

for some values of $x_1, x_2, \ldots, x_k$. This equation would not necessarily give us a one-to-one parameterization. For it to do so, the x_j would need to be uniquely determined.

DEFINITION

(Spanning set) A collection of vectors $\{\mathbf{v}_1, \mathbf{v}_2, \ldots, \mathbf{v}_k\}$ in a subspace S of $\mathbb{R}^n$ *spans* S if every vector in S can be written as a linear combination

$$\mathbf{w} = t_1\mathbf{v}_1 + t_2\mathbf{v}_2 + \cdots + t_k\mathbf{v}_k \tag{3.1}$$

in at least one way. In this case, $\{\mathbf{v}_1, \mathbf{v}_2, \ldots, \mathbf{v}_k\}$ is a *spanning set* for S.

EXAMPLE 18

(Spanning sets) The vectors $\{\mathbf{e}_1, \mathbf{e}_2, \mathbf{e}_3\}$ in $\mathbb{R}^3$ span $\mathbb{R}^3$ since certainly any vector $\mathbf{w}$ in $\mathbb{R}^3$ can be expressed as a linear combination of them. We have been using this fact all along; the only thing new here is another way to talk about it.

We can find other examples of spanning sets in parameterizations. Let $A = \begin{bmatrix} 1 & 2 & 1 & 2 \\ 2 & 3 & 2 & 3 \\ 1 & 1 & 1 & 1 \\ 0 & 1 & 0 & 1 \end{bmatrix}$, the matrix considered in Example 2 in Section 1. We found there that $\text{Img}(A)$ is the subspace S of $\mathbb{R}^4$ parameterized by

$$t_1 \begin{bmatrix} 1 \\ 2 \\ 1 \\ 0 \end{bmatrix} + t_2 \begin{bmatrix} 1 \\ 1 \\ 0 \\ 1 \end{bmatrix} = t_1 \mathbf{u}_1 + t_2 \mathbf{u}_2, \tag{3.2}$$

where this equality defines $\mathbf{u}_1$ and $\mathbf{u}_2$. Hence, $\{\mathbf{u}_1, \mathbf{u}_2\}$ spans $S = \text{Img}(A)$.

In general, given any set $\{\mathbf{v}_1, \mathbf{v}_2, \ldots, \mathbf{v}_k\}$ of vectors in $\mathbb{R}^n$, we can form the set of all linear combinations of these vectors.

DEFINITION	**(Span)** Given a set $\{\mathbf{v}_1, \mathbf{v}_2, \ldots, \mathbf{v}_k\}$ of vectors in $\mathbb{R}^n$, the *span* of $\{\mathbf{v}_1, \mathbf{v}_2, \ldots, \mathbf{v}_k\}$ is the set of all linear combinations of these vectors. It is denoted by $$\text{Sp}(\mathbf{v}_1, \mathbf{v}_2, \ldots, \mathbf{v}_k). \tag{3.3}$$

$\text{Sp}(\mathbf{v}_1, \mathbf{v}_2, \ldots, \mathbf{v}_k)$ is always a subspace. If we form the $n \times k$ matrix $A = [\mathbf{v}_1, \mathbf{v}_2, \ldots, \mathbf{v}_k]$, then

$$A \begin{bmatrix} t_1 \\ t_2 \\ \vdots \\ t_k \end{bmatrix} = t_1 \mathbf{v}_1 + t_2 \mathbf{v}_2 + \cdots + t_k \mathbf{v}_k,$$

the "general" linear combination of these vectors. Hence, the image of A is exactly the span of $\{\mathbf{v}_1, \mathbf{v}_2, \ldots, \mathbf{v}_k\}$. We know that $\text{Img}(A)$ is a subspace, and hence, $\text{Sp}(\mathbf{v}_1, \mathbf{v}_2, \ldots, \mathbf{v}_k)$ is a subspace.

THEOREM 4   **(Image of A and the span of its columns)** *Let A be any matrix. Then the image of A is the span of the columns of A.*

PROOF: As we just saw,

$$\text{Img}([\mathbf{v}_1, \mathbf{v}_2, \ldots, \mathbf{v}_k]) = \text{Sp}(\mathbf{v}_1, \mathbf{v}_2, \ldots, \mathbf{v}_k). \tag{3.4}$$

∎

3.3 Linear independence

Suppose that $\{\mathbf{v}_1, \mathbf{v}_2, \ldots, \mathbf{v}_k\}$ is any set of k vectors in $\mathbb{R}^m$ that spans a subspace S. Then by definition, any vector $\mathbf{w}$ in S is a linear combination of the $\{\mathbf{v}_1, \mathbf{v}_2, \ldots, \mathbf{v}_k\}$:

$$\mathbf{w} = x_1 \mathbf{v}_1 + x_2 \mathbf{v}_2 + \cdots + x_k \mathbf{v}_k \tag{3.5}$$

for some values of $x_1, x_2, \ldots, x_k$. We now ask the question

- *Under what circumstances does (3.5) provide us with a one-to-one parameterization of S?*

Since the parameterization in (3.5) is one-to-one if and only if for each $\mathbf{w}$ the coefficients $x_1, x_2, \ldots, x_k$ are uniquely determined, this question is about whether or not a linear system has more than one solution. This we know.

To rephrase the question in matrix form, introduce the $m \times k$ matrix $A = [\mathbf{v}_1, \mathbf{v}_2, \ldots, \mathbf{v}_k]$ and the vector

$$\mathbf{x} = \begin{bmatrix} x_1 \\ x_2 \\ \vdots \\ x_k \end{bmatrix}.$$

Then we can rewrite (3.5) as

$$A\mathbf{x} = \mathbf{w}. \tag{3.6}$$

Since the fact that $\mathbf{w}$ belongs to S means that (3.6) has at least one solution, Theorem 5 of Chapter 2 tells us that (3.6) has a unique solution if and only if $\text{Ker}(A) = 0$.

In other words, the parameterization (3.5) is one-to-one if and only if $\text{Ker}(A) = 0$, that is, if and only if $A\mathbf{x} = 0$ has only the zero solution. Rephrased yet again, the parameterization (3.5) is one-to-one if and only if

$$x_1\mathbf{v}_1 + x_2\mathbf{v}_2 + \cdots + x_k\mathbf{v}_k = 0 \iff x_j = 0 \qquad \text{for } j = 1, 2, \ldots, k. \tag{3.7}$$

DEFINITION

(Linear independence) A set of vectors $\{\mathbf{v}_1, \mathbf{v}_2, \ldots, \mathbf{v}_k\}$ in $\mathbb{R}^m$ is *linearly independent* if and only if

$$x_1\mathbf{v}_1 + x_2\mathbf{v}_2 + \cdots + x_k\mathbf{v}_k = 0 \Rightarrow x_j = 0 \qquad \text{for each } j = 1, 2, \ldots, k. \tag{3.8}$$

THEOREM 5

(Independence and uniqueness) *For any set of vectors* $\{\mathbf{v}_1, \mathbf{v}_2, \ldots, \mathbf{v}_k\}$ *in* $\mathbb{R}^n$, *the following are equivalent:*

1. $\{\mathbf{v}_1, \mathbf{v}_2, \ldots, \mathbf{v}_k\}$ *is linearly independent.*

2. $(x_1\mathbf{v}_1 + x_2\mathbf{v}_2 + \cdots + x_k\mathbf{v}_k) = (y_1\mathbf{v}_1 + y_2\mathbf{v}_2 + \cdots + y_k\mathbf{v}_k) \Rightarrow x_j = y_j$ *for each* j.

3. $\text{Ker}([\mathbf{v}_1, \mathbf{v}_2, \ldots, \mathbf{v}_k]) = 0$.

4. *The rank of the* $n \times k$ *matrix* $[\mathbf{v}_1, \mathbf{v}_2, \ldots, \mathbf{v}_k]$ *is* k.

PROOF: The proof works by showing that each of statements (1)–(3) is equivalent to statement (4). We know from Theorem 7 of Chapter 2 that statement (4) holds if and only if $[\mathbf{v}_1, \mathbf{v}_2, \ldots, \mathbf{v}_k]\mathbf{x} = 0$ has only the zero solution, just another way of expressing statement (3). So statements (3) and (4) are equivalent.

Next, notice that with $A = [\mathbf{v}_1, \mathbf{v}_2, \ldots, \mathbf{v}_k]$, statement (2) is just a longer way of writing

$$A\mathbf{x} = A\mathbf{y} \Rightarrow \mathbf{x} = \mathbf{y}.$$

So statement (2) says that solutions of $A\mathbf{x} = \mathbf{b}$ are unique whenever they exist. Again by Theorem 7 of Chapter 2, we know that this case is so if and only if the rank of A is k, the number of columns. Hence, statements (2) and (4) are equivalent.

Finally, (3.8) can be rewritten in terms of A as $A\mathbf{x} = 0 \Rightarrow \mathbf{x} = 0$, which means that $A\mathbf{x} = 0$ has only the zero solution. Again by Theorem 7 of Chapter 2, this case is so if and only if the rank of A is k, the number of columns. Thus, statements (1) and (4) are equivalent. ∎

Theorem 5 provides a useful means of checking for independence since we know how to use row reduction to check whether or not $\text{Ker}(A) = 0$.

EXAMPLE 19   **(Checking for linear independence)** Let

$$\mathbf{v}_1 = \begin{bmatrix} 1 \\ 2 \\ 0 \end{bmatrix}, \qquad \mathbf{v}_2 = \begin{bmatrix} -1 \\ 0 \\ 1 \end{bmatrix}, \qquad \text{and} \qquad \mathbf{v}_3 = \begin{bmatrix} 0 \\ 2 \\ 1 \end{bmatrix}.$$

Is $\{\mathbf{v}_1, \mathbf{v}_2, \mathbf{v}_3\}$ a linearly independent set of vectors? To check, form

$$A = [\mathbf{v}_1, \mathbf{v}_2, \mathbf{v}_3] = \begin{bmatrix} 1 & -1 & 0 \\ 2 & 0 & 2 \\ 0 & 1 & 1 \end{bmatrix}.$$

Row reducing yields

$$\begin{bmatrix} 1 & -1 & 0 \\ 0 & 2 & 2 \\ 0 & 0 & 0 \end{bmatrix}.$$

Evidently, the rank of A is 2, not 3, so there is a nonpivotal variable, and $\{\mathbf{v}_1, \mathbf{v}_2, \mathbf{v}_3\}$ is not a linearly independent set of vectors. On the other hand,

$$[\mathbf{v}_1, \mathbf{v}_2] = \begin{bmatrix} 1 & -1 \\ 2 & 0 \\ 0 & 1 \end{bmatrix}$$

row reduces to

$$\begin{bmatrix} 1 & -1 \\ 0 & 2 \\ 0 & 0 \end{bmatrix},$$

which has rank 2. This time there are just two columns—so $\{\mathbf{v}_1, \mathbf{v}_2\}$ is a linearly independent set of vectors.

Before moving on, we point out that it would not be right to summarize the result of Example 19 as being "$\mathbf{v}_1$ and $\mathbf{v}_2$ are independent, but $\mathbf{v}_3$ is not."

• *Independence is not a property of individual vectors; it is a property of sets of vectors. Therefore, it makes no sense to talk about an individual vector being independent or not.*

Indeed, there is nothing at all "wrong" with $\mathbf{v}_3$ itself in this example; you can and should check that both $\{\mathbf{v}_1, \mathbf{v}_3\}$ and $\{\mathbf{v}_2, \mathbf{v}_3\}$ are independent. Be sure you understand this point. A lot of first-time-around confusion comes from trying to think about linear independence as a property of individual vectors.

3.4 Bases

Putting "span" and "linearly independent" together brings us to the central definition of this section.

DEFINITION	**(Basis)** A collection of vectors $\{\mathbf{v}_1, \mathbf{v}_2, \ldots, \mathbf{v}_k\}$ in a subspace S of $\mathbb{R}^n$ is a *basis* for S if it is linearly independent and it spans S.

EXAMPLE 20   **(The standard basis vectors)** The standard basis vectors $\{\mathbf{e}_1, \mathbf{e}_2 \ldots, \mathbf{e}_n\}$ are a basis for $\mathbb{R}^n$ since

$$\begin{bmatrix} x_1 \\ x_2 \\ \vdots \\ x_n \end{bmatrix} = x_1\mathbf{e}_1 + x_2\mathbf{e}_2 + \cdots + x_n\mathbf{e}_n,$$

showing at once that we can write any $\mathbf{x}$ in $\mathbb{R}^n$ as a linear combination of standard basis elements and that the linear combination is the zero vector if and only if each $x_i = 0$. So this collection is independent, and it spans $\mathbb{R}^n$. This is what it means to be a basis in the proper technical sense, which is the only way we will use the word from now on.

EXAMPLE 21 **(Bases and parameterizations)** Continuing with Example 18, let $t_1\mathbf{u}_1 + t_2\mathbf{u}_2$ be the one-to-one parameterization (3.2) of the subspace S. The fact that every vector in S can be written in the form $t_1\mathbf{u}_1 + t_2\mathbf{u}_2$ in just one way means that $\{\mathbf{u}_1, \mathbf{u}_2\}$ is linearly independent. In Example 16, we saw that $\{\mathbf{u}_1, \mathbf{u}_2\}$ spans S, so it is a basis for S.

The connection between bases and one-to-one parameterizations that we used in Example 21 is a useful general fact.

THEOREM 6 **(Bases and one-to-one parameterizations)** *Let S be a subspace of $\mathbb{R}^n$. Then*

$$t_1\mathbf{v}_1 + t_2\mathbf{v}_2 + \cdots + t_k\mathbf{v}_k \tag{3.9}$$

is a one-to-one parameterization of S if and only if $\{\mathbf{v}_1, \mathbf{v}_2, \ldots, \mathbf{v}_k\}$ is a basis for S. In particular, if A is any matrix, the pivotal columns of A are a basis for Img(A).

PROOF: Suppose that (3.9) is a parameterization of S. Then every $\mathbf{v}$ in S can be written in the form (3.9), so $\{\mathbf{v}_1, \mathbf{v}_2, \ldots, \mathbf{v}_k\}$ spans S. Since the parameterization is one-to-one, statement (2) of Theorem 5 is true, so $\{\mathbf{v}_1, \mathbf{v}_2, \ldots, \mathbf{v}_k\}$ is linearly independent. Hence, it is a basis for S.

Conversely, if $\{\mathbf{v}_1, \mathbf{v}_2, \ldots, \mathbf{v}_k\}$ spans S, then every $\mathbf{v}$ in S can be written in the form (3.9), so (3.9) is a parameterization of S. Moreover, since $\{\mathbf{v}_1, \mathbf{v}_2, \ldots, \mathbf{v}_k\}$ is independent, there cannot be two different ways to write $\mathbf{v}$ as a linear combination of the $\{\mathbf{v}_1, \mathbf{v}_2, \ldots, \mathbf{v}_k\}$. Hence, statement (2) of Theorem 5 is true, so the parameterization is one-to-one. ∎

Theorem 6 tells us how to find a basis for not only Img(A) but Ker(A) as well, since we know how to find a one-to-one parameterization of Ker(A).

EXAMPLE 22 **(A basis for Ker(A))** Let

$$A = \begin{bmatrix} 1 & 1 & 0 & 1 \\ 4 & 2 & 6 & 0 \\ 4 & 3 & 1 & 2 \\ 1 & 2 & 0 & 3 \end{bmatrix},$$

and let us find bases for its image and kernel. Row reducing A, we find

$$\begin{bmatrix} 1 & 1 & 0 & 1 \\ 0 & -2 & 6 & -4 \\ 0 & 0 & 2 & 0 \\ 0 & 0 & 0 & 0 \end{bmatrix}.$$

Only the final variable is nonpivotal. Back substitution then gives

$$t \begin{bmatrix} 1 \\ -2 \\ 0 \\ 1 \end{bmatrix}$$

as the general solution of $A\mathbf{x} = 0$, so $\{\mathbf{w}_1\}$ where

$$\mathbf{w}_1 = \begin{bmatrix} 1 \\ -2 \\ 0 \\ 1 \end{bmatrix}$$

is a basis for $\text{Ker}(A)$.

The first three columns of A are the pivotal columns, so $\{\mathbf{v}_1, \mathbf{v}_2, \mathbf{v}_3\}$ is a basis for $\text{Img}(A)$, where $\mathbf{v}_1$, $\mathbf{v}_2$, and $\mathbf{v}_3$ are the first, second, and third columns of A, respectively.

3.5 Coordinates and bases

DEFINITION

(Coordinates) Let S be a subspace of $\mathbb{R}^n$, and let $\{\mathbf{v}_1, \mathbf{v}_2, \ldots, \mathbf{v}_k\}$ be a basis for S. Any vector $\mathbf{b}$ in S can be written in one and only one way as a linear combination of the vectors in $\{\mathbf{v}_1, \mathbf{v}_2, \ldots, \mathbf{v}_k\}$:
$$\mathbf{b} = x_1\mathbf{v}_1 + x_2\mathbf{v}_2 + \cdots x_k\mathbf{v}_k. \tag{3.10}$$
The vector $\mathbf{x}$ in $\mathbb{R}^k$ with entries $x_1, x_2, \ldots, x_k$ is called the *coordinate vector* of $\mathbf{b}$ with respect to the basis $\{\mathbf{v}_1, \mathbf{v}_2, \ldots, \mathbf{v}_k\}$, and $x_1, x_2, \ldots, x_k$ are called the *coordinates* of $\mathbf{b}$ with respect to this basis.

EXAMPLE 23   **(Computing coordinates)** We saw in Example 1 that the image of

$$A = \begin{bmatrix} 1 & -1 & 0 \\ 2 & 0 & 2 \\ 0 & 1 & 1 \end{bmatrix}$$

is the plane of vectors

$$\begin{bmatrix} x \\ y \\ z \end{bmatrix}$$

in $\mathbb{R}^3$ satisfying $2x - y + 2z = 0$. We saw also that the first two columns were pivotal, so by Theorem 6, a basis for $\text{Img}(A)$ is $\{\mathbf{v}_1, \mathbf{v}_2\}$, where

$$\mathbf{v}_1 = \begin{bmatrix} 1 \\ 2 \\ 0 \end{bmatrix} \quad \text{and} \quad \mathbf{v}_2 = \begin{bmatrix} -1 \\ 0 \\ 1 \end{bmatrix}.$$

Now let $S = \text{Img}(A)$, and let

$$\mathbf{b} = \begin{bmatrix} 2 \\ 2 \\ -1 \end{bmatrix}.$$

You can easily check that $\mathbf{b}$ satisfies the equation $2x - y + 2z = 0$ and hence belongs to $S = \text{Img}(A)$. What are the coordinates of $\mathbf{b}$ with respect to the basis $\{\mathbf{v}_1, \mathbf{v}_2\}$?

We have to find x_1 and x_2 so that $x_1\mathbf{v}_1 + x_2\mathbf{v}_2 = \mathbf{b}$. Introduce $V = [\mathbf{v}_1, \mathbf{v}_2]$ so that

$$V = \begin{bmatrix} 1 & -1 \\ 2 & 0 \\ 0 & 1 \end{bmatrix}.$$

Let

$$\mathbf{x} = \begin{bmatrix} x_1 \\ x_2 \end{bmatrix}$$

so that, from the very definition of matrix–vector multiplication, $V\mathbf{x} = x_1\mathbf{v}_1 + x_2\mathbf{v}_2$. Hence, to find $\mathbf{x}$, we only need to solve $V\mathbf{x} = \mathbf{b}$.

To do so, we row reduce $[V|\mathbf{b}]$ to find

$$\begin{bmatrix} 1 & -1 & 2 \\ 0 & 2 & -2 \\ 0 & 0 & 0 \end{bmatrix}.$$

The unique solution is $x_2 = -1$ and $x_1 = 1$. These are the coordinates of $\mathbf{b}$ with respect to the basis $\{\mathbf{v}_1, \mathbf{v}_2\}$. We can easily check that

$$\mathbf{v}_1 - \mathbf{v}_2 = \begin{bmatrix} 1 \\ 2 \\ 0 \end{bmatrix} - \begin{bmatrix} -1 \\ 0 \\ 1 \end{bmatrix} = \begin{bmatrix} 2 \\ 2 \\ -1 \end{bmatrix}.$$

The procedure used in Example 23 works in general:

- *Given a basis $\{\mathbf{v}_1, \mathbf{v}_2, \ldots, \mathbf{v}_k\}$ of a subspace S, and given a vector $\mathbf{b}$ in S, you can compute the coordinate vector $\mathbf{x}$ of $\mathbf{b}$ with respect to this basis by solving $V\mathbf{x} = \mathbf{b}$, where $V = [\mathbf{v}_1, \mathbf{v}_2, \ldots, \mathbf{v}_k]$.*

Notice that since $\{\mathbf{v}_1, \mathbf{v}_2, \ldots, \mathbf{v}_k\}$ is a basis for S, the columns of V are linearly independent, and they span S. Thus, for any $\mathbf{b}$ in S, $V\mathbf{x} = \mathbf{b}$ has a unique solution.

A good way to think of the coordinates of $\mathbf{b}$ with respect to a basis $\{\mathbf{v}_1, \mathbf{v}_2, \ldots, \mathbf{v}_k\}$ is as *a set of instructions for getting to $\mathbf{b}$ from the origin.* Suppose that you live in the vector space S, and you are throwing a party to be held at the location $\mathbf{b}$. If $\mathbf{b}$ is given by (3.10) so that its coordinate vector is

$$\mathbf{x} = \begin{bmatrix} x_1 \\ x_2 \\ \vdots \\ x_k \end{bmatrix},$$

you could tell your friends to get to the party from the origin by going x_1 multiples of $\mathbf{v}_1$ along the direction of $\mathbf{v}_1$, then x_2 multiples of $\mathbf{v}_2$, and so on. If $k = 2$, we would be done. These directions are not far from telling your friends to start from the town center, go 3 blocks west, then go 2 blocks north. Coordinate descriptions of a vector are very concise and useful. This is especially true when S is a k-dimensional subspace of $\mathbb{R}^n$ and k is much less than n. In terms of a given basis, it takes only k numbers to describe a vector $\mathbf{b}$ in S, namely, the coordinates of $\mathbf{b}$ with respect to the given basis, while it takes n numbers to specify $\mathbf{b}$ considered as a vector in $\mathbb{R}^n$.

Some bases are easier to work with than others, and *orthonormal bases* are particularly amenable. Recall from Chapter 1 that a set $\{\mathbf{u}_1, \mathbf{u}_2, \ldots, \mathbf{u}_k\}$ of vectors in $\mathbb{R}^n$ is *orthonormal* if and only if

$$\mathbf{u}_i \cdot \mathbf{u}_j = \begin{cases} 1 & \text{if } i = j \\ 0 & \text{if } i \neq j. \end{cases} \tag{3.11}$$

Note in particular that $|\mathbf{u}_j| = 1$ for each j.

EXAMPLE 24 **(An orthonormal set in $\mathbb{R}^3$)** Let

$$\mathbf{u}_1 = \frac{1}{\sqrt{2}} \begin{bmatrix} -1 \\ 1 \\ 0 \end{bmatrix} \quad \text{and} \quad \mathbf{u}_2 = \frac{1}{\sqrt{6}} \begin{bmatrix} -1 \\ -1 \\ 2 \end{bmatrix}.$$

We can check that $\mathbf{u}_1 \cdot \mathbf{u}_1 = \mathbf{u}_2 \cdot \mathbf{u}_2 = 1$ and $\mathbf{u}_1 \cdot \mathbf{u}_2 = 0$. Hence, $\{\mathbf{u}_1, \mathbf{u}_2\}$ is an orthonormal set in $\mathbb{R}^3$.

One nice thing about orthonormal sets of vectors is that they are always linearly independent.

THEOREM 7   **(Orthonormality and independence)** *Any set* $\{\mathbf{u}_1, \mathbf{u}_2, \ldots, \mathbf{u}_k\}$ *of orthonormal vectors is linearly independent.*

PROOF: Let $\mathbf{b} = x_1\mathbf{u}_1 + x_2\mathbf{u}_2 + \cdots + x_k\mathbf{u}_k$. We must show that if $\mathbf{b} = 0$, then $x_j = 0$ for each j. But

$$
|\mathbf{b}|^2 = \mathbf{b} \cdot \mathbf{b} = \left(\sum_{i=1}^{k} x_i\mathbf{u}_i \right) \cdot \left(\sum_{j=1}^{k} x_j\mathbf{u}_j \right)
$$
$$
= \sum_{i,j=1}^{k} x_i x_j (\mathbf{u}_i \cdot \mathbf{u}_j) = \sum_{i=1}^{k} x_i^2 , \tag{3.12}
$$

where we used (3.11) in the last line. Clearly, $\mathbf{b} = 0$ if and only if each $x_j = 0$. ∎

We can draw a number of useful conclusions from the theorem and from its proof. First, suppose that S is a subspace of $\mathbb{R}^n$. Then any orthonormal set of k vectors in S that spans S is a basis for S. Such bases are particularly nice to work with, so they get a special name.

DEFINITION   **(Orthonormal bases)** A basis for a subspace of $\mathbb{R}^n$ that is an orthonormal set of vectors is called an *orthonormal basis*.

EXAMPLE 25   **(An orthonormal basis for a subspace of $\mathbb{R}^3$)** Consider the orthonormal set $\{\mathbf{u}_1, \mathbf{u}_2\}$ from Example 24. Notice that $\mathbf{a} \cdot \mathbf{u}_1 = 0$ and $\mathbf{a} \cdot \mathbf{u}_2 = 0$, where

$$
\mathbf{a} = \begin{bmatrix} 1 \\ 1 \\ 1 \end{bmatrix} .
$$

Hence both $\mathbf{u}_1$ and $\mathbf{u}_2$ lie in the subspace S in $\mathbb{R}^3$ that is the plane given by $\mathbf{a} \cdot \mathbf{x} = 0$, or in other words, $x + y + z = 0$. Since S is a plane, the pair of vectors $\{\mathbf{u}_1, \mathbf{u}_2\}$ is an orthonormal basis for S.

Suppose that S is a subspace of $\mathbb{R}^n$ and that $\{\mathbf{u}_1, \mathbf{u}_2, \ldots, \mathbf{u}_k\}$ is an orthonormal basis for S. Then any vector $\mathbf{b}$ in S can be written as a linear combination

$$
\mathbf{b} = x_1\mathbf{u}_1 + x_2\mathbf{u}_2 + \cdots + x_k\mathbf{u}_k. \tag{3.13}
$$

Taking the dot product of both sides of (3.13) with $\mathbf{u}_j$,

$$
x_j = \mathbf{b} \cdot \mathbf{u}_j \tag{3.14}
$$

since only the jth term in the sum on the right in (3.13) contributes, according to (3.11). So it is easy to compute coordinates with respect to an orthonormal basis.

There is another way to look at this that is quite useful. Our general method for computing coordinates is to compute $\mathbf{x}$ by solving $U\mathbf{x} = \mathbf{b}$ where $U = [\mathbf{u}_1, \mathbf{u}_2, \dots, \mathbf{u}_k]$. But since the columns of U are orthonormal, U is an isometry, so U^t is a left inverse of U. Hence, multiplying both sides of $U\mathbf{x} = \mathbf{b}$ by U^t, we get $\mathbf{x} = U^t\mathbf{b}$. Hence, by **V.I.F.2**,

$$\begin{bmatrix} x_2 \\ x_2 \\ \vdots \\ x_k \end{bmatrix} = \mathbf{x} = U^t\mathbf{b} = \begin{bmatrix} \mathbf{u}_1 \cdot \mathbf{b} \\ \mathbf{u}_2 \cdot \mathbf{b} \\ \vdots \\ \mathbf{u}_k \cdot \mathbf{b} \end{bmatrix},$$

and we arrive at the same result: $x_i = \mathbf{u}_i \cdot \mathbf{b}$.

- *Finding coordinates for a general basis means solving a matrix equation. When the basis is orthonormal, coordinates can be found simply by taking dot products.*

EXAMPLE 26

(Finding coordinates for an orthonormal basis) In Example 25, we found that $\{\mathbf{u}_1, \mathbf{u}_2\}$ is an orthonormal basis for the subspace S of $\mathbb{R}^3$ given by the equation $x + y + z = 0$.

Now consider the vector

$$\mathbf{b} = \begin{bmatrix} -3 \\ 2 \\ 1 \end{bmatrix}.$$

Evidently, $\mathbf{a} \cdot \mathbf{b} = 0$, where $\mathbf{b}$ satisfies the equation $x + y + z = 0 = 0$, so $\mathbf{b}$ belongs to S. Let x_1 and x_2 be its coordinates with respect to the basis $\{\mathbf{u}_1, \mathbf{u}_2\}$. What are x_1 and x_2? Just take dot products:

$$x_1 = \mathbf{b} \cdot \mathbf{u}_1 = 5/\sqrt{2} \quad \text{and} \quad x_2 = 3/\sqrt{6}.$$

Better yet, the coordinates of $\mathbf{b}$ with respect to an orthonormal basis have an immediate geometric meaning. As we see from (3.12), $|\mathbf{b}| = |\mathbf{x}|$, where $\mathbf{x}$ is the coordinate vector of $\mathbf{b}$ with respect to an orthonormal basis.

Exercises

3.1 Let $\mathbf{v}_1 = \begin{bmatrix} 1 \\ 2 \\ 3 \\ 4 \end{bmatrix}$, $\mathbf{v}_2 = \begin{bmatrix} 4 \\ 3 \\ 2 \\ 1 \end{bmatrix}$, and $\mathbf{v}_3 = \begin{bmatrix} 2 \\ 1 \\ 4 \\ 3 \end{bmatrix}$. Is $\{\mathbf{v}_1, \mathbf{v}_2, \mathbf{v}_3\}$ linearly independent?

3.2 Let $\mathbf{v}_1 = \begin{bmatrix} 1 \\ 2 \\ 3 \\ 4 \end{bmatrix}$, $\mathbf{v}_2 = \begin{bmatrix} 4 \\ 3 \\ 2 \\ 1 \end{bmatrix}$, $\mathbf{v}_3 = \begin{bmatrix} 2 \\ 1 \\ 4 \\ 3 \end{bmatrix}$, and $\mathbf{v}_4 = \begin{bmatrix} 3 \\ 4 \\ 1 \\ 2 \end{bmatrix}$. Is $\{\mathbf{v}_1, \mathbf{v}_2, \mathbf{v}_3, \mathbf{v}_4\}$ linearly independent?

3.3 Let $\mathbf{v}_1 = \begin{bmatrix} 1 \\ 2 \\ 3 \\ 4 \end{bmatrix}$, $\mathbf{v}_2 = \begin{bmatrix} 2 \\ 3 \\ 4 \\ 1 \end{bmatrix}$, $\mathbf{v}_3 = \begin{bmatrix} 3 \\ 4 \\ 1 \\ 2 \end{bmatrix}$, and $\mathbf{v}_4 = \begin{bmatrix} 4 \\ 1 \\ 2 \\ 3 \end{bmatrix}$. Is $\{\mathbf{v}_1, \mathbf{v}_2, \mathbf{v}_3, \mathbf{v}_4\}$ linearly independent?

3.4 Let A and B be $n \times n$ matrices. Suppose that the kernel of B is the zero subspace and that the columns of A are linearly independent. Then is AB necessarily invertible? Either explain why it is so, or give a counterexample.

3.5 Let A be an $m \times n$ matrix, let B be an $n \times p$ matrix, and let $C = AB$. If the columns of A are linearly independent and the columns of B are linearly independent, must the columns of C also be linearly independent? Justify your answer.

3.6 Let $\mathbf{v}_1 = \begin{bmatrix} 1 \\ 1 \\ a \end{bmatrix}$, $\mathbf{v}_2 = \begin{bmatrix} 1 \\ a \\ 1 \end{bmatrix}$, and $\mathbf{v}_3 = \begin{bmatrix} a \\ 1 \\ 1 \end{bmatrix}$. For which values of the parameter a is $\{\mathbf{v}_1, \mathbf{v}_2, \mathbf{v}_3\}$ a basis for $\mathbb{R}^3$?

3.7 (a) Find all values a for which $\begin{bmatrix} 1 \\ 1 \\ 1 \end{bmatrix}$ is in the span of $\{\mathbf{v}_1, \mathbf{v}_2, \mathbf{v}_3\}$, where

$$\mathbf{v}_1 = \begin{bmatrix} 1 \\ 0 \\ -1 \end{bmatrix}, \qquad \mathbf{v}_2 = \begin{bmatrix} 1 \\ -1 \\ 0 \end{bmatrix}, \qquad \text{and} \qquad \mathbf{v}_3 = \begin{bmatrix} a \\ 1 \\ 1 \end{bmatrix}.$$

(b) Find all values of a for which $\{\mathbf{v}_1, \mathbf{v}_2, \mathbf{v}_3\}$ is linearly independent.

3.8 (a) Find all values a for which $\begin{bmatrix} 1 \\ 1 \\ 1 \end{bmatrix}$ is in the span of $\{\mathbf{v}_1, \mathbf{v}_2, \mathbf{v}_3\}$, where

$$\mathbf{v}_1 = \begin{bmatrix} 1 \\ a \\ a \end{bmatrix}, \qquad \mathbf{v}_2 = \begin{bmatrix} a \\ 1 \\ a \end{bmatrix}, \qquad \text{and} \qquad \mathbf{v}_3 = \begin{bmatrix} a \\ a \\ 1 \end{bmatrix}.$$

(b) Find all values of a for which $\{\mathbf{v}_1, \mathbf{v}_2, \mathbf{v}_3\}$ is linearly independent.

3.9 Consider the following matrices:

$$A = \begin{bmatrix} 1 & 1 & 0 & 1 \\ 4 & 2 & 6 & 0 \\ 4 & 3 & 1 & 2 \\ 1 & 2 & 0 & 3 \end{bmatrix} \quad B = \begin{bmatrix} 0 & -1 & -1 \\ 1 & 0 & -1 \\ 1 & 1 & 0 \end{bmatrix} \quad C = \begin{bmatrix} 1 & 2 & 3 & 4 \\ 1 & 0 & 1 & 2 \\ 3 & 1 & 4 & 7 \\ 1 & 1 & 2 & 3 \end{bmatrix} \quad D = \begin{bmatrix} 1 & 1 & 3 & 1 \\ 2 & 0 & 1 & 1 \\ 3 & 1 & 4 & 2 \\ 4 & 2 & 7 & 3 \end{bmatrix}.$$

In each case, find a basis for the image and the kernel.

3.10 Consider the following matrices:

$$A = \begin{bmatrix} 1 & 2 & 1 & 2 \\ 2 & 1 & 2 & 1 \\ 4 & 1 & 4 & 1 \\ 1 & 4 & 4 & 1 \end{bmatrix} \quad B = \begin{bmatrix} 1 & 2 & 4 & 1 \\ 2 & 1 & 1 & 4 \\ 1 & 2 & 4 & 4 \\ 2 & 1 & 1 & 1 \end{bmatrix} \quad C = \begin{bmatrix} 1 & 2 & 4 \\ 1 & 3 & 9 \\ 1 & 4 & 16 \end{bmatrix} \quad D = \begin{bmatrix} 1 & 0 & 1 & 0 \\ 0 & 1 & 0 & 1 \\ 2 & 3 & 2 & 3 \end{bmatrix}.$$

In each case, find a basis for the image and the kernel.

3.11 Let S be the set of vectors $\mathbf{x}$ in $\mathbb{R}^3$ with $|\mathbf{x}| \leq 1$. Is S a subspace?

3.12 Let S and $\tilde{S}$ be two subspaces of $\mathbb{R}^n$. Is their intersection $S \cap \tilde{S}$ necessarily a subspace? Explain why, or give a counterexample.

3.13 Let S and $\tilde{S}$ be two subspaces of $\mathbb{R}^n$. Define $S + \tilde{S}$ to be the set of all vectors $\mathbf{v}$ in $\mathbb{R}^n$ so that $\mathbf{v} = \mathbf{x} + \mathbf{y}$ for some vectors $\mathbf{x}$ in S and $\mathbf{y}$ in $\tilde{S}$. Show that $S + \tilde{S}$ is a subspace of $\mathbb{R}^n$.

3.14 Let S be the subspace of $\mathbb{R}^3$ given by the equation $x + 2y + z = 0$. Show that $\{\mathbf{v}_1, \mathbf{v}_2\}$ is a basis for S, where $\mathbf{v}_1 = \begin{bmatrix} 1 \\ -1 \\ 1 \end{bmatrix}$ and $\mathbf{v}_2 = \begin{bmatrix} -1 \\ 0 \\ 1 \end{bmatrix}$. Then show that $\mathbf{w} = \begin{bmatrix} 1 \\ -2 \\ 3 \end{bmatrix}$ belongs to S, and find its coordinate vector with respect to the basis $\{\mathbf{v}_1, \mathbf{v}_2\}$.

3.15 Let S be the subspace of $\mathbb{R}^3$ given by the equation $x - 2y + 3z = 0$. Show that $\{\mathbf{v}_1, \mathbf{v}_2\}$ is a basis for S, where $\mathbf{v}_1 = \begin{bmatrix} 2 \\ 1 \\ 0 \end{bmatrix}$ and $\mathbf{v}_2 = \begin{bmatrix} 0 \\ 3 \\ 2 \end{bmatrix}$. Then show that $\mathbf{w} = \begin{bmatrix} 1 \\ 2 \\ 1 \end{bmatrix}$ belongs to S, and find its coordinate vector with respect to the basis $\{\mathbf{v}_1, \mathbf{v}_2\}$.

3.16 Let S be the subspace of $\mathbb{R}^3$ given by the equation $2x - 2y - z = 0$. Show that $\{\mathbf{v}_1, \mathbf{v}_2\}$ is an orthonormal basis for S, where $\mathbf{v}_1 = \dfrac{1}{3} \begin{bmatrix} 1 \\ 2 \\ -2 \end{bmatrix}$ and $\mathbf{v}_2 = \dfrac{1}{3} \begin{bmatrix} 2 \\ 1 \\ 2 \end{bmatrix}$. Then show that $\mathbf{w} = \begin{bmatrix} 1 \\ 1 \\ 0 \end{bmatrix}$ belongs to S, and find its coordinate vector with respect to the basis $\{\mathbf{v}_1, \mathbf{v}_2\}$.

3.17 Let S be the subspace of $\mathbb{R}^3$ given by the equation $x + 2y - 2z = 0$. Show that $\{\mathbf{v}_1, \mathbf{v}_2\}$ is an orthonormal basis for S, where $\mathbf{v}_1 = \dfrac{1}{3} \begin{bmatrix} 2 \\ -2 \\ -1 \end{bmatrix}$ and $\mathbf{v}_2 = \dfrac{1}{3} \begin{bmatrix} 2 \\ 1 \\ 2 \end{bmatrix}$. Then show that $\mathbf{w} = \begin{bmatrix} 2 \\ 4 \\ 5 \end{bmatrix}$ belongs to S, and find its coordinate vector with respect to the basis $\{\mathbf{v}_1, \mathbf{v}_2\}$.

SECTION 4 | Dimension

4.1 Bases and dimension

We have seen a number of ways to find a one-to-one parameterization of a subspace S of $\mathbb{R}^n$. By Theorem 6, finding such a parameterization is the same thing as finding a basis for S.

Finding a basis means finding a set of vectors in S that *spans S* and is *linearly independent*. The following theorem summarizes some important facts about sets of vectors that span or are linearly independent or both.

THEOREM 8

(**The basis theorem**) *Suppose that S is a nonzero subspace of $\mathbb{R}^n$.*

1. *No linearly independent set of vectors in S contains more vectors than any set that spans S.*

2. *If $\{\mathbf{v}_1, \mathbf{v}_2, \ldots, \mathbf{v}_k\}$ is a linearly independent subset of S and $\mathbf{v}_{k+1}$ is any vector in S that is not in the span of $\{\mathbf{v}_1, \mathbf{v}_2, \ldots, \mathbf{v}_k\}$, then $\{\mathbf{v}_1, \mathbf{v}_2, \ldots, \mathbf{v}_k, \mathbf{v}_{k+1}\}$ is a linearly independent subset of S.*

3. *S has a basis, and in fact, every set of vectors in S that spans S contains a basis for S.*

Theorem 8 has many consequences; here are some of them.

COROLLARY 1   *Any two bases of a subspace S of $\mathbb{R}^n$ always consist of the same number of vectors.*

PROOF: Consider two bases $\{\mathbf{v}_1, \mathbf{v}_2, \ldots, \mathbf{v}_k\}$ and $\{\mathbf{w}_1, \mathbf{w}_2, \ldots, \mathbf{w}_\ell\}$ of S. Since both sets are bases, the first set is linearly independent and the second spans. It then follows from statement (1) of Theorem 8 that $k \leq \ell$. Reversing the roles of the two bases, we conclude in the same way that $\ell \leq k$. Once again, two inequalities make an equality: $k = \ell$. ■

EXAMPLE 27 **(Every subspace is an image)** We know that the image of every matrix is a subspace. Is it true that every subspace of $\mathbb{R}^n$ is the image of some matrix A with n rows? The answer is yes. By Corollary 1, every subspace S has a basis $\{\mathbf{v}_1, \mathbf{v}_2, \ldots, \mathbf{v}_k\}$. By Theorem 4, the image of $A = [\mathbf{v}_1, \mathbf{v}_2, \ldots, \mathbf{v}_k]$ is the span of $\{\mathbf{v}_1, \mathbf{v}_2, \ldots, \mathbf{v}_k\}$, which is S.

Even more significantly, Corollary 1 states that no matter which basis we use to parameterize a subspace, we will need the *same* number of parameters to do it. This number has a name.

DEFINITION | **(Dimension)** The number of elements in a basis for a subspace S of $\mathbb{R}^n$ is called the *dimension* of S, and it is denoted by $\dim(S)$.

COROLLARY 2 *Let S be a subspace of $\mathbb{R}^n$, and let $d = \dim(S)$. Then*

1. *Any set of d linearly independent vectors in S is a basis of S.*
2. *Any set of d vectors in S that spans S is a basis of S.*

PROOF: Let $\{\mathbf{v}_1, \mathbf{v}_2, \ldots, \mathbf{v}_d\}$ be any d linearly independent vectors in S. It must also span, since otherwise, according to statement (3) of Theorem 8, we could add a vector $\mathbf{v}_{d+1}$ to the set, keeping it independent. But by statement (1) of Theorem 8, no set of $d + 1$ vectors in S can be independent.

For statement (2), let $\{\mathbf{v}_1, \mathbf{v}_2, \ldots, \mathbf{v}_d\}$ be a set of d vectors that spans S. According to statement (3) of Theorem 8, if it is not a basis, we can delete vectors from it to obtain a basis. But we would produce a basis with fewer than d elements, which Corollary 1 says is impossible. ■

To make use of Corollary 2, we need to know the dimension of the subspace S in question. We are most likely interested in subspaces that are given as images or kernels. According to the next theorem, we already have a name for the dimension of the image; let us now give one to the dimension of the kernel.

DEFINITION | **(Nullity)** The *nullity* of a matrix is the dimension of its kernel.

THEOREM 9 **(Dimension formula)** *Let B be an $m \times n$ matrix of rank r. Then*

$$\dim(\text{Ker}(B)) = n - r \quad \text{and} \quad \dim(\text{Img}(B)) = r \tag{4.1}$$

so that, in particular,

$$\text{nullity}(B) + \text{rank}(B) = n. \tag{4.2}$$

PROOF: By Theorem 6, the pivotal columns of B are a basis for Img(B), and there are r of them by the definition of rank. Likewise, Theorem 6 also tells us that Ker(B) has a basis consisting of the $n - r$ vectors, one for each nonpivotal variable, arising in a parameterization of the solution set of $B\mathbf{x} = 0$ produced using row reduction. ∎

EXAMPLE 28

(Finding bases) Let S be the plane through the origin in $\mathbb{R}^3$ given by $x_1 + x_2 + x_3 = 0$. In other words S is the kernel of the 1×3 matrix $B = [1 \quad 1 \quad 1]$. Clearly rank(B) = 1 so that dim(S) = dim(Ker(B)) = $3 - 1 = 2$ by Theorem 9. Of course the dimension of a plane should be 2. If it were not, there would have been something wrong with our definition.

By Corollary 2, *any* two linearly independent vectors in Ker(B) are a basis for S. Let us look for a vector $\mathbf{w}_1$ in S of the form

$$\mathbf{w}_1 = \begin{bmatrix} 1 \\ t \\ 0 \end{bmatrix}.$$

Then $B\mathbf{w}_1 = 1 + t = 0$ if $t = -1$. Making this choice,

$$\mathbf{w}_1 = \begin{bmatrix} 1 \\ -1 \\ 0 \end{bmatrix}$$

belongs to Ker(B) = S.

Now look for a vector $\mathbf{w}_2$ in S of the form

$$\mathbf{w}_2 = \begin{bmatrix} 0 \\ t \\ 1 \end{bmatrix}.$$

Then $B\mathbf{w}_2 = 1 + t = 0$ if $t = -1$. Making this choice,

$$\mathbf{w}_2 = \begin{bmatrix} 0 \\ -1 \\ 1 \end{bmatrix}$$

belongs to Ker(B) = S. Notice that because $\mathbf{w}_1$ has a zero in the third place and $\mathbf{w}_2$ does not, these vectors are not proportional and so are linearly independent. Hence, by Corollary 2, $\{\mathbf{w}_1, \mathbf{w}_2\}$ is a basis for S.

Finally, we prove the basis theorem.

PROOF OF THEOREM 8: First, we prove statement (1). Suppose $\{\mathbf{v}_1, \mathbf{v}_2, \ldots, \mathbf{v}_m\}$ spans S, and $\{\mathbf{u}_1, \mathbf{u}_2, \ldots, \mathbf{u}_\ell\}$ is linearly independent. We will show that $\ell \leq m$.

Let V denote the $n \times m$ matrix $[\mathbf{v}_1, \mathbf{v}_2, \ldots, \mathbf{v}_m]$. Since $\{\mathbf{v}_1, \mathbf{v}_2, \ldots, \mathbf{v}_m\}$ spans S, every vector in S belongs to Img(V). In particular, for each j, $\mathbf{u}_j$ belongs to S and hence to Img(V). So there is a vector $\mathbf{r}_j$ in $\mathbb{R}^m$ with $V\mathbf{r}_j = \mathbf{u}_j$. Let U be the $n \times \ell$ matrix $U = [\mathbf{u}_1, \mathbf{u}_2, \ldots, \mathbf{u}_\ell]$, and let R be the $m \times \ell$ matrix $R = [\mathbf{r}_1, \mathbf{r}_2, \ldots, \mathbf{r}_\ell]$. In fact, by V.I.F.3,

$$VR = [V\mathbf{r}_1, V\mathbf{r}_2, \ldots, V\mathbf{r}_\ell] = [\mathbf{u}_1, \mathbf{u}_2, \ldots, \mathbf{u}_\ell] = U. \tag{4.3}$$

Now if $\mathbf{x}$ is any vector in $\mathbb{R}^\ell$ with $R\mathbf{x} = 0$, then $V(R\mathbf{x}) = 0$ too. But by (4.3),

$$V(R\mathbf{x}) = (VR)\mathbf{x} = U\mathbf{x}.$$

Therefore, $U\mathbf{x} = 0$. But since $\{\mathbf{u}_1, \mathbf{u}_2, \ldots, \mathbf{u}_\ell\}$ is linearly independent, Theorem 5 tells us that Ker(U) = 0. Hence, the only solution of $R\mathbf{x} = 0$ is the zero vector, so Ker(R) = 0.

Since R is an $m \times \ell$ matrix, this shows that $\ell \leq m$. Indeed, by Theorem 7 of Chapter 2, for any matrix A with more columns than rows, $A\mathbf{x} = 0$ always has a nonzero solution. But R does not have a nonzero solution, so it cannot have more columns than rows. Hence $\ell \leq m$.

To prove statement (2), suppose that $\{\mathbf{v}_1, \mathbf{v}_2, \ldots, \mathbf{v}_k\}$ is a linearly independent subset of S and that $\mathbf{v}_{k+1}$ is not in its span. If

$$x_1\mathbf{v}_1 + x_2\mathbf{v}_2 + \cdots + x_k\mathbf{v}_k + x_{k+1}\mathbf{v}_{k+1} = 0, \tag{4.4}$$

it must be the case that $x_{k+1} = 0$. Otherwise, from (4.4), we would have

$$\mathbf{v}_{k+1} = -\left(\frac{x_1}{x_{k+1}} \mathbf{v}_1 + \frac{x_2}{x_{k+1}} \mathbf{v}_2 + \cdots + \frac{x_v}{x_{k+1}} \mathbf{v}_n \right),$$

which would put $\mathbf{v}_{k+1}$ in the span of $\{\mathbf{v}_1, \mathbf{v}_2, \ldots, \mathbf{v}_k\}$. It is not there, so $x_{k+1} = 0$, and hence

$$x_1 \mathbf{v}_1 + x_2 \mathbf{v}_2 + \cdots + x_k \mathbf{v}_k = 0.$$

But since $\{\mathbf{v}_1, \mathbf{v}_2, \ldots, \mathbf{v}_k\}$ is linearly independent, $x_1 = x_2 + \cdots = x_k = 0$. Hence, necessarily $x_j = 0$ for each $j = 1, \ldots, k + 1$ in (4.4), so $\{\mathbf{v}_1, \mathbf{v}_2, \ldots, \mathbf{v}_k, \mathbf{v}_{k+1}\}$ is linearly independent.

To prove statement (3), let T be the given spanning set so that $S = \mathrm{Sp}(T)$. Pick any nonzero vector $\mathbf{v}_1$ in T. If $\{\mathbf{v}_1\}$ spans S, we have our basis. If not, $\mathrm{Sp}(\{\mathbf{v}_1\})$ does not contain all of T,[*] so we can pick a vector $\{\mathbf{v}_2\}$ in T that is not in $\mathrm{Sp}(\{\mathbf{v}_1\})$. By statement (2), $\{\mathbf{v}_1, \mathbf{v}_2\}$ will be linearly independent.

If $\{\mathbf{v}_1, \mathbf{v}_2\}$ spans S, we have our basis. Otherwise, just as earlier, we can add a vector $\mathbf{v}_3$ chosen from T so that $\{\mathbf{v}_1, \mathbf{v}_2, \mathbf{v}_3\}$ is linearly independent. Continuing, this procedure either terminates when a basis is found or it keeps going, producing an ever larger linearly independent subset of S.

But an arbitrarily large linearly independent subset of S cannot be produced: $\mathbb{R}^n$ is spanned by the n standard basis vectors, so statement (1) of Theorem 8 says that no collection of $n + 1$ (or more) vectors in $\mathbb{R}^n$ can possibly be linearly independent. Since S is a subspace of $\mathbb{R}^n$, no collection of $n + 1$ (or more) vectors in S can possibly be linearly independent. Hence, the procedure must terminate and produce a basis in n steps at most. The same selection argument shows that any nonzero subspace contains a finite spanning set and hence a basis. ∎

4.2 The dimension principle

Dimension is a useful concept in part because it provides an easy way to decide when two subspaces S and $\tilde{S}$ are equal.

Recall that sets, and in particular subspaces, are equal if and only if they have the same members. This means that S is a subset of $\tilde{S}$ and $\tilde{S}$ is a subset of S, or, using the standard symbols, $S \subset \tilde{S}$ and $\tilde{S} \subset S$. So one way to show that two subspaces are the same is to show that both

$$S \subset \tilde{S} \quad \text{and} \quad \tilde{S} \subset S. \tag{4.5}$$

Sometimes one is easier to show than the other, and dimension provides a way around that.

THEOREM 10   **(Dimension principle)** *Suppose S and $\tilde{S}$ are two subspaces of $\mathbb{R}^n$. Then*

1. *If $\tilde{S} \subset S$, then $\dim(\tilde{S}) \leq \dim(S)$.*

2. *If*

$$\tilde{S} \subset S \quad \text{and} \quad \dim(\tilde{S}) = \dim(S), \tag{4.6}$$

then $\tilde{S} = S$.

---

[*]Any *subspace* that contains T contains $\mathrm{Sp}(T) = S$.

PROOF: Let $\{v_1, v_2, \ldots, v_d\}$ be any basis for $\tilde{S}$. Since $\tilde{S} \subset S$, this set of vectors in S is linearly independent. If it spans S, then every vector in S is a linear combination of $\{v_1, v_2, \ldots, v_d\}$ and hence is in $\tilde{S}$. Thus, in this case, $S \subset \tilde{S}$ and $\tilde{S} \subset S$, so $S = \tilde{S}$. Hence $S = \tilde{S}$. Clearly the converse is true: If $S = \tilde{S}$, then $\{v_1, v_2, \ldots, v_d\}$ spans S.

Hence, $S \neq \tilde{S}$ if and only if $\{v_1, v_2, \ldots, v_d\}$ does not span S. In this case, we obtain a basis for S by adding vectors to $\{v_1, v_2, \ldots, v_d\}$, and hence the dimension of S exceeds that of $\tilde{S}$. ■

To see the dimension principle in action, let A and B be any two matrices whose sizes permit the product BA to be formed. Notice that if y belongs to $\text{Img}(BA)$, then there is a vector x so that $y = BAx$. But then $y = Bz$, where $z = Ax$, so y belongs to $\text{Img}(B)$. Hence, whenever y belongs to $\text{Img}(BA)$, y belongs to $\text{Img}(B)$. In other words,

$$\text{Img}(BA) \subset \text{Img}(B). \tag{4.7}$$

Then, by the dimension principle and Theorem 9,

$$\text{rank}(BA) \leq \text{rank}(B). \tag{4.8}$$

Likewise, if $Ax = 0$, then $BAx = B(Ax) = B0 = 0$. Hence, whenever x belongs to $\text{Ker}(A)$, x belongs to $\text{Ker}(BA)$. In other words,

$$\text{Ker}(A) \subset \text{Ker}(BA). \tag{4.9}$$

Then, by the dimension principle and Theorem 9,

$$\text{nullity}(BA) \geq \text{nullity}(A). \tag{4.10}$$

However, BA and A have the same number of columns, say n, and so by (4.2), we can restate (4.10) in terms of rank as

$$\text{rank}(BA) \leq \text{rank}(A). \tag{4.11}$$

We summarize in the following corollary.

COROLLARY 1   *Let A and B be any two matrices whose sizes permit the product BA to be formed. Then*

$$\text{rank}(BA) \leq \min\{\text{rank}(A), \text{rank}(B)\}. \tag{4.12}$$

In particular, the corollary says that for square matrices A, $\text{rank}(A^2) \leq \text{rank}(A)$. The inequality can be strict.

EXAMPLE 29   **(The rank of A^2 can be less than the rank of A)** Let

$$A = \begin{bmatrix} 0 & 1 \\ 0 & 0 \end{bmatrix}.$$

Clearly, $\text{rank}(A) = \text{nullity}(A) = 1$. In fact, both the image and the kernel of A are the line spanned by e_1. Computing

$$A^2 = \begin{bmatrix} 0 & 0 \\ 0 & 0 \end{bmatrix},$$

we see that $\text{rank}(A^2) = 0$, and $\text{nullity}(A^2) = 2$.

Now you can see why the nullity of BA is at least as large as the nullity of A and sometimes larger. Compute $BA\mathbf{x}$ in two stages: $B(A\mathbf{x})$. Every $\mathbf{x}$ that is in the kernel of A gets "zeroed out" at the first stage. But even if $\mathbf{x}$ is not in $\text{Ker}(A)$, A can move it into the $\text{Ker}(B)$, where it gets zeroed out in the second stage. On the other hand, if

$$A = \begin{bmatrix} 1 & 1 \\ 0 & 0 \end{bmatrix},$$

then $A^2 = A$, so in this case $\text{rank}(A^2) = \text{rank}(A)$.

We see from the example that the inequality in (4.12) is sometimes strict and sometimes not.

The corollary of Theorem 2 gives an important example in which there is equality in (4.9) and hence in (4.11) as well: the case $B = A^t$. Indeed, according to (2.20),

$$\text{rank}(A^t A) = \text{rank}(A^t) \quad \text{and} \quad \text{nullity}(A^t A) = \text{nullity}(A).$$

But $A^t A$ and A have the same number of columns, so by Theorem 9, $\text{nullity}(A^t A) = \text{nullity}(A)$ is equivalent to $\text{rank}(A^t A) = \text{rank}(A)$, which leads to a remarkable conclusion.

COROLLARY 2   *Let A be any $m \times n$ matrix. Then*

$$\text{rank}(A^t) = \text{rank}(A) = \text{rank}(A^t A). \tag{4.13}$$

We say that this conclusion is remarkable because the span of the columns of A is a subspace of $\mathbb{R}^m$ and the span of the rows is a subspace of $\mathbb{R}^n$, and nonetheless, these two subspaces always have the same dimension.

Connecting the concepts of rank and dimension gives us new insight into linear systems. If you want to compute the rank of a particular matrix, row reduction is certainly still the way to proceed. But if you think only in terms of row reduction, it is not so easy to see why, when you row reduce AB, you can never find more pivots than you would if you row reduce either A or B; nor is it so easy to see why, when you row reduce A^t, you always find the same number of pivots as when you row reduce A.

EXAMPLE 30   **(When are the rows of A independent?)** The fact that $\text{rank}(A^t) = \text{rank}(A)$ has many consequences, as we have said. Here is one. We already know from Theorem 3 that if A is an $m \times n$ matrix, the columns of A are independent if and only if $\text{rank}(A) = n$. When are the rows independent? The rows of A are the columns of A^t, so the columns of A^t are independent when $\text{rank}(A^t) = m$. Since $\text{rank}(A^t) = \text{rank}(A)$, the rows of A are independent if and only if $\text{rank}(A) = m$.

COROLLARY 3   **(More dimension formulas)** *Let A be an $m \times n$ matrix, and let r be the number of pivotal columns. Then*

$$\begin{aligned} \dim(\text{Img}(A)) &= r & \dim(\text{Ker}(A)) &= n - r \\ \dim(\text{Img}(A^t)) &= r & \dim(\text{Ker}(A^t)) &= m - r. \end{aligned} \tag{4.14}$$

PROOF: The dimensions of $\text{Img}(A)$ and $\text{Ker}(A)$ are already given in Theorem 10. The identity (4.13) says that A^t and A have the same rank, namely, r, so another application of Theorem 9 gives the formulas for $\text{Img}(A^t)$ and $\text{Ker}(A^t)$. ∎

EXAMPLE 31   **(Finding bases for Img(A), Ker(A), and so on)** Find bases for $\text{Img}(A)$, $\text{Ker}(A)$, $\text{Img}(A^t)$, and $\text{Ker}(A^t)$, where

$$A = \begin{bmatrix} 1 & 0 & -1 \\ 0 & 1 & -1 \\ 1 & -1 & 0 \end{bmatrix}.$$

We row reduce A in two steps to

$$\begin{bmatrix} 1 & 0 & -1 \\ 0 & 1 & -1 \\ 0 & 0 & 0 \end{bmatrix}$$

and see that $\text{rank}(A) = 2$. It follows from Theorem 9 that

$$\dim(\text{Img}(A)) = 2, \quad \dim(\text{Ker}(A)) = 1, \quad \dim(\text{Img}(A^t)) = 2, \quad \text{and} \quad \dim(\text{Ker}(A^t)) = 1.$$

Writing $A = [\mathbf{v}_1, \mathbf{v}_2, \mathbf{v}_3]$, the first two columns of A are pivotal, and by Theorem 6, $\{\mathbf{v}_1, \mathbf{v}_2\}$ is a basis for $\text{Img}(A)$.

To get a basis for $\text{Img}(A^t)$, we do not have to row reduce A^t. Since we know that $\dim(\text{Img}(A^t)) = 2$, Corollary 2 of Theorem 8 says that any pair of independent vectors in $\text{Img}(A^t)$ are a basis for $\text{Img}(A^t)$. Writing $A^t = [\mathbf{w}_1, \mathbf{w}_2, \mathbf{w}_3]$, we see that no pair of these vectors is proportional, so *any* pair of vectors chosen from among $\{\mathbf{w}_1, \mathbf{w}_2, \mathbf{w}_3\}$ will do. For example, $\{\mathbf{w}_2, \mathbf{w}_3\}$ is a basis for $\text{Img}(A^t)$.

The same reasoning could have been applied to $\text{Img}(A)$. We are not restricted to the pivotal columns of A. By Corollary 2 of Theorem 8 and the fact that any pair chosen from $\{\mathbf{v}_1, \mathbf{v}_2, \mathbf{v}_3\}$ is independent, we see that any such pair is a basis. For example, $\{\mathbf{v}_2, \mathbf{v}_3\}$ is a basis for $\text{Img}(A)$.

Since both kernels are one-dimensional, we just need to find a nonzero vector in each. If we happen to notice that the rows of A sum to zero, then we see that $A\mathbf{a} = 0$, where

$$a = \begin{bmatrix} 1 \\ 1 \\ 1 \end{bmatrix}.$$

Hence, $\{\mathbf{a}\}$ is a basis for $\text{Ker}(A)$.

Now for A^t. We are looking for only one nonzero vector $\mathbf{x}$ satisfying $A^t\mathbf{x} = 0$. Let us look at A^t and see if we can avoid a row reduction and back substitution:

$$A^t = \begin{bmatrix} 1 & 0 & 1 \\ 0 & 1 & -1 \\ -1 & -1 & 0 \end{bmatrix},$$

and any vector in $\text{Ker}(A^t)$ must be orthogonal to each row of A^t. If

$$\mathbf{x} = \begin{bmatrix} x \\ y \\ z \end{bmatrix}$$

is orthogonal to the second row, $y = z$. And if it is orthogonal to the first row, $x = -y$. Hence,

$$\mathbf{x} = \begin{bmatrix} -1 \\ 1 \\ 1 \end{bmatrix}$$

is a nonzero vector in $\text{Ker}(A^t)$ and is therefore a basis for it.

Notice that we avoided ever row reducing A^t. In this example, that does not really save us much. But this kind of reasoning can often be applied in situations with larger matrices in which it would save us a great amount of work.

There are yet more ways to use the theorems of this section to find bases. Some ways may be more work than others, but if they lead to a particularly convenient basis, the work may be worthwhile. After all, if you are computing a basis, presumably you are going to be doing some calculations with it. If you are going to be doing a lot of calculating, it may pay to do a bit of work before you start to get a really nice basis.

For example, let A be an $m \times n$ matrix. Theorem 6 tells us that the set of pivotal columns of A is a basis for $\text{Img}(A)$. To find out which columns are pivotal, you row reduce A, producing the row-reduced form U. But then the basis is given by the pivotal columns of the original matrix A, not the columns of U. (Row operations preserve the kernel but not the image of a matrix.) This is too bad since generally U is a much simpler matrix than A.

There is a way to keep this simplification in a basis for $\mathrm{Img}(A)$. First, take the transpose of A, and row reduce A^t all the way to the reduced row echelon form. (Note that you reduce A^t, not A.)

If the rank of A is r, the rank of A^t is r, too, by (4.13), so there will be r rows in $\mathrm{rref}\,(A^t)$. We claim that these r rows are a basis of $\mathrm{Img}(A)$:

- *If A is any $m \times n$ matrix with rank r, and $W = \mathrm{rref}\,(A^t)$, the reduced row echelon form of A^t, then W has r nonzero rows $\{\mathbf{w}_1, \mathbf{w}_2, \ldots, \mathbf{w}_r\}$, and these are a basis of $\mathrm{Img}(A)$.*

To see why, first note that the number of rows of W will be the rank of A^t. But by (4.13), $\mathrm{rank}(A^t) = \mathrm{rank}(A) = r$. The $r \times m$ matrix, W, is clearly in echelon form, and it has rank r. By Example 30, its rows, $\{\mathbf{w}_1, \mathbf{w}_2, \ldots, \mathbf{w}_r\}$, are linearly independent.

Next, each row of W is a linear combination of the rows of A^t by the nature of row reduction. But the rows of A^t are the columns of A, so $\{\mathbf{w}_1, \mathbf{w}_2, \ldots, \mathbf{w}_r\}$ lies in the span of the columns of A, that is, $\mathrm{Img}(A)$.

But by Theorem 6, $\dim(\mathrm{Img}(A)) = r$, and by Corollary 2 of Theorem 8, any independent set of r vectors in $\mathrm{Img}(A)$ is a basis for $\mathrm{Img}(A)$. So $\{\mathbf{w}_1, \mathbf{w}_2, \ldots, \mathbf{w}_r\}$ is a basis for $\mathrm{Img}(A)$. The basis you get in this way is particularly nice. To see what is so nice, let us look at an example.

EXAMPLE 32　　**(Finding a nice base for Img(A))**　Let

$$A = \begin{bmatrix} 1 & 4 & 3 & 1 \\ 1 & 1 & 3 & 3 \\ 3 & 3 & 9 & 1 \\ 1 & 4 & 3 & 3 \end{bmatrix}.$$

This row reduces to

$$\begin{bmatrix} 1 & 4 & 3 & 1 \\ 0 & -3 & 0 & 2 \\ 0 & 0 & 0 & -8 \\ 0 & 0 & 0 & 0 \end{bmatrix}.$$

From this we see that columns 1, 2, and 4 are pivotal, so $\{\mathbf{v}_1, \mathbf{v}_2, \mathbf{v}_3\}$ is a basis of A, where

$$\mathbf{v}_1 = \begin{bmatrix} 1 \\ 1 \\ 3 \\ 1 \end{bmatrix}, \qquad \mathbf{v}_2 = \begin{bmatrix} 4 \\ 1 \\ 3 \\ 4 \end{bmatrix}, \qquad \text{and} \qquad \mathbf{v}_3 = \begin{bmatrix} 1 \\ 3 \\ 1 \\ 3 \end{bmatrix}.$$

To find the "nice" basis, row reduce A^t all the way to the reduced row echelon form:

$$W = \begin{bmatrix} 1 & 0 & 0 & 1 \\ 0 & 1 & 0 & 3/4 \\ 0 & 0 & 1 & -1/4 \end{bmatrix}.$$

The basis is $\{\mathbf{w}_1, \mathbf{w}_2, \mathbf{w}_3\}$, where

$$\mathbf{w}_1 = \begin{bmatrix} 1 \\ 0 \\ 0 \\ 1 \end{bmatrix}, \qquad \mathbf{w}_2 = \begin{bmatrix} 0 \\ 4 \\ 0 \\ 3 \end{bmatrix}, \qquad \text{and} \qquad \mathbf{w}_3 = \begin{bmatrix} 0 \\ 0 \\ 4 \\ -1 \end{bmatrix}.$$

(We have multiplied through by 4 in two cases to eliminate fractions.)

The vectors in the second basis are much simpler than in the "pivotal" basis. This will be helpful when we compute with this basis, as we do in later sections.

Here is one more example using the identities of Corollary 3.

EXAMPLE 33 **(Rank and independence of columns and rows)** For the matrices

$$A = \begin{bmatrix} 1 & 2 & 1 \\ 2 & 3 & 1 \\ 2 & 2 & 0 \end{bmatrix}, \quad B = \begin{bmatrix} 1 & -2 & -3 & 1 \\ 4 & -3 & -5 & 3 \\ 1 & 3 & 4 & 0 \end{bmatrix}, \quad C = \begin{bmatrix} 1 & 2 & 3 \\ 3 & 1 & 2 \\ 2 & 3 & 1 \end{bmatrix}, \quad \text{and} \quad D = \begin{bmatrix} 0 & 0 & 1 \\ 0 & 1 & 0 \\ 1 & 3 & 6 \\ 2 & 5 & 2 \\ 1 & 0 & 4 \end{bmatrix},$$

answer the following questions: (a) For which of them, if any, are the columns independent? (b) For which of them, if any, are the rows independent? (c) For which of them, if any, do the columns span $\mathbb{R}^3$? (d) For which of them, if any, do the rows span $\mathbb{R}^3$?

We first compute the ranks of all of the matrices. Once the ranks are known, theorems in this section make it easy to answer all questions of this sort.

Checking the theorems for easy ones, we see that D^t is already row reduced. There are three pivots, so $\text{rank}(D) = \text{rank}(D^t) = 3$. Next, A row reduces to

$$\begin{bmatrix} 1 & 2 & 2 \\ 0 & 1 & 1 \\ 0 & 0 & 0 \end{bmatrix},$$

so $\text{rank}(A) = 2$. And B row reduces to

$$\begin{bmatrix} 1 & -2 & -3 & 1 \\ 0 & 5 & 7 & -1 \\ 0 & 0 & 0 & 0 \end{bmatrix},$$

so $\text{rank}(B) = 2$. Finally, C row reduces to

$$\begin{bmatrix} 1 & 2 & 3 \\ 0 & -5 & -7 \\ 0 & 0 & -18/5 \end{bmatrix},$$

so $\text{rank}(C) = 3$. Let us summarize:

$$\text{rank}(A) = 2, \quad \text{rank}(B) = 2, \quad \text{rank}(C) = 3, \quad \text{and} \quad \text{rank}(D) = 3.$$

Now the answers: (a) The columns are independent when the rank equals the number of columns (Theorem 6). Hence, only for C and D are the columns independent. (b) The rows are independent when the rank equals the number of rows (Example 30). Hence, only C. (c) The columns span $\mathbb{R}^3$ when the columns are vectors in $\mathbb{R}^3$ and the dimension of the image is 3. By Theorem 6, this requires $r = 3$; hence, only C. (d) The span of the rows is the image of the transpose, so the rows span $\mathbb{R}^3$ when the rows are vectors in $\mathbb{R}^3$ and the dimension of the image of the transpose is 3. By Theorem 6, this requires $r = 3$; hence, only C and D.

4.3 The given dot products problem

Here is another type of problem that often comes up, and in it the fact that $\text{rank}(A^t A) = \text{rank}(A)$ is important.

Suppose $\{\mathbf{v}_1, \mathbf{v}_2, \ldots, \mathbf{v}_k\}$ is a basis for a subspace S of $\mathbb{R}^n$. Suppose $\mathbf{w}$ is a vector in S and you are given the values of $a_1, a_2, \ldots, a_k$, where

$$a_j = \mathbf{v}_j \cdot \mathbf{w} \quad \text{for} \quad j = 1, 2, \ldots, k. \tag{4.15}$$

Does this problem determine the coordinates of $\mathbf{w}$ with respect to $\{\mathbf{v}_1, \mathbf{v}_2, \ldots, \mathbf{v}_k\}$, and if so, how do we find them? In other words, we would like to find $x_1, x_2, \ldots, x_k$ so that

$$\mathbf{w} = x_1 \mathbf{v}_1 + x_2 \mathbf{v}_2 + \cdots + x_k \mathbf{v}_k. \tag{4.16}$$

Let $A = [\mathbf{v}_1, \mathbf{v}_2, \ldots, \mathbf{v}_k]$. Since the columns of A are linearly independent, $\text{rank}(A) = k$.

Let $\mathbf{x} = \begin{bmatrix} x_1 \\ x_2 \\ \vdots \\ x_k \end{bmatrix}$ so that $\mathbf{w} = A\mathbf{x}$. Let $\mathbf{a} = \begin{bmatrix} a_1 \\ a_2 \\ \vdots \\ a_k \end{bmatrix}$. Then $A^t = \begin{bmatrix} \mathbf{v}_1 \\ \mathbf{v}_2 \\ \vdots \\ \mathbf{v}_k \end{bmatrix}$, so $A^t \mathbf{w} = \begin{bmatrix} \mathbf{v}_1 \cdot \mathbf{w} \\ \mathbf{v}_2 \cdot \mathbf{w} \\ \vdots \\ \mathbf{v}_k \cdot \mathbf{w} \end{bmatrix} = \mathbf{a}.$

Substituting $\mathbf{w} = A\mathbf{x}$, we have $A^t A\mathbf{x} = \mathbf{a}$. Since $A^t A$ is a $k \times k$ matrix, and since $\text{rank}(A^t A) = k$, $A^t A$ is invertible. Therefore, $A^t A\mathbf{x} = \mathbf{a}$ has a unique solution no matter what $\mathbf{a}$ is.

EXAMPLE 34

(Finding a vector with given dot products) Let A be the 4×4 matrix from Example 13. There, we found two one-to-one parameterizations of $\text{Img}(A)$. We now know that the vectors in these parameterizations are a basis for $\text{Img}(A)$. In particular, $\{\mathbf{v}_1, \mathbf{v}_2\}$ is a basis for $\text{Img}(A)$, where

$$\mathbf{v}_1 = \begin{bmatrix} -1 \\ 1 \\ 7 \\ 0 \end{bmatrix} \quad \text{and} \quad \mathbf{v}_2 = \begin{bmatrix} 2 \\ 5 \\ 0 \\ 7 \end{bmatrix}.$$

What we have just seen assures us that there is exactly one vector $\mathbf{w}$ in $\text{Img}(A)$ with

$$\mathbf{w} \cdot \mathbf{v}_1 = 2 \quad \text{and} \quad \mathbf{w} \cdot \mathbf{v}_2 = 3.$$

The values 2 and 3 are arbitrary choices. With these choices made, we can now find $\mathbf{w}$.

First, form

$$A^t A = \begin{bmatrix} 53 & 3 \\ 3 & 78 \end{bmatrix}.$$

Then

$$(A^t A)^{-1} \begin{bmatrix} 2 \\ 3 \end{bmatrix} = (1/27) \begin{bmatrix} 1 \\ 1 \end{bmatrix}.$$

The right-hand side is the coordinate vector $\mathbf{w}$ in this basis. It follows that

$$\mathbf{w} = \frac{1}{27} \begin{bmatrix} -1 \\ 1 \\ 7 \\ 0 \end{bmatrix} + \frac{1}{27} \begin{bmatrix} 2 \\ 5 \\ 0 \\ 7 \end{bmatrix} = \frac{1}{27} \begin{bmatrix} 1 \\ 6 \\ 7 \\ 7 \end{bmatrix}.$$

Exercises

4.1 Let S be the subspace of $\mathbb{R}^4$ spanned by the vectors $\{\mathbf{v}_1, \mathbf{v}_2, \mathbf{v}_3, \mathbf{v}_4\}$ from Exercise 3.3. What is the dimension of S?

4.2 Let S be the subspace of $\mathbb{R}^4$ spanned by the vectors $\{\mathbf{v}_1, \mathbf{v}_2, \mathbf{v}_3, \mathbf{v}_4\}$ from Exercise 3.2. What is the dimension of S?

4.3 Consider the matrices

$$A = \begin{bmatrix} 1 & 1 & 1 \\ 2 & 0 & 2 \\ 0 & 0 & 1 \\ 3 & 2 & 1 \end{bmatrix}, \quad B = \begin{bmatrix} 1 & 2 & 3 \\ 2 & 1 & 3 \\ 1 & 1 & 2 \end{bmatrix}, \quad C = \begin{bmatrix} 1 & 2 & 0 & 1 \\ 1 & 3 & 1 & 2 \\ 1 & 2 & 2 & 2 \end{bmatrix}, \quad \text{and} \quad D = \begin{bmatrix} 1 & 2 & 4 \\ 2 & 2 & 4 \\ 0 & 2 & -1 \end{bmatrix}.$$

Answer the following questions, and justify your answers.

(a) For which of these matrices, if any, are the columns linearly independent?

(b) For which of these matrices, if any, are the rows linearly independent?

(c) For which of these matrices, if any, is the kernel the zero subspace?

(d) For which of these matrices, if any, is the dimension of the image of the transpose equal to 3?

(e) For which of these matrices, if any, is the dimension of the image equal to 3?

4.4 Let $A = \begin{bmatrix} 1 & 2 & 3 \\ 2 & 0 & 2 \\ 0 & 1 & 1 \\ 1 & 2 & 3 \end{bmatrix}$, $B = \begin{bmatrix} 1 & 0 & 0 \\ 2 & 1 & 0 \\ 1 & 1 & 2 \end{bmatrix}$, $C = \begin{bmatrix} 1 & 0 & 0 & 1 \\ 0 & 2 & 1 & 1 \\ 0 & 0 & 2 & 2 \end{bmatrix}$, and $D = \begin{bmatrix} 1 & 2 & 1 \\ 1 & 0 & 1 \\ 0 & 2 & 0 \end{bmatrix}$. Answer questions (a) through (e) from Exercise 4.3, and justify your answers.

4.5 Consider the matrices

$$A = \begin{bmatrix} 1 & 2 & 3 \\ 2 & 1 & 1 \\ 3 & 3 & 4 \end{bmatrix}, \qquad B = \begin{bmatrix} 1 & 7 & -3 & -4 \\ 3 & -2 & 5 & 1 \\ 0 & 2 & 1 & 1 \end{bmatrix}, \qquad C = \begin{bmatrix} 1 & 2 & 4 \\ 1 & 3 & 9 \\ 1 & 4 & 16 \end{bmatrix}, \qquad \text{and} \qquad D = \begin{bmatrix} 1 & 0 & 1 & 0 \\ 0 & 1 & 0 & 1 \\ 2 & 3 & 2 & 3 \end{bmatrix}.$$

Answer questions (a) through (e) from Exercise 4.3, and justify your answers.

4.6 Let A be an $m \times n$ matrix with $n > m$, and suppose that A has linearly independent rows. Consider the square matrices

$$C = A^t A \qquad \text{and} \qquad D = AA^t.$$

(a) Must C be invertible? Justify your answer.

(b) Must D be invertible? Justify your answer.

4.7 Let $v_1 = \begin{bmatrix} 1 \\ -1 \\ 0 \end{bmatrix}$, $v_2 = \begin{bmatrix} 0 \\ 1 \\ -1 \end{bmatrix}$, and $v_3 = \begin{bmatrix} 1 \\ 1 \\ 1 \end{bmatrix}$. Find numbers t_1, t_2, and t_3 so that the vector $b = \begin{bmatrix} 1 \\ 2 \\ 3 \end{bmatrix}$ can be written $b = t_1 v_1 + t_2 v_2 + t_3 v_3$. Is $\{v_1, v_2, v_3\}$ a basis for $\mathbb{R}^3$?

4.8 Let $v_1 = \begin{bmatrix} 1 \\ 3 \\ 9 \end{bmatrix}$, $v_2 = \begin{bmatrix} 1 \\ 2 \\ 4 \end{bmatrix}$, and $v_3 = \begin{bmatrix} 1 \\ 1 \\ 1 \end{bmatrix}$. Find numbers t_1, t_2, and t_3 so that the vector $b = \begin{bmatrix} 1 \\ 2 \\ 3 \end{bmatrix}$ can be written $b = t_1 v_1 + t_2 v_2 + t_3 v_3$. Is $\{v_1, v_2, v_3\}$ a basis for $\mathbb{R}^3$?

4.9 Let $v_1 = \begin{bmatrix} 1 \\ -1 \\ 0 \end{bmatrix}$, $v_2 = \begin{bmatrix} 0 \\ 1 \\ -1 \end{bmatrix}$, and $v_3 = \begin{bmatrix} 1 \\ -2 \\ 1 \end{bmatrix}$. Find numbers t_1, t_2, and t_3 so that the vector $b = \begin{bmatrix} 1 \\ 2 \\ -3 \end{bmatrix}$ can be written $b = t_1 v_1 + t_2 v_2 + t_3 v_3$. How many ways can this be done? Is $\{v_1, v_2, v_3\}$ a basis for $\mathbb{R}^3$?

4.10 Find bases, if possible,* for $\text{Img}(A)$, $\text{Ker}(A)$, $\text{Img}(A^t)$, and $\text{Ker}(A^t)$ where $A = \begin{bmatrix} 1 & 2 & 3 \\ 0 & 3 & 2 \\ 2 & 0 & 1 \end{bmatrix}$.

4.11 Find bases, if possible, for $\text{Img}(A)$, $\text{Ker}(A)$, $\text{Img}(A^t)$, and $\text{Ker}(A^t)$ where $A = \begin{bmatrix} 1 & 2 & 4 & 1 \\ 0 & 2 & 2 & 0 \\ 2 & 3 & 7 & 1 \\ 1 & 1 & 3 & 0 \end{bmatrix}$.

4.12 Find bases, if possible, for $\text{Img}(A)$, $\text{Ker}(A)$, $\text{Img}(A^t)$, and $\text{Ker}(A^t)$ where $A = \begin{bmatrix} 1 & 0 \\ 1 & 1 \\ 0 & 1 \end{bmatrix}$.

*It will not be possible to find bases for the zero subspace, which has no basis.

4.13 Find bases, if possible, for $\mathrm{Img}(A)$, $\mathrm{Ker}(A)$, $\mathrm{Img}(A^t)$, and $\mathrm{Ker}(A^t)$, where $A = \begin{bmatrix} 1 & 1 & 2 \\ 1 & -1 & 0 \\ 0 & 1 & 1 \end{bmatrix}$.

4.14 Find bases, if possible, for $\mathrm{Img}(A)$, $\mathrm{Ker}(A)$, $\mathrm{Img}(A^t)$, and $\mathrm{Ker}(A^t)$, where $A = \begin{bmatrix} 1 & 0 \\ 1 & 1 \\ 0 & 1 \end{bmatrix}$.

4.15 Let $\mathbf{v}_1 = \begin{bmatrix} -1 \\ -2 \\ 2 \end{bmatrix}$ and $\mathbf{v}_2 = \begin{bmatrix} 1 \\ 1 \\ -1 \end{bmatrix}$. Let $\mathbf{w}$ be a vector of the form $\mathbf{w} = x\mathbf{v}_1 + y\mathbf{v}_2$ such that $\mathbf{w} \cdot \mathbf{v}_1 = 2$ and $\mathbf{w} \cdot \mathbf{v}_2 = 4$. Find x and y.

4.16 Let $\mathbf{v}_1 = \begin{bmatrix} 1 \\ 3 \\ 2 \\ 1 \end{bmatrix}$, $\mathbf{v}_2 = \begin{bmatrix} 2 \\ 1 \\ 3 \\ -11 \end{bmatrix}$, and $\mathbf{v}_3 = \begin{bmatrix} 3 \\ 2 \\ 1 \\ 1 \end{bmatrix}$. Let $\mathbf{w}$ be a vector of the form $\mathbf{w} = x\mathbf{v}_1 + y\mathbf{v}_2 + z\mathbf{v}_3$ such that $\mathbf{w} \cdot \mathbf{v}_1 = 27$, $\mathbf{w} \cdot \mathbf{v}_2 = 135$, and $\mathbf{w} \cdot \mathbf{v}_3 = 27$. Find x, y, and z.

4.17 Let $\mathbf{v}_1 = \begin{bmatrix} -1 \\ -2 \\ 2 \\ 1 \end{bmatrix}$, $\mathbf{v}_2 = \begin{bmatrix} -2 \\ 1 \\ -1 \\ 2 \end{bmatrix}$, and $\mathbf{v}_3 = \begin{bmatrix} 2 \\ 0 \\ 1 \\ -3 \end{bmatrix}$. Let $\mathbf{w}$ be a vector of the form $\mathbf{w} = x\mathbf{v}_1 + y\mathbf{v}_2 + z\mathbf{v}_3$ such that $\mathbf{w} \cdot \mathbf{v}_1 = 2$, $\mathbf{w} \cdot \mathbf{v}_2 = 4$, and $\mathbf{w} \cdot \mathbf{v}_3 = 1$. Find x, y, and z.

4.18 Let $\mathbf{v}_1 = \begin{bmatrix} 1 \\ 1 \\ 1 \\ 1 \end{bmatrix}$, $\mathbf{v}_2 = \begin{bmatrix} 2 \\ -2 \\ 3 \\ -3 \end{bmatrix}$, and $\mathbf{v}_3 = \begin{bmatrix} 4 \\ 4 \\ 9 \\ 9 \end{bmatrix}$. Let $\mathbf{w}$ be a vector of the form $\mathbf{w} = x\mathbf{v}_1 + y\mathbf{v}_2 + z\mathbf{v}_3$ such that $\mathbf{w} \cdot \mathbf{v}_1 = 50$, $\mathbf{w} \cdot \mathbf{v}_2 = 26$, and $\mathbf{w} \cdot \mathbf{v}_3 = 50$. Find x, y, and z.

4.19 Let $A = \begin{bmatrix} 1 & 2 \\ 2 & 1 \\ 3 & 3 \end{bmatrix}$ and $B = \begin{bmatrix} 7 & 10 \\ 5 & 8 \\ 12 & 18 \end{bmatrix}$.

(a) Compute the ranks of A and B.

(b) Find an equation for $\mathrm{Img}(A)$. Does each column of B belong to $\mathrm{Img}(A)$?

(c) Is it true that $\mathrm{Img}(B) \subset \mathrm{Img}(A)$? Justify your answer.

(d) Is it true that $\mathrm{Img}(B) = \mathrm{Img}(A)$? Justify your answer.

4.20 Let S and $\tilde{S}$ be subsets of $\mathbb{R}^n$, and let $S + \tilde{S}$ be the subspace of $\mathbb{R}^n$ defined in Exercise 3.13 of this chapter.

(a) Show that $\dim(S + \tilde{S}) \leq \dim(S) + \dim(\tilde{S})$.

(b) Give an example in which $\dim(S + \tilde{S}) = \dim(S) + \dim(\tilde{S})$.

(c) Give an example in which $\dim(S + \tilde{S}) < \dim(S) + \dim(\tilde{S})$.

Another look at the dimension formula

The following exercises provide another look at the dimension formula

$$\mathrm{rank}(A) + \mathrm{nullity}(A) = n$$

and the fact that

$$\mathrm{rank}(A) = \mathrm{rank}(A^t).$$

In the text, we have proved the dimension formula by a careful analysis of row reduction, and then we have shown (in Theorem 2) that the normal equations always have a solution, which amounts to saying that $\mathrm{Img}(A^tA) = \mathrm{Img}(A^t)$. From this demonstration and the dimension formula, a few simple steps took us to the conclusion that $\mathrm{rank}(A) = \mathrm{rank}(A^t)$.

Although we have chosen to base our proofs on a line of reasoning that is clearly connected with solving equations and thus manifestly practical, other ways of proceeding are more direct, if also somewhat more abstract.

4.21 Let A be an $m \times n$ matrix. Let $\{\mathbf{w}_1, \ldots, \mathbf{w}_r\}$ be a basis for $\text{Img}(A)$. By the definition of the image, there are vectors $\{\mathbf{v}_1, \ldots, \mathbf{v}_r\}$ in $\mathbb{R}^n$ so that $\mathbf{w}_j = A\mathbf{v}_j$ for each $j = 1, \ldots, n$. Show that $\{\mathbf{v}_1, \ldots, \mathbf{v}_r\}$ is linearly independent.

4.22 Let A be an $m \times n$ matrix. Let $\{\mathbf{w}_1, \ldots, \mathbf{w}_r\}$ be a basis for $\text{Img}(A)$.

(a) Show that $\{A^t\mathbf{w}_1, \ldots, A^t\mathbf{w}_r\}$ is linearly independent. Hint: Consider a set of vectors $\{\mathbf{v}_1, \ldots, \mathbf{v}_r\}$ in $\mathbb{R}^n$ such that $\mathbf{w}_j = A\mathbf{v}_j$ for each $j = 1, \ldots, n$.

(b) Using part (a), give another proof of the fact that $\text{Img}(A)$ and $\text{Img}(A^t)$ have the same dimension.

4.23 Let A be an $m \times n$ matrix. Let $\{\mathbf{u}_1, \ldots, \mathbf{u}_k\}$ be a basis for $\text{Ker}(A)$. Since any linearly independent set in $\mathbb{R}^n$ can be extended to a basis of $\mathbb{R}^n$, we can extend this set to a basis $\{\mathbf{u}_1, \ldots, \mathbf{u}_k, \mathbf{u}_{k+1}, \ldots, \mathbf{u}_n\}$ of $\mathbb{R}^n$.

(a) Show that $\{A\mathbf{u}_{k+1}, \ldots, A\mathbf{u}_n\}$ spans $\text{Img}(A)$.

(b) Show that $\{A\mathbf{u}_{k+1}, \ldots, A\mathbf{u}_n\}$ is linearly independent.

(c) Use the results of parts (a) and (b) to give another proof that $\text{rank}(A) + \text{nullity}(A) = n$.

4.24 Let A and $\tilde{A}$ be $m \times n$ matrices, and suppose that they have the same kernel. Let $\{\mathbf{w}_1, \ldots, \mathbf{w}_r\}$ be a basis for $\text{Img}(A)$. By the definition of the image, there are vectors $\{\mathbf{v}_1, \ldots, \mathbf{v}_r\}$ in $\mathbb{R}^n$ so that $\mathbf{w}_j = A\mathbf{v}_j$ for each $j = 1, \ldots, n$. Define $\{\tilde{\mathbf{w}}_1, \ldots, \tilde{\mathbf{w}}_r\}$ by $\tilde{\mathbf{w}}_j = \tilde{A}\mathbf{v}_j$ for $j = 1, \ldots, r$.

(a) Show that $\{\tilde{\mathbf{w}}_1, \ldots, \tilde{\mathbf{w}}_r\}$ is a basis for $\text{Img}(\tilde{A})$.

(b) Extend $\{\mathbf{w}_1, \ldots, \mathbf{w}_r\}$ and $\{\tilde{\mathbf{w}}_1, \ldots, \tilde{\mathbf{w}}_r\}$ to bases of $\mathbb{R}^m$: $\{\mathbf{w}_1, \ldots, \mathbf{w}_r, \mathbf{w}_{r+1}, \ldots, \mathbf{w}_m\}$ and $\{\tilde{\mathbf{w}}_1, \ldots, \tilde{\mathbf{w}}_r, \tilde{\mathbf{w}}_{r+1}, \ldots, \tilde{\mathbf{w}}_m\}$, respectively. Define the $m \times m$ matrices W and $\tilde{W}$ by

$$W = [\mathbf{w}_1, \ldots, \mathbf{w}_r, \mathbf{w}_{r+1}, \ldots, \mathbf{w}_m] \quad \text{and} \quad \tilde{W} = [\tilde{\mathbf{w}}_1, \ldots, \tilde{\mathbf{w}}_r, \tilde{\mathbf{w}}_{r+1}, \ldots, \tilde{\mathbf{w}}_m] .$$

Show that these matrices are invertible and that with $B = \tilde{W}W^{-1}$,

$$\tilde{\mathbf{w}}_j = B\mathbf{w}_j \quad \text{for} \quad j = 1, \ldots, m .$$

(c) Show that with B defined as in part (b), $\tilde{A} = BA$. This equality gives another proof that whenever two $m \times n$ matrices A and $\tilde{A}$ have the same kernel, there is an invertible $m \times m$ matrix B so that $\tilde{A} = BA$.

4.25 Let A be an $m \times n$ matrix with rank r. Suppose $r < n$, so A has a nonzero kernel.

(a) Suppose that there are ℓ pivotal columns to the left of the kth nonpivotal column of A. Then the kth nonpivotal column of $\text{rref}(A)$ has the form

$$\begin{bmatrix} y_1 \\ \vdots \\ y_\ell \\ 0 \\ \vdots \\ 0 \end{bmatrix} .$$

This is a vector in $\mathbb{R}^r$, and every entry after the ℓth is zero, as indicated (provided $\ell < r$ or there are no such entries). Define a new vector $\mathbf{w}_k$ by

$$\mathbf{w}_k = y_1\mathbf{e}_{j(1)} + \cdots + y_\ell\mathbf{e}_{j(\ell)} - \mathbf{e}_{(k+\ell)} .$$

Show that this vector is in $\text{Ker}(A)$.

(b) Let $\{\mathbf{w}_1, \ldots, \mathbf{w}_{n-r}\}$ be the set of vectors obtained by doing the same thing for each of the $n - r$ nonpivotal columns. Show that $\{\mathbf{w}_1, \ldots, \mathbf{w}_{n-r}\}$ is a basis for $\text{Ker}(A)$. *This is one more of the things that computing* $\text{rref}(A)$ *is good for: It allows you to read off a basis for* $\text{Ker}(A)$. *Moreover, this basis consists of vectors with lots of zero entries, so it will be easy to work with.*

SECTION 5 | Orthogonal Complements and Projections

5.1 What orthogonal complements are

Here is a diagram of a plane S through the origin in $\mathbb{R}^3$. In it, you see a vector $\mathbf{x}$ written as the sum of a vector $\mathbf{y}$ in the plane S and another vector $\mathbf{z}$ in the normal line to the plane, which is denoted by $S^\perp$ in the diagram. (The symbol $\perp$ denotes orthogonality or perpendicularity and is called "perp." One reads $S^\perp$ aloud as "S perp.")

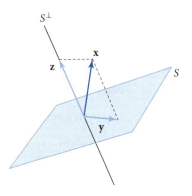

The vector $\mathbf{y}$ is the vector in S that is closest to $\mathbf{x}$. With this choice of $\mathbf{y}$, we will soon see that $\mathbf{z} = \mathbf{x} - \mathbf{y}$ is orthogonal to $\mathbf{y}$.

This "decomposition" of $\mathbf{x}$ into the "components"* $\mathbf{y}$ and $\mathbf{z}$ is useful for many purposes.

When S is a plane through the origin in $\mathbb{R}^3$, the normal line is the line consisting of all of the vectors in $\mathbb{R}^3$ that are orthogonal to every vector in the plane. Look at the picture again and visualize this description. It gives us something we can generalize: For any subspace S of $\mathbb{R}^n$, let $S^\perp$ be the set of all vectors in $\mathbb{R}^n$ that are orthogonal to *every* vector in S.

DEFINITION	**(Orthogonal complement)** For any subspace S of $\mathbb{R}^n$, the *orthogonal complement of S, $S^\perp$,* is the set of all vectors $\mathbf{x}$ in $\mathbb{R}^n$ such that $\mathbf{v} \cdot \mathbf{x} = 0$ for every $\mathbf{v}$ in S.

The first thing to observe is that $S^\perp$ is itself a subspace of $\mathbb{R}^n$. Indeed, if $\mathbf{x}$ is in $S^\perp$ and t is any number, $\mathbf{v} \cdot (t\mathbf{x}) = t(\mathbf{v} \cdot \mathbf{x}) = 0$ for all $\mathbf{v}$ in S, so $S^\perp$ is closed under multiplication by numbers. Next, if $\mathbf{x}_1$ and $\mathbf{x}_2$ belong to $S^\perp$, $\mathbf{v} \cdot (\mathbf{x}_1 + \mathbf{x}_2) = \mathbf{v} \cdot \mathbf{x}_1 + \mathbf{v} \cdot \mathbf{x}_2 = 0$, so $S^\perp$ is closed under addition. This means it is a subspace.

*We give a proper definition of the term "component" shortly, but look at the diagram and see if you can guess what it must be.

Let us find a system of equations for $S^\perp$, that is, a system of equations whose solution set is $S^\perp$. In some sense we already have one: A vector $\mathbf{x}$ belongs to $S^\perp$ if and only if

$$\mathbf{v} \cdot \mathbf{x} = 0 \quad \text{for all vectors } \mathbf{v} \text{ in } S.$$

The problem is that this is a system of *infinitely* many equations, one for each $\mathbf{v}$ in S. But we can reduce it to an *equivalent* system of finitely many equations. Here is how.

Suppose that $\{\mathbf{v}_1, \mathbf{v}_2, \ldots, \mathbf{v}_k\}$ is any spanning set for S. Then any vector $\mathbf{v}$ in S can be written

$$\mathbf{v} = t_1\mathbf{v}_1 + t_2\mathbf{v}_2 + \cdots + t_k\mathbf{v}_k.$$

Since by definition $\mathbf{x}$ belongs to $S^\perp$ if and only if $\mathbf{v} \cdot \mathbf{x} = 0$ for all $\mathbf{v}$ in S, $\mathbf{x}$ belongs to $S^\perp$ if and only if

$$\begin{aligned}
\mathbf{v} \cdot \mathbf{x} &= (t_1\mathbf{v}_1 + t_2\mathbf{v}_2 + \cdots + t_k\mathbf{v}_k) \cdot \mathbf{x} \\
&= t_1(\mathbf{v}_1 \cdot \mathbf{x}) + t_2(\mathbf{v}_2 \cdot \mathbf{x}) + \cdots + t_k(\mathbf{v}_k \cdot \mathbf{x}) = 0 \qquad \text{for } all \; t_1, t_2, \ldots, t_k.
\end{aligned} \tag{5.1}$$

If we choose $t_1 = 1$ and $t_j = 0$ for all other j, this says $\mathbf{v}_1 \cdot \mathbf{x} = 0$. Making similar choices, we see that (5.1) holds for every choice of $t_1, t_2, \ldots, t_k$ if and only if $\mathbf{v}_j \cdot \mathbf{x} = 0$ for all $j = 1, 2, \ldots, k$. Now we have a system of *finitely* many equations.

Introduce $V = [\mathbf{v}_1, \mathbf{v}_2, \ldots, \mathbf{v}_k]$ so that $\text{Img}(V) = S$. Then

$$V^t\mathbf{x} = \begin{bmatrix} \mathbf{v}_1 \cdot \mathbf{x} \\ \mathbf{v}_2 \cdot \mathbf{x} \\ \vdots \\ \mathbf{v}_m \cdot \mathbf{x} \end{bmatrix}.$$

Hence,

$$\mathbf{x} \text{ belongs to } S^\perp \iff V^t\mathbf{x} = 0,$$

and we have our system of equations for $S^\perp$. In other words, $S^\perp = \text{Ker}(V^t)$.

Summarizing, whenever $\{\mathbf{v}_1, \mathbf{v}_2, \ldots, \mathbf{v}_k\}$ spans S, $S = \text{Img}(V)$ and $S^\perp = \text{Ker}(V^t)$. In particular, if A is any $m \times n$ matrix and $S = \text{Img}(A)$, we can take the columns of A as the spanning set, in which case $V = A$, so

$$(\text{Img}(A))^\perp = \text{Ker}(A^t). \tag{5.2}$$

If we take $\{\mathbf{v}_1, \mathbf{v}_2, \ldots, \mathbf{v}_k\}$ to be a basis of S, not just a spanning set, we can compute the dimension of $S^\perp$.

Notice that $V = [\mathbf{v}_1, \mathbf{v}_2, \ldots, \mathbf{v}_k]$ is an $n \times k$ matrix and $\text{rank}(V) = k$. It follows from Corollary 2 of Theorem 10 that V^t is a $k \times n$ matrix and $\text{rank}(V^t) = k$. Then from the dimension formula, Theorem 9, $\dim(\text{Ker}(V^t)) = n - k$. Hence, if $\dim(S) = k$, then $\dim(S^\perp) = n - k$.

EXAMPLE 35   **(An orthogonal complement)** Let S be the plane in $\mathbb{R}^3$ given by the equation $x + y + z = 0$. Since S is two-dimensional, any two linearly independent vectors in S are a basis for it. Clearly,

$$\mathbf{v}_1 = \begin{bmatrix} 1 \\ -1 \\ 0 \end{bmatrix} \quad \text{and} \quad \mathbf{v}_2 = \begin{bmatrix} 0 \\ 1 \\ -1 \end{bmatrix}$$

both satisfy $x + y + z = 0$, so they both belong to S. They are not proportional, so they are linearly independent. Hence, $\{\mathbf{v}_1, \mathbf{v}_2\}$ is a basis for S.

We therefore let

$$V = [\mathbf{v}_1, \mathbf{v}_2] = \begin{bmatrix} 1 & 0 \\ -1 & 1 \\ 0 & -1 \end{bmatrix}$$

so that $S^\perp = \text{Ker}(V^t)$, where

$$V^t = \begin{bmatrix} 1 & -1 & 0 \\ 0 & 1 & -1 \end{bmatrix}.$$

Since V^t is already in row-reduced form—a lucky break—we can find the solution set of $V^t\mathbf{x} = 0$ by back substitution. The variable z is nonpivotal, and expressing x and y in terms of z, we find $y = z$ and then $x = y = z$. The solution set is

$$z \begin{bmatrix} 1 \\ 1 \\ 1 \end{bmatrix},$$

so the single vector

$$\begin{bmatrix} 1 \\ 1 \\ 1 \end{bmatrix}$$

is a basis for $S^\perp$. Indeed, this vector specifies the direction of the normal line, so $S^\perp$ is the normal line to the plane S, as we knew it should be. Notice also that $\dim(S) = 2$ and $\dim(S^\perp) = 1$, so $\dim(S) + \dim(S^\perp) = 3$, as we knew it should.

The following theorem summarizes what we have learned so far and a bit more.

THEOREM 11   **(Finding orthogonal complements)** *Let S be any subspace of $\mathbb{R}^n$. Then $S^\perp$ is also a subspace of $\mathbb{R}^n$.*

1. *Let $\{\mathbf{v}_1, \mathbf{v}_2, \ldots, \mathbf{v}_k\}$ be a set of vectors that spans S. Then a vector $\mathbf{x}$ in $\mathbb{R}^n$ belongs to $S^\perp$ if and only if $\mathbf{v}_j \cdot \mathbf{x} = 0$ for each $j = 1, 2, \ldots, n$. In other words, with $V = [\mathbf{v}_1, \mathbf{v}_2, \ldots, \mathbf{v}_k]$ so that $S = \text{Img}(V)$, $S^\perp = \text{Ker}(V^t)$.*

    In particular, if A is any $m \times n$ matrix,

    $$(\text{Img}(A))^\perp = \text{Ker}(A^t). \tag{5.3}$$

2. *The following relations between S and $S^\perp$ are true:*

    $$\dim(S) + \dim(S^\perp) = n \tag{5.4}$$

    $$S \cap S^\perp = 0 \tag{5.5}$$

    $$(S^\perp)^\perp = S. \tag{5.6}$$

PROOF: Everything has been explained in the preceding text except (5.5) and (5.6). To prove (5.5), suppose that $\mathbf{x}$ belongs to S and $S^\perp$. Then $\mathbf{x}$ is orthogonal to itself, so $|\mathbf{x}|^2 = \mathbf{x} \cdot \mathbf{x} = 0$. Hence $\mathbf{x} = 0$.

To prove (5.6), we use the dimension principle. First, observe that $\mathbf{y}$ belongs to $(S^\perp)^\perp$ if and only if $\mathbf{y} \cdot \mathbf{v} = 0$ for every $\mathbf{v}$ in $S^\perp$. But whenever $\mathbf{v}$ belongs to $S^\perp$ and $\mathbf{y}$ belongs to S, then $\mathbf{y} \cdot \mathbf{v} = 0$, so every $\mathbf{y}$ in S belongs to $(S^\perp)^\perp$. In other words,

$$S \subset (S^\perp)^\perp. \tag{5.7}$$

By (5.4), $\dim(S) + \dim(S^\perp) = n = \dim(S^\perp) + \dim((S^\perp)^\perp)$. Canceling $\dim(S^\perp)$ from both sides, we see $\dim(S) = \dim((S^\perp)^\perp)$. Together with (5.7) and the dimension principle, this proves (5.6). ■

COROLLARY TO
THEOREM 11

(Kernels, images, and orthogonal complements) *For any matrix A,*

$$(\text{Img}(A))^{\perp} = \text{Ker}(A^t) \qquad \text{and} \qquad \text{Img}(A) = (\text{Ker}(A^t))^{\perp} \qquad (5.8)$$

$$(\text{Ker}(A))^{\perp} = \text{Img}(A^t) \qquad \text{and} \qquad \text{Ker}(A) = (\text{Img}(A^t))^{\perp}. \qquad (5.9)$$

PROOF: The first equality is (5.3). To get the second, take the orthogonal complements of both sides, and use (5.6). Then substitute A^t for A in (5.8), and use the fact that $(A^t)^t = A$. The result is (5.9) written in a different order. ∎

In Section 1 of this chapter, we learned how to find a system of equations for $\text{Img}(A)$ by row reduction. Now we have another description of $\text{Img}(A)$, not as the solution set of a system of equations, but as the *orthogonal complement* of $\text{Ker}(A)$, the solution set of $A\mathbf{x} = 0$. This description provides a useful criterion for solvability of $A\mathbf{x} = \mathbf{b}$.

• **(Fredholm criterion)** *Suppose* $\{\mathbf{v}_1, \mathbf{v}_2, \ldots, \mathbf{v}_k\}$ *is a basis of* $\text{Ker}(A^t)$. *Then*

$$A\mathbf{x} = \mathbf{b} \quad \text{has at least one solution} \quad \Longleftrightarrow \quad \mathbf{v}_j \cdot \mathbf{b} = 0 \quad \text{for } j = 1, 2, \ldots, k. \qquad (5.10)$$

The reason this works is that $A\mathbf{x} = \mathbf{b}$ is solvable exactly when $\mathbf{b}$ belongs to $\text{Img}(A)$. By (5.8), this is exactly the case when $\mathbf{b}$ belongs to $(\text{Ker}(A^t))^{\perp}$. By Theorem 11, this is the case if and only if $\mathbf{b} \cdot \mathbf{v}_j = 0$ for each $\mathbf{v}_j$ in a spanning set for $\text{Ker}(A^t)$.

This criterion is useful since it turns out that for many very large matrices A that come up in applications, $\text{Ker}(A^t)$ has a low dimension, so k will be a small number.

EXAMPLE 36 **(Fredholm criterion)** Consider

$$A = \begin{bmatrix} 1 & 0 & 0 & 0 & \cdots & 0 & -1 \\ -1 & 1 & 0 & 0 & \cdots & 0 & 0 \\ 0 & -1 & 1 & 0 & \cdots & 0 & 0 \\ 0 & 0 & -1 & 1 & \cdots & 0 & 0 \\ & & \vdots & & \vdots & & \\ 0 & 0 & 0 & 0 & \cdots & -1 & 1 \end{bmatrix}.$$

That is, for each $i = 1, 2, \ldots, n$, $(A\mathbf{x})_i = x_i - x_{i-1}$, where x_0 is interpreted as x_n. Some thought should convince you that A^t does something pretty similar:

$$(A^t\mathbf{x})_i = x_i - x_{i+1},$$

where x_{n+1} is interpreted as x_1.

Therefore,

$$|A^t\mathbf{x}|^2 = \sum_{i=1}^{n}(x_i - x_{i+1})^2$$

so that $A^t\mathbf{x} = 0$ if and only if $x_i = x_i + 1$ for each i. Hence, $\mathbf{x}$ is in the kernel of A^t if and only if x_i is independent of i so that $\mathbf{x}$ is a multiple of

$$\mathbf{v} = \begin{bmatrix} 1 \\ 1 \\ \vdots \\ 1 \end{bmatrix}.$$

Hence, $\{\mathbf{v}\}$ is a basis for $\text{Ker}(A^t)$, so $\mathbf{v} \cdot \mathbf{x} = 0$ is an equation for $\text{Img}(A)$. (We could write it in matrix form, but with only one row, there is very little point.)

This equation is very easy to check: Because of the specific form of $\mathbf{v}$ in this example, $\mathbf{b} \cdot \mathbf{v} = 0$ is equivalent to

$$\sum_{j=1}^{n} b_j = 0.$$

So checking for solvability of $A\mathbf{x} = \mathbf{b}$ reduces to summing the entries of $\mathbf{b}$.

The matrices A and A^t are not artificial; you see that they perform "finite differences," and they come up in many numerical treatments of differential equations. There is more on this in Chapter 6.

We conclude with a comment on terminology. You may be wondering why $S^{\perp}$ is called the "orthogonal complement" of S. It is clear what the role of orthogonality is, but in what sense is $S^{\perp}$ *complementary* to S?

This question is important, and let us begin by saying what the answer is not: $S^{\perp}$ *is not complementary to S in the set-theoretic sense.*

Recall that the set-theoretic complement of any subset S of $\mathbb{R}^n$ is the set S^c of vectors that do not belong to S. For one thing, all subspaces contain the zero vector, so the set-theoretic intersection of S and $S^{\perp}$ is not empty, while the set-theoretic intersection of S and S^c is empty by definition. The analogy with the set-theoretic complement is that $(S^c)^c = S$, which is similar to (5.6). In general, a "complementarity relation" has the property that the complement of a complement is what you started with.

5.2 Orthogonal projections and closest vectors

Let S be any subspace of $\mathbb{R}^n$, and let $\mathbf{x}$ be any vector in $\mathbb{R}^n$. Let $\{\mathbf{v}_1, \mathbf{v}_2, \ldots, \mathbf{v}_d\}$ be a basis for S, and let A be the $n \times d$ matrix $A = [\mathbf{v}_1, \mathbf{v}_2, \ldots, \mathbf{v}_d]$. Then $S = \mathrm{Img}(A)$.

In Section 2 we learned how to find the vector $\mathbf{y}$ in $\mathrm{Img}(A) = S$ that is closest to $\mathbf{x}$: Since $\{\mathbf{v}_1, \mathbf{v}_2, \ldots, \mathbf{v}_d\}$ is a basis for $\mathrm{Img}(A)$, $t_1\mathbf{v}_1 + t_2\mathbf{v}_2 + \cdots + t_d\mathbf{v}_d$ is a one-to-one parameterization of $\mathrm{Img}(A)$, so Theorem 2 tells us that

$$\mathbf{y} = A(A^t A)^{-1} A^t \mathbf{x}. \tag{5.11}$$

Notice that

$$A^t \mathbf{y} = A^t (A(A^t A)^{-1} A^t \mathbf{x}) = (A^t A)(A^t A)^{-1} A^t \mathbf{x} = A^t \mathbf{x}.$$

Therefore, $A^t(\mathbf{x} - \mathbf{y}) = A^t \mathbf{x} - A^t \mathbf{y} = A^t \mathbf{x} - A^t \mathbf{x} = 0$, which means that $\mathbf{x} - \mathbf{y}$ belongs to $\mathrm{Ker}(A^t)$. By Theorem 11, $\mathrm{Ker}(A^t) = (\mathrm{Img}(A))^{\perp} = S^{\perp}$.

Hence, with $\mathbf{z}$ defined by $\mathbf{z} = \mathbf{x} - \mathbf{y}$,

$$\mathbf{x} = \mathbf{y} + \mathbf{z} \qquad \text{where } \mathbf{y} \text{ belongs to } S \text{ and } \mathbf{z} \text{ belongs to } S^{\perp}. \tag{5.12}$$

Now suppose that $\mathbf{x} = \mathbf{y}_1 + \mathbf{z}_1 = \mathbf{y}_2 + \mathbf{z}_2$ with both $\mathbf{y}_1$ and $\mathbf{y}_2$ in S and both $\mathbf{z}_1$ and $\mathbf{z}_2$ in $S^{\perp}$. Then $\mathbf{y}_1 - \mathbf{y}_2 = \mathbf{z}_2 - \mathbf{z}_1$. But the left-hand side belongs to $S^{\perp}$, the right-hand side to S. Since both sides are equal, they are both in $S \cap S^{\perp} = 0$. So $\mathbf{y}_1 - \mathbf{y}_2 = \mathbf{z}_2 - \mathbf{z}_1 = 0$. Hence, $\mathbf{y}_1 = \mathbf{y}_2$ and $\mathbf{z}_1 = \mathbf{z}_2$, so the decomposition in (5.12) is unique.

The fact that, given S and $\mathbf{x}$, the vectors $\mathbf{y}$ and $\mathbf{z}$ in (5.12) are uniquely determined justifies the following definition.

DEFINITION	**(Components)** With $\mathbf{x}$, $\mathbf{y}$, and $\mathbf{z}$ related as in (5.12), $\mathbf{y}$ is called the *component of* $\mathbf{x}$ *in S* and $\mathbf{z}$ is called the *component of* $\mathbf{x}$ *in* $S^{\perp}$.

Now let $\mathbf{w}$ be the vector in $S^{\perp}$ that is closest to $\mathbf{x}$. Define $\mathbf{u} = \mathbf{x} - \mathbf{w}$. Then $\mathbf{x} = \mathbf{w} + \mathbf{u}$ and $\mathbf{w}$ is in $S^{\perp}$. By the preceding argument, $\mathbf{u}$ is in $(S^{\perp})^{\perp} = S$. Because there is just one such decomposition, $\mathbf{w} = \mathbf{z}$ and $\mathbf{u} = \mathbf{y}$. In particular, $\mathbf{z}$ is the vector in $S^\perp$ that is closest to $\mathbf{x}$. We summarize in the following theorem.

THEOREM 12   **(Orthogonal complements and closest vectors)** *Let S be a k-dimensional subspace of $\mathbb{R}^n$. Then every vector $\mathbf{x}$ in $\mathbb{R}^n$ can be written*

$$\mathbf{x} = \mathbf{y} + \mathbf{z} \qquad \text{with } \mathbf{y} \text{ in } S \text{ and } \mathbf{z} \text{ in } S^\perp, \tag{5.13}$$

and there is just one such decomposition: $\mathbf{y}$ is the vector in S closest to $\mathbf{x}$, and $\mathbf{z}$ is the vector in $S^\perp$ closest to $\mathbf{x}$. In particular, if $\{\mathbf{v}_1, \mathbf{v}_2, \ldots, \mathbf{v}_k\}$ is a basis for S, and $V = [\mathbf{v}_1, \mathbf{v}_2, \ldots, \mathbf{v}_k]$, then $\mathbf{y} = P_S\mathbf{x}$, where the $n \times n$ matrix P_S is given by

$$P_S = V(V^t V)^{-1} V^t. \tag{5.14}$$

It is important to notice that although the *formula* for P_S depends on the choice of the basis for S, the resulting matrix P_S does not depend on this choice. The reason is that without making any choice of a basis, $P_S\mathbf{x}$ is uniquely determined as the vector in S that is closest to $\mathbf{x}$. In particular, for any j, $P_S\mathbf{e}_j$ is independent of the choice of basis, so $(P_S)_{i,j} = \mathbf{e}_i \cdot P_S\mathbf{e}_j$ is independent of the choice of basis.

DEFINITION   **(Orthogonal projection)** The orthogonal projection onto a subspace S of $\mathbb{R}^n$ is the matrix P_S such that $P_S\mathbf{x} = \mathbf{y}$, where $\mathbf{x} = \mathbf{y} + \mathbf{z}$ with $\mathbf{y}$ in S and $\mathbf{z}$ in $S^\perp$.

It follows directly from the definition that if $\mathbf{x} = \mathbf{y} + \mathbf{z}$ with $\mathbf{y}$ in S and $\mathbf{z}$ in $S^\perp$, then $P_{S^\perp}\mathbf{x} = \mathbf{z}$, which, together with $P_S\mathbf{x} = \mathbf{y}$, gives $\mathbf{x} = P_S\mathbf{x} + P_{S^\perp}\mathbf{x}$. Since this identity holds for all $\mathbf{x}$, $P_S + P_{S^\perp} = I$. This is a very useful formula. We summarize as follows.

THEOREM 13   **(Properties of orthogonal projections)** *Let S be a subspace of $\mathbb{R}^n$. Given basis $\{\mathbf{v}_1, \mathbf{v}_2, \ldots, \mathbf{v}_k\}$ for S, let $A = [\mathbf{v}_1, \mathbf{v}_2, \ldots, \mathbf{v}_k]$ so that $S = \text{Img}(A)$ and A has independent columns. Then P_S, the orthogonal projection onto S, is given by*

$$P_S = A(A^t A)^{-1} A^t. \tag{5.15}$$

Moreover,

1. $$P_S + P_{S^\perp} = I. \tag{5.16}$$

2. $$P = P^t, \qquad P^2 = P \tag{5.17}$$

3. $$\text{Img}(P) = S \qquad \text{and} \qquad \text{Ker}(P) = S^\perp. \tag{5.18}$$

PROOF: We have already explained everything except for parts (2) and (3). Using the facts that $(BC)^t = C^t B^t$ and $(B^t)^t = B$, together with the formula (5.15) for P_S, we see that $P^t = P$ and $P^2 = A(A^t A)^{-1} A^t A(A^t A)^{-1} A^t = A(A^t A)^{-1} A^t = P$.

Next, if $\mathbf{b}$ is in S, then by definition, $P\mathbf{b} = \mathbf{b}$, so $\mathbf{b}$ is in $\text{Img}(P)$. But also by the definition of P, $\text{Img}(P) \subset S$. Hence $\text{Img}(P) = S$.

It now follows, using Theorem 11, that $\text{Ker}(P) = (\text{Img}(P^t))^\perp = (\text{Img}(P))^\perp = S^\perp$. ∎

EXAMPLE 37   **(Computing an orthogonal projection)** Let

$$B = \begin{bmatrix} 2 & 1 \\ 1 & -1 \\ 2 & 1 \end{bmatrix}.$$

Then

$$B^t B = \begin{bmatrix} 9 & 3 \\ 3 & 3 \end{bmatrix},$$

so

$$(B^t B)^{-1} = (1/6) \begin{bmatrix} 1 & -1 \\ -1 & 3 \end{bmatrix}.$$

By (5.15),

$$P_S = \frac{1}{2} \begin{bmatrix} 1 & 0 & 1 \\ 0 & 2 & 0 \\ 1 & 0 & 1 \end{bmatrix}.$$

Next, by Theorem 11, $S^\perp = \text{Ker}(B^t)$. It is easy to see that the vector

$$\mathbf{v} = \begin{bmatrix} 1 \\ 0 \\ -1 \end{bmatrix}$$

belongs to $\text{Ker}(B^t)$. Since $\dim(S^\perp) = 1$ by Theorem 11, $\{\mathbf{v}\}$ is a basis for $S^\perp$. Let $\mathbf{v}$ also denote the 3×1 matrix whose single column is $\mathbf{v}$. That is, we are just writing $\mathbf{v}$ to denote $[\mathbf{v}]$. It may look like abuse of notation, but it is useful and quite standard.

With this notational convention, $\mathbf{v}^t \mathbf{v} = \mathbf{v} \cdot \mathbf{v} = 2$, so

$$P_{S^\perp} = \mathbf{v}(\mathbf{v}^t \mathbf{v})^{-1} \mathbf{v}^t = \frac{1}{2} \begin{bmatrix} 1 & 0 & -1 \\ 0 & 0 & 0 \\ -1 & 0 & 1 \end{bmatrix}.$$

Again, in this computation we are regarding $\mathbf{v}$ as a 3×1 matrix.

Notice that $P_S + P_{S^\perp} = I$, as Theorem 11 says. In fact, we could have used this statement to simplify the computation of P_S. Here is the idea: Since $S^\perp$ is one-dimensional, it is relatively easy to use (5.15) to compute it. But once it is computed, you can get P_S from $P_S = I - P_{S^\perp}$. This indirect route can be much faster. It is likely to be faster whenever $\dim(S^\perp) < \dim(S)$.

Partly for the reason discussed at the end of the last example, the one-dimensional case is very important.

EXAMPLE 38   **(Orthogonal projection onto a line)** Let S be a one-dimensional subspace of $\mathbb{R}^n$ spanned by some vector $\mathbf{v}$. Again, letting $\mathbf{v}$ denote the $n \times 1$ matrix $[\mathbf{v}]$ whose single column is $\mathbf{v}$, we have $\mathbf{v}^t \mathbf{v} = \mathbf{v} \cdot \mathbf{v}$, so $(\mathbf{v}^t \mathbf{v})^{-1} = |\mathbf{v}|^{-2}$, regarded as a 1×1 matrix. From (5.15),

$$(P_S)_{i,j} = v_i |\mathbf{v}|^{-2} v_j$$

$$= \frac{1}{|\mathbf{v}|^2} v_i v_j. \tag{5.19}$$

For example, if

$$\mathbf{v} = \begin{bmatrix} 2 \\ 0 \\ 1 \\ 2 \end{bmatrix},$$

then

$$P_S = \frac{1}{9} \begin{bmatrix} 4 & 0 & 2 & 4 \\ 0 & 0 & 0 & 0 \\ 2 & 0 & 1 & 2 \\ 4 & 0 & 2 & 4 \end{bmatrix}.$$

The computation of orthogonal projections is easy if we use a basis consisting of orthonormal vectors* to compute P_S. Suppose that $\{\mathbf{u}_1, \mathbf{u}_2, \ldots, \mathbf{u}_k\}$ is an orthonormal basis for S. Then let $Q = [\mathbf{u}_1, \mathbf{u}_2, \ldots, \mathbf{u}_k]$ so that $S = \text{Img}(Q)$. We must have $P_S = Q(Q^t Q)^{-1} Q^t$. Since Q has orthonormal columns, Q is an isometry, so $Q^t Q = I$. There is no inverse to compute! The formula (5.15) simplifies to

$$P_S = QQ^t.$$

5.3 Outer products and orthogonal projections

Let $\mathbf{v}$ and $\mathbf{w}$ be any two vectors in $\mathbb{R}^n$, and think of them as $n \times 1$ matrices—as matrices with just one column. Then their transposes, $\mathbf{v}^t$ and $\mathbf{w}^t$, are $1 \times n$ matrices—matrices with just one row. The matrix product $\mathbf{v}^t \mathbf{w}$ is then defined, and it is a 1×1 matrix—just a number. In fact,

$$\mathbf{v}^t \mathbf{w} = \mathbf{v} \cdot \mathbf{w}.$$

The dot product is sometimes called the *inner product*. With that in mind, we observe that the matrix product $\mathbf{v} \mathbf{w}^t$ is also defined and is an $n \times n$ matrix. We call it the *outer product* of $\mathbf{v}$ and $\mathbf{w}$.

DEFINITION

(Outer product) Let $\mathbf{v}$ and $\mathbf{w}$ be any two vectors in $\mathbb{R}^n$ considered as $n \times 1$ matrices. The *outer product* of $\mathbf{v}$ and $\mathbf{w}$ is the $n \times n$ matrix

$$\mathbf{v} \mathbf{w}^t.$$

EXAMPLE 39   (An outer product) Let

$$\mathbf{v} = \begin{bmatrix} 1 \\ 2 \\ 3 \end{bmatrix} \quad \text{and} \quad \mathbf{w} = \begin{bmatrix} 3 \\ 2 \\ 1 \end{bmatrix}.$$

Then

$$\mathbf{v} \mathbf{w}^t = \begin{bmatrix} 1 \\ 2 \\ 3 \end{bmatrix} \begin{bmatrix} 3 & 2 & 1 \end{bmatrix} = \begin{bmatrix} 3 \begin{bmatrix} 1 \\ 2 \\ 3 \end{bmatrix} & 2 \begin{bmatrix} 1 \\ 2 \\ 3 \end{bmatrix} & 1 \begin{bmatrix} 1 \\ 2 \\ 3 \end{bmatrix} \end{bmatrix} = \begin{bmatrix} 3 & 2 & 1 \\ 6 & 4 & 2 \\ 9 & 6 & 3 \end{bmatrix}.$$

Let $\mathbf{x}$ be any vector in $\mathbb{R}^n$. To compute $(\mathbf{v} \mathbf{w}^t)\mathbf{x}$, think of $\mathbf{x}$ as an $n \times 1$ matrix, and use the associativity of matrix multiplication:

$$(\mathbf{v} \mathbf{w}^t)\mathbf{x} = \mathbf{v}(\mathbf{w}^t \mathbf{x}) = \mathbf{v}(\mathbf{w} \cdot \mathbf{x}) = (\mathbf{w} \cdot \mathbf{x})\mathbf{v}.$$

Now for any matrix A, $A_{i,j} = \mathbf{e}_i \cdot A\mathbf{e}_j$, so

$$(\mathbf{v} \mathbf{w}^t)_{i,j} = \mathbf{e}_i \cdot ((\mathbf{w} \cdot \mathbf{e}_j)\mathbf{v}) = (\mathbf{e}_i \cdot \mathbf{v})(\mathbf{w} \cdot \mathbf{e}_j).$$

That is,

$$(\mathbf{v} \mathbf{w}^t)_{i,j} = v_i w_j. \tag{5.20}$$

We bring this up now because there is a convenient way to write orthogonal projections in terms of outer products. It is popular for good reason, and you are likely to see a lot of it. For example, using (5.20), the formula (5.19) can be written in terms of an outer product as

$$P_S = \frac{1}{|\mathbf{v}|^2} \mathbf{v} \mathbf{v}^t.$$

---

*We will see in Section 7 that every subspace has such a basis; there is even an algorithm for turning any basis into a basis consisting of orthonormal vectors.

More generally, if $\{\mathbf{u}_1, \mathbf{u}_2, \ldots, \mathbf{u}_k\}$ is an orthonormal basis for S, then P_S, the orthogonal projection onto S, is given by

$$P_S = \mathbf{u}_1\mathbf{u}_1^t + \mathbf{u}_2\mathbf{u}_2^t + \cdots + \mathbf{u}_k\mathbf{u}_k^t. \tag{5.21}$$

Exercises

5.1 Let A be a 4×4 matrix such that $A^t\mathbf{x} = 0$ only when $\mathbf{x}$ is a multiple of $\begin{bmatrix} 1 \\ -2 \\ 1 \\ 0 \end{bmatrix}$. Let $\mathbf{b} = \begin{bmatrix} 1 \\ 1 \\ 1 \\ 1 \end{bmatrix}$. Does $A\mathbf{x} = \mathbf{b}$ have a solution? Does $A\mathbf{x} = \mathbf{b}$ have a *unique* solution? Justify your answers.

5.2 Let A be a 4×4 matrix such that $A\mathbf{x} = 0$ only when $\mathbf{x}$ is a multiple of $\begin{bmatrix} 1 \\ -2 \\ 1 \\ 0 \end{bmatrix}$. Let $\mathbf{b} = \begin{bmatrix} 1 \\ 1 \\ 1 \\ 1 \end{bmatrix}$. Does $A\mathbf{x} = \mathbf{b}$ necessarily have a solution? Does $A\mathbf{x} = \mathbf{b}$ have a *unique* solution? How would your answers change if A were symmetric? Justify your answers.

5.3 Find the orthogonal projection onto S where S is the span of $\mathbf{v}_1 = \begin{bmatrix} 1 \\ 2 \\ 3 \end{bmatrix}$ and $\mathbf{v}_2 = \begin{bmatrix} 3 \\ 2 \\ 1 \end{bmatrix}$. Also find the orthogonal projection $P^\perp$ onto $S^\perp$, and give a one-to-one parametric representation for $S^\perp$.

5.4 Let A be the matrix $A = \begin{bmatrix} 1 & 0 & 2 \\ 2 & 2 & 1 \\ 3 & 2 & 3 \end{bmatrix}$.

(a) Find the orthogonal projection P_c onto $\mathrm{Img}(A)$, and $P_c^\perp$. ($\mathrm{Img}(A)$ is spanned by the column of A, hence the c.)

(b) Find the orthogonal projection P_r onto $\mathrm{Img}(A^t)$, and $P_r^\perp$. ($\mathrm{Img}(A^t)$ is spanned by the rows of A, hence the r.)

(c) Find one-to-one parametric representations of $\mathrm{Img}(A)$, and of $\mathrm{Ker}(A)$.

5.5 Let A be the matrix $A = \begin{bmatrix} 1 & 1 & 2 \\ 1 & -1 & 0 \\ 0 & 1 & 1 \end{bmatrix}$.

(a) Find the orthogonal projection onto $\mathrm{Ker}(A^t)$.

(b) Find the orthogonal projection onto $\mathrm{Img}(A)$.

(c) How are your answers for parts (a) and (b) related?

(d) Find the orthogonal projection onto $\mathrm{Img}(A^t)$.

5.6 Let A be the matrix $A = \begin{bmatrix} 1 & 1 & 2 \\ 1 & 0 & 1 \\ 0 & 1 & 1 \end{bmatrix}$.

(a) Find the orthogonal projection P_c onto $\mathrm{Img}(A)$.

(b) Find the orthogonal projection P_r onto $\mathrm{Img}(A^t)$.

(c) Find the orthogonal projection P_c onto $\mathrm{Ker}(A)$.

(d) Find the orthogonal projection P_r onto $\mathrm{Ker}(A^t)$.

5.7 Let S_1 and S_2 be two subspaces of $\mathbb{R}^n$, with $S_1 \subset S_2$. Show that $S_2^\perp \subset S_1^\perp$.

| Gram-Schmidt and QR Factorization

6.1 From a basis to an orthonormal basis

We closed Section 2 with the following question:

- *Given an $m \times n$ matrix A, can we always find an isometry Q so that* $\mathrm{Img}(Q) = \mathrm{Img}(A)$*, and if so, how?*

This question is important in the theory of least squares solutions. We can now answer it, but it will help to rephrase it first. If we can find an orthonormal basis $\{u_1, u_2, \ldots, u_k\}$ for $\mathrm{Img}(A)$, then $Q = [u_1, u_2, \ldots, u_k]$ will be an isometry, and

$$\mathrm{Img}(Q) = \mathrm{Sp}(\{u_1, u_2, \ldots, u_k\}) = \mathrm{Img}(A).$$

Therefore, we may as well ask whether or not we can always find an orthonormal basis for the image of a matrix. But the image of every matrix is a subspace and every subspace is the image of some matrix, so we can rephrase our question:

- *Given a subspace S of $\mathbb{R}^n$, can we always find an orthonormal basis of S, and if so, how?*

Phrased this way, the question is now easy to answer, using what we have learned in the last few sections.

The answer is useful for many things besides finding least squares solutions. Many computations are much easier if you do them using an orthonormal basis. Unfortunately, the methods we have for finding bases do not usually give us an orthonormal basis. Fortunately, there is a simple procedure for producing an orthonormal basis for S out of any spanning set for S. This procedure is called the *Gram-Schmidt orthonormalization procedure*. Let us begin by stepping through it in a simple example. We do so in $\mathbb{R}^2$ so that we can draw diagrams.

Let

$$v_1 = \begin{bmatrix} 1 \\ 2 \end{bmatrix} \quad \text{and} \quad v_2 = \begin{bmatrix} 2 \\ 1 \end{bmatrix}.$$

These vectors are linearly independent, so $\{v_1, v_2\}$ is a basis for $\mathbb{R}^2$, though not an orthonormal basis.

Step One: Define the first element, u_1, of the new basis to be the *normalization* of v_1, the first element of the old basis. That is, we rescale v_1 to make it a unit vector:

$$u_1 = \frac{1}{|v_1|} v_1 = \frac{1}{\sqrt{5}} \begin{bmatrix} 1 \\ 2 \end{bmatrix}. \tag{6.1}$$

Step Two: When we get to the second vector, we have to take the requirement for orthogonality into account, not just normalization. Here is what to do: Project v_2 onto the subspace $S^{\perp}$ orthogonal to the subspace S spanned by u_1. Both subspaces are lines, as indicated in the following diagram, and S is also the subspace spanned by v_1.

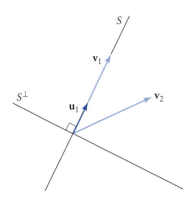

By (5.21), the orthogonal projection P_S onto S is given by $P = \mathbf{u}_1\mathbf{u}_1^t$, so the projection $P_{S^\perp}$ is given by $P_{S^\perp} = I - \mathbf{u}_1\mathbf{u}_1^t$. The projection of $\mathbf{v}_2$ onto $S^\perp$ is

$$\mathbf{w}_2 = P_{S^\perp}\mathbf{v}_2 = (I - \mathbf{u}_1\mathbf{u}_1^t)\mathbf{v}_2 = \mathbf{v}_2 - (\mathbf{v}_2 \cdot \mathbf{u}_1)\mathbf{u}_1. \tag{6.2}$$

Since $\mathbf{v}_2 \cdot \mathbf{u}_1 = 4/\sqrt{5}$,

$$\mathbf{w}_2 = \begin{bmatrix} 2 \\ 1 \end{bmatrix} - \frac{4}{5}\begin{bmatrix} 1 \\ 2 \end{bmatrix} = \frac{3}{5}\begin{bmatrix} 2 \\ -1 \end{bmatrix}.$$

Step Three: Normalize $\mathbf{w}_2$: Define

$$\mathbf{u}_2 = \frac{1}{|\mathbf{w}_2|}\mathbf{w}_2 = \frac{1}{\sqrt{5}}\begin{bmatrix} 2 \\ -1 \end{bmatrix}.$$

Then $\{\mathbf{u}_1, \mathbf{u}_2\}$ is the new orthonormal basis.

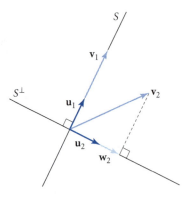

Study the diagrams carefully, and visualize the process.

Now let us apply the process to find an orthonormal basis of a two-dimensional subspace of $\mathbb{R}^3$. This example is more interesting than the previous one since we already know one orthonormal basis for $\mathbb{R}^2$, namely, $\{\mathbf{e}_1, \mathbf{e}_2\}$. However, it is harder to draw diagrams this time, and we do not try.

Let S be spanned by $\{\mathbf{v}_1, \mathbf{v}_2\}$ where

$$\mathbf{v}_1 = \begin{bmatrix} 1 \\ 1 \\ 0 \end{bmatrix} \quad \text{and} \quad \mathbf{v}_2 = \begin{bmatrix} 2 \\ 1 \\ 1 \end{bmatrix}.$$

Since these vectors are not proportional, they are linearly independent, so $\{v_1, v_2\}$ is actually a basis, not just a spanning set for S, which is therefore a two-dimensional subspace of $\mathbb{R}^3$.

We now step through the Gram-Schmidt procedure for turning this basis into an orthonormal basis $\{u_1, u_2\}$.

Step One: Define the first element, u_1, by normalizing v_1.

$$u_1 = \frac{1}{|v_1|}v_1. \tag{6.3}$$

Notice that u_1 belongs to S since it is a multiple of a vector in S, namely v_1.

Step Two: Project v_2 onto the orthogonal complement of the span of v_1, which is also the span of u_1. The projection w_2 is given by

$$\begin{aligned} w_2 &= (I - u_1 u_1^t)v_2 \\ &= v_2 - (v_2 \cdot u_1)u_1. \end{aligned} \tag{6.4}$$

- *Notice that w_2 is a linear combination of u_1 and v_2, which are both vectors in S. Since S is a subspace, w_2 also belongs to S, by Theorem 3.*

Step Three: Divide w_2 by its length to obtain a unit vector u_2:

$$u_2 = \frac{1}{|w_2|}w_2. \tag{6.5}$$

We now have a pair $\{u_1, u_2\}$ of orthonormal vectors in S. By Theorem 7, any orthonormal set is linearly independent, and since the dimension of S is 2, $\{u_1, u_2\}$ must also span S. So it is indeed an orthonormal basis.

Of course, we could not do the third step if it involved dividing by zero. But $w_2 = 0$ if and only if

$$0 = v_2 - (v_2 \cdot u_1)u_1 = v_2 - \frac{v_2 \cdot v_1}{|v_1|^2}v_1,$$

which would mean that v_2 is a multiple of v_1, which cannot happen if $\{v_1, v_2\}$ is linearly independent. The procedure evidently does what it is supposed to, and it easily generalizes to higher dimensions for S.

EXAMPLE 40 **(Gram-Schmidt)** Now let us work through the case just discussed and do the actual computations. First, $|v_1| = \sqrt{2}$ so that by (6.3),

$$u_1 = \frac{1}{\sqrt{2}}\begin{bmatrix} 1 \\ 1 \\ 0 \end{bmatrix}.$$

Second, we compute $v_2 \cdot u_1 = 3/\sqrt{2}$, and hence by (6.4),

$$w_2 = \begin{bmatrix} 2 \\ 1 \\ 1 \end{bmatrix} - \frac{3}{\sqrt{2}}\frac{1}{\sqrt{2}}\begin{bmatrix} 1 \\ 1 \\ 0 \end{bmatrix} = \frac{1}{2}\begin{bmatrix} 1 \\ -1 \\ 2 \end{bmatrix}.$$

Finally, by (6.5),

$$u_2 = \frac{1}{|w_2|}w_2 = \frac{1}{\sqrt{6}}\begin{bmatrix} 1 \\ -1 \\ 2 \end{bmatrix},$$

and the orthonormal basis is $\{u_1, u_2\}$, where

$$u_1 = \frac{1}{\sqrt{2}}\begin{bmatrix} 1 \\ 1 \\ 0 \end{bmatrix} \quad \text{and} \quad u_2 = \frac{1}{\sqrt{6}}\begin{bmatrix} 1 \\ -1 \\ 2 \end{bmatrix}.$$

Now that we have found an orthonormal basis for the subspace S of $\mathbb{R}^3$ spanned by $\{\mathbf{v}_1, \mathbf{v}_2\}$, we can easily write down the orthogonal projection P_S onto S. To do so, form the matrix $Q = [\mathbf{u}_1, \mathbf{u}_2]$. Then by what was explained following Example 38, $P_S = QQ^t$.

EXAMPLE 41   **(Computing a projection)** Let us compute the orthogonal projection P_S onto the span of $\{\mathbf{v}_1, \mathbf{v}_2\}$, where $\mathbf{v}_1$ and $\mathbf{v}_2$ are the vectors considered in Example 40. First we form

$$Q = [\mathbf{u}_1, \mathbf{u}_2] = \frac{1}{\sqrt{6}} \begin{bmatrix} \sqrt{3} & 1 \\ \sqrt{3} & -1 \\ 0 & 2 \end{bmatrix}.$$

Then

$$P_S = QQ^t = \frac{1}{3} \begin{bmatrix} 2 & 1 & 1 \\ 1 & 2 & -1 \\ 1 & -1 & 2 \end{bmatrix}.$$

Now that we have the orthogonal projection in hand, we can easily answer a host of geometric questions. For example: What is the vector $\mathbf{c}$ in S that is closest to $\mathbf{b} = \begin{bmatrix} 1 \\ 2 \\ 3 \end{bmatrix}$, and what is the distance from $\mathbf{b}$ to S?

The closest vector $\mathbf{c}$ is the orthogonal projection of $\mathbf{b}$ onto S and therefore is

$$\mathbf{c} = P_S \mathbf{b} = \frac{1}{3} \begin{bmatrix} 2 & 1 & 1 \\ 1 & 2 & -1 \\ 1 & -1 & 2 \end{bmatrix} \begin{bmatrix} 1 \\ 2 \\ 3 \end{bmatrix} = \frac{1}{3} \begin{bmatrix} 7 \\ 2 \\ 5 \end{bmatrix}.$$

The distance from $\mathbf{b}$ to S is $|\mathbf{b} - \mathbf{c}| = 4/\sqrt{3}$.

Next, what about more than two vectors? Let $\{\mathbf{v}_1, \mathbf{v}_2, \ldots, \mathbf{v}_k\}$ be a basis for a subspace S of $\mathbb{R}^n$. Define $\mathbf{u}_1$ and $\mathbf{u}_2$ just as previously, using (6.3), (6.4), and (6.5). As before, the vector $\mathbf{w}_2$ given in (6.4) is a nonzero linear combination of $\mathbf{v}_1$ and $\mathbf{v}_2$ and hence is in S. Also, it cannot be zero since $\{\mathbf{v}_1, \mathbf{v}_2\}$ is linearly independent.

Now we just keep going in the same way. The orthogonal projection onto the orthogonal complement of the span of $\{\mathbf{u}_1, \mathbf{u}_2\}$ is $I - \mathbf{u}_1\mathbf{u}_1^t - \mathbf{u}_2\mathbf{u}_2^t$, so we define

$$\begin{aligned} \mathbf{w}_3 &= (I - \mathbf{u}_1\mathbf{u}_1^t - \mathbf{u}_2\mathbf{u}_2^t)\mathbf{v}_3 \\ &= \mathbf{v}_3 - (\mathbf{v}_3 \cdot \mathbf{u}_1)\mathbf{u}_1 - (\mathbf{v}_3 \cdot \mathbf{u}_2)\mathbf{u}_2. \end{aligned}$$

Perhaps more directly, if we define

$$\mathbf{w}_3 = \mathbf{v}_3 - (\mathbf{v}_3 \cdot \mathbf{u}_1)\mathbf{u}_1 - (\mathbf{v}_3 \cdot \mathbf{u}_2)\mathbf{u}_2 \tag{6.6}$$

and keep in mind that $\mathbf{u}_1 \cdot \mathbf{u}_1 = \mathbf{u}_2 \cdot \mathbf{u}_2 = 1$ and $\mathbf{u}_1 \cdot \mathbf{u}_2 = 0$, we can easily check that

$$\mathbf{w}_3 \cdot \mathbf{u}_1 = \mathbf{w}_3 \cdot \mathbf{u}_2 = 0.$$

Hence, $\mathbf{w}_3$ is orthogonal to $\mathbf{u}_1$ and $\mathbf{u}_2$.

Also, since $\mathbf{u}_1$ and $\mathbf{u}_2$ are linear combinations of $\mathbf{v}_1$ and $\mathbf{v}_2$, it is clear that $\mathbf{w}_3$ is a linear combination of $\mathbf{v}_1$, $\mathbf{v}_2$, and $\mathbf{v}_3$. Therefore it is in S, and since these vectors are linearly independent, $|\mathbf{w}_3| > 0$, and we can divide by it to define

$$\mathbf{u}_3 = \frac{1}{|\mathbf{w}_3|}\mathbf{w}_3,$$

where $\mathbf{w}_3$ is given by (6.6).

By now we see the pattern, and we make the following definition.

DEFINITION

(**Gram-Schmidt procedure**) Given a basis $\{\mathbf{v}_1, \mathbf{v}_2, \ldots, \mathbf{v}_r\}$ of a subspace S of $\mathbb{R}^n$, its *Gram-Schmidt orthonormalization* is the basis $\{\mathbf{u}_1, \mathbf{u}_2, \ldots, \mathbf{u}_r\}$, where the $\mathbf{u}_j$ are defined recursively by

$$\mathbf{u}_1 = \frac{1}{|\mathbf{v}_1|}\mathbf{v}_1$$

and then, for $j \geq 2$, by

$$\mathbf{u}_j = \frac{1}{|\mathbf{w}_j|}\mathbf{w}_j, \qquad \text{where } \mathbf{w}_j = \mathbf{v}_j - \sum_{i=1}^{j-1}(\mathbf{v}_j \cdot \mathbf{u}_i)\mathbf{u}_i. \tag{6.7}$$

Notice that the sum on the right in (6.7) involves only $\mathbf{u}_1$ through $\mathbf{u}_{j-1}$, so once they are found, we can use (6.7) to find $\mathbf{u}_j$, and so on. Thus, this definition is a valid recursive definition.

THEOREM 14

(**Gram-Schmidt procedure**) *The Gram-Schmidt procedure always produces an orthonormal basis $\{\mathbf{u}_1, \mathbf{u}_2, \ldots, \mathbf{u}_r\}$ for a subspace S out of an arbitrary basis $\{\mathbf{v}_1, \mathbf{v}_2, \ldots, \mathbf{v}_r\}$ for S.*

PROOF: First of all, we have explained that the formula on the right in (6.7) defines $\mathbf{w}_j$ as a linear combination of the $\mathbf{v}_i$ for $i \leq j$. Since $\{\mathbf{v}_1, \ldots, \mathbf{v}_r\}$ is linearly independent, and since the coefficient of $\mathbf{v}_j$ in this linear combination is not zero, $\mathbf{w}_j$ is never zero. Hence, the division by $|\mathbf{w}_j|$ on the left-hand side of (6.7) is always well defined, and the algorithm always produces a new sequence of vectors $\{\mathbf{u}_1, \ldots, \mathbf{u}_r\}$, without fail.

To see that $\{\mathbf{u}_1, \ldots, \mathbf{u}_r\}$ is orthonormal, we use induction—which is natural in discussing a recursive algorithm. We shall show that for each $j = 2, \ldots r$, if $\{\mathbf{u}_1, \ldots, \mathbf{u}_{j-1}\}$ is orthonormal, then $\{\mathbf{u}_1, \ldots, \mathbf{u}_j\}$ is orthonormal. Since $\{\mathbf{u}_1\}$ is evidently orthonormal, we then learn that $\{\mathbf{u}_1, \mathbf{u}_2\}$ is orthonormal, and so on, up to $\{\mathbf{u}_1, \ldots, \mathbf{u}_r\}$.

Therefore, let us suppose that $\{\mathbf{u}_1, \ldots, \mathbf{u}_{j-1}\}$ is orthonormal. Since $\mathbf{u}_j$ is clearly a unit vector, if we show that $\mathbf{u}_k \cdot \mathbf{u}_j = 0$ for all $j < k$, it will follow that $\{\mathbf{u}_1, \ldots, \mathbf{u}_j\}$ is orthonormal. But since $\mathbf{u}_j$ is a nonzero multiple of $\mathbf{w}_j$, $\mathbf{u}_k \cdot \mathbf{u}_j = 0$ if and only if $\mathbf{u}_k \cdot \mathbf{w}_j = 0$.

Now let us compute $\mathbf{u}_k \cdot \mathbf{w}_j$ using the definition of $\mathbf{w}_j$ on the right-hand side of (6.7):

$$\mathbf{u}_k \cdot \mathbf{w}_j = \mathbf{u}_k \cdot \left(\mathbf{v}_j - \sum_{i=1}^{j-1}(\mathbf{v}_j \cdot \mathbf{u}_i)\mathbf{u}_i\right)$$

$$= \mathbf{u}_k \cdot \mathbf{v}_j - \sum_{i=1}^{j-1}(\mathbf{v}_j \cdot \mathbf{u}_i)\mathbf{u}_k \cdot \mathbf{u}_i$$

$$= \mathbf{u}_k \cdot \mathbf{v}_j - \mathbf{u}_k \cdot \mathbf{v}_j = 0 \,.$$

(Only the $i = k$ term in the the sum is not zero by the inductive hypothesis). ∎

6.2 Gram-Schmidt and QR factorization

Let $\{\mathbf{v}_1, \mathbf{v}_2, \ldots, \mathbf{v}_k\}$ be a basis for a subspace S of $\mathbb{R}^n$, and form the matrix $A = [\mathbf{v}_1, \mathbf{v}_2, \ldots, \mathbf{v}_k]$. Then $S = \text{Img}(A)$. In many applications, the matrix A will be the actual starting point. If we now apply the Gram-Schmidt procedure to $\{\mathbf{v}_1, \mathbf{v}_2, \ldots, \mathbf{v}_k\}$, the columns of A, we get a new, orthonormal basis $\{\mathbf{u}_1, \mathbf{u}_2, \ldots, \mathbf{u}_k\}$ for $S = \text{Img}(A)$.

Here is a useful way to express the relation between the original basis and the new orthonormal basis in matrix terms: Organize the vectors $\mathbf{u}_1, \mathbf{u}_2, \ldots, \mathbf{u}_k$ into a matrix $Q = [\mathbf{u}_1, \mathbf{u}_2, \ldots, \mathbf{u}_k]$. Since the columns of Q are orthonormal, Q is an isometry, so $Q^t Q = I$. Also, since $\{\mathbf{u}_1, \mathbf{u}_2, \ldots, \mathbf{u}_k\}$ is an orthonormal basis for $\text{Img}(A)$, formula (5.15) of Theorem 13 tells us that $P_{\text{Img}(A)} = Q(Q^t Q)^{-1} Q^t$. But since $Q^t Q = I$, this simplifies to

$$P_{\text{Img}(A)} = QQ^t.$$

Define a new matrix R:

$$R = Q^t A. \tag{6.8}$$

Then since $P_{\text{Img}(A)} \mathbf{v} = \mathbf{v}$ for every vector $\mathbf{v}$ in $\text{Img}(A)$,

$$\begin{aligned}
QR = Q(Q^t A) &= (QQ^t)A \\
&= P_{\text{Img}(A)} A \\
&= [P_{\text{Img}(A)} \mathbf{v}_1, P_{\text{Img}(A)} \mathbf{v}_2, \ldots, P_{\text{Img}(A)} \mathbf{v}_k] \\
&= [\mathbf{v}_1, \mathbf{v}_2, \ldots, \mathbf{v}_k] = A.
\end{aligned}$$

That is,

$$A = QR. \tag{6.9}$$

- *This formula (6.9) gives us a factorization of A as a product of two matrices, Q and R. Q is an isometry, which is nice and simple. We will now see that R is nice and simple, too.*

First of all, both A and Q are $n \times k$ matrices, so from the definition (6.8), R is a $k \times k$ matrix. Next, since the k columns of A are linearly independent, $\text{rank}(A) = k$. But by (6.9) and Corollary 1 of Theorem 10,

$$k = \text{rank}(A) = \text{rank}(QR) \leq \min\{\text{rank}(Q), \ \text{rank}(R)\}.$$

It follows that $\text{rank}(R) \geq k$, and since R is $k \times k$, $\text{rank}(R) \leq k$. Hence the rank of R is k, and by the basic criterion for invertibility of square matrices, R is invertible.

This is good news: R is always invertible. But that is not all. To see the rest, rewrite (6.8) in terms of the bases $\{\mathbf{v}_1, \mathbf{v}_2, \ldots, \mathbf{v}_k\}$ and $\{\mathbf{u}_1, \mathbf{u}_2, \ldots, \mathbf{u}_k\}$. Here is how: From one of the fundamental formulas for matrix multiplication,

$$\begin{aligned}
R_{i,j} = (Q^t A)_{i,j} &= (\text{row } i \text{ of } Q^t) \cdot (\text{column } j \text{ of } A) \\
&= (\text{column } i \text{ of } Q) \cdot (\text{column } j \text{ of } A) \\
&= \mathbf{u}_i \cdot \mathbf{v}_j.
\end{aligned}$$

In short,

$$R_{i,j} = \mathbf{u}_i \cdot \mathbf{v}_j. \tag{6.10}$$

Notice several things about this formula. First of all, the dot products $\mathbf{u}_i \cdot \mathbf{v}_j$ for $i < j$ all are computed in the course of doing the Gram-Schmidt orthogonalization.

Even better,

$$\mathbf{u}_i \cdot \mathbf{v}_j = 0 \qquad \text{for } i > j. \tag{6.11}$$

Indeed, from the definition of $\mathbf{w}_j$ in (6.7),

$$\mathbf{v}_j = \mathbf{w}_j + \sum_{i=1}^{j-1} (\mathbf{v}_j \cdot \mathbf{u}_i)\mathbf{u}_i = |\mathbf{w}_j|\mathbf{u}_j + \sum_{i=1}^{j-1} (\mathbf{v}_j \cdot \mathbf{u}_i)\mathbf{u}_i \tag{6.12}$$

since $\mathbf{w}_j = |\mathbf{w}_j|\mathbf{u}_j$ by the left side of (6.7). This equation displays $\mathbf{v}_j$ as a linear combination of $\{\mathbf{u}_1, \mathbf{u}_2, \ldots, \mathbf{u}_j\}$. But for $i > j$, $\mathbf{u}_i$ is orthogonal to each of these, so for $i > j$, $\mathbf{u}_i$ is orthogonal to $\mathbf{v}_j$. Therefore, (6.11) is always true. This means that R will always be upper-triangular. In particular,

- *The matrix R is always already in row-reduced form, so any equation of the form $R\mathbf{x} = \mathbf{c}$ can be solved by back substitution. Moreover, all except for the diagonal entries of R are computed in the course of working out a Gram-Schmidt orthonormalization, so R is easy to compute.*

Since we can write (6.9) and (6.8) as

$$[\mathbf{v}_1, \mathbf{v}_2, \ldots, \mathbf{v}_k] = [\mathbf{u}_1, \mathbf{u}_2, \ldots, \mathbf{u}_k]R \qquad \text{and} \qquad [\mathbf{u}_1, \mathbf{u}_2, \ldots, \mathbf{u}_k] = [\mathbf{v}_1, \mathbf{v}_2, \ldots, \mathbf{v}_k]R^{-1},$$
$$(6.13)$$

the matrix R relates the two bases $\{\mathbf{v}_1, \mathbf{v}_2, \ldots, \mathbf{v}_k\}$ and $\{\mathbf{u}_1, \mathbf{u}_2, \ldots, \mathbf{u}_k\}$. Suppose that $\mathbf{b}$ is any vector in $S = \text{Img}(A)$ and the coordinate vector of $\mathbf{b}$ with respect to the basis $\{\mathbf{v}_1, \mathbf{v}_2, \ldots, \mathbf{v}_k\}$ is

$$\mathbf{x} = \begin{bmatrix} x_1 \\ x_2 \\ \vdots \\ x_k \end{bmatrix}.$$

Then, using the first equality in (6.13),

$$\begin{aligned} \mathbf{b} &= x_1\mathbf{v}_1 + x_2\mathbf{v}_2 + \cdots + x_k\mathbf{v}_k \\ &= [\mathbf{v}_1, \mathbf{v}_2, \ldots, \mathbf{v}_k]\mathbf{x} \\ &= ([\mathbf{u}_1, \mathbf{u}_2, \ldots, \mathbf{u}_k]R)\mathbf{x} \\ &= [\mathbf{u}_1, \mathbf{u}_2, \ldots, \mathbf{u}_k]R\mathbf{x}. \end{aligned}$$

Defining $\mathbf{y} = R\mathbf{x}$, we have $\mathbf{b} = y_1\mathbf{u}_1 + y_2\mathbf{u}_2 + \cdots + y_k\mathbf{u}_k$. In other words, if $\mathbf{x}$ is the coordinate vector of $\mathbf{b}$ with respect to $\{\mathbf{v}_1, \mathbf{v}_2, \ldots, \mathbf{v}_k\}$, then $\mathbf{y} = R\mathbf{x}$ is the coordinate vector of $\mathbf{b}$ with respect to $\{\mathbf{u}_1, \mathbf{u}_2, \ldots, \mathbf{u}_k\}$. For this reason, R is sometimes referred to as the *change-of-basis matrix* for these two bases.

We summarize in the following theorem.

THEOREM 15   **(Change of basis for Gram-Schmidt)** *Let $\{\mathbf{v}_1, \mathbf{v}_2, \ldots, \mathbf{v}_k\}$ be a basis for a subspace S of $\mathbb{R}^n$, and let $\{\mathbf{u}_1, \mathbf{u}_2, \ldots, \mathbf{u}_k\}$ be its Gram-Schmidt orthonormalization. Let $Q = [\mathbf{u}_1, \mathbf{u}_2, \ldots, \mathbf{u}_k]$ and let $A = [\mathbf{v}_1, \mathbf{v}_2, \ldots, \mathbf{v}_k]$. Then the $k \times k$ change-of-basis matrix R with $A = QR$ is given by (6.10) and is upper-triangular and invertible. In particular, every $n \times k$ matrix A with rank k can be factored into a product of an isometry Q and an invertible upper-triangular matrix R.*

Our original motivation for seeking orthonormal bases was to facilitate the solution of least squares problems. We now explain that if you want least squares solutions of $A\mathbf{x} = \mathbf{b}$, it is very useful to have A factored as in Theorem 15.

Since the orthogonal projection onto $\text{Img}(A)$ is QQ^t, it follows that the vector in $\text{Img}(A)$ that is closest to $\mathbf{b}$ is $QQ^t\mathbf{b}$. The least squares solution of $A\mathbf{x} = \mathbf{b}$ is then given by

$$A\mathbf{x} = QQ^t\mathbf{b}.$$

Substituting $A = QR$, the equation is $QR\mathbf{x} = QQ^t\mathbf{b}$. Since Q is an isometry, Q^t is a left inverse for Q, so if we multiply both sides by Q^t,

$$R\mathbf{x} = Q^t\mathbf{b}. \qquad (6.14)$$

Since R is invertible, our least squares solution is

$$\mathbf{x} = R^{-1}Q^t\mathbf{b}.$$

In practice, we would not compute the inverse of R; we would solve (6.14) by back substitution. The QR factorization of A as in (6.9) provides a good way to solve least squares problems. It is easy to use when working with a computer, since all computer programs for linear algebra have built-in methods for computing the Q and the R. Hence, we can proceed directly to (6.14).

EXAMPLE 42   **(Finding and using a QR factorization)** Use the Gram-Schmidt procedure to compute an orthonormal basis for $\text{Img}(A)$, where A is the 4×3 matrix

$$A = \begin{bmatrix} 1 & 1 & 1 \\ 0 & 1 & 1 \\ 2 & 4 & 1 \\ 2 & 0 & 3 \end{bmatrix},$$

find a QR factorization of A, and then, with

$$\mathbf{b} = \begin{bmatrix} 1 \\ 1 \\ 2 \\ 2 \end{bmatrix},$$

solve the least squares problem $A\mathbf{x} = \mathbf{b}$.

Let us proceed by assuming that the columns of A are linearly independent, which they are. We could check this by doing a row reduction, but it is not necessary. As explained in the next section, if we just start into the procedure, we will discover any dependence and be able to deal with it. So let us just proceed.

Write $A = [\mathbf{v}_1, \mathbf{v}_2, \mathbf{v}_3]$. Then $|\mathbf{v}_1| = 3$, so

$$\mathbf{u}_1 = \frac{1}{3} \begin{bmatrix} 1 \\ 0 \\ 2 \\ 2 \end{bmatrix}.$$

We compute $\mathbf{v}_2 \cdot \mathbf{u}_1 = 3$, and hence

$$\mathbf{w}_2 = \begin{bmatrix} 1 \\ 1 \\ 4 \\ 0 \end{bmatrix} - \begin{bmatrix} 1 \\ 0 \\ 2 \\ 2 \end{bmatrix} = \begin{bmatrix} 0 \\ 1 \\ 2 \\ -2 \end{bmatrix},$$

so

$$\mathbf{u}_2 = \frac{1}{3}\mathbf{w}_2 = \frac{1}{3} \begin{bmatrix} 0 \\ 1 \\ 2 \\ -2 \end{bmatrix}.$$

Next, $\mathbf{v}_3 \cdot \mathbf{u}_1 = 3$ and $\mathbf{v}_3 \cdot \mathbf{u}_2 = -1$. Therefore,

$$\mathbf{w}_3 = \begin{bmatrix} 1 \\ 1 \\ 1 \\ 3 \end{bmatrix} - \begin{bmatrix} 1 \\ 0 \\ 2 \\ 2 \end{bmatrix} + \frac{1}{3} \begin{bmatrix} 0 \\ 1 \\ 2 \\ -2 \end{bmatrix} = \frac{1}{3} \begin{bmatrix} 0 \\ 4 \\ -1 \\ 1 \end{bmatrix}.$$

Normalizing,

$$\mathbf{u}_3 = \frac{1}{\sqrt{18}} \begin{bmatrix} 0 \\ 4 \\ -1 \\ 1 \end{bmatrix}$$

is the third element of our orthonormal basis, so

$$Q = [\mathbf{u}_1, \mathbf{u}_2, \mathbf{u}_3] = \frac{1}{6} \begin{bmatrix} 2 & 0 & 0 \\ 0 & 2 & 4\sqrt{2} \\ 4 & 4 & -\sqrt{2} \\ 4 & -4 & \sqrt{2} \end{bmatrix}.$$

The change of basis matrix R is given by

$$R = \begin{bmatrix} \mathbf{u}_1 \cdot \mathbf{v}_1 & \mathbf{u}_1 \cdot \mathbf{v}_2 & \mathbf{u}_1 \cdot \mathbf{v}_3 \\ 0 & \mathbf{u}_2 \cdot \mathbf{v}_2 & \mathbf{u}_2 \cdot \mathbf{v}_3 \\ 0 & 0 & \mathbf{u}_3 \cdot \mathbf{v}_3 \end{bmatrix}.$$

We have already computed $\mathbf{u}_1 \cdot \mathbf{v}_2 = 3$, $\mathbf{u}_1 \cdot \mathbf{v}_3 = 3$, and $\mathbf{u}_2 \cdot \mathbf{v}_3 = -1$, so to find R, we need to compute only the three diagonal entries $\mathbf{u}_1 \cdot \mathbf{v}_1 = 3$, $\mathbf{u}_2 \cdot \mathbf{v}_2 = 3$, and $\mathbf{u}_3 \cdot \mathbf{v}_3 = \sqrt{7/6}$. Hence,

$$R = \begin{bmatrix} 3 & 3 & 3 \\ 0 & 3 & -1 \\ 0 & 0 & \sqrt{2} \end{bmatrix}.$$

Next,

$$Q^t \mathbf{b} = \frac{1}{3} \begin{bmatrix} 9 \\ 1 \\ 2\sqrt{2} \end{bmatrix},$$

so we find our least squares solution to $A\mathbf{x} = \mathbf{b}$ by solving

$$\begin{bmatrix} 3 & 3 & 3 \\ 0 & 3 & -1 \\ 0 & 0 & \sqrt{2} \end{bmatrix} \begin{bmatrix} x \\ y \\ z \end{bmatrix} = \frac{1}{3} \begin{bmatrix} 9 \\ 1 \\ 2\sqrt{2} \end{bmatrix}.$$

By back substitution,

$$\begin{bmatrix} x \\ y \\ z \end{bmatrix} = \begin{bmatrix} 0 \\ 1/3 \\ 2/3 \end{bmatrix}.$$

If we multiply out, we see that this is an exact solution.

6.3 How things change if $\{\mathbf{v}_1, \mathbf{v}_2, \ldots, \mathbf{v}_k\}$ only spans S

What happens if we do not bother to extract a basis from $\{\mathbf{v}_1, \mathbf{v}_2, \ldots, \mathbf{v}_r\}$? Let us see.

EXAMPLE 43 **(QR with dependent columns)** Let A be the 3×3 matrix

$$A = \begin{bmatrix} 1 & 0 & 1 \\ 0 & 1 & 1 \\ 1 & 1 & 2 \end{bmatrix}.$$

Writing this matrix as $A = [\mathbf{v}_1, \mathbf{v}_2, \mathbf{v}_3]$, from Theorem 4 $\{\mathbf{v}_1, \mathbf{v}_2, \mathbf{v}_3\}$ spans $\mathrm{Img}(A)$. Let us try to transform it into an orthonormal basis using the Gram-Schmidt procedure.

First, we normalize $\mathbf{v}_1$:

$$\mathbf{u}_1 = \frac{1}{\sqrt{2}} \begin{bmatrix} 1 \\ 0 \\ 1 \end{bmatrix}.$$

Then

$$\mathbf{w}_2 = \mathbf{v}_2 - (\mathbf{u}_1 \cdot \mathbf{v}_2)\mathbf{u}_1 = \begin{bmatrix} 0 \\ 1 \\ 1 \end{bmatrix} - \frac{1}{2}\begin{bmatrix} 1 \\ 0 \\ 1 \end{bmatrix} = \frac{1}{2}\begin{bmatrix} -1 \\ 2 \\ 1 \end{bmatrix}.$$

Normalizing,

$$\mathbf{u}_2 = \frac{1}{\sqrt{6}} \begin{bmatrix} -1 \\ 2 \\ 1 \end{bmatrix}.$$

So far, so good. Next,

$$\mathbf{w}_3 = \mathbf{v}_3 - (\mathbf{u}_1 \cdot \mathbf{v}_3)\mathbf{u}_1 - (\mathbf{u}_2 \cdot \mathbf{v}_3)\mathbf{u}_2 = \begin{bmatrix} 1 \\ 1 \\ 2 \end{bmatrix} - \frac{3}{2}\begin{bmatrix} 1 \\ 0 \\ 1 \end{bmatrix} - \frac{1}{2}\begin{bmatrix} -1 \\ 2 \\ 1 \end{bmatrix} = \begin{bmatrix} 0 \\ 0 \\ 0 \end{bmatrix}.$$

This time we cannot normalize, since $\mathbf{w}_3 = 0$. What should we do? Nothing! This result tells us that

$$\mathbf{v}_3 = (\mathbf{u}_1 \cdot \mathbf{v}_3)\mathbf{u}_1 + (\mathbf{u}_2 \cdot \mathbf{v}_3)\mathbf{u}_2,$$

which means that $\mathbf{v}_3$ is in the span of $\{\mathbf{u}_1, \mathbf{u}_2\}$.

Since $\mathbf{v}_1$ and $\mathbf{v}_2$ are also in the span of $\{\mathbf{u}_1, \mathbf{u}_2\}$, we see that $\{\mathbf{u}_1, \mathbf{u}_2\}$ spans $\text{Img}(A)$. Since it is linearly independent—orthonormal sets always are—it is a basis for $\text{Img}(A)$.

There is a relation between $A = [\mathbf{v}_1, \mathbf{v}_2, \mathbf{v}_3]$ and $Q = [\mathbf{u}_1, \mathbf{u}_2]$ that is very much like the situation described in Theorem 15. We have known that $A = [\mathbf{v}_1, \mathbf{v}_2, \mathbf{v}_3]$ from the start, and by the preceding computations,

$$Q = \begin{bmatrix} 1/\sqrt{2} & -1/\sqrt{6} \\ 0 & 2/\sqrt{6} \\ 1/\sqrt{2} & 1/\sqrt{6} \end{bmatrix}. \tag{6.15}$$

Since $\{\mathbf{u}_1, \mathbf{u}_2\}$ is a basis for $\text{Img}(A)$, there is a unique vector $\mathbf{r}_j$ with

$$Q\mathbf{r}_j = [\mathbf{u}_1, \mathbf{u}_2]\mathbf{r}_j = \mathbf{v}_j$$

for $j = 1, 2, 3$. The entries of $\mathbf{r}_j$ are the coordinates of $\mathbf{v}_j$ in the basis $\{\mathbf{u}_1, \mathbf{u}_2\}$. But we know how to compute coordinates with respect to an orthonormal basis:

$$\mathbf{v}_j = (\mathbf{v}_j \cdot \mathbf{u}_1)\mathbf{u}_1 + (\mathbf{v}_j \cdot \mathbf{u}_2)\mathbf{u}_2.$$

Hence

$$\mathbf{r}_j = \begin{bmatrix} \mathbf{u}_1 \cdot \mathbf{v}_j \\ \mathbf{u}_2 \cdot \mathbf{v}_j \end{bmatrix}.$$

Therefore, if R is the 2×3 matrix with $R_{i,j} = \mathbf{u}_i \cdot \mathbf{v}_j$,

$$A = [\mathbf{v}_1, \mathbf{v}_2, \mathbf{v}_3] = [Q\mathbf{r}_1, Q\mathbf{r}_2, Q\mathbf{r}_3] = QR. \tag{6.16}$$

From this formula,

$$R = \frac{1}{\sqrt{2}} \begin{bmatrix} 2 & 1 & 3 \\ 0 & \sqrt{3} & \sqrt{3} \end{bmatrix}. \tag{6.17}$$

Notice that R is in row-reduced form. Since Q is an isometry, we have just found a way to write A as the product of two nice matrices:

$$A = QR,$$

where Q is an isometry and where R is in row-reduced form.

This "factorization" of A into these nice factors is called the *QR decomposition*. It can always be done for any $m \times n$ matrix A just as in the example. All you do is apply the Gram-Schmidt procedure to the columns of $A = [\mathbf{v}_1, \mathbf{v}_2, \ldots, \mathbf{v}_n]$. As you go along, you might find that certain of the $\mathbf{w}_j$ vectors are zero. *When this happens, it means that $\mathbf{v}_j$ is already in the span of $\{\mathbf{u}_1, \mathbf{u}_2, \ldots, \mathbf{u}_{j-1}\}$.* Simply skip $\mathbf{v}_j$, and go on to the next vector. In the end you still get an orthonormal basis $\{\mathbf{u}_1, \mathbf{u}_2, \ldots, \mathbf{u}_r\}$ for $\text{Img}(A)$, where r is the rank of A. (You know that $r = \dim(\text{Img}(A))$). If you define $Q = [\mathbf{u}_1, \mathbf{u}_2, \ldots, \mathbf{u}_r]$, then Q is an isometry, and if R is the $r \times n$ matrix with $R_{i,j} = \mathbf{u}_i \cdot \mathbf{v}_j$,

$$A = QR.$$

It is important to notice that whenever we do so,

$$\text{rank}(R) = \text{rank}(Q) = \text{rank}(A).$$

To see it, let r be the rank of A. Then $\text{Img}(A)$ is an r-dimensional subspace. This means that Q has r columns, since they are a basis for $\text{Img}(A)$. Since the columns are independent, $\text{rank}(Q) = r$. Finally, since Q is an isometry, $|A\mathbf{x}| = |QR\mathbf{x}| = |R\mathbf{x}|$, so $A\mathbf{x} = 0$ if and only if $R\mathbf{x} = 0$, so $\text{Ker}(A) = \text{Ker}(R)$. Since A and R both have n columns,

$$n - \text{rank}(R) = \dim(\text{Ker}(R)) = \dim(\text{Ker}(A)) = n - r.$$

Hence, $\text{rank}(R) = r$, too.

Now since R has r rows, $R\mathbf{x} = \mathbf{c}$ will have a solution for every $\mathbf{c}$ in $\mathbb{R}^m$, which means that the factorization $A = QR$ is useful for solving equations even when the columns of A are not independent.

In general, just as when the columns of A are independent, QQ^t is the orthogonal projection onto $\text{Img}(A)$, so any least squares solution to $A\mathbf{x} = \mathbf{b}$ is a solution to $A\mathbf{x} = QQ^t\mathbf{b}$. Writing $A = QR$ and multiplying on the left by Q^t,

$$R\mathbf{x} = Q^t\mathbf{b}, \qquad (6.18)$$

as before. By what we have observed, when $\text{rank}(A) = r$, Q has r columns, so $Q^t\mathbf{b}$ is in $\mathbb{R}^r$. But since $\text{rank}(R) = r$, (6.18) is always solvable. And since R is already in row-reduced form, this equation is easy to solve. Let us summarize this in a theorem.

THEOREM 16 (**QR Factorization**) *Let A be any $m \times n$ matrix with rank r. Then there is a matrix factorization*

$$A = QR$$

where:

1. *Q is an $m \times r$ matrix that has orthonormal columns. The columns of Q are an orthonormal basis for $\text{Img}(A)$.*

2. *R is an $r \times n$ matrix in row-reduced form that has rank r.*

3. *For any $\mathbf{b}$ in $\mathbb{R}^m$, the set of least squares solutions of $A\mathbf{x} = \mathbf{b}$ is exactly the solution set of $R\mathbf{x} = Q^t\mathbf{b}$.*

Moreover, Q and R can be found by applying the Gram-Schmidt procedure to the columns of A, as previously explained.

The factorization can also be found by using a built-in routine on any computer program for doing linear algebra, which is much more convenient. So in practice, finding Q and R is easy, and it is very useful if you know what to do with them!

EXAMPLE 44 (**QR and least squares**) Let A be the matrix from Example 42, and consider the equation $A\mathbf{x} = \mathbf{b}$, where

$$\mathbf{b} = \begin{bmatrix} 1 \\ 1 \\ 1 \end{bmatrix}.$$

This equation can be written as $QR\mathbf{x} = \mathbf{b}$. Multiplying on the left by Q^t and using the explicit form of Q found in (6.15),

$$R\mathbf{x} = Q^t\mathbf{b} = \begin{bmatrix} 1/\sqrt{2} & 0 & 1/\sqrt{2} \\ -1/\sqrt{6} & 2/\sqrt{6} & \sqrt{1/6} \end{bmatrix} \begin{bmatrix} 1 \\ 1 \\ 1 \end{bmatrix}$$

$$= \begin{bmatrix} \sqrt{2} \\ 2/\sqrt{6} \end{bmatrix}.$$

Next, we solve

$$R\mathbf{x} = \begin{bmatrix} \sqrt{2} \\ 2/\sqrt{6} \end{bmatrix}.$$

Using the explicit form of R found in (6.17),

$$\begin{bmatrix} 2 & 1 & 3 \\ 0 & \sqrt{3} & \sqrt{3} \end{bmatrix} \begin{bmatrix} x_1 \\ x_2 \\ x_3 \end{bmatrix} = \begin{bmatrix} 2 \\ 2/\sqrt{3} \end{bmatrix}.$$

The variable x_3 is nonpivotal; set $x_3 = t$. Back substitution then gives

$$x_2 = \frac{2}{3} - t \qquad \text{and} \qquad x_1 = \frac{2}{3} - t.$$

We have found the one-parameter family of what Theorem 16 says are least squares solutions:

$$\begin{bmatrix} x_1 \\ x_2 \\ x_3 \end{bmatrix} = \begin{bmatrix} 2/3 - t \\ 2/3 - t \\ t \end{bmatrix} = \frac{2}{3}\begin{bmatrix} 1 \\ 1 \\ 0 \end{bmatrix} + t\begin{bmatrix} -1 \\ -1 \\ 1 \end{bmatrix}.$$

Let us check this out. Multiplying,

$$A\left(\frac{2}{3}\begin{bmatrix} 1 \\ 1 \\ 0 \end{bmatrix} + t\begin{bmatrix} -1 \\ -1 \\ 1 \end{bmatrix}\right) = \frac{2}{3}A\begin{bmatrix} 1 \\ 1 \\ 0 \end{bmatrix} + tA\begin{bmatrix} -1 \\ -1 \\ 1 \end{bmatrix} = \frac{2}{3}\begin{bmatrix} 1 \\ 1 \\ 2 \end{bmatrix}, \tag{6.19}$$

since

$$A\begin{bmatrix} -1 \\ -1 \\ 1 \end{bmatrix} = 0.$$

We claim that the right-hand side is the vector in Img(A) that is closest to **b**. (The theorems we have proved assure us of it, but let us check it out directly.) To do so, we find the equation for Img(A). Row reduce

$$\begin{bmatrix} 1 & 0 & 1 & \bigm| & x_1 \\ 0 & 1 & 1 & \bigm| & x_2 \\ 1 & 1 & 2 & \bigm| & x_3 \end{bmatrix}$$

to find

$$\begin{bmatrix} 1 & 0 & 1 & \bigm| & x_1 \\ 0 & 1 & 1 & \bigm| & x_2 \\ 0 & 0 & 2 & \bigm| & x_3 - x_2 - x_1 \end{bmatrix}.$$

Hence, Img(A) is the plane given by $x_1 + x_2 - x_3 = 0$. Notice that the entries of **b** do not satisfy this equation, so **b** is not in the image of A. Since Img(A) is the plane given by $\mathbf{a} \cdot \mathbf{x} = 0$ where

$$\mathbf{a} = \begin{bmatrix} 1 \\ 1 \\ -1 \end{bmatrix},$$

we can subtract the component of **b** that is orthogonal to the plane to deduce that the vector in this plane that is closest to **b** is

$$\mathbf{b} - \frac{\mathbf{a} \cdot \mathbf{b}}{|\mathbf{a}|^2}\mathbf{a} = \begin{bmatrix} 1 \\ 1 \\ 1 \end{bmatrix} - \frac{1}{3}\begin{bmatrix} 1 \\ 1 \\ -1 \end{bmatrix} = \frac{2}{3}\begin{bmatrix} 1 \\ 1 \\ 2 \end{bmatrix} = \mathbf{c}.$$

Comparing with (6.19), we see that the solution set we have found,

$$\frac{2}{3}\begin{bmatrix} 1 \\ 1 \\ 0 \end{bmatrix} + t\begin{bmatrix} -1 \\ -1 \\ 1 \end{bmatrix},$$

is the solution set of $A\mathbf{x} = \mathbf{c}$, in which the right-hand side is "corrected" to belong to Img(A).

Exercises

6.1 Find an orthonormal basis for Img(A), where $A = \begin{bmatrix} 1 & 2 & 4 & 1 \\ 0 & 2 & 2 & 0 \\ 2 & 3 & 7 & 1 \\ 1 & 1 & 3 & 0 \end{bmatrix}$.

6.2 Find an orthonormal basis for Img(A^t), where A is the matrix from Exercise 6.1.

6.3 Let A be the matrix $A = \begin{bmatrix} 1 & 2 & 3 \\ 0 & 3 & 2 \\ 2 & 0 & 1 \end{bmatrix}$. Find the QR decomposition of A. Check your result by computing the product QR.

6.4 Let A be the matrix $A = \begin{bmatrix} 1 & 1 & 2 \\ 1 & -1 & 0 \\ 0 & 1 & 1 \end{bmatrix}$.

(a) Find an orthonormal basis for $\text{Img}(A)$, and find a matrix Q with orthonormal columns and another matrix R so that $A = QR$.

(b) Express the second column of A as a linear combination of the columns of Q.

(c) Express the second column of Q as a linear combination of the columns of A.

(d) Give parametric descriptions of $\text{Img}(A)$ and $\text{Ker}(A)$. If either one of these is a plane, give the equation of the plane.

(e) Are the rows of A linearly independent? Are the columns of A linearly independent?

6.5 Let A be the matrix $A = \begin{bmatrix} 1 & 1 & 2 \\ 1 & 0 & 1 \\ 0 & 1 & 1 \end{bmatrix}$.

(a) Find an orthonormal basis for $\text{Img}(A)$, and find a matrix Q with orthonormal columns and another matrix R so that $A = QR$.

(b) Find an orthonormal basis for $\text{Img}(A^t)$, and find the orthogonal projection P_r onto $\text{Img}(A^t)$.

(c) Give parametric descriptions of $\text{Img}(A)$ and $\text{Ker}(A)$. If either one of these is a plane, give the equation of the plane.

(d) What are the dimensions of $\text{Img}(A)$ and $\text{Img}(A^t)$?

(e) Are the rows of A linearly independent? Are the columns of A linearly independent?

6.6 Let A be the matrix $A = \begin{bmatrix} 1 & 0 \\ 1 & 1 \\ 0 & 1 \end{bmatrix}$.

(a) Find an orthonormal basis for $\text{Img}(A)$, and find a matrix Q with orthonormal columns and another matrix R so that $A = QR$.

(b) $\text{Img}(A)$ is a plane. Find the equation of this plane.

(c) Let $\mathbf{b}$ be the vector $\mathbf{b} = \begin{bmatrix} 1 \\ 1 \\ 1 \end{bmatrix}$. Find the least squares solution to $A\mathbf{x} = \mathbf{b}$. What is the minimum value of $|A\mathbf{x} - \mathbf{b}|^2$ as $\mathbf{x}$ varies over R^2?

6.7 Let A be the matrix $A = \begin{bmatrix} 1 & 0 \\ 0 & 1 \\ 1 & 1 \end{bmatrix}$.

(a) Find an orthonormal basis for Img(A), and find a matrix Q with orthonormal columns and another matrix R so that $A = QR$.

(b) For this matrix A, Img(A) is a plane. Find the equation of the plane.

6.8 Let A be the matrix $A = \begin{bmatrix} 1 & 1 \\ 0 & 1 \\ 1 & 2 \end{bmatrix}$.

(a) Find an orthonormal basis for Img(A), and find a matrix Q with orthonormal columns and another matrix R so that $A = QR$.

(b) For this matrix A, Img(A) is a plane. Find the equation of the plane.

6.9 Consider the vectors

$$\mathbf{v}_1 = \begin{bmatrix} -1 \\ -1 \\ 0 \\ 0 \end{bmatrix}, \qquad \mathbf{v}_2 = \begin{bmatrix} 2 \\ 0 \\ 1 \\ 1 \end{bmatrix}, \qquad \mathbf{v}_3 = \begin{bmatrix} 1 \\ 0 \\ 0 \\ 1 \end{bmatrix}, \qquad \text{and} \qquad \mathbf{v}_4 = \begin{bmatrix} 0 \\ -1 \\ 1 \\ 0 \end{bmatrix}.$$

(a) Let S be the span of $\{\mathbf{v}_1, \mathbf{v}_2, \mathbf{v}_3, \mathbf{v}_4\}$. Find an orthonormal basis for S. What is the dimension of S?

(b) Find an orthonormal basis for $S^\perp$, the orthogonal complement to S. What is the dimension of $S^\perp$?

(c) Find the orthogonal projections onto $S^\perp$ and onto S.

(d) Consider the vector $\mathbf{x} = \begin{bmatrix} 1 \\ 2 \\ 3 \\ 4 \end{bmatrix}$. Find vectors $\mathbf{w}$ and $\mathbf{v}$ with $\mathbf{w}$ in S and $\mathbf{v}$ in $S^\perp$, so that $\mathbf{x} = \mathbf{w} + \mathbf{v}$.

6.10 Consider the vectors

$$\mathbf{v}_1 = \begin{bmatrix} 1 \\ -1 \\ 0 \\ 0 \end{bmatrix}, \qquad \mathbf{v}_2 = \begin{bmatrix} 2 \\ 0 \\ -1 \\ -1 \end{bmatrix}, \qquad \mathbf{v}_3 = \begin{bmatrix} 1 \\ 0 \\ 0 \\ -1 \end{bmatrix}, \qquad \text{and} \qquad \mathbf{v}_4 = \begin{bmatrix} 0 \\ -1 \\ 0 \\ 1 \end{bmatrix}.$$

(a) Let S be the span of $\{\mathbf{v}_1, \mathbf{v}_2, \mathbf{v}_3, \mathbf{v}_4\}$. Find an orthonormal basis for S. What is the dimension of S?

(b) Find an orthonormal basis for $S^\perp$, the orthogonal complement to S. What is the dimension of $S^\perp$?

(c) Find the orthogonal projections onto $S^\perp$ and onto S.

(d) Consider the vector $\mathbf{x} = \begin{bmatrix} 1 \\ 2 \\ 3 \\ 4 \end{bmatrix}$. Find vectors $\mathbf{w}$ and $\mathbf{v}$ with $\mathbf{w}$ in S and $\mathbf{v}$ in $S^\perp$ so that $\mathbf{x} = \mathbf{w} + \mathbf{v}$.

6.11 Let $C = \begin{bmatrix} 1 & 1 & 2 \\ 2 & 0 & 2 \\ 0 & 1 & 1 \end{bmatrix}$ and let $\mathbf{b} = \begin{bmatrix} 1 \\ 2 \\ 3 \end{bmatrix}$.

(a) Find vectors $\mathbf{w}$ and $\mathbf{v}$ with $\mathbf{w}$ in S and $\mathbf{v}$ in $S^\perp$ so that $\mathbf{b} = \mathbf{w} + \mathbf{v}$.

(b) Find the distance from $\mathbf{b}$ to Img(C).

(c) Find the distance from $\mathbf{b}$ to $(\text{Img}(C))^\perp$.

6.12 Compute the distance from $\mathbf{b} = \begin{bmatrix} 1 \\ 0 \\ 0 \\ 1 \end{bmatrix}$ to the line in $\mathbb{R}^4$ through the origin and $\begin{bmatrix} 0 \\ 1 \\ 2 \\ 0 \end{bmatrix}$.

6.13 Compute the distance from $\mathbf{b} = \begin{bmatrix} 1 \\ 0 \\ 0 \\ 1 \end{bmatrix}$ to the line in $\mathbb{R}^4$ through $\begin{bmatrix} 0 \\ 1 \\ 1 \\ 0 \end{bmatrix}$ and $\begin{bmatrix} 0 \\ 1 \\ 0 \\ 2 \end{bmatrix}$.

SECTION 7 | # QR Factorization and Least Squares Solutions

7.1 Least squares: The advantage of the QR method

In this section we return to least squares problems and apply the QR decomposition to solve them. As explained, this method is likely to yield more reliable results than an application of the normal equations. To demonstrate, we return to the "dangerous case" discussed in Section 2.6. The matrix and vector considered there were

$$A = \begin{bmatrix} 1 & 1 \\ 1 & 1+a \\ 1 & 1-a \end{bmatrix} \quad \text{and} \quad \mathbf{b} = \begin{bmatrix} 1 \\ 2 \\ 1 \end{bmatrix}.$$

Recall that if $1 + a^2$ is rounded off to 1, which would happen if you were calculating with 16 accurate digits and $a = 10^{-9}$, say, then round-off error destroys the invertability of $A^t A$, and the normal equation method breaks down. How does the QR method do? Much better!

Write $A = [\mathbf{v}_1, \mathbf{v}_2]$. Applying the Gram-Schmidt method, we first form

$$\mathbf{u}_1 = \frac{1}{|\mathbf{v}_1|}\mathbf{v}_1 - \frac{1}{\sqrt{3}}\begin{bmatrix} 1 \\ 1 \\ 1 \end{bmatrix}.$$

Then

$$\mathbf{u}_1 \cdot \mathbf{v}_2 = \sqrt{3}, \tag{7.1}$$

so

$$\mathbf{w}_2 = \begin{bmatrix} 1 \\ 1+a \\ 1-a \end{bmatrix} - \begin{bmatrix} 1 \\ 1 \\ 1 \end{bmatrix} = a\begin{bmatrix} 0 \\ 1 \\ -1 \end{bmatrix}.$$

Normalizing, we neglect the a in front of $\mathbf{w}_2$, as usual, and find

$$\mathbf{u}_2 = \frac{1}{\sqrt{2}}\begin{bmatrix} 0 \\ 1 \\ -1 \end{bmatrix}.$$

Hence,

$$Q = [\mathbf{u}_1, \mathbf{u}_2] = \begin{bmatrix} 1/\sqrt{3} & 0 \\ 1/\sqrt{3} & 1/\sqrt{2} \\ 1/\sqrt{3} & -1/\sqrt{2} \end{bmatrix}.$$

To determine R, we need compute only $\mathbf{u}_1 \cdot \mathbf{v}_1$ and $\mathbf{u}_2 \cdot \mathbf{v}_2$, since $\mathbf{u}_1 \cdot \mathbf{v}_2$ is already computed in (7.1). We find $\mathbf{u}_1 \cdot \mathbf{v}_1 = \sqrt{3}$ and $\mathbf{u}_2 \cdot \mathbf{v}_2 = \sqrt{2}a$. Hence,

$$R = \begin{bmatrix} \sqrt{3} & \sqrt{3} \\ 0 & \sqrt{2}a \end{bmatrix}.$$

We have gone through these computations in detail to bring out an important point: We never squared a at any stage and, more significantly, never produced any quantities like $1 + a^2$ that would get rounded off to 1. The QR method computes Q and R correctly, with no error from round off. We can now proceed to find the least squares solution:

$$Q^t \mathbf{b} = \begin{bmatrix} 4/\sqrt{3} \\ 1/\sqrt{2} \end{bmatrix}.$$

We then solve $R\mathbf{x} = Q^t\mathbf{b}$, or

$$\begin{bmatrix} \sqrt{3} & \sqrt{3} \\ 0 & \sqrt{2}a \end{bmatrix} \begin{bmatrix} x \\ y \end{bmatrix} = \begin{bmatrix} 4/\sqrt{3} \\ 1/\sqrt{2} \end{bmatrix},$$

which gives

$$y = \frac{1}{2a} \qquad \text{and} \qquad x = \frac{4}{3} - \frac{1}{2a}.$$

One reason the QR factorization is used often in practical applications is that it is much less susceptible to round-off error than direct use of the normal equations. It may *look* much more complicated, but if you are using a computer, all the ugly square roots are given in a simple decimal form, and you can compute Q and R with a single command—no more work than computing $A^t A$, as far as typing the commands goes.

To demonstrate, let us look at another curve-fitting problem: Let us find a good polynomial fit to $\sin(x)$ on $0 \le x \le \pi/2$. Since this function is odd, we use a linear combination of odd powers of x. If we use the first four odd powers, we must find values of a, b, c, and d so that

$$\sin(x) \approx ax + bx^3 + cx^5 + dx^7 \qquad \text{for } 0 \le x \le \frac{\pi}{2}.$$

Let us pick some values of x for which we know the exact value of $\sin(x)$ by elementary trigonometry. For example, we know that

$$\sin(\pi/6) = \frac{1}{2}, \qquad \sin(\pi/4) = \frac{1}{\sqrt{2}}, \qquad \sin(\pi/3) = \frac{\sqrt{3}}{2}, \qquad \text{and} \qquad \sin(\pi/2) = 1.$$

From the angle addition formula, $\sin(5\pi/12) = \sin(\pi/3 + \pi/4) = (\sqrt{3}+1)/(2\sqrt{2})$ and $\sin(\pi/12) = \sin(\pi/3 - \pi/4) = (\sqrt{3}-1)/(2\sqrt{2})$, giving four data points. Notice in this example that there are no experimental errors. The error comes in through our attempt to treat $\sin(x)$ like a polynomial of seventh degree, which it is not.

In any case, we have our data, and we define

$$x_1 = \frac{\pi}{12}, \qquad x_2 = \frac{\pi}{6}, \qquad x_3 = \frac{\pi}{4}, \qquad x_4 = \frac{\pi}{3}, \qquad x_5 = \frac{5\pi}{12}, \qquad \text{and} \qquad x_6 = \frac{\pi}{2}$$

and

$$y_1 = \sin\frac{\pi}{12}, \qquad y_2 = \sin\frac{\pi}{6}, \qquad y_3 = \sin\frac{\pi}{4}, \qquad y_4 = \sin\frac{\pi}{3}, \qquad y_5 = \sin\frac{5\pi}{12}, \qquad \text{and} \qquad y_6 = \sin\frac{\pi}{2}.$$

The matrix A is now

$$A = \begin{bmatrix} x_1 & x_1^3 & x_1^5 & x_1^7 \\ x_2 & x_2^3 & x_2^5 & x_2^7 \\ x_3 & x_3^3 & x_3^5 & x_3^7 \\ x_4 & x_4^3 & x_4^5 & x_4^7 \\ x_5 & x_5^3 & x_5^5 & x_5^7 \\ x_6 & x_6^3 & x_6^5 & x_6^7 \end{bmatrix}.$$

It would be very tedious to do the calculation by hand, but with a computer program that finds *QR* decompositions, it is easy. The answer, kept to 10 digits, is

$$\begin{bmatrix} a \\ b \\ c \\ d \end{bmatrix} = \begin{bmatrix} .9999976966 \\ -.1666522644 \\ .008309725100 \\ -.0001844086724 \end{bmatrix},$$

yielding

$$f(x) = (.9999976966)x - (.1666522644)x^3 + (.008309725100)x^5 - (.0001844086724)x^7.$$

If we plot this polynomial from 0 to $\pi/2$, the graph is indistinguishable to the eye from that of $\sin(x)$, but if we plot from 0 to π, we can see the two graphs pull apart.

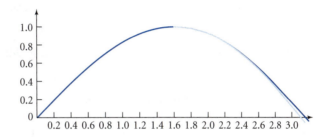

The same is true of the seventh-degree Taylor polynomial approximation to $\sin(x)$. Which one is better? Let us use both to approximate

$$\int_0^{\pi/2} \sin(x)\, dx = 1.$$

We easily work out

$$\int_0^{\pi/2} f(x)\, dx = 1.000000117,$$

which is almost accurate to one part in 10^7.

Now let $g(x)$ be the seventh-degree Taylor approximation to $\sin(x)$:

$$g(x) = x - \frac{1}{6}x^3 + \frac{1}{120}x^5 - \frac{1}{5040}x^7.$$

We easily work out

$$\int_0^{\pi/2} g(x)\, dx = .9999752637,$$

which is off by more than two parts in 10^5. In fact, the ratio of the errors, Taylor to least square, is

$$\frac{1 - .9999752637}{1.000000117 - 1} = 211.426934.$$

The least squares approach to curve fitting can provide extremely accurate fits that can then be used for purposes such as numerical integration.

7.2 The least length solution and the generalized inverse

When a system of equations $A\mathbf{x} = \mathbf{b}$ has infinitely many solutions, which of them might we want to actually use in an application? Very often, it is natural to focus on the solution $\mathbf{x}_0$ with the property that

$$|\mathbf{x}_0|^2 < |\mathbf{x}|^2$$

for any other solution $\mathbf{x}$. This is the *least length solution*.

- *Even when $A\mathbf{x} = \mathbf{b}$ has infinitely many solutions, there is exactly one solution in $\text{Img}(A^t)$ and it is always the least length solution.*

The space $\text{Img}(A^t)$ is spanned by the columns of A^t, which are the rows of A. Therefore, it is often called the *row space* of A, just as $\text{Img}(A)$ is often called the *column space* of A.

To see why the claim made in the bulleted point is true, suppose that $\mathbf{x}$ is *any* vector in $\mathbb{R}^n$ with $A\mathbf{x} = \mathbf{b}$. Let $\mathbf{x} = \mathbf{y} + \mathbf{z}$ be the decomposition of $\mathbf{x}$ into its components in $\text{Img}(A^t)$ and $\text{Ker}(A)$, respectively (by Theorem 11, these two spaces are orthogonal complements). Then

$$\mathbf{b} = A\mathbf{x} = A(\mathbf{y} + \mathbf{z}) = A\mathbf{y} + A\mathbf{z} = A\mathbf{y}$$

since $\mathbf{z}$ is in $\text{Ker}(A)$, so $A\mathbf{z} = 0$. Cutting out the middle,

$$A\mathbf{y} = \mathbf{b}\,.$$

Therefore, $\mathbf{y}$, the orthogonal projection of $\mathbf{x}$ onto the row space of A, is also a solution, and clearly it lies in the row space.

Moreover, since $\mathbf{y}$ and $\mathbf{z}$ are orthogonal components of $\mathbf{x}$,

$$|\mathbf{x}|^2 = |\mathbf{y}|^2 + |\mathbf{z}|^2\,. \tag{7.2}$$

Thus, $|\mathbf{x}| > |\mathbf{y}|$ unless $\mathbf{z} = 0$, which would mean that $\mathbf{x}$ already belonged to the row space. We conclude that solutions in the row space are shorter vectors than solutions that are not in the row space.

To show that there is never more than one solution in the row space, suppose that $\mathbf{y}_1$ and $\mathbf{y}_2$ are two solutions in the row space. Then

$$0 = \mathbf{b} - \mathbf{b} = A\mathbf{y}_1 - A\mathbf{y}_2 = A(\mathbf{y}_1 - \mathbf{y}_2)\,.$$

Thus, $\mathbf{y}_1 - \mathbf{y}_2$ is in the kernel of A. But it is in a linear combination of vectors in the row space of A, so it is also in the row space. Since the kernel and the row space are orthogonal complements, (5.5) of Theorem 11 tells us that $\mathbf{y}_1 - \mathbf{y}_2 = 0$. That is, $\mathbf{y}_1$ and $\mathbf{y}_2$ must actually be the same.

Hence if $A\mathbf{x} = \mathbf{b}$ has solutions, it has one special solution $\mathbf{x}_0$ that lies in the row space. To find this special solution, we can take any other solution and project it orthogonally onto the row space.

Moreover, even if $A\mathbf{x} = \mathbf{b}$ does not have solutions, it always has a least squares solution. All least squares solutions are exact solutions of the normal equations $A^t A\mathbf{x} = A^t\mathbf{b}$. Hence, to get the least length, least squares solution, take *any* least squares solution and project it orthogonally onto the row space of $A^t A$.

However, $A^t A$ and A have the same row space, so it is the same thing to take *any* least squares solution and project it orthogonally onto the row space of A. Exact solution or least squares solution, one always projects orthogonally onto the row space of A.

To see why $A^t A$ and A have the same row space, recall from the Corollary to Theorem 2 that $\text{Ker}(A^t A) = \text{Ker}(A)$. By Theorem 11, the row space is the orthogonal complement of the kernel, and clearly the orthogonal complements of identical subspaces are identical.

Now recall that every true solution of $A\mathbf{x} = \mathbf{b}$ is a least squares solution—one with zero residual. Thus, we can express our conclusion more succinctly by referring only to least squares solutions: If they happen to be true solutions, so much the better.

Let us summarize in a theorem.

THEOREM 17   **(Projection onto the row space and least length solutions)** *Let A be an $m \times n$ matrix. Let P_r be the orthogonal projection onto $\text{Img}(A^t)$, the row space of A.*

There is exactly one least squares solution of $A\mathbf{x} = \mathbf{b}$ in the row space of A. Moreover, this special solution, $\mathbf{x}_0$, is the least length, least squares solution in that

$$|\mathbf{x}_0| < |\mathbf{x}|$$

for any other least squares solution $\mathbf{x}$. It is given by

$$\mathbf{x}_0 = P_r\mathbf{x}$$

for any least squares solution $\mathbf{x}$.

There is a system of equations characterizing the least length, least squares solution of $A\mathbf{x} = \mathbf{b}$. To see it, let P_c denote the orthogonal projection onto $\text{Img}(A)$, the column space of A. Recall that $\mathbf{x}$ is a least squares solution to $A\mathbf{x} = \mathbf{b}$ if and only if $A\mathbf{x} = P_c\mathbf{b}$ since $P_c\mathbf{b}$ is the vector in the column space that is closest to $\mathbf{b}$. Furthermore, a least squares solution $\mathbf{x}$ has the least length if and only if it lies in the row space, that is, if and only if $\mathbf{x} = P_r\mathbf{x}$. To summarize:

- *Let A be an $m \times n$ matrix. Let P_r and P_c be the orthogonal projections onto its row and column spaces, respectively. Then a vector $\mathbf{x}$ in $\mathbb{R}^n$ is the least length, least squares solution of $A\mathbf{x} = \mathbf{b}$ if and only if*

$$A\mathbf{x} = P_c\mathbf{b} \qquad \text{and} \qquad \mathbf{x} = P_r\mathbf{x} \,. \tag{7.3}$$

There are several ways we could go about computing the least length, least squares solution to $A\mathbf{x} = \mathbf{b}$. One three-step procedure follows: (1) Use the QR decomposition to compute P_r. (2) Use the QR decomposition to find some least squares solution of $A\mathbf{x} = \mathbf{b}$. (3) Orthogonally project this solution onto the row space using P_r.

As you see, the QR factorization plays a starring role in this approach. Unfortunately, we need a QR factorization of A^t in step 1 and a QR factorization of A in step 2. There is a much better way to proceed. It still uses two QR factorizations, but it has many advantages. Here it is:

(The double QR factorization) *Let A be an $m \times n$ matrix, and let r be the rank of A. Then*

1. *Let $A = Q_c R$ be a QR factorization of A. (That is, compute a QR factorization of A, and call the isometry part Q_r.)*
2. *Let $R^t = Q_r T^t$ be a QR factorization of R^t. (That is, compute a QR factorization of R^t, and call the isometry part Q_r and the row-reduced part T^t.)*
3. *Put it all back together: $A = Q_c R = Q_c(R^t)^t = Q_c(Q_r T^t)^t = Q_c T Q_r^t$. Then*

$$A = Q_c T Q_r^t$$

is the double QR factorization of A.

The factors have some nice properties:

1. *Q_c is an $m \times r$ isometry whose columns are an orthonormal basis for the column space of A. Hence $P_c = Q_c Q$.*
2. *T is an invertible, lower-triangular $r \times r$ matrix.*
3. *Q_r is an $n \times r$ isometry whose columns are an orthonormal basis for the row space of A. Hence $P_r = Q_r Q$.*

The fact that statement (1) is true follows directly from Theorem 16, which also says that R is an $r \times n$ matrix of rank r.

Since the transpose of a matrix has the same rank (Corollary 2 of Theorem 10), R^t is an $n \times r$ matrix with rank r. By Theorem 16 once again, Q_r is an $n \times r$ isometry, and T^t is an $r \times r$ upper-triangular matrix with rank r. By Corollary 2 of Theorem 10 again, it follows that T itself is an $r \times r$ lower-triangular matrix of rank r. Every $r \times r$ matrix with rank r is invertible, so T is invertible, which proves statement (2).

Finally, $A^t = (Q_c T Q_r^t)^t = Q_r(T^t Q_c^t)$. By (4.7),

$$\text{Img}(A^t) \subset \text{Img}(Q_r) \,.$$

Since the transpose of a matrix has the same rank (Corollary 2 of Theorem 10 again), A^t has rank r, and by Theorem 16, Q_r has rank r. Then by Theorem 9, since A^t and Q_r both have rank r, the dimensions of $\text{Img}(A^t)$ and $\text{Img}(Q_r)$ are equal. Thus, by the dimension principle, Theorem 10,

$$\text{Img}(A^t) = \text{Img}(Q_r) \,.$$

Therefore, the columns of Q_r are an orthonormal basis for $\text{Img}(A^t)$, which is the row space. Hence, statement (3) is proved.

To see what these properties are good for and how they help us to compute least length, least squares solutions, define the $n \times m$ matrix A^+ by

$$A^+ = Q_r T^{-1} Q_c^t \,.$$

(Notice that now Q_r is on the left and Q_c^t is on the right.) This matrix is called the *generalized inverse* of A. It has the following properties, which justify the name:

$$A^+ A = P_r \qquad \text{and} \qquad A A^+ = P_c \tag{7.4}$$

and

$$P_r A^+ = A^+ \qquad \text{and} \qquad A^+ P_c = A^+ \,. \tag{7.5}$$

For the moment, let us assume that these identities are true and focus on understanding what they are telling us. First, from (7.4), if the rank of A is n, the dimension of the row space is n, so it is all of $\mathbb{R}^n$. In this case, $P_r = I_{n \times n}$ and A^+ is a left inverse of A.

Likewise, if the rank of A is m, the dimension of the column space is m, so it is all of $\mathbb{R}^m$. In this case, $P_c = I_{m \times m}$ and A^+ is a right inverse of A.

Evidently, if A is an invertible $n \times n$ matrix, then A^+ is just A^{-1}. What we have just explained justifies calling A^+ the generalized inverse of A. But what does it do in general, when the rank of A is neither m nor n? Here is the answer:

- *For every* $\mathbf{b}$ *in* $\mathbb{R}^m$, $A^+\mathbf{b}$ *is the unique least length, least squares solution to* $A\mathbf{x} = \mathbf{b}$.

This is good news: To compute the least length, least squares solution to $A\mathbf{x} = \mathbf{b}$, just compute $A^+\mathbf{b}$. Since most computer programs for working with linear algebra have a built-in command for computing A^+, it is extremely easy to compute least length, least squares solutions—now that you know about the generalized inverse.

To see why the claim in the bulleted point is true, we just have to check that $A^+\mathbf{b}$ satisfies (7.3). Since by (7.4), $A A^+ = P_c$,

$$A(A^+\mathbf{b}) = (A A^+)\mathbf{b} = P_c\mathbf{b} \,.$$

Next, from (7.5),

$$A^+\mathbf{b} = (P_r A^+)\mathbf{b} = P_r(A^+\mathbf{b}) \,.$$

Thus, both equations in (7.3) are satisfied, and $A^+\mathbf{b}$ is the least length, least squares solution of $A\mathbf{x} = \mathbf{b}$.

In fact, if you think about it, you will see that we only used the second equation in (7.4) and the first one in (7.5). There can be only one $n \times m$ matrix that satisfies these two equations. To see why, let B be any $n \times m$ matrix such that

$$AB = P_c \qquad \text{and} \qquad P_r B = B \,. \tag{7.6}$$

Then, by what we have just seen, $B\mathbf{e}_j$ is the least length, least squares solution to $A\mathbf{x} = \mathbf{e}_j$. By V.I.F.1, $B\mathbf{e}_j$ is the jth column of B. But it is also true that $A^+\mathbf{e}_j$ is the least length, least squares solution to $A\mathbf{x} = \mathbf{e}_j$ and the jth column of A. Since there is just one least length, least squares solution, B and A^+ have the same jth column for every j. Hence if B is an $n \times m$ matrix satisfying (7.6), $B = A^+$. Let us summarize.

THEOREM 18 **(Generalized inverses)** *Let A be an $m \times n$ matrix. Let P_r and P_c be the orthogonal projections the row space and the column space of A, respectively.*

Then there is a unique $n \times m$ matrix A^+ satisfying the identities in (7.4) and (7.5). If $\mathbf{b}$ is any vector in $\mathbb{R}^m$, then $A^+\mathbf{b}$ is the least length, least squares solution of $A\mathbf{x} = \mathbf{b}$.

PROOF: For the existence part, we have to check that if A^+ is defined by $A^+ = Q_r T^{-1} Q_c^t$ using the double QR factorization, then A^+ does indeed satisfy the identities in (7.4) and (7.5).

First,

$$AA^+ = (Q_c T Q_r^t)(Q_r T^{-1} Q_c^t) = Q_c(T(Q_r^t Q_r)T^{-1})Q_c^t = Q_c Q_c^t = P_c \,,$$

where we have used $Q_r^t Q_r = I$ and $Q_r Q_r^t = P_r$. This proves the second identity in (7.4), and the first one follows in the same way.

Second,

$$P_r A^+ = (Q_r Q_r^t)(Q_r T^{-1} Q_c^t) = Q_r(Q_r^t Q_r)T^{-1}Q_c^t = Q_r T^{-1} Q_c^t = A^+ \,,$$

again using $Q_r^t Q_r = I$. This proves the first identity in (7.5), and the second one follows in the same way.

Hence such a matrix always exists, and the double QR factorization provides one way of computing it.

Moreover, we have seen that any $n \times m$ matrix B satisfying (7.6) must equal A^+, so there is only one matrix that satisfies the identities (7.4) and (7.5)—or even half of them in (7.6). We have also seen why these identities ensure that then $A^+\mathbf{b}$ is the least length, least squares solution of $A\mathbf{x} = \mathbf{b}$. ∎

DEFINITION **(Generalized inverse)** Let A be an $m \times n$ matrix. Let P_r and P_c be the orthogonal projections the row space and the column space of A, respectively. The *generalized inverse* of A is the unique $n \times m$ matrix A^+ satisfying the identities (7.4) and (7.5).

Now some examples.

EXAMPLE 45 **(Computing a double QR factorization and a generalized inverse)** Let $A = \begin{bmatrix} 2 & 1 \\ 1 & 2 \\ 2 & 2 \end{bmatrix}$.

Computing a QR factorization of A, $A = Q_c R$ where

$$Q_c = \frac{1}{3\sqrt{17}}\begin{bmatrix} 2\sqrt{17} & -7 \\ 1\sqrt{17} & 10 \\ 2\sqrt{17} & 2 \end{bmatrix} \quad \text{and} \quad R = \frac{1}{3}\begin{bmatrix} 9 & 8 \\ 0 & \sqrt{17} \end{bmatrix}.$$

Next, computing a QR factorization of R^t, $R^t = Q_r T^t$ where

$$Q_r = \frac{1}{\sqrt{145}} \begin{bmatrix} 9 & -8 \\ 8 & 9 \end{bmatrix} \quad \text{and} \quad T^t = \frac{1}{3} \begin{bmatrix} \sqrt{145} & 8\sqrt{17/145} \\ 0 & 9\sqrt{17/145} \end{bmatrix},$$

so

$$T = \frac{1}{3} \begin{bmatrix} \sqrt{145} & 0 \\ 8\sqrt{17/145} & 9\sqrt{17/145} \end{bmatrix}.$$

Thus A has the double QR factorization

$$A = Q_c T Q_r^t$$

with the matrices Q_c, T, and Q_r given.

Now let us compute the generalized inverse, A^+. First we compute

$$T^{-1} = \frac{1}{3\sqrt{145}} \begin{bmatrix} 9 & 0 \\ -8 & 145/\sqrt{17} \end{bmatrix}.$$

Next, multiplying out $A^+ = Q_r T^{-1} Q_c^t$,

$$A^+ = \frac{1}{17} \begin{bmatrix} 10 & -7 & 2 \\ -7 & 10 & 2 \end{bmatrix}.$$

We can now multiply out A^+A and AA^+:

$$A^+A = \begin{bmatrix} 1 & 0 \\ 0 & 1 \end{bmatrix} \quad \text{and} \quad AA^+ = \frac{1}{17} \begin{bmatrix} 13 & -4 & 6 \\ -4 & 13 & 6 \\ 6 & 6 & -8 \end{bmatrix}.$$

We can also easily check that

$$P_r = Q_r Q_r^t = \begin{bmatrix} 1 & 0 \\ 0 & 1 \end{bmatrix} \quad \text{and} \quad P_c = Q_c Q_c^t = \frac{1}{17} \begin{bmatrix} 13 & -4 & 6 \\ -4 & 13 & 6 \\ 6 & 6 & -8 \end{bmatrix}$$

so that (7.4) is satisfied.

Notice something nice: When we reassemble the pieces to obtain A^+, all the square roots in the individual factors canceled out. The square roots in this (or any other) example may make the procedure look ugly and cumbersome. But do not let radicals mislead you! After all, if you were doing your computations on a computer in decimal form, you would not see any radicals.

In fact, it is mainly the generalized inverse itself that we are interested in, not the double QR factorization. In Chapter 6, we will meet a variant of the double QR factorization in which the matrix in the middle is not only triangular but is even diagonal. It is called the *singular value factorization*, or more commonly, the *singular value decomposition*. It too can be used to compute generalized inverses. However, the method presented here is much simpler and is quite effective on a computer.

EXAMPLE 46   **(Computing a second generalized inverse)** Let $B = \frac{1}{17} \begin{bmatrix} 10 & -7 & 2 \\ -7 & 10 & 2 \end{bmatrix}$. Notice that the matrix B is just the generalized inverse of the matrix A from the previous example. Let us compute the generalized inverse of B.

Applying the procedure from the previous example,

$$B^+ = \begin{bmatrix} 2 & 1 \\ 1 & 2 \\ 2 & 2 \end{bmatrix}.$$

We do not show the intermediate steps because the matrices involve horrible-looking square roots. In fact, we recommend that you check the computations, but do so using a computer to work out the QR factorizations. Be practical!

The important thing to notice from this example is that $(A^+)^+ = A$. This is generally true; see Exercise 7.15. Taking the generalized inverse twice gets you back to where you started.

In the next example, we focus on using generalized inverses to compute least length, least squares solutions. In many ways, it is more important to understand what generalized inverses are and how to use them than to be able to compute them because computer programs for doing linear algebra have built-in commands for computing generalized inverses. We suggest that you use such a program when checking the computations in the next example and when doing the corresponding exercises at the end of this section.

EXAMPLE 47   **(Using a generalized inverse)** Let $A = \begin{bmatrix} 1 & 5 & 2 & 7 & 9 \\ 0 & -3 & -1 & -5 & -4 \\ -2 & -1 & -1 & 1 & 0 \\ 1 & 2 & 1 & 2 & 1 \end{bmatrix}$.

Before we begin with the generalized inverse, let us compute rref(A), the reduced row echelon form of A, using a computer, of course:

$$\text{rref}(A) = \begin{bmatrix} 1 & 0 & 1/3 & -4/3 & 0 \\ 0 & 1 & 1/3 & 5/3 & 0 \\ 0 & 0 & 0 & 0 & 1 \end{bmatrix}.$$

(Recall that by definition we throw out any "all-zero" rows in the reduced row echelon form. Your computer program may or may not do this. If you get an "all-zero" row, just ignore it.

We see from the computation of rref(A) that the rank of A is 3. Therefore, solutions of $A\mathbf{x} = \mathbf{b}$ will not exist for all $\mathbf{b}$, and when they do exist, they will never be unique.

Let us find the least length, least squares solution of $A\mathbf{x} = \mathbf{b}$ for

$$\mathbf{b} = \begin{bmatrix} 1 \\ 2 \\ -2 \\ -1 \end{bmatrix}.$$

Using a computer to do the computation, the generalized inverse of A is

$$A^+ = \frac{1}{2806} \begin{bmatrix} 51 & 176 & -657 & 245 \\ -171 & -260 & -273 & 499 \\ -40 & -28 & -310 & 248 \\ -353 & -668 & 421 & 505 \\ 671 & 610 & -61 & -793 \end{bmatrix}.$$

The least length, least squares solution of $A\mathbf{x} = \mathbf{b}$ is therefore

$$\mathbf{x}_0 = A^+\mathbf{b} = \frac{1}{61} \begin{bmatrix} 32 \\ -14 \\ 6 \\ -66 \\ 61 \end{bmatrix}.$$

Is this an exact solution? Let us check: Multiplying out $A\mathbf{x}_0$, we find that indeed $A\mathbf{x}_0 = \mathbf{b}$. So this is an exact solution. There are infinitely many others.

Since we have found rref(A), it is easy to give a one-to-one parameterization of the kernel of A. Letting $\mathbf{v}_j$ denote the jth column of A, from rref(A),

$$\mathbf{v}_3 = \frac{1}{3}(\mathbf{v}_1 + \mathbf{v}_2) \quad \text{and} \quad \mathbf{v}_4 = \frac{1}{3}(-4\mathbf{v}_1 + 5\mathbf{v}_2).$$

Therefore, a basis for the kernel of A is $\{\mathbf{w}_1, \mathbf{w}_2\}$, where

$$\mathbf{w}_1 = \begin{bmatrix} 1 \\ 1 \\ -3 \\ 0 \\ 0 \end{bmatrix} \quad \text{and} \quad \mathbf{w}_2 = \begin{bmatrix} -4 \\ 5 \\ 0 \\ -3 \\ 0 \end{bmatrix}.$$

Hence, the general solution of $A\mathbf{x} = \mathbf{b}$ is

$$\mathbf{x}_0 + t_1\mathbf{w}_1 + t_2\mathbf{w}_2.$$

Let us try this parameterization again but with

$$\mathbf{b} = \begin{bmatrix} 1 \\ 2 \\ 2 \\ 1 \end{bmatrix}.$$

The least length, least squares solution of $A\mathbf{x} = \mathbf{b}$ is

$$\mathbf{x}_0 = A^+\mathbf{b} = \frac{1}{1403} \begin{bmatrix} -333 \\ -369 \\ -234 \\ -171 \\ 488 \end{bmatrix}.$$

Is this an exact solution? Let us check: Multiplying out $A\mathbf{x}_0$,

$$A\mathbf{x}_0 = \frac{1}{23} \begin{bmatrix} 9 \\ 4 \\ 18 \\ -19 \end{bmatrix}.$$

This is not even close to $\mathbf{b}$, so this is not an exact solution. It is, however, a least squares solution—with a rather large residual at that—but at least with a lesser length than that of any other least squares solution.

There are infinitely many others. The general least squares solution is $\mathbf{x}_0 + t_1\mathbf{w}_1 + t_2\mathbf{w}_2$ with the newly computed $\mathbf{x}_0$ but the same vectors $\mathbf{w}_1$ and $\mathbf{w}_2$.

Exercises

7.1 Consider the matrix $A = \begin{bmatrix} 0 & 1 & 1 \\ 1 & 0 & 1 \\ 0 & 1 & 1 \\ 0 & 0 & 1 \end{bmatrix}$.

(a) Give the QR factorization of A.

(b) Find the least squares solution to $A\mathbf{x} = \mathbf{b}$ where $\mathbf{b} = \begin{bmatrix} 1 \\ 2 \\ 3 \\ 1 \end{bmatrix}$.

(c) Find the least squares solution to $A\mathbf{x} = \mathbf{c}$ where $\mathbf{c} = \begin{bmatrix} 1 \\ 0 \\ -1 \\ 0 \end{bmatrix}$.

7.2 Find the values of a, b, and c that give the best least squares fit of the quadratic

$$y = a + bx + cx^2$$

to the data points

$$(-1, 1), \qquad (0, 1), \qquad (1, 4), \qquad \text{and} \qquad (2, 6).$$

7.3 (A computer is recommended to help with the computations here.)

(a) Find the values of a, b, c, and d that give the best least squares fit of the even polynomial

$$y = ax^6 + bx^4 + cx^2 + d$$

to the function $f(x) = e^{-x^2}$, using as data points

$$(0, f(0)), \quad (1/2, f(1/2)), \quad (1, f(1)), \quad (3/2, f(3/2)), \quad (2, f(2)), \quad \text{and} \quad (5/2, f(5/2)).$$

(b) Use your result from part (a) to approximate the integral

$$\int_0^{5/2} e^{-x^2} \, dx.$$

7.4 Let A be the matrix $A = \begin{bmatrix} 1 & 0 & 2 & 0 \\ 0 & 1 & -1 & -1 \end{bmatrix}$.

(a) Find an orthonormal basis for the $\mathrm{Ker}(A)$, and find the orthogonal projection onto $\mathrm{Ker}(A)$.

(b) For this matrix A, $\mathrm{Img}(A)$ is $\mathbb{R}^2$. Explain why, and find the least length solution of

$$A\mathbf{x} = \begin{bmatrix} 1 \\ 2 \end{bmatrix}.$$

(c) Find two linearly independent vectors $\mathbf{v}_1$ and $\mathbf{v}_2$ in $\mathbb{R}^4$ that are orthogonal to

$$\begin{bmatrix} 1 \\ 0 \\ 2 \\ 0 \end{bmatrix} \quad \text{and} \quad \begin{bmatrix} 0 \\ 1 \\ -1 \\ -1 \end{bmatrix}.$$

7.5 Let A be the matrix $A = \begin{bmatrix} 1 & 1 & 2 \\ 1 & -1 & 0 \\ 0 & 1 & 1 \end{bmatrix}$. Let $\mathbf{b} = \begin{bmatrix} 1 \\ 2 \\ 3 \end{bmatrix}$. Find a least squares solution to $A\mathbf{x} = \mathbf{b}$. Which vector in $\mathrm{Img}(A)$ is closest to $\mathbf{b}$?

7.6 Let A be the matrix $A = \begin{bmatrix} 1 & 1 & 2 \\ 1 & 0 & 1 \\ 0 & 1 & 1 \end{bmatrix}$. Let $\mathbf{b} = \begin{bmatrix} 1 \\ 1 \\ 1 \end{bmatrix}$. Find a least squares solution to $A\mathbf{x} = \mathbf{b}$. Which vector in $\mathrm{Img}(A)$ is closest to $\mathbf{b}$?

7.7 Let A be the matrix $A = \begin{bmatrix} 1 & 0 \\ 1 & 1 \\ 0 & 1 \end{bmatrix}$. Let $\mathbf{b} = \begin{bmatrix} 1 \\ 1 \\ 1 \end{bmatrix}$. Find the least squares solution to $A\mathbf{x} = \mathbf{b}$. What is the minimum value of $|A\mathbf{x} - \mathbf{b}|^2$ as $\mathbf{x}$ varies over $\mathbb{R}^2$?

7.8 Let A be the matrix $B = \begin{bmatrix} 1 & 0 & 2 \\ 1 & 1 & -1 \end{bmatrix}$.

(a) Find an orthonormal basis for $\mathrm{Ker}(A)$.

(b) Find the orthogonal projection onto $\mathrm{Img}(A^t)$.

(c) Find the solution of the system

$$\begin{aligned} x + 2z &= 1 \\ x + y - z &= 2 \end{aligned}$$

for which $x^2 + y^2 + z^2$ has the minimum value.

7.9 Let $A = \begin{bmatrix} 1 & 1 & 2 \\ 0 & 1 & 1 \\ 1 & 0 & 1 \end{bmatrix}$.

(a) Find an orthonormal basis for $\mathrm{Img}(A)$.

(b) Find P_c, the orthogonal projection onto $\text{Img}(A)$.

(c) Find a least squares solution to $A\mathbf{x} = \mathbf{b}$, where $\mathbf{b} = \begin{bmatrix} 1 \\ 1 \\ 1 \end{bmatrix}$.

(d) What is the dimension of $\text{Img}(A)$? If the $\text{Img}(A)$ is a plane, give the equation of the plane. If it is a line, give a parametric representation of the line.

7.10 Let $A = \begin{bmatrix} 2 & 1 & 2 \\ 1 & 2 & 2 \end{bmatrix}$.

(a) Find a double QR factorization of A.

(b) Find the generalized inverse A^+ of A and the least length, least squares solution to $A\mathbf{x} = \begin{bmatrix} 1 \\ 3 \end{bmatrix}$.

7.11 Let $A = \begin{bmatrix} 4 & 0 \\ 2 & 6 \\ 5 & 3 \end{bmatrix}$.

(a) Find a double QR factorization of A.

(b) Find the generalized inverse A^+ of A and the least length, least squares solution to $A\mathbf{x} = \begin{bmatrix} 1 \\ 1 \\ 1 \end{bmatrix}$.

7.12 Let $A = \begin{bmatrix} 2 & 2 \\ 1 & -1 \\ -1 & 1 \end{bmatrix}$.

(a) Find a double QR factorization of A.

(b) Find the generalized inverse A^+ of A and the least length, least squares solution to $A\mathbf{x} = \begin{bmatrix} 1 \\ 1 \\ 1 \end{bmatrix}$.

7.13 Let $A = \begin{bmatrix} 1 & 1 & 1 \\ 1 & 1 & 1 \\ 1 & 1 & 1 \end{bmatrix}$.

(a) Find a double QR factorization of A.

(b) Find the generalized inverse A^+ of A and the least length, least squares solution to $A\mathbf{x} = \begin{bmatrix} 1 \\ 2 \\ 3 \end{bmatrix}$.

7.14 Let $A = \begin{bmatrix} 2 & 1 & 3 \\ 2 & 2 & 4 \\ 1 & 0 & 1 \end{bmatrix}$.

(a) Find a double QR factorization of A.

(b) Find the generalized inverse A^+ of A and the least length, least squares solution to $A\mathbf{x} = \begin{bmatrix} 1 \\ 2 \\ 3 \end{bmatrix}$.

7.15 Let A be an $m \times n$ matrix.
(a) Show that the row space of A^+ is the column space of A and that the column space of A^+ is the row space of A.
(b) Using the result of part (a), show that $(A^+)^+ = A$.

7.16 Let A be an $m \times n$ matrix, and let A^+ be its generalized inverse. Show that

$$A^+ A A^+ = A^+ \qquad \text{and} \qquad A A^+ A = A .$$

7.17 Let A be an $m \times n$ matrix. The four identities in (7.4) and (7.5) characterize its generalized inverse A^+. However, we have seen that this list is somewhat redundant: Any matrix B that satisifies the two identities in (7.6) must be A^+ and hence must satisfy the other two identities as well. Can we draw similar conclusions from other pairs of these four identities? In particular,

(a) Suppose an $n \times m$ matrix B satisfies just the identities in (7.4), that is,

$$BA = P_r \qquad \text{and} \qquad AB = P_c ,$$

where P_r and P_c are the orthogonal projections the row space and the column space of A, respectively. Does this satisfaction imply that $B = A^+$? Prove it or give a counterexample.

(b) Suppose an $n \times m$ matrix B satisfies just the identities in (7.5), that is,

$$P_r B = B \quad \text{and} \quad BP_c = B,$$

where P_r and P_c are the orthogonal projections the row space and the column space of A, respectively. Does this imply that $B = A^+$? Prove it or give a counterexample.

7.18 Let A be an $m \times n$ matrix with rank r. What is the rank of A^+?

7.19 Let $a > 0$, and let $A(a)$ and B be the matrices

$$A(a) = \begin{bmatrix} 1 & 1 \\ 1 & 1+a \\ 1 & 1-a \end{bmatrix} \quad \text{and} \quad B = \begin{bmatrix} 1 & 1 \\ 1 & 1 \\ 1 & 1 \end{bmatrix}.$$

Notice that $A(a)$ is the "dangerous" matrix discussed in Section 7.1.

(a) Compute double QR factorizations and then generalized inverses for each of these matrices. Use the QR decomposition of $A(a)$ already computed in Section 7.1.

(b) Show that although

$$\lim_{a \to 0} A(a) = B$$

in the obvious sense of taking limits entry by entry, it is *not* the case that $\lim_{a \to 0} A^+(a) = B^+$. Indeed, $\lim_{a \to 0} A^+(a)$ does not even exist.

7.20 Let A be a symmetric square matrix. Show that $(A^2)^+ = (A^+)^2$. Is this true in general, assuming A is symmetric?

7.21 Show that for any $m \times n$ matrix A, $(A^t)^+ = (A^+)^t$.

SECTION 8 | # Changing Bases

8.1 Relating the coordinates for two different bases

Suppose $\{v_1, v_2, \ldots, v_k\}$ and $\{w_1, w_2, \ldots, w_k\}$ are two different bases for a subspace S of $\mathbb{R}^n$. Suppose that b is any vector in S. Let x be the coordinate vector of b with respect to the first basis, and let y be the coordinate vector of b with respect to the second basis. How are x and y related?

Before going into this question, it will pay to introduce some notation for bases so that we can refer to them briefly by name. Lowercase Greek letters will denote bases. Let $\beta = \{v_1, v_2, \ldots, v_k\}$ and let $\gamma = \{w_1, w_2, \ldots, w_k\}$. The question now is "How are coordinates with respect to β related to coordinates with respect to γ?"

To answer the question, let $V = [v_1, v_2, \ldots, v_k]$ and let $W = [w_1, w_2, \ldots, w_k]$. As we saw when we introduced the notion of coordinates,

$$V x = b \quad \text{and} \quad W y = b,$$

which means that

$$V x = W y. \tag{8.1}$$

If W were invertible, we would have $y = W^{-1} V x$. However, W is an $n \times k$ matrix, so in general it is not even square, let alone invertible. All is not lost: Corollary 2 of Theorem 10 tells us that when the columns of W are linearly independent so that W has rank k, the $k \times k$ matrix

$W^t W$ has rank k and hence is invertible. To put this fact to work, multiply both sides of (8.1) by W^t:

$$W^t W \mathbf{y} = W^t V \mathbf{x};$$

hence

$$\mathbf{y} = (W^t W)^{-1} W^t V \mathbf{x}.$$

Since $W^t V$ is a $k \times k$ matrix, so is $(W^t W)^{-1} W^t V$. This matrix transforms coordinates with respect to β into coordinates with respect to γ. Let $[\gamma, \beta]$ denote the $k \times k$ matrix $(W^t W)^{-1} W^t V$. That is,

$$[\gamma, \beta] = (W^t W)^{-1} W^t V. \tag{8.2}$$

DEFINITION | **(Change of basis matrix)** Let $\beta = \{\mathbf{v}_1, \mathbf{v}_2, \ldots, \mathbf{v}_k\}$ and $\gamma = \{\mathbf{w}_1, \mathbf{w}_2, \ldots, \mathbf{w}_k\}$ be two bases of a subspace S of $\mathbb{R}^n$. Let $V = [\mathbf{v}_1, \mathbf{v}_2, \ldots, \mathbf{v}_k]$ and let $W = [\mathbf{w}_1, \mathbf{w}_2, \ldots, \mathbf{w}_k]$. The $k \times k$ matrix $[\gamma, \beta]$ defined by (8.2) is called the *change-of-basis matrix from β to γ*.

EXAMPLE 48   **(Computing a change-of-basis matrix)** Consider the image of the matrix

$$A = \begin{bmatrix} 1 & 4 & 3 & 1 \\ 1 & 1 & 3 & 3 \\ 3 & 3 & 9 & 1 \\ 1 & 4 & 3 & 3 \end{bmatrix}$$

that was studied in Example 32. We found two bases for it there: $\{\mathbf{w}_1, \mathbf{w}_2, \mathbf{w}_3\}$ and $\{\mathbf{v}_1, \mathbf{v}_2, \mathbf{v}_3\}$ where

$$\mathbf{w}_1 = \begin{bmatrix} 1 \\ 0 \\ 0 \\ 1 \end{bmatrix}, \qquad \mathbf{w}_2 = \begin{bmatrix} 0 \\ 4 \\ 0 \\ 3 \end{bmatrix}, \qquad \text{and} \qquad \mathbf{w}_3 = \begin{bmatrix} 0 \\ 0 \\ 4 \\ -1 \end{bmatrix}$$

and

$$\mathbf{v}_1 = \begin{bmatrix} 1 \\ 1 \\ 3 \\ 1 \end{bmatrix}, \qquad \mathbf{v}_2 = \begin{bmatrix} 4 \\ 1 \\ 3 \\ 4 \end{bmatrix}, \qquad \text{and} \qquad \mathbf{v}_3 = \begin{bmatrix} 1 \\ 3 \\ 1 \\ 3 \end{bmatrix}.$$

Therefore, we have

$$W = \begin{bmatrix} 1 & 0 & 0 \\ 0 & 4 & 0 \\ 0 & 0 & 4 \\ 1 & 3 & -1 \end{bmatrix} \qquad \text{and} \qquad V = \begin{bmatrix} 1 & 4 & 1 \\ 1 & 1 & 3 \\ 3 & 3 & 1 \\ 1 & 4 & 3 \end{bmatrix}.$$

It is now a fairly simple matter to compute $[\gamma, \beta] = (W^t W)^{-1} W^t V$:

$$[\gamma, \beta] = \frac{1}{4} \begin{bmatrix} 4 & 16 & 4 \\ 1 & 1 & 3 \\ 3 & 3 & 1 \end{bmatrix}.$$

The formula (8.2) for $[\gamma, \beta]$ may look a bit complicated, but it should also look a bit familiar. We saw in Section 5 that the orthogonal projection onto S, P_S, is given by $P_S = W(W^t W)^{-1} W^t$ since γ is a basis for S. Therefore,

$$W[\gamma, \beta] = W((W^t W)^{-1} W^t V) = (W(W^t W)^{-1} W^t) V = P_S V.$$

But since each $\mathbf{v}_j$ belongs to S, $P_S \mathbf{v}_j = \mathbf{v}_j$,

$$P_S V = P_S[\mathbf{v}_1, \mathbf{v}_2, \ldots, \mathbf{v}_k] = [P_S \mathbf{v}_1, P_S \mathbf{v}_2, \ldots, P_S \mathbf{v}_k] = [\mathbf{v}_1, \mathbf{v}_2, \ldots, \mathbf{v}_k] = V.$$

Combining the last two results, $$W[\gamma, \beta] = V. \tag{8.3}$$

This identity gives us a useful way to think about the change-of-basis matrix in terms of its columns. Let $\mathbf{c}_j$ denote the jth column of $[\gamma, \beta]$ so that

$$[\gamma, \beta] = [\mathbf{c}_1, \mathbf{c}_2, \ldots, \mathbf{c}_k].$$

By (8.3),

$$W[\mathbf{c}_1, \mathbf{c}_2, \ldots, \mathbf{c}_k] = [W\mathbf{c}_1, W\mathbf{c}_2, \ldots, W\mathbf{c}_k] = [\mathbf{v}_1, \mathbf{v}_2, \ldots, \mathbf{v}_k].$$

In other words, for each j,

$$W\mathbf{c}_j = \mathbf{v}_j,$$

which means, as in Example 23, that $\mathbf{c}_j$ is the coordinate vector of $\mathbf{v}_j$ with respect to the basis γ. In other words,

- *Let* $\beta = \{\mathbf{v}_1, \mathbf{v}_2, \ldots, \mathbf{v}_k\}$ *and* $\gamma = \{\mathbf{w}_1, \mathbf{w}_2, \ldots, \mathbf{w}_k\}$ *be any two bases of some subspace of* $\mathbb{R}^n$. *Then the jth column of* $[\gamma, \beta]$, $\mathbf{c}_j$, *is the coordinate vector of* $\mathbf{v}_j$ *with respect to* γ:

$$\mathbf{v}_j = (\mathbf{c}_j)_1\mathbf{w}_1 + (\mathbf{c}_j)_2\mathbf{w}_2 + \cdots + (\mathbf{c}_j)_k\mathbf{w}_k = W\mathbf{c}_j. \tag{8.4}$$

As we will soon see, this provides a way to compute the matrix $[\gamma, \beta]$ without using formula (8.2). Since (8.2) can be a bit cumbersome, it is good to have an alternative. First, let us continue to become familiar with $[\gamma, \beta]$.

The change-of-basis matrix $[\gamma, \beta]$ is invertible. In fact, since it changes coordinates for β into coordinates for γ, the inverse should do the reverse: change coordinates for γ into coordinates for β, suggesting that

$$[\gamma, \beta]^{-1} = [\beta, \gamma]. \tag{8.5}$$

To verify, we compute

$$
\begin{aligned}
[\beta, \gamma][\gamma, \beta] &= ((V^tV)^{-1}V^tW)[\gamma, \beta] \\
&= ((V^tV)^{-1}V^t)(W[\gamma, \beta]) \\
&= ((V^tV)^{-1}V^t)V \\
&= (V^tV)^{-1}(V^tV) = I,
\end{aligned}
\tag{8.6}
$$

where the third equality is from (8.3). The same sort of argument shows that $[\gamma, \beta][\beta, \gamma] = I$, so $[\gamma, \beta]$ is invertible and $[\beta, \gamma]$ is its inverse.

The following theorem summarizes what we have just learned.

THEOREM 19   **(Changing bases)** *Let S be a subspace of* $\mathbb{R}^n$, *and let* $\beta = \{\mathbf{v}_1, \mathbf{v}_2, \ldots, \mathbf{v}_k\}$ *and* $\gamma = \{\mathbf{w}_1, \mathbf{w}_2, \ldots, \mathbf{w}_k\}$ *be any two bases of S. Let* $V = [\mathbf{v}_1, \mathbf{v}_2, \ldots, \mathbf{v}_k]$ *and* $W = [\mathbf{w}_1, \mathbf{w}_2, \ldots, \mathbf{w}_k]$. *Then the $k \times k$ matrix* $[\gamma, \beta]$ *given by*

$$[\gamma, \beta] = (W^tW)^{-1}W^tV \tag{8.7}$$

is invertible, and for any $\mathbf{b}$ *in S, if the coordinate vector of* $\mathbf{b}$ *with respect to* β *is* $\mathbf{x}$ *and the coordinate vector of* $\mathbf{b}$ *with respect to* γ *is* $\mathbf{y}$, *then*

$$\mathbf{y} = [\gamma, \beta]\mathbf{x}. \tag{8.8}$$

For each j, the jth column of $[\gamma, \beta]$ *is the coordinate vector of* $\mathbf{v}_j$ *with respect to* γ.

Let us apply this theorem to work out some relations between coordinate vectors.

EXAMPLE 49   **(Coordinates in different bases)** Let **b** be the third column of the matrix A that we considered in Example 48. Evidently $\mathbf{b} = A\mathbf{e}_3$, so it is in $\text{Img}(A)$. We will now find its coordinate vectors with respect to each of the bases for $\text{Img}(A)$ studied in Example 48.

The coordinate vector

$$\mathbf{x} = \begin{bmatrix} x_1 \\ x_2 \\ x_3 \end{bmatrix}$$

for the basis $\beta = \{\mathbf{v}_1, \mathbf{v}_2, \mathbf{v}_k\}$ is very easy to find. Notice that $\mathbf{b} = 3\mathbf{v}_1$. Therefore,

$$\mathbf{x} = \begin{bmatrix} 3 \\ 0 \\ 0 \end{bmatrix}.$$

Let **y** be the coordinate vector of **b** with respect to $\gamma = \{\mathbf{w}_1, \mathbf{w}_2, \mathbf{w}_3\}$. Now using the relation $\mathbf{y} = [\gamma, \beta]\mathbf{x}$ and the result from Example 48 that

$$[\gamma, \beta] = \frac{1}{4}\begin{bmatrix} 4 & 16 & 4 \\ 1 & 1 & 3 \\ 3 & 3 & 1 \end{bmatrix},$$

we have

$$\mathbf{y} = [\gamma, \beta]\mathbf{x} = \frac{1}{4}\begin{bmatrix} 4 & 16 & 4 \\ 1 & 1 & 3 \\ 3 & 3 & 1 \end{bmatrix}\begin{bmatrix} 3 \\ 0 \\ 0 \end{bmatrix} = \frac{3}{4}\begin{bmatrix} 4 \\ 1 \\ 3 \end{bmatrix}.$$

The vector on the right is the coordinate vector of **b** for the basis γ. To check, compute

$$3\mathbf{w}_1 + \frac{3}{4}\mathbf{w}_2 + \frac{9}{4}\mathbf{w}_3 = 3\begin{bmatrix} 1 \\ 0 \\ 0 \\ 1 \end{bmatrix} + \frac{3}{4}\begin{bmatrix} 0 \\ 4 \\ 0 \\ 3 \end{bmatrix} + \frac{9}{4}\begin{bmatrix} 0 \\ 0 \\ 4 \\ -1 \end{bmatrix} = \begin{bmatrix} 3 \\ 3 \\ 9 \\ 3 \end{bmatrix} = \mathbf{b}.$$

8.2 Computing the change-of-basis matrix efficiently

The fact that the jth column of $[\gamma, \beta]$ is the unique solution of $W\mathbf{x} = \mathbf{v}_j$ gives us an efficient way to compute $[\gamma, \beta]$. To compute the jth column, just row reduce the augmented matrix $[W|\mathbf{v}_j]$. We *could* do this for each j separately, but it would be more efficient to do it in parallel for all $j = 1, 2, \ldots, k$ at once, just as when we compute an inverse by row reduction. Therefore, we form the augmented matrix $[W|V]$ and row reduce it. As when we are computing inverses, we may as well proceed with the row reduction all the way to the reduced row echelon form of $[W|V]$. From it we can easily read off the solutions of $W\mathbf{x} = \mathbf{v}_j, j = 1, 2, \ldots, k$. Since these are the columns of $[\gamma, \beta]$, we easily find $[\gamma, \beta]$ in this way.

EXAMPLE 50   **(Calculating $[\gamma, \beta]$ by row reduction)** Consider the equation

$$x_1 + x_2 - x_3 + x_4 = 0$$

in $\mathbb{R}^4$. Its solution set S is a three-dimensional subspace of $\mathbb{R}^4$. Let $\{\mathbf{w}_1, \mathbf{w}_2, \mathbf{w}_3\}$ and $\{\mathbf{v}_1, \mathbf{v}_2, \mathbf{v}_3\}$ be given by

$$\mathbf{w}_1 = \begin{bmatrix} 2 \\ 0 \\ 1 \\ -1 \end{bmatrix}, \quad \mathbf{w}_2 = \begin{bmatrix} 0 \\ -1 \\ 0 \\ 1 \end{bmatrix}, \quad \text{and} \quad \mathbf{w}_3 = \begin{bmatrix} 0 \\ 0 \\ 1 \\ 1 \end{bmatrix}$$

and

$$\mathbf{v}_1 = \begin{bmatrix} 1 \\ 0 \\ 0 \\ -1 \end{bmatrix}, \quad \mathbf{v}_2 = \begin{bmatrix} 0 \\ 1 \\ 0 \\ -1 \end{bmatrix}, \quad \text{and} \quad \mathbf{v}_3 = \begin{bmatrix} 0 \\ 0 \\ 1 \\ 1 \end{bmatrix}.$$

We can see that each of these vectors satisfies the equation defining S and that each set is linearly independent. Therefore, $\beta = \{\mathbf{v}_1, \mathbf{v}_2, \mathbf{v}_3\}$ and $\gamma = \{\mathbf{w}_1, \mathbf{w}_2, \mathbf{w}_3\}$ are both bases for S.

We now compute $[\gamma, \beta]$ by row reducing $[W|V]$.

$$[W|V] = \begin{bmatrix} 2 & 0 & 0 & | & 1 & 0 & 0 \\ 0 & -1 & 0 & | & 0 & 1 & 0 \\ 1 & 0 & 1 & | & 0 & 0 & 1 \\ -1 & 1 & 1 & | & -1 & -1 & 1 \end{bmatrix}.$$

Row reducing all the way to reduced row echelon form, as in computing an inverse, we find

$$\begin{bmatrix} 1 & 0 & 0 & | & 1/2 & 0 & 0 \\ 0 & 1 & 0 & | & 0 & -1 & 0 \\ 0 & 0 & 1 & | & -1/2 & 0 & 1 \\ 0 & 0 & 0 & | & 0 & 0 & 0 \end{bmatrix},$$

which tells us that

$$W \begin{bmatrix} 1/2 \\ 0 \\ -1/2 \end{bmatrix} = \mathbf{v}_1, \qquad W \begin{bmatrix} 0 \\ -1 \\ 0 \end{bmatrix} = \mathbf{v}_2, \qquad \text{and} \qquad W \begin{bmatrix} 0 \\ 0 \\ 1 \end{bmatrix} = \mathbf{v}_3.$$

Hence, the change-of-basis matrix is

$$[\gamma, \beta] = \begin{bmatrix} 1/2 & 0 & 0 \\ 0 & -1 & 0 \\ -1/2 & 0 & 1 \end{bmatrix}.$$

This is just the upper right 3×3 matrix in our row reduction of $[V|U]$.

8.3 When one basis is orthonormal

If γ is an orthonormal basis, then the formula (8.2) for $[\gamma, \beta]$ simplifies, no matter what kind of basis β is.

THEOREM 20

(Change of basis for orthonormal bases) *Let $\beta = \{\mathbf{v}_1, \mathbf{v}_2, \ldots, \mathbf{v}_k\}$ and $\gamma = \{\mathbf{u}_1, \mathbf{u}_2, \ldots, \mathbf{u}_k\}$ be two bases for a subspace S of $\mathbb{R}^n$, and suppose that γ is orthonormal. Then the entries of the change-of-basis matrix $[\gamma, \beta]$ are given by*

$$[\gamma, \beta]_{i,j} = \mathbf{u}_i \cdot \mathbf{v}_j. \tag{8.9}$$

PROOF: Let $V = [\mathbf{v}_1, \mathbf{v}_2, \ldots, \mathbf{v}_k]$ and $U = [\mathbf{u}_1, \mathbf{u}_2, \ldots, \mathbf{u}_k]$, as usual. Since γ is orthonormal, U is an isometry, and so $U^t U = I$. Then, of course, $(U^t U)^{-1} = I$, so $[\gamma, \beta] = (U^t U)^{-1} U^t V$ simplifies to

$$[\gamma, \beta] = U^t V. \tag{8.10}$$

By, **V.I.F.4**,

$$(U^t V)_{i,j} = (\text{row } i \text{ of } U^t) \cdot (\text{column } j \text{ of } V)$$
$$= (\text{column } i \text{ of } U) \cdot (\text{column } j \text{ of } V)$$
$$= \mathbf{u}_i \cdot \mathbf{v}_j,$$

which gives us (8.10). ∎

EXAMPLE 51

(Change of basis with an orthonormal basis) Let $\gamma = \{\mathbf{u}_1, \mathbf{u}_2\}$ be the orthonormal bases for the plane $x + y + z = 0$ considered in Examples 24 and 25 back in Section 3. Recall that

$$\mathbf{u}_1 = \frac{1}{\sqrt{2}} \begin{bmatrix} -1 \\ 1 \\ 0 \end{bmatrix} \qquad \text{and} \qquad \mathbf{u}_2 = \frac{1}{\sqrt{6}} \begin{bmatrix} -1 \\ -1 \\ 2 \end{bmatrix}.$$

If we parameterize the solution set equation $x + y + z = 0$ in the usual way, we find the basis $\beta = \{\mathbf{v}_1, \mathbf{v}_2\}$ for S where

$$\mathbf{v}_1 = \begin{bmatrix} -1 \\ 1 \\ 0 \end{bmatrix} \quad \text{and} \quad \mathbf{v}_2 = \begin{bmatrix} -1 \\ 0 \\ 1 \end{bmatrix}.$$

Since γ is orthonormal,

$$[\gamma, \beta] = \begin{bmatrix} \mathbf{u}_1 \cdot \mathbf{v}_1 & \mathbf{u}_1 \cdot \mathbf{v}_2 \\ \mathbf{u}_2 \cdot \mathbf{v}_1 & \mathbf{u}_2 \cdot \mathbf{v}_2 \end{bmatrix} = \begin{bmatrix} \sqrt{2} & 1/\sqrt{2} \\ 0 & \sqrt{3/2} \end{bmatrix}.$$

What about changing in the other direction? We can compute $[\beta, \gamma]$ using the fact that $[\beta, \gamma] = [\gamma, \beta]^{-1}$. Hence

$$[\beta, \gamma] = \begin{bmatrix} 1/\sqrt{2} & -1/\sqrt{6} \\ 0 & \sqrt{2/3} \end{bmatrix}.$$

8.4 Change of basis in $\mathbb{R}^n$

There is another important case in which the formula (8.2) for $[\gamma, \beta]$ simplifies, namely, when $S = \mathbb{R}^n$. Then both W and V are square, $n \times n$ matrices with zero kernels, so they are both invertible. When W itself is invertible, so is W^t, and

$$(W^t W)^{-1} = W^{-1}(W^t)^{-1}.$$

Therefore, $[\gamma, \beta] = W^{-1}(W^t)^{-1} W^t V = W^{-1} V$. That is,

- If $\beta = \{\mathbf{v}_1, \mathbf{v}_2, \ldots, \mathbf{v}_n\}$ *and* $\gamma = \{\mathbf{w}_1, \mathbf{w}_2, \ldots, \mathbf{w}_n\}$ *are bases for* $\mathbb{R}^n$,

$$[\gamma, \beta] = W^{-1} V. \tag{8.11}$$

In particular, when β is the standard basis of $\mathbb{R}^n$, that is, $\beta = \{\mathbf{e}_1, \mathbf{e}_2, \ldots, \mathbf{e}_n\}$, then $V = I$ and $[\gamma, \beta] = W$.

Furthermore, if $\gamma = \{\mathbf{u}_1, \mathbf{u}_2, \ldots, \mathbf{u}_n\}$ is an orthonormal basis of $\mathbb{R}^n$, $U = [\mathbf{u}_1, \mathbf{u}_2, \ldots, \mathbf{u}_n]$ is more than just an isometry; it is an *invertible* isometry, that is, an orthogonal matrix.

You may be wondering why a practical person would ever want to use any basis for $\mathbb{R}^n$ except the standard basis $\{\mathbf{e}_1, \mathbf{e}_2, \ldots, \mathbf{e}_n\}$? Would not this basis *always* be the simplest basis to use in $\mathbb{R}^n$?

The answer depends on what you are trying to do. If you are trying to solve problems involving some given $m \times n$ matrix A, it might be better to use bases for $\mathbb{R}^m$ and $\mathbb{R}^n$ that "simplify" A.

For any $m \times n$ matrix with rank r, there are always bases

$$\beta = \{\mathbf{v}_1, \ldots, \mathbf{v}_n\} \quad \text{and} \quad \gamma = \{\mathbf{w}_1, \ldots, \mathbf{w}_m\}$$

of $\mathbb{R}^n$ and $\mathbb{R}^m$, respectively, so that if

$$V = [\mathbf{v}_1, \ldots, \mathbf{v}_n] \quad \text{and} \quad W = [\mathbf{w}_1, \ldots, \mathbf{w}_m],$$

then

$$W^{-1} A V = \begin{bmatrix} I_{r \times r} & 0 \\ 0 & 0 \end{bmatrix}_{m \times n}, \tag{8.12}$$

where

$$\begin{bmatrix} I_{r \times r} & 0 \\ 0 & 0 \end{bmatrix}_{m \times n} \tag{8.13}$$

denotes the $m \times n$ matrix with $I_{r \times r}$ filling out the upper left $r \times r$ block and all other entries being zero.

Another way to write (8.12) is

$$A = W \begin{bmatrix} I_{r \times r} & 0 \\ 0 & 0 \end{bmatrix}_{m \times n} V^{-1}. \tag{8.14}$$

This factorization allows us to write A as the product of nice and simple matrices, and it makes answering all sorts of questions about A very easy. For instance, suppose we are trying to solve $A\mathbf{x} = \mathbf{b}$. Using (8.14), we see that

$$A\mathbf{x} = \mathbf{b} \quad \text{is equivalent to} \quad \begin{bmatrix} I_{r \times r} & 0 \\ 0 & 0 \end{bmatrix}_{m \times n} (V^{-1}\mathbf{x}) = (W^{-1}\mathbf{b}) \,.$$

Now, $V^{-1}\mathbf{x}$ and $W^{-1}\mathbf{b}$ are just the coordinate vectors of $\mathbf{x}$ and $\mathbf{b}$ for the β and γ bases, respectively. Call them $\tilde{\mathbf{x}}$ and $\tilde{\mathbf{y}}$ so that the last equation becomes

$$\begin{bmatrix} I_{r \times r} & 0 \\ 0 & 0 \end{bmatrix}_{m \times n} \tilde{\mathbf{x}} = \tilde{\mathbf{b}} \,.$$

This equation—equivalent to the original one—is very easy!

- *Using these special bases, the linear transformation described by A does something very, very simple: It just "wipes out" the last $n - r$ entries of the input coordinate vector.*

In particular, $A\mathbf{x} = \mathbf{b}$ will have a solution if and only if $\tilde{\mathbf{b}} = W^{-1}\mathbf{b}$ is all zeros after the rth entry (assuming that $m > r$, otherwise it is always solvable). Practically *any* question about A of the kind we have been asking so far can easily be answered using the factorization (8.14).

Think back to Chapter 1. Let f be any linear transformation from $\mathbb{R}^n$ to $\mathbb{R}^m$. We saw that using the standard bases for $\mathbb{R}^n$ and $\mathbb{R}^m$—the only bases we even knew about back then—we could write down a matrix A representing this linear transformation. If the matrix A has a complicated structure, it might be difficult to work with. But if we change the bases in $\mathbb{R}^n$ and $\mathbb{R}^m$ properly, then we can arrange that the matrix representing our linear transformation has the simple form in (8.13).

- *A major theme in the remaining chapters of the book is to find "good bases" that allow us to write*

$$A = W[\text{some simple matrix}]V^{-1} \,.$$

The simple matrix in the middle will not always have the form (8.13), but it will have some nice properties adapted to the kind of problem we are trying to solve.

For example, in Chapter 5, we shall see how such a choice of bases enables us to solve an important class of systems of differential equations. However, back to the matter at hand. How does one choose the bases β and γ so that (8.12) is true?

- *One chooses the bases β and γ "respecting the four subspaces"* $\text{Ker}(A)$, $\text{Img}(A)$, $\text{Ker}(A^t)$ *and* $\text{Img}(A^t)$.

Here is how:

1. Choose *any* basis $\{\mathbf{u}_1, \ldots, \mathbf{u}_{n-r}\} \subset \mathbb{R}^n$ for $\text{Ker}(A)$

2. Choose *any* basis $\{\mathbf{z}_1, \ldots, \mathbf{z}_{m-r}\} \subset \mathbb{R}^m$ for $\text{Ker}(A^t)$.

3. Choose *any* basis $\{\mathbf{a}_1, \ldots, \mathbf{a}_r\} \subset \mathbb{R}^m$ for $\text{Img}(A^t)$, the row space. For example, one natural choice for (3) would be the pivotal columns of A^t.

There is one more basis to choose, but we do not choose it arbitrarily. For each $\mathbf{a}_j$, define $\mathbf{b}_j = A\mathbf{a}_j$. Then $\{\mathbf{b}_1, \ldots, \mathbf{b}_r\}$ is clearly a subset of $\text{Img}(A)$. It is also linearly independent with the maximal number of elements and therefore a basis of $\text{Img}(A)$.

To see why, suppose that $\sum_{j=1}^{r} c_j \mathbf{b}_j = 0$, but then

$$0 = \sum_{j=1}^{r} c_j \mathbf{b}_j = \sum_{j=1}^{r} c_j A\mathbf{a}_j = A\left(\sum_{j=1}^{r} c_j \mathbf{a}_j\right),$$

which means that $\sum_{j=1}^{r} c_j \mathbf{a}_j$ is in the kernel of A as well as in $\text{Img}(A^t)$, the row space. But $\text{Img}(A^t)$ and $\text{Ker}(A)$ are orthogonal complements, so their only vector in common is 0. Thus, $\sum_{j=1}^{r} c_j \mathbf{a}_j = 0$, and since $\{\mathbf{a}_1, \ldots, \mathbf{a}_r\}$ is linearly independent, each c_j is zero. Showing that $c_j = 0$ for each j whenever $\sum_{j=1}^{r} c_j \mathbf{b}_j = 0$ proves that $\{\mathbf{b}_1, \ldots, \mathbf{b}_r\}$ is linearly independent.

Therefore, with this special choice $\mathbf{b}_j = A\mathbf{a}_j$,

4. $\{\mathbf{b}_1, \ldots, \mathbf{b}_r\} \subset \mathbb{R}^n$ is a basis for $\text{Img}(A)$.

Now let us combine the bases as follows. Define

$$\beta = \{\mathbf{v}_1, \ldots, \mathbf{v}_n\} = \{\mathbf{a}_1, \ldots, \mathbf{a}_r, \mathbf{u}_1, \ldots, \mathbf{u}_{n-r}\}$$

and

$$\gamma = \{\mathbf{w}_1, \ldots, \mathbf{w}_n\} = \{\mathbf{b}_1, \ldots, \mathbf{b}_r, \mathbf{z}_1, \ldots, \mathbf{z}_{m-r}\}.$$

It is easy to see that these sets are both linearly independent sets with the maximal number of elements and therefore are both bases. For example, suppose that $\sum_{j=1}^{n} c_j \mathbf{v}_j = 0$. Then

$$\left(\sum_{j=1}^{r} c_j \mathbf{v}_j\right) = -\left(\sum_{j=r+1}^{n} c_j \mathbf{v}_j\right).$$

The vector on the left belongs to $\text{Img}(A^t)$, the vector on the right to $\text{Img}(A)$. These subspaces are orthogonal complements, so their only vector in common is the zero vector. Thus

$$\left(\sum_{j=1}^{r} c_j \mathbf{v}_j\right) = 0 = \left(\sum_{j=r+1}^{n} c_j \mathbf{v}_j\right).$$

Now since $\{\mathbf{v}_1, \ldots, \mathbf{v}_r\}$ is linearly independent, $c_j = 0$ for $j = 1, \ldots, r$. Next, since $\{\mathbf{v}_{r+1}, \ldots, \mathbf{v}_n\}$ is linearly independent, $c_j = 0$ for $j = r + 1, \ldots, n$. Hence $c_j = 0$ for each j, which is what we had to show.

We will now show that these bases "do the trick." That is nice, because we learned early on in the chapter how to find the required bases of the four subspaces!

By the choices made in the definition of the bases,

$$A\mathbf{v}_j = \mathbf{w}_j \quad \text{for } j \leq r \quad \text{while} \quad A\mathbf{v}_j = 0 \quad \text{for } j > r. \tag{8.15}$$

Then, by **V.I.F.3** and (8.15),

$$AV = [\mathbf{w}_1, \ldots, \mathbf{w}_r, 0 \ldots, 0]. \tag{8.16}$$

Next, by **V.I.F.4**

$$(W^{-1}AV)_{i,j} = (\text{row } i \text{ of } W^{-1}) \cdot (\text{column } j \text{ of } AV).$$

By (8.16), if $j > r$, $(\text{column } j \text{ of } AV) = 0$ so that

$$(W^{-1}AV)_{i,j} = 0 \quad \text{for } j > r.$$

Also, by (8.16), if $j \leq r$,

$$(\text{column } j \text{ of } AV) = (\text{column } j \text{ of } W),$$

so

$$(\text{row } i \text{ of } W^{-1}) \cdot (\text{column } j \text{ of } AV) = (\text{row } i \text{ of } W^{-1}) \cdot (\text{column } j \text{ of } W) = (I_{m \times m})_{i,j}.$$

Thus the left r columns are those of $I_{m \times m}$, which proves that (8.12) holds with this choice of V and W.

There are further nice properties of these bases that are worth noting. Let P_c denote the orthogonal projection onto $\text{Img}(A)$, the column space of A. We claim that

$$P_c = W \begin{bmatrix} I_{r \times r} & 0 \\ 0 & 0 \end{bmatrix}_{m \times m} W^{-1}. \tag{8.17}$$

To see it, let $\mathbf{b}$ be any vector in $\mathbb{R}^m$, and let $\tilde{\mathbf{b}} = W^{-1}$ be its coordinate vector for the basis γ. Then (using **V.I.F.1** in the last step)

$$\left(W \begin{bmatrix} I_{r \times r} & 0 \\ 0 & 0 \end{bmatrix}_{m \times m} W^{-1} \right) \mathbf{b} = W \begin{bmatrix} I_{r \times r} & 0 \\ 0 & 0 \end{bmatrix}_{m \times m} \tilde{\mathbf{b}} = \sum_{j=1}^{r} (\tilde{\mathbf{b}})_j \mathbf{w}_j \,,$$

since the matrix in the middle "wipes out" all entries above the rth.

But note that

$$P_c \mathbf{w}_j = \mathbf{w}_j \quad \text{for } j \le r \qquad \text{while} \qquad P_c \mathbf{w}_j = 0 \quad \text{for } j > r \,,$$

since for $j \le r$, $\mathbf{w}_j$ is already in the column space, while for $j > r$, $\mathbf{w}_j$ is in $\text{Ker}(A^t)$ and thus is orthogonal to the column space. Therefore,

$$P_c \mathbf{b} = P_c \left(\sum_{j=1}^{m} (\tilde{\mathbf{b}})_j \mathbf{w}_j \right) = \sum_{j=1}^{m} (\tilde{\mathbf{b}})_j P_c \mathbf{w}_j = \sum_{j=1}^{r} (\tilde{\mathbf{b}})_j \mathbf{w}_j \,.$$

We get the same result applying P_c as we did applying $W \begin{bmatrix} I_{r \times r} & 0 \\ 0 & 0 \end{bmatrix}_{m \times m} \cdot W^{-1}$, and since $\mathbf{b}$ was any vector in $\mathbb{R}^m$, the two matrices must therefore be the same.

The same reasoning shows that if P_r denotes the orthogonal projection onto $\text{Img}(A^t)$, the row space of A, then

$$P_r = V \begin{bmatrix} I_{r \times r} & 0 \\ 0 & 0 \end{bmatrix}_{n \times n} V^{-1}. \tag{8.18}$$

The demonstration of (8.18) is the same as that of (8.17).

We next claim that the generalized inverse of A, A^+, is given by

$$A^+ = V \begin{bmatrix} I_{r \times r} & 0 \\ 0 & 0 \end{bmatrix}_{n \times m} W^{-1}. \tag{8.19}$$

Let B denote the matrix on the right in (8.19). Then, using (8.17) and (8.18), we can easily check that

$$AB = P_c \qquad \text{and} \qquad BA = P_r$$

and

$$P_c B = B \qquad \text{and} \qquad BP_r = B \,.$$

Comparing with the identities (7.4) and (7.5), we see from Theorem 18 that B is indeed the generalized inverse of A.

The ideas we have just explained provide a way to compute generalized inverses that is especially simple if either $\text{rank}(A) = m$ or $\text{rank}(A) = n$. This approach is developed in the exercises. The important thing to bear in mind, which is far more important than computing generalized inverses, is that *if we choose bases for $\mathbb{R}^n$ and $\mathbb{R}^m$ that are somehow "adapted" to a* *matrix A, we can often greatly simplify the solution of a problem involving A.* The next chapters further develop this idea.

Exercises

8.1 The sets of vectors $\{w_1, w_2\}$ and $\{v_1, v_2\}$, where

$$w_1 = \begin{bmatrix} 1 \\ -1 \\ 0 \end{bmatrix}, \qquad w_2 = \begin{bmatrix} 0 \\ 1 \\ -1 \end{bmatrix} \quad \text{and} \quad v_1 = \begin{bmatrix} 2 \\ -1 \\ -1 \end{bmatrix}, \qquad v_2 = \begin{bmatrix} 1 \\ 1 \\ -2 \end{bmatrix}$$

are both bases for S, where S is the plane in $\mathbb{R}^3$ given by $x_1 + x_2 + x_3 = 0$.

(a) Find the 2×2 matrix $[\gamma, \beta]$ that is the change-of-basis matrix from the basis $\beta = \{v_1, v_2\}$ to the basis $\gamma = \{w_1, w_2\}$.

(b) Express $2v_1 - 3v_2$ as a linear combination of w_1 and w_2.

8.2 The sets of vectors $\{w_1, w_2\}$ and $\{v_1, v_2\}$, where

$$w_1 = \begin{bmatrix} 1 \\ 2 \\ 3 \end{bmatrix}, \qquad w_2 = \begin{bmatrix} 3 \\ 2 \\ 1 \end{bmatrix} \quad \text{and} \quad v_1 = \begin{bmatrix} 2 \\ 1 \\ 0 \end{bmatrix}, \qquad v_2 = \begin{bmatrix} 0 \\ 1 \\ 2 \end{bmatrix}$$

are both bases for S, where S is the plane in $\mathbb{R}^3$ given by $x_1 - 2x_2 + x_3 = 0$.

(a) Find the 2×2 matrix $[\gamma, \beta]$ that is the change-of-basis matrix from the basis $\beta = \{v_1, v_2\}$ to the basis $\gamma = \{w_1, w_2\}$.

(b) Express $3v_1 - v_2$ as a linear combination of w_1 and w_2.

8.3 Let $A = \begin{bmatrix} 1 & 2 & 3 \\ 2 & 1 & 3 \\ 1 & 1 & 2 \end{bmatrix}$. Let the columns of V be the pivotal column basis for $\text{Img}(A)$. Let the columns of W be the "nice" basis found using the method of Example 32. Find the corresponding change-of-basis matrix.

8.4 Let $A = \begin{bmatrix} 1 & 2 & 1 \\ 2 & 0 & 1 \\ 3 & 2 & 2 \end{bmatrix}$. Let the columns of V be the pivotal column basis for $\text{Img}(A)$. Let the columns of W be the "nice" basis found using the method of Example 32. Find the corresponding change-of-basis matrix.

8.5 Let $A = \begin{bmatrix} 1 & 1 & 0 & 1 \\ 4 & 2 & 6 & 0 \\ 4 & 3 & 1 & 2 \\ 1 & 2 & 0 & 3 \end{bmatrix}$. Let the columns of V be the pivotal column basis for $\text{Img}(A)$. Let the columns of W be the "nice" basis found using the method of Example 32. Find the corresponding change-of-basis matrix.

8.6 Let $A = \begin{bmatrix} 1 & 3 & 1 & 4 \\ 0 & 1 & 2 & 1 \\ 4 & 7 & -6 & 11 \\ 2 & 1 & -8 & 3 \end{bmatrix}$. Let the columns of V be the pivotal column basis for $\text{Img}(A)$. Let the columns of W be the "nice" basis found using the method of Example 32. Find the corresponding change-of-basis matrix.

8.7 Give the simplest example of a 3×3 orthogonal matrix that you can think of.

Computing generalized inverses: Independent rows

The crucial thing about the special bases used in Section 8.4 is that for $j < r$, $\mathbf{v}_j$ lies in the row space of j, and $\mathbf{w}_j = A\mathbf{v}_j$. There is an easy way to arrange this in the special case that the rows of A are linearly independent: Then the rows of A themselves are a basis for the row space, so we take $\mathbf{v}_j$ to be the jth row, and then $\mathbf{w}_j = A\mathbf{v}_j$.

8.8 (See the note before this exercise.) Let A be the matrix $A = \begin{bmatrix} 1 & 2 & 2 \\ 2 & 1 & 2 \end{bmatrix}$. The two rows are not proportional, so they are linearly independent. Define a basis $\{\mathbf{v}_1, \mathbf{v}_2, \mathbf{v}_3\}$ for $\mathbb{R}^3$ by taking $\mathbf{v}_1$ to be the first row of A, $\mathbf{v}_2$ to be the second row of A, and $\mathbf{v}_3$ to be a nonzero vector in the kernel of A. Define $\mathbf{w}_1$ and $\mathbf{w}_2$ by

$$\mathbf{w}_1 = A\mathbf{v}_1 \quad \text{and} \quad \mathbf{w}_2 = A\mathbf{v}_2 .$$

(a) Show that for $j = 1, 2$, $\mathbf{v}_j$ is the least length solution of $A\mathbf{x} = \mathbf{w}_j$.

(b) Show that $\{\mathbf{w}_1, \mathbf{w}_2\}$ is a basis for $\text{Img}(A) = \mathbb{R}^2$.

(c) Let $V = [\mathbf{v}_1, \mathbf{v}_2, \mathbf{v}_3]$ and $W = [\mathbf{w}_1, \mathbf{w}_2]$. Show that

$$W^{-1}AV = \begin{bmatrix} 1 & 0 & 0 \\ 0 & 1 & 0 \end{bmatrix} .$$

(d) Compute A^+, the generalized inverse of A.

(e) Compute the least length solution of $A\mathbf{x} = \begin{bmatrix} 1 \\ 2 \end{bmatrix}$.

8.9 (See the note before Exercise 8.8.) Let A be the matrix $A = \begin{bmatrix} 1 & 2 & 3 \\ 3 & 2 & 1 \end{bmatrix}$. The two rows are not proportional, so they are linearly independent. Define a basis $\{\mathbf{v}_1, \mathbf{v}_2, \mathbf{v}_3\}$ for $\mathbb{R}^3$ by taking $\mathbf{v}_1$ to be the first row of A, $\mathbf{v}_2$ to be the second row of A, and $\mathbf{v}_3$ to be a nonzero vector in the kernel of A. Define $\mathbf{w}_1$ and $\mathbf{w}_2$ by

$$\mathbf{w}_1 = A\mathbf{v}_1 \quad \text{and} \quad \mathbf{w}_2 = A\mathbf{v}_2 .$$

(a) Show that for $j = 1, 2$, $\mathbf{v}_j$ is the least length solution of $A\mathbf{x} = \mathbf{w}_j$.

(b) Show that $\{\mathbf{w}_1, \mathbf{w}_2\}$ is a basis for $\text{Img}(A) = \mathbb{R}^2$.

(c) Let $V = [\mathbf{v}_1, \mathbf{v}_2, \mathbf{v}_3]$ and $W = [\mathbf{w}_1, \mathbf{w}_2]$. Show that

$$W^{-1}AV = \begin{bmatrix} 1 & 0 & 0 \\ 0 & 1 & 0 \end{bmatrix} .$$

(d) Compute A^+, the generalized inverse of A.

(e) Compute the least length solution of $A\mathbf{x} = \begin{bmatrix} 1 \\ 3 \end{bmatrix}$.

Computing generalized inverses: Independent columns

As we have seen, it is easy to find the "special bases" if the rows of A are linearly independent. What if only the columns of A are linearly independent? This case arises often—overdetermined problems arise more frequently than underdetermined ones in most applications.

There is a simple cure: If A has linearly independent columns, A^t has linearly independent rows. Apply the method used in the two previous exercises to arrange

$$A^t = V \begin{bmatrix} I & 0 \\ 0 & 0 \end{bmatrix} W^{-1} .$$

But then

$$A = (W^{-1})^t \begin{bmatrix} I & 0 \\ 0 & 0 \end{bmatrix}^t V^t ,$$

which is what we want for A: $(W^{-1})^t$ is the "V matrix" for A, and $(V^{-1})^t$ is the "W matrix" for A.

8.10 (See the note just before this exercise.) Let $A = \begin{bmatrix} 1 & 2 \\ 2 & 1 \\ 2 & 2 \end{bmatrix}$.

(a) Find invertible matrices V and W so that

$$W^{-1}AV = \begin{bmatrix} 1 & 0 \\ 0 & 1 \\ 0 & 0 \end{bmatrix} .$$

(b) Compute A^+, the generalized inverse of A.

8.11 (See the note before Exercise 8.10.) Let $A = \begin{bmatrix} 1 & 2 \\ 2 & 1 \\ 2 & 2 \end{bmatrix}$.

(a) Find invertible matrices V and W so that

$$W^{-1}AV = \begin{bmatrix} 1 & 0 \\ 0 & 1 \\ 0 & 0 \end{bmatrix} .$$

(b) Compute A^+, the generalized inverse of A.

Determinants

Overview of Chapter 4

We are now at a major milestone in the development of our subject. It is time for a brief look back and ahead. We began Chapter 3 with the problem of computing least squares solutions. In the course of developing a general and effective approach to this problem, we learned that the "natural" coordinates on $\mathbb{R}^n$ might not always be the best ones to use in a linear algebra computation. We saw that using a coordinate system that was adapted to the matrix A in question could be very advantageous. These other systems of coordinates are based on the use of other bases for $\mathbb{R}^n$ besides the "natural" one. At the end of Chapter 3, we saw how to construct special bases for $\mathbb{R}^n$ and $\mathbb{R}^m$, paying attention to the "four subspaces" of the $m \times n$ matrix A—$\text{Ker}(A)$, $\text{Ker}(A^t)$, $\text{Img}(A)$, and $\text{Img}(A^t)$—and that the action of A was so simple in these coordinates that we could easily compute the generalized inverse of A.

The main theme of the rest of the book is a further development of this one: *If you choose your coordinate system carefully, you can make matrix computations easy and numerically reliable.* Choosing a coordinate system amounts to choosing the basis on which computations can be made. How do we find good bases?

One important answer to this question is given in Chapter 5: *By looking for eigenvectors.* An eigenvector of an $n \times n$ matrix A, as we shall explain in Chapter 5, is simply a nonzero vector $\mathbf{v}$ in $\mathbb{R}^n$ such that $A\mathbf{v}$ is a multiple of $\mathbf{v}$ itself. In other words, it is a nonzero vector whose "direction" is unchanged by A. We shall explain what makes eigenvectors so useful when we get to Chapter 5, but for problems involving an $n \times n$ matrix A, finding a basis of $\mathbb{R}^n$ consisting of eigenvectors of A is often the key to progress.

But to use eigenvectors, we need to be able to find them. This chapter develops a subject that is very useful for finding them and for other things as well: the theory of determinants. We will be taking a slight detour away from our main new theme—choosing effective coordinate systems—but we are paving the way for a return to this in Chapters 5, 6, and 7.

There are also geometric reasons for studying determinants: Determinants can be used to compute areas and volumes. The geometric theorems in this chapter, as well as the ideas leading up to them, significantly strengthen the bond that we have built up so far between algebra and geometry.

SECTION 1 | Determinants

1.1 The determinant of a 2 × 2 matrix

Consider the general 2×2 matrix

$$A = \begin{bmatrix} a & b \\ c & d \end{bmatrix}.$$

We saw in Chapter 1 that the quantity $ad - bc$ conveys important algebraic and geometric information about A. In an algebraic context, we saw that A is invertible if and only if $ad - bc \neq 0$, in which case

$$A^{-1} = \frac{1}{ad - bc} \begin{bmatrix} d & -b \\ -c & a \end{bmatrix}.$$

In a geometric context, we saw that $|ad - bc|$ is the area of the image of the unit square under the linear transformation of $\mathbb{R}^2$ induced by A.

This motivates defining a numerically valued function on the set of 2×2 matrices through the rule

$$\begin{bmatrix} a & b \\ c & d \end{bmatrix} \rightarrow ad - bc.$$

This function is called the *determinant function*, and we denote its value at A by writing $\det(A)$. In other words,

$$\det\left(\begin{bmatrix} a & b \\ c & d \end{bmatrix}\right) = ad - bc. \tag{1.1}$$

We can now rephrase the two results from Chapter 1 just mentioned:

(i)   *The 2×2 matrix A is invertible if and only if $\det(A) \neq 0$.*

(ii)   *The area of the image of the unit square under the linear transformation of $\mathbb{R}^2$ induced by a 2×2 matrix A is $|\det(A)|$.*

It would be nice to generalize the determinant function to $n \times n$ matrices for arbitrary n. However, it is not so easy, just looking at (1.1), to guess a generalization of the determinant function to $n \times n$ matrices for which results analogous to (i) and (ii) would be valid.

To find the generalization, we need a different way of looking at the determinant function. The way forward is to think of A in terms of its row vectors:

$$A = \begin{bmatrix} \mathbf{r}_1 \\ \mathbf{r}_2 \end{bmatrix}.$$

Instead of thinking of the determinant as a function on 2×2 matrices, we can think of it as a function on ordered pairs of vectors in $\mathbb{R}^2$. To emphasize this point of view, we write

$$\det(\mathbf{r}_1, \mathbf{r}_2) = \det\left(\begin{bmatrix} \mathbf{r}_1 \\ \mathbf{r}_2 \end{bmatrix}\right).$$

EXAMPLE 1 **(The 2×2 determinant as a function of two vectors)** Let

$$\mathbf{v} = \begin{bmatrix} 1 \\ 2 \end{bmatrix} \qquad \text{and} \qquad \mathbf{w} = \begin{bmatrix} 3 \\ 4 \end{bmatrix}.$$

Then

$$\begin{bmatrix} \mathbf{v} \\ \mathbf{w} \end{bmatrix} = \begin{bmatrix} 1 & 2 \\ 3 & 4 \end{bmatrix}$$

so that

$$\det(\mathbf{v}, \mathbf{w}) = \det\left(\begin{bmatrix} 1 & 2 \\ 3 & 4 \end{bmatrix}\right) = -2.$$

More generally, if

$$\mathbf{v} = \begin{bmatrix} a \\ b \end{bmatrix} \qquad \text{and} \qquad \mathbf{w} = \begin{bmatrix} c \\ d \end{bmatrix},$$

then

$$\det(\mathbf{v}, \mathbf{w}) = \det\left(\begin{bmatrix} a & b \\ c & d \end{bmatrix}\right) = ad - bc.$$

We can see from this that the order of $\mathbf{v}$ and $\mathbf{w}$ matters:

$$\det(\mathbf{w}, \mathbf{v}) = \det\left(\begin{bmatrix} c & d \\ a & b \end{bmatrix}\right) = cb - da.$$

In other words,

$$\det(\mathbf{w}, \mathbf{v}) = -\det(\mathbf{v}, \mathbf{w}). \tag{1.2}$$

EXAMPLE 2 **(Swapping rows and the determinant)** Let

$$A = \begin{bmatrix} 1 & 2 \\ 3 & 4 \end{bmatrix} \qquad \text{and} \qquad B = \begin{bmatrix} 3 & 4 \\ 1 & 2 \end{bmatrix}$$

so that B is obtained from A by swapping the rows. We compute

$$\det(A) = -2 \qquad \det(B) = 2,$$

so swapping the rows changed the sign of the determinant, as (1.2) says it must.

The fact that swapping the rows of a matrix changes the sign of its determinant has the following important consequence: If A has two identical rows and we swap them, we still have A.

So it must be that $\det(A) = -\det(A)$, which means that $\det(A) = 0$. Therefore, the determinant function has the following property:

(iii)   *Swapping the rows in a 2×2 matrix simply changes the sign of the determinant, so if A has two identical rows,* $\det(A) = 0$.

Here is the next key property: The determinant function $\det(\mathbf{v}, \mathbf{w})$ is a linear function of both $\mathbf{v}$ and $\mathbf{w}$.

(iv)   *When we say it is linear in* $\mathbf{v}$, *we just mean that for any two vectors* $\mathbf{v}_1$ *and* $\mathbf{v}_2$ *in* $\mathbb{R}^2$ *and any numbers s and t,*

$$\det(s\mathbf{v}_1 + t\mathbf{v}_2, \mathbf{w}) = s\det(\mathbf{v}_1, \mathbf{w}) + t\det(\mathbf{v}_2, \mathbf{w}). \tag{1.3}$$

Linearity in $\mathbf{w}$ means the analogous thing.

The point is that although $ad - bc$ is a second-degree polynomial in the variables a, b, c, and d, if we keep c and d fixed and regard only a and b as variables, then it is first-degree in a and b. That this implies (1.3) may be quite clear, but it is still a good idea to write out (1.3) in terms of the components of the vectors and check it directly. Please do this now.

Notice that with $A = \begin{bmatrix} \mathbf{v} \\ \mathbf{w} \end{bmatrix}$,

$$\det(tA) = \det(t\mathbf{v}, t\mathbf{w}) = t\det(\mathbf{v}, t\mathbf{w}) = t^2\det(\mathbf{v}, \mathbf{w}) = t^2\det(A).$$

Taking $t = 2$ and $B = A$, we see that it is *not true in general* that for 2×2 matrices A and B and numbers s and t

$$\det(sA + tB) = s\det(A) + t\det(B) \qquad \text{False, in general!}$$

The determinant function "respects" linear combinations in each row separately but not in the matrix as a whole.

To make use of the linearity in the rows, let us add a multiple of the first row to the second. That is, we perform the row operation

$$\begin{bmatrix} \mathbf{v} \\ \mathbf{w} \end{bmatrix} \to \begin{bmatrix} \mathbf{v} \\ \mathbf{w} + t\mathbf{v} \end{bmatrix}.$$

How does this operation affect the determinant? Not at all! By linearity in the second row,

$$\det(\mathbf{v}, \mathbf{w} + t\mathbf{v}) = \det(\mathbf{v}, \mathbf{w}) + t\det(\mathbf{v}, \mathbf{v}).$$

By (1.2),

$$\det(\mathbf{v}, \mathbf{v}) = -\det(\mathbf{v}, \mathbf{v}),$$

so $\det(\mathbf{v}, \mathbf{v}) = 0$. Hence,

$$\det(\mathbf{v}, \mathbf{w} + t\mathbf{v}) = \det(\mathbf{v}, \mathbf{w}). \tag{1.4}$$

EXAMPLE 3   **(Adding a multiple of one row to the other and the determinant)** Let

$$A = \begin{bmatrix} 1 & 2 \\ 3 & 4 \end{bmatrix}.$$

Let us form B by subtracting 3 times the first row from the second. We are just adding a negative multiple, so it does not affect the determinant. We find

$$B = \begin{bmatrix} 1 & 2 \\ 0 & -2 \end{bmatrix}, \qquad \text{so} \qquad \det(A) = -2 \qquad \text{and} \qquad \det(B) = -2.$$

The row operation changed the matrix but not its determinant.

The same argument applies if we add a multiple of the second row to the first, and we conclude that if a 2×2 matrix A is transformed into B by adding any multiple of one row to another, then $\det(B) = \det(A)$. In other words, the determinant function is invariant under this type of row operation. Also, we have already seen how swapping rows affects the determinant, so we deduce another property of the determinant:

(v) *If A is transformed into B through any sequence of row operations in which a multiple of one row is added to the other or rows are swapped, then*

$$\det(B) = (-1)^N \det(A),$$

where N is the number of row swaps.

We have produced a list of (i) through (v) properties of the determinant function for 2×2 matrices. What is this long list good for? It is pretty clear what the original properties, namely, (i) and (ii), are good for, but what are the new properties good for? Computing!

Let f be *some* numerically valued function on the set of 2×2 matrices. Suppose we are asked to compute

$$f\left(\begin{bmatrix} 1 & 2 \\ 3 & 4 \end{bmatrix}\right),$$

but we are not given any formula for f. Instead, we are told three properties of f:

1. If B is obtained from A by swapping rows, $f(B) = -f(A)$; in particular, if A has two identical rows, $f(A) = 0$.

2. The function $f(A)$ is linear in each of the rows of A.

3. The function f takes the value 1 at the identity. That is, $f(I_{2\times 2}) = 1$.

How is this list related to the previous list? Note that (1) and (2) are just properties (iii) and (iv) from previous list of properties. As we saw, the property (v) is an automatic consequence of these two and therefore need not be listed to be used. Finally, (3) just gives the value of the function at one particular input.

Can we do the computation? Yes, even though we are not given the formula for f and we are only given the value of f for one particular 2×2 matrix, namely $I_{2\times 2}$!

To do the computation, first of all note that properties (1) and (2) imply that $f(B) = f(A)$ when B is obtained by adding any multiple of one row of A to the other, as we have just seen. Therefore, if we subtract 3 times the first row from the second row of

$$\begin{bmatrix} 1 & 2 \\ 3 & 4 \end{bmatrix},$$

we obtain

$$\begin{bmatrix} 1 & 2 \\ 0 & -2 \end{bmatrix}.$$

It follows that

$$f\left(\begin{bmatrix} 1 & 2 \\ 3 & 4 \end{bmatrix}\right) = f\left(\begin{bmatrix} 1 & 2 \\ 0 & -2 \end{bmatrix}\right). \tag{1.5}$$

Adding the second row of

$$\begin{bmatrix} 1 & 2 \\ 0 & -2 \end{bmatrix}$$

to the first, we obtain the diagonal matrix

$$\begin{bmatrix} 1 & 0 \\ 0 & -2 \end{bmatrix}.$$

By the invariance of f under such row operations,

$$f\left(\begin{bmatrix} 1 & 2 \\ 0 & -2 \end{bmatrix}\right) = f\left(\begin{bmatrix} 1 & 0 \\ 0 & -2 \end{bmatrix}\right). \tag{1.6}$$

Finally, by the linearity in the second row and by property (3),

$$f\left(\begin{bmatrix} 1 & 0 \\ 0 & -2 \end{bmatrix}\right) = -2f\left(\begin{bmatrix} 1 & 0 \\ 0 & 1 \end{bmatrix}\right) = -2. \tag{1.7}$$

Combining (1.5), (1.6), and (1.7), we finish our computation:

$$f\left(\begin{bmatrix} 1 & 2 \\ 3 & 4 \end{bmatrix}\right) = -2.$$

If this is the first time you have seen such a computation, you may find it quite remarkable. We computed the value of a function without using a formula for it, but instead we used a short list of properties of the function. This method of computation can be very effective.

Of course, for 2×2 determinants, we have the simple formula (1.1), so there is no point in getting fancy just for 2×2 matrices. But for larger matrices, this "property-based" approach to computation beats the "formula-based" approach hands down, as we will see. Therefore, let us further familiarize ourselves with it in the 2×2 case.

Suppose we are asked to compute a formula for

$$f\left(\begin{bmatrix} a & b \\ c & d \end{bmatrix}\right),$$

given only that f has the properties (1), (2), and (3) just listed. Can we do it? Yes, in essentially the same way we computed

$$f\left(\begin{bmatrix} 1 & 2 \\ 3 & 4 \end{bmatrix}\right) = -2.$$

The only difference is that we have to consider several cases separately.

First, consider the case in which both $a = 0$ and $c = 0$. Then by the linearity in the rows,

$$f\left(\begin{bmatrix} 0 & b \\ 0 & d \end{bmatrix}\right) = bf\left(\begin{bmatrix} 0 & 1 \\ 0 & d \end{bmatrix}\right) = bdf\left(\begin{bmatrix} 0 & 1 \\ 0 & 1 \end{bmatrix}\right),$$

and by property (1),*

$$f\left(\begin{bmatrix} 0 & 1 \\ 0 & 1 \end{bmatrix}\right) = 0.$$

Hence,

$$f\left(\begin{bmatrix} a & b \\ c & d \end{bmatrix}\right) = 0 = ad - bc \qquad \text{when both } a = 0 \text{ and } c = 0. \tag{1.8}$$

Second, suppose that $a \neq 0$. Then, by the linearity in the first row,

$$f\left(\begin{bmatrix} a & b \\ c & d \end{bmatrix}\right) = af\left(\begin{bmatrix} 1 & b/a \\ c & d \end{bmatrix}\right). \tag{1.9}$$

Subtracting c times the first row from the second row,

$$f\left(\begin{bmatrix} 1 & b/a \\ c & d \end{bmatrix}\right) = f\left(\begin{bmatrix} 1 & b/a \\ 0 & d - (cb)/a \end{bmatrix}\right).$$

If $d - (cb)/a = 0$, then the bottom row of

$$\begin{bmatrix} 1 & b/a \\ 0 & d - (cb)/a \end{bmatrix}$$

---

*Remember the key fact that 0 is the only solution of $x = -x$.

is zero. Without affecting the value of f, we can then add the first row to it and get a matrix with two rows the same. For such a matrix, $f = 0$, and since $d - (cb)/a = 0$ implies $ad - bc = 0$, we see.

$$f\left(\begin{bmatrix} a & b \\ c & d \end{bmatrix}\right) = 0 = ad - bc \qquad \text{when } a \neq 0 \text{ and } d - b/a = 0. \tag{1.10}$$

On the other hand, if $d - (cb)/a \neq 0$, then by linearity in the second row,

$$f\left(\begin{bmatrix} 1 & b/a \\ 0 & d - (cb)/a \end{bmatrix}\right) = (d - (cb)/a)f\left(\begin{bmatrix} 1 & (cb)/a \\ 0 & 1 \end{bmatrix}\right). \tag{1.11}$$

Subtracting $(cb)/a$ times the second row from the first, we see that

$$f\left(\begin{bmatrix} 1 & (cb)/a \\ 0 & 1 \end{bmatrix}\right) = f(I_{2\times 2}) = 1.$$

Combining this equation with (1.9) and (1.11) together with (1.10), we see that

$$f\left(\begin{bmatrix} a & b \\ c & d \end{bmatrix}\right) = ad - bc \tag{1.12}$$

whenever $a \neq 0$.

The remaining case $a = 0$ and $c \neq 0$ is reduced to the case $a \neq 0$ by swapping rows,* so the properties (1), (2), and (3) imply that in all cases,

$$f\left(\begin{bmatrix} a & b \\ c & d \end{bmatrix}\right) = ad - bc = \det\left(\begin{bmatrix} a & b \\ c & d \end{bmatrix}\right). \tag{1.13}$$

We have just deduced the formula for the 2×2 determinant from the list of properties (1), (2), and (3), while earlier we derived this list of properties from (1.1). *Hence these properties are equivalent to the formula, and the determinant function is the only function of 2×2 matrices with these properties.* This is the jumping-off point into higher dimensions.

1.2 Determinants of $n \times n$ matrices

We have just seen that for 2×2 matrices, the list of properties (1), (2), and (3) is essentially equivalent to the formula for the determinant—one can be derived from the other.

This equivalence is helpful, since it is not so clear how to make a *useful* generalization of the formula (1.1) even to 3×3 matrices. By a useful generalization, we mean one that can be used to check for invertibility at the very least. Some "obvious" analogs of (1.1) fail this test; see the exercises.

On the other hand, it is easy to generalize the list of properties (1), (2), and (3) to $n \times n$ matrices for arbitrary n, and doing so turns out to provide a determinant in higher dimensions that can be used to check for invertibility and to compute volume.

It is not obvious at this point that generalizing the properties (1), (2), and (3) to $n \times n$ matrices determines a computable function $f(A)$ for $n > 2$, much less that this function would

---

* Swapping rows switches a and c.

have any interesting algebraic or geometric properties for $n > 2$. But let us try this strategy out and see where it takes us.

DEFINITION

(Determinant function for $n \times n$ matrices) The determinant function on the set of $n \times n$ matrices is the uniquely determined function f on this set with the properties that

1. $f(A)$ changes sign when any two rows of A are interchanged.
2. $f(A)$ is linear in each row of A.
3. $f(I) = 1$.

We denote $f(A)$ by $\det(A)$.

For $n > 2$, this definition is "unjustified" as it stands. We do not yet know that there is *any* such function for $n \times n$ matrices, except when $n = 2$. Mathematical definitions are not allowed to refer to objects that do not exist. For example, we cannot define an infinite sequence $\{a_n\}$ by defining a_n to be the nth even prime number—the only even prime number is 2. So for all we know right now, two things might go wrong with this definition. There might be *no* function of $n \times n$ matrices with these properties for all n, or there might be *more than one,* in which case we could not talk about "the" function with these properties.

As we shall see, though, it all works out in the end. First, let us derive some *rules of computation* from the definition.

As in the 2×2 case, it will help to think of $\det(A)$ as a function of the rows of A. Let $\mathbf{r}_i$ denote the ith row of A so that

$$A = \begin{bmatrix} \mathbf{r}_1 \\ \mathbf{r}_2 \\ \vdots \\ \mathbf{r}_n \end{bmatrix}.$$

As in the 2×2 case, we write

$$\det(A) = \det(\mathbf{r}_1, \mathbf{r}_2, \ldots, \mathbf{r}_n).$$

If $\mathbf{r}_i = \mathbf{r}_j$ for any $i \neq j$ and B is obtained by swapping the i and jth rows, then $B = A$ since these rows are the same. Therefore, by property (1),

$$\det(A) = -\det(B) = -\det(A),$$

so $\det(A) = 0$, as before.

Likewise, if any row of A is all zeros, we can add any other row to it by property (2) without changing the value of the determinant. The resulting matrix has two identical rows, so its determinant is zero by the argument just made. Hence, if $\mathbf{r}_j = 0$ for any j, then $\det(A) = 0$.

- *If A has two identical rows or a row that is zero, $\det(A) = 0$.*

Now suppose that $k \neq j$ and we add α times the jth row to the kth. By the linearity property,

$$\det(\mathbf{r}_1, \ldots \mathbf{r}_j, \ldots, \mathbf{r}_k + \alpha \mathbf{r}_j \ldots, \mathbf{r}_n) = \det(\mathbf{r}_1, \ldots \mathbf{r}_j, \ldots, \mathbf{r}_k \ldots, \mathbf{r}_n)$$
$$+ \alpha \det(\mathbf{r}_1, \ldots \mathbf{r}_j, \ldots, \mathbf{r}_j \ldots, \mathbf{r}_n).$$

By what we just saw, $\det(\mathbf{r}_1, \ldots \mathbf{r}_j, \ldots, \mathbf{r}_j \ldots, \mathbf{r}_n) = 0$, so

$$\det(\mathbf{r}_1, \ldots \mathbf{r}_j, \ldots, \mathbf{r}_k + \alpha \mathbf{r}_j \ldots, \mathbf{r}_n) = \det(\mathbf{r}_1, \ldots \mathbf{r}_j, \ldots, \mathbf{r}_k \ldots, \mathbf{r}_n).$$

In other words,

- *Adding a multiple of one row of A to another row does not change the value of* $\det(A)$.

We can use this *invariance property* of the determinant to compute $\det(A)$.

EXAMPLE 4   **(Row reduction to echelon form and determinants)** Consider the Van der Monde matrix

$$A = \begin{bmatrix} 1 & 1 & 1 \\ 1 & 2 & 4 \\ 1 & 3 & 9 \end{bmatrix}.$$

Subtracting the first row from the second and third, we get

$$\begin{bmatrix} 1 & 1 & 1 \\ 0 & 1 & 3 \\ 0 & 2 & 8 \end{bmatrix}.$$

This does not change the determinant. Subtracting twice the second row from the third takes us to row-reduced form, and we still have not affected the value of the determinant:

$$\det(A) = \begin{bmatrix} 1 & 1 & 1 \\ 0 & 1 & 3 \\ 0 & 0 & 2 \end{bmatrix}. \tag{1.14}$$

Multiplying the final row through by $1/2$, we get, using the linearity in the rows,

$$\frac{1}{2}\det(A) = \begin{bmatrix} 1 & 1 & 1 \\ 0 & 1 & 3 \\ 0 & 0 & 1 \end{bmatrix}.$$

Three more row operations—all of them adding multiples of one row to another—"clean out" the upper-right entries of

$$\begin{bmatrix} 1 & 1 & 1 \\ 0 & 1 & 3 \\ 0 & 0 & 1 \end{bmatrix},$$

giving the identity. Since these operations do not affect the value of the determinant,

$$\det\left(\begin{bmatrix} 1 & 1 & 1 \\ 0 & 1 & 3 \\ 0 & 0 & 1 \end{bmatrix}\right) = \det(I_{3\times 3}) = 1.$$

Therefore, $\frac{1}{2}\det(A) = 1$, or, in other words,

$$\det(A) = 2.$$

This first computation was not as efficient as it could have been. To compute more efficiently and easily, we need one more observation.

Recall that a square matrix A is *upper triangular* if every entry below the diagonal is zero. The key observation is that

- *If A is an upper-triangular matrix, then* $\det(A)$ *is the product of the diagonal entries of A.*

Going back to the matrix on the right in (1.14), we see right away that $\det(A) = 2$. In the next example, we give another check on the claim we have made for upper-triangular matrices.

EXAMPLE 5   **(Determinant of an upper-triangular matrix)** Consider the matrix

$$A = \begin{bmatrix} 2 & 6 & 4 \\ 0 & 3 & 6 \\ 0 & 0 & 4 \end{bmatrix}.$$

Then by linearity in the rows and by property (3),

$$\det\left(\begin{bmatrix} 2 & 6 & 4 \\ 0 & 3 & 6 \\ 0 & 0 & 4 \end{bmatrix}\right) = 2\det\left(\begin{bmatrix} 1 & 3 & 2 \\ 0 & 3 & 6 \\ 0 & 0 & 4 \end{bmatrix}\right) = 6\det\left(\begin{bmatrix} 1 & 3 & 2 \\ 0 & 1 & 2 \\ 0 & 0 & 4 \end{bmatrix}\right) = 24\det\left(\begin{bmatrix} 1 & 3 & 2 \\ 0 & 1 & 2 \\ 0 & 0 & 1 \end{bmatrix}\right) = 24.$$

Now we can reduce

$$\begin{bmatrix} 1 & 3 & 2 \\ 0 & 1 & 2 \\ 0 & 0 & 1 \end{bmatrix}$$

to the identity just by adding multiples of rows to other rows. Adding -2 times the third row to the second and then -2 times the third row to the first, we get

$$\begin{bmatrix} 1 & 3 & 0 \\ 0 & 1 & 0 \\ 0 & 0 & 1 \end{bmatrix}.$$

Next, adding -3 times the second row to the first, we get the identity. Since none of these operations affects the value of the determinant,

$$\det\left(\begin{bmatrix} 1 & 3 & 0 \\ 0 & 1 & 0 \\ 0 & 0 & 1 \end{bmatrix}\right) = \det(I_{3\times 3}) = 1.$$

Therefore,

$$\det(A) = 24,$$

which is the product of the diagonal entries.

To see why the claim is true in general, proceed as in Example 5. Consider any $n \times n$ upper-triangular matrix A. If any diagonal entry is zero, then A has fewer than n pivots. If we row reduce—using row operations that do not affect the determinant—we get a matrix that has at least one zero row at the bottom. Hence, if $A_{i,i} = 0$ for any i, then $\det(A) = 0$.

Next, suppose that $A_{i,i} \neq 0$ for any i. Using the linearity property, we can divide the ith row through by $A_{i,i}$. Doing this for each i, we get a *unit upper-triangular* matrix B. That is, B is upper triangular, and every diagonal entry is 1.

By the linearity property of the determinant,

$$\det(A) = \left(\prod_{i=1}^{n} A_{i,i}\right)\det(B).$$

But we can clearly reduce any unit upper-triangular matrix B to the identity just by adding multiples of rows to other rows, as in Example 5. None of these operations affects the determinant, so $\det(B) = \det(I) = 1$. Thus,

$$\det(A) = \prod_{i=1}^{n} A_{i,i} \qquad \text{for any } n \times n \text{ upper-triangular matrix.} \tag{1.15}$$

EXAMPLE 6   **(Determinant of an upper-triangular matrix the fast way)** Consider

$$A = \begin{bmatrix} 1 & 2 & 3 \\ 0 & 9 & 3 \\ 0 & 0 & 2 \end{bmatrix}.$$

By (1.15) we conclude immediately that $\det(A) = 18$.

We now have a good strategy for computing determinants:

- **(Row-reduction method for computing determinants)** *Let A be an $n \times n$ matrix. Row reduce to the upper-triangular form B, keeping track of row swaps. If N row swaps are used, then*

$$\det(A) = (-1)^N \det(B). \tag{1.16}$$

Then use (1.15) to evaluate $\det(B)$.

EXAMPLE 7   **(Determinants via row operations)** Consider the matrix A from Example 4. Three simple row operations (two to "clean" out the first column, one for the second) reduce it to upper triangular with no swapping of rows:

$$\begin{bmatrix} 1 & 1 & 1 \\ 1 & 2 & 4 \\ 1 & 3 & 9 \end{bmatrix} \rightarrow \begin{bmatrix} 1 & 1 & 1 \\ 0 & 1 & 3 \\ 0 & 2 & 8 \end{bmatrix} \rightarrow \begin{bmatrix} 1 & 1 & 1 \\ 0 & 1 & 3 \\ 0 & 0 & 2 \end{bmatrix}.$$

From the right-hand side, we set $\det(A) = 2$, as we found in Example 4.

EXAMPLE 8   **(Another determinant via row operations)** Consider

$$A = \begin{bmatrix} 3 & 1 & 0 & 1 \\ 1 & 3 & 1 & 0 \\ 0 & 1 & 3 & 1 \\ 1 & 0 & 1 & 3 \end{bmatrix}.$$

If we swap rows 1 and 4 and then swap rows 2 and 3, the result is

$$\begin{bmatrix} 1 & 0 & 1 & 3 \\ 0 & 1 & 3 & 1 \\ 1 & 3 & 1 & 0 \\ 3 & 1 & 0 & 1 \end{bmatrix}.$$

The good thing is that the first two rows are now in upper-triangular form. And since we used two row swaps, we did not change the sign of the determinant. Now simple row operations "clean out" the first two columns, resulting in the matrix

$$\begin{bmatrix} 1 & 0 & 1 & 3 \\ 0 & 1 & 3 & 1 \\ 0 & 0 & -9 & -6 \\ 0 & 0 & -6 & -9 \end{bmatrix}.$$

None of these row operations changes the value of the determinant, so

$$\det(A) = \det\left(\begin{bmatrix} 1 & 0 & 1 & 3 \\ 0 & 1 & 3 & 1 \\ 0 & 0 & -9 & -6 \\ 0 & 0 & -6 & -9 \end{bmatrix}\right).$$

Now multiply the third row through by -6 and the last row through by -9 to get

$$54\det(A) = \det\left(\begin{bmatrix} 1 & 0 & 1 & 3 \\ 0 & 1 & 3 & 1 \\ 0 & 0 & 54 & 36 \\ 0 & 0 & 54 & 81 \end{bmatrix}\right).$$

Now, subtracting the third row from the fourth,

$$54\det(A) = \det\left(\begin{bmatrix} 1 & 0 & 1 & 3 \\ 0 & 1 & 3 & 1 \\ 0 & 0 & 54 & 36 \\ 0 & 0 & 0 & 45 \end{bmatrix}\right).$$

Therefore, using (1.15), $54\det(A) = 1 \cdot 1 \cdot 54 \cdot 45$, or $\det(A) = 45$.

Now that we know how to compute determinants, we return to a key question: What does the number $\det(A)$ tell us about A? Here is a first answer, which shows we are on the right track:

- *An $n \times n$ matrix A is invertible if and only if $\det(A) \neq 0$.*

This is an easy consequence of (1.16): Row reduce A to upper row-reduced form. The result is upper triangular, and the diagonal entries are all nonzero if and only if there are n pivots, which is the case if and only if A is invertible.

We summarize what we have learned so far in this section.

THEOREM 1   (**Determinants and row reduction**) *Let A be an $n \times n$ matrix, and let B be an upper-triangular matrix obtained from A by repeatedly adding multiples of one row to another and swapping pairs of rows. Suppose that there were N interchanges of pairs of rows used. Then*

$$\det(A) = (-1)^N \left(B_{1,1} B_{2,2} \cdots B_{n,n} \right). \tag{1.17}$$

Moreover, A is invertible if and only if $\det(A) \neq 0$.

We are now in a position to partially justify our definition. We see that there is *at most* one function f possessing the properties we used to define the determinant because just using these properties, we can use row reduction to determine the value of $f(A)$ for any A. That is, if f and g are two such functions and A is any $n \times n$ matrix, then we could use the exact same sequence of row operations for both f and g, and we would arrive at the exact same result. Hence, we would have $f(A) = g(A)$. Therefore, there cannot be two different functions with these properties.

However, even though we have learned how to compute $\det(A)$, we do not yet know that there is any function at all with the properties that we used to define the determinant, except, of course, when $n = 2$.

The point is that a function, to be a function, must return exactly one output for each input on which it is defined. But there are choices to be made in doing a row reduction. If different ways of doing the row reduction led to different values of output, our definition would not define a function at all. We have run into this issue before, for example with the definition of *rank*, which is a numerical valued function on the set of matrices that we compute by row reduction. Everything will work out, as before, though the reasons are a bit more intricate this time around.

EXAMPLE 9   (**Different row reductions**) Consider the matrix A from Example 7. First, swapping rows 1 and 2 and then subtracting multiples of rows from rows below,

$$\begin{bmatrix} 1 & 1 & 1 \\ 1 & 2 & 4 \\ 1 & 3 & 9 \end{bmatrix} \to \begin{bmatrix} 1 & 1 & 1 \\ 1 & 3 & 9 \\ 1 & 2 & 4 \end{bmatrix} \to \begin{bmatrix} 1 & 1 & 1 \\ 0 & 2 & 8 \\ 0 & 1 & 3 \end{bmatrix} \to \begin{bmatrix} 1 & 1 & 1 \\ 0 & 2 & 8 \\ 0 & 0 & -1 \end{bmatrix}.$$

Since we used one row swap,

$$\det(A) = \det \left(\begin{bmatrix} 1 & 1 & 1 \\ 0 & 2 & 8 \\ 0 & 0 & -1 \end{bmatrix} \right) = -(-2) = 2,$$

as we found in Example 7.

The two row reductions were different, but the final answer was the same. If it were not, we would have found a counterexample to the existence of the determinant function as we have defined it.

The differences in the upper-triangular matrices produced by different choices in the row reduction can be even more striking. Consider

$$A = \begin{bmatrix} 1 & 5 & 2 \\ 2 & 3 & 4 \\ 3 & 1 & 3 \end{bmatrix}.$$

Then, just adding multiples of one row to another,

$$\begin{bmatrix} 1 & 5 & 2 \\ 2 & 3 & 4 \\ 3 & 1 & 3 \end{bmatrix} \rightarrow \begin{bmatrix} 1 & 5 & 2 \\ 0 & -7 & 0 \\ 0 & -14 & -3 \end{bmatrix} \rightarrow \begin{bmatrix} 1 & 5 & 2 \\ 0 & -7 & 0 \\ 0 & 0 & -3 \end{bmatrix}.$$

We conclude that $\det(A) = 1 \cdot (-7) \cdot (-3) = 21$.

Next, swap the second and third rows:

$$\begin{bmatrix} 1 & 5 & 2 \\ 2 & 3 & 4 \\ 3 & 1 & 3 \end{bmatrix} \rightarrow \begin{bmatrix} 1 & 5 & 2 \\ 3 & 1 & 3 \\ 2 & 3 & 4 \end{bmatrix}.$$

Then, just adding multiples of one row to another,

$$\begin{bmatrix} 1 & 5 & 2 \\ 3 & 1 & 3 \\ 2 & 3 & 4 \end{bmatrix} \rightarrow \begin{bmatrix} 1 & 5 & 2 \\ 0 & -14 & -3 \\ 0 & -7 & 0 \end{bmatrix} \rightarrow \begin{bmatrix} 1 & 5 & 2 \\ 0 & -14 & -3 \\ 0 & 0 & 3/2 \end{bmatrix}.$$

Since there was one row swap, we conclude that

$$\det(A) = (-1)^1 \cdot 1 \cdot (-14) \cdot (3/2) = 21.$$

The row reductions are different, but both give the same value for the determinant.

We claim that things always work out like they did in Example 9: No matter how you go about row reducing any $n \times n$ matrix A to upper-triangular form B using only row swaps and adding multiples of one row to another, you always find exactly the same value for $(-1)^N (B_{1,1}B_{2,2} \cdots B_{n,n})$, even though the possible row reductions can themselves be quite different.

1.3 The determinant and block matrices

In Example 8 we encountered the 4×4 matrix

$$\begin{bmatrix} 1 & 0 & 1 & 3 \\ 0 & 1 & 3 & 1 \\ 0 & 0 & -9 & -6 \\ 0 & 0 & -6 & -9 \end{bmatrix}.$$

This matrix can be considered as a *block matrix*

$$\begin{bmatrix} A & B \\ C & D \end{bmatrix} \tag{1.18}$$

where A, B, C, and D are the 2×2 matrices

$$A = \begin{bmatrix} 1 & 0 \\ 0 & 1 \end{bmatrix}, \qquad B = \begin{bmatrix} 1 & 3 \\ 3 & 1 \end{bmatrix}, \qquad C = \begin{bmatrix} 0 & 0 \\ 0 & 0 \end{bmatrix}, \qquad \text{and} \qquad D = \begin{bmatrix} -9 & -6 \\ -6 & -9 \end{bmatrix}.$$

Notice that $C = 0$, so the block matrix in (1.18) is upper triangular. In this case, the determinant is the product of the determinants of the diagonal blocks:

$$\det\left(\begin{bmatrix} 1 & 0 & 1 & 3 \\ 0 & 1 & 3 & 1 \\ 0 & 0 & -9 & -6 \\ 0 & 0 & -6 & -9 \end{bmatrix}\right) = \det\left(\begin{bmatrix} 1 & 0 \\ 0 & 1 \end{bmatrix}\right) \det\left(\begin{bmatrix} -9 & -6 \\ -6 & -9 \end{bmatrix}\right) = 1 \cdot 45,$$

as we found in Example 8.

This result is general, and it is very useful: You do not need to row reduce to clean out everything below the diagonal. If you "clean out" the lower-left corner, the problem splits into two much simpler problems!

In general, an $(m + n) \times (m + n)$ block matrix is a matrix of the type $\begin{bmatrix} A & B \\ C & D \end{bmatrix}$, where A is $m \times m$, D is $n \times n$, B is $m \times n$, and C is $n \times m$. Matrix (1.18) is an example with $m = n = 2$.

THEOREM 2   **(Determinants and block matrices)** *Let*

$$\begin{bmatrix} A & B \\ 0 & D \end{bmatrix}$$

be an $(m + n) \times (m + n)$ block matrix whose lower-left block is the $n \times m$ zero matrix. Then

$$\det\left(\begin{bmatrix} A & B \\ 0 & D \end{bmatrix}\right) = \det(A)\det(D).$$

PROOF: To compute

$$\det\left(\begin{bmatrix} A & B \\ 0 & D \end{bmatrix}\right),$$

we row reduce from left to right as usual. Since the first m columns are zero below the mth row, the row operations affect only the first m rows when we "clean out" the first m columns. So, after doing that, A has become an upper-triangular matrix T, B has been changed into some other $m \times n$ matrix, and the bottom n rows are untouched. We now have

$$\begin{bmatrix} T & \tilde{B} \\ 0 & D \end{bmatrix}.$$

The key observation is that our choice of row operations up to now has depended only on A. Any sequence of row operations that row reduces A itself will get us to this point.

To finish the row reduction, we need to work on only D. That is, we proceed with row operations on the last n rows until D has become an upper-triangular matrix U. We then have

$$\begin{bmatrix} T & \tilde{B} \\ 0 & U \end{bmatrix}.$$

This matrix is an upper-triangular matrix, so its determinant is the product of the diagonal elements. But these are the diagonal elements of T and U, so

$$\det\left(\begin{bmatrix} T & \tilde{B} \\ 0 & U \end{bmatrix}\right) = \det(T)\det(U).$$

Now T is a row reduction of A to upper-triangular form, and if we used M row swaps in the reduction, we would have

$$\det(A) = (-1)^M \det(T).$$

Likewise, if we used N row swaps in reducing D to U, we would have

$$\det(D) = (-1)^N \det(U).$$

But then

$$\begin{bmatrix} A & B \\ 0 & D \end{bmatrix}$$

row reduces to

$$\begin{bmatrix} T & \tilde{B} \\ 0 & U \end{bmatrix}$$

with $M + N$ row swaps, so

$$\det \left(\begin{bmatrix} A & B \\ 0 & D \end{bmatrix} \right) = (-1)^{M+N} \det \left(\begin{bmatrix} T & \tilde{B} \\ 0 & U \end{bmatrix} \right)$$

$$= (-1)^M \det(T)(-1)^N \det(U)$$

$$= \det(A)\det(D).$$

∎

Exercises

1.1 Compute the determinant of $A = \begin{bmatrix} 2 & 1 & 1 \\ 1 & 2 & 1 \\ 1 & 1 & 2 \end{bmatrix}$.

1.2 Compute the determinant of $A = \begin{bmatrix} 3 & 2 & 1 \\ 1 & 3 & 2 \\ 2 & 1 & 3 \end{bmatrix}$.

1.3 Compute the determinant of $A = \begin{bmatrix} 1 & 2 & 4 & 8 \\ 1 & -2 & 4 & -8 \\ 1 & 3 & 9 & 27 \\ 1 & -3 & 9 & -27 \end{bmatrix}$.

1.4 Compute the determinant of $A = \begin{bmatrix} a & 2 & 0 & 0 \\ 2 & a & 2 & 0 \\ 0 & 2 & a & 2 \\ 0 & 0 & 2 & a \end{bmatrix}$. For which values of a is A invertible?

1.5 Compute the determinant of $A = \begin{bmatrix} 0 & 0 & 2 & a \\ 0 & 2 & a & 2 \\ 2 & a & 2 & 0 \\ a & 2 & 0 & 0 \end{bmatrix}$. For which values of a is A invertible?

Exercises on block matrices

In the following exercises, consider the $(2n) \times (2n)$ block matrix $\begin{bmatrix} A & B \\ C & D \end{bmatrix}$, where A, B, C, and D are $n \times n$ matrices. In the next few exercises, we will see that using Theorem 2 enables us to do the row reduction for computing determinants in a very efficient way. We will see that the block matrix point of view can be quite helpful in computation.

1.6 Consider the case in which $C = 0$. Show that $\begin{bmatrix} A & B \\ 0 & D \end{bmatrix}$ is invertible if and only if both A and D are invertible and that in this case,

$$\begin{bmatrix} A & B \\ 0 & D \end{bmatrix}^{-1} = \begin{bmatrix} A^{-1} & -A^{-1}BD^{-1} \\ 0 & D^{-1} \end{bmatrix}.$$

1.7 (a) Show that if A is invertible,

$$\begin{bmatrix} I & 0 \\ -CA^{-1} & I \end{bmatrix} \begin{bmatrix} A & B \\ C & D \end{bmatrix} = \begin{bmatrix} A & B \\ 0 & F \end{bmatrix},$$

where $F = D - CA^{-1}B$.

(b) Use part (a) to show that if A is invertible; then

$$\det\left(\begin{bmatrix} A & B \\ C & D \end{bmatrix}\right) = \det(A)\det(D - CA^{-1}B).$$

(c) The 4×4 matrices in Exercises 1.3 and 1.4 can be viewed as a block matrix of 2×2 matrices. Use part (b) to compute the determinants of the matrices in these exercises.

(d) Must the square matrices A and D in $\begin{bmatrix} A & B \\ C & D \end{bmatrix}$ necessarily have the same size? Can you generalize to the case in which A is $n \times n$ and D is $m \times m$?

1.8 Use part (a) of the previous exercise to show that provided A and $F = D - CA^{-1}B$ are invertible, so is $\begin{bmatrix} A & B \\ C & D \end{bmatrix}$, and

$$\begin{bmatrix} A & B \\ C & D \end{bmatrix}^{-1} = \begin{bmatrix} A^{-1} & 0 \\ -A^{-1}BF^{-1} & F^{-1} \end{bmatrix} \begin{bmatrix} I & 0 \\ -CA^{-1} & I \end{bmatrix}.$$

Use this equation to compute the inverse of the matrix in Exercise 1.4 for all values of a for which it is invertible.

SECTION 2 | Properties of Determinants

2.1 The product formula

In Section 1 of this chapter, we defined the determinant function in terms of three properties we required it to have. As we saw there, these properties are as good as a formula—using them we can compute the determinant of any square matrix. Since these properties are all we need to evaluate the determinant, there can be at most one function on the set of square matrices that has these properties: If f and g were any two such functions, we could use the properties to evaluate $f(A)$ and $g(A)$. Using the same properties in the same way, we would find $f(A) = g(A)$, so the two functions would be the same. Thus, there is at most one function on the set of square matrices that has all three properties. We summarize this in the following theorem.

THEOREM 3   **(Uniqueness of the determinant)** *Let f be any numerically valued function on the $n \times n$ matrices with the properties that*

1.   *$f(A)$ changes sign when any two rows of A are interchanged.*

2.   *$f(A)$ is linear in each row of A.*

3.   *$f(I) = 1$.*

Then $f(A) = \det(A)$.

This theorem has an important consequence. For any two $n \times n$ matrices A and B, $\det(AB) = \det(A)\det(B)$.

To see it, fix any invertible $n \times n$ matrix B and consider the function $f_B(A)$ defined by

$$f_B(A) = \frac{1}{\det(B)} \det(AB).$$

From **V.I.F.4**,

$$(AB)_{i,j} = (\text{row } i \text{ of } A) \cdot (\text{col } j \text{ of } B), \tag{2.1}$$

we see that every entry in the ith row of AB is a linear combination of the entries in the ith row of A.

We then see two things: First, swapping two rows of A swaps the same two rows of AB. Second, since the composition of linear functions is a linear function, $f_B(A) = \det(AB)$ is a linear function of the rows of A. Also, it is evident that $f_B(I) = \det(B)/\det(B) = 1$. Then from Theorem 3 we have $f_B(A) = \det(A)$, so

$$\det(A) = f_B(A) = \frac{1}{\det(B)} \det(AB).$$

Therefore, $\det(AB) = \det(A)\det(B)$. If B is not invertible, then neither is AB, so $\det(AB) = \det(B) = 0$ and $\det(AB) = \det(A)\det(B)$ is still true. Taking $B = A^{-1}$, we get $1 = \det(I) = \det(A)\det(A^{-1})$ so that $\det(A^{-1}) = 1/\det(A)$. We summarize.

THEOREM 4   **(Product property of the determinant)** *For any two $n \times n$ matrices A and B,*

$$\det(AB) = \det(A)\det(B), \tag{2.2}$$

and if A is invertible,

$$\det(A^{-1}) = (\det(A))^{-1}. \tag{2.3}$$

EXAMPLE 10   **(The determinant of a product)** Let us consider the following matrices:

$$A = \begin{bmatrix} 1 & 2 & 3 & 4 \\ 0 & 1 & 2 & 1 \\ 0 & 0 & 5 & 1 \\ 0 & 0 & 0 & 0 \end{bmatrix} \qquad B = \begin{bmatrix} 1 & 2 & 1 & 4 \\ 0 & 1 & 7 & 1 \\ 0 & 0 & 1 & 6 \\ 0 & 0 & 0 & 7 \end{bmatrix}.$$

Using the previous theorems, we can immediately say $\det(AB) = 0$.

The product property has many important consequences. Here is the first one: For any $n \times n$ matrix A,

$$\det(A^t) = \det(A). \tag{2.4}$$

This identity is not so easy to see directly from the definition, but combining the product property with things we learned in Chapter 2, we can give a fairly simple indirect proof.

First, consider a special case. We saw in the last section that for any $n \times n$ upper-triangular matrix A, $\det(A)$ is simply the product of the diagonal elements of A:

$$\det(A) = \prod_{i=1}^{n} A_{i,i}.$$

This equation is just (1.15), and the proof that we gave there is easily adapted to apply to *lower-triangular matrices* as well. The proof is left as an exercise.

Now, the transpose of an upper-triangular matrix is lower triangular, and vice versa, but the diagonals are the same. Thus,

- *Whenever A is upper triangular or lower triangular, $\det(A^t) = \det(A)$.*

Moving beyond the special case of triangular matrices, recall from Section 2.5 of Chapter 3 that whenever it is possible to row reduce an $n \times n$ matrix A without swapping rows, we can factor A as a product

$$A = LU,$$

where L is unit lower triangular and U is row reduced and hence upper triangular since it is square. By the product property and what we just saw,

$$\det(A) = \det(L)\det(U) = \det(U^t)\det(L^t) = \det(U^t L^t) = \det(A^t).$$

This proves (2.4) in this case, which is already quite broad.

In general, though, we have not $A = LU$ but rather

$$PA = LU$$

where P is an $n \times n$ permutation matrix. In this case, by the product property,

$$\det(P)\det(A) = \det(L)\det(U). \tag{2.5}$$

But from $PA = LU$,

$$A^t P^t = U^t L^t.$$

Applying the product property and what we have seen earlier about triangular matrices,

$$\det(A^t)\det(P^t) = \det(U^t)\det(L^t) = \det(U)\det(L). \tag{2.6}$$

Comparing (2.5) and (2.6), we see that

$$\det(P^t)\det(A^t) = \det(P)\det(A). \tag{2.7}$$

We also know from Section 5.4 of Chapter 2 that every permutation matrix is invertible with $P = P^t$, so neither $\det(P)$ nor $\det(P^t)$ is zero. Therefore, if we can show that $\det(P) = \det(P^t)$ for every permutation matrix P, then we can cancel $\det(P) = \det(P^t)$ from both sides in (2.7) and conclude that $\det(A^t) = \det(A)$.

Therefore, all that remains to be done to prove (2.4) in general is to prove it for every permutation matrix P. To do so, we again consider a special case. Suppose P is the permutation matrix obtained from $I_{n \times n}$ by swapping rows i and j.

DEFINITION

(Pair-permutation matrix) An $n \times n$ matrix obtained from $I_{n \times n}$ by swapping rows i and j for some $i \neq j$ is called a *pair-permutation matrix*.

As explained in Section 5.4 of Chapter 2, any permutation P is invertible and P^t is the inverse of P. Therefore, when P is a pair permutation, $P^t = P^{-1}$. But if rows i and j an swapped twice, they end up in their original places, so for a pair permutation P, $P^2 = I$, so $P = P^{-1} = P^t$. Clearly, then, for a pair-permutation matrix P,

$$\det(P^t) = \det(P). \tag{2.8}$$

To conclude the argument, we need to observe only that every permutation matrix P can be written as a product

$$P = P_k P_{k-1} \cdots P_1$$

of k pair-permutation matrices for some $k \leq n$. Indeed, since P is just the identity matrix with the row reordered, we can obtain P from the identity by swapping a pair of rows to get the first row in the right place (if it is not already) and then swap again to get the second row in the right place (if it is not already) and so on, finishing in at most n steps. Let P_j denote the pair permutation that does the swapping at the jth step. Then $P = P_k P_{k-1} \cdots P_1 I$.

Since each P_j is a pair-permutation matrix,

$$
\begin{aligned}
\det(P^t) &= \det((P_k P_{k-1} \cdots P_1)^t) \\
&= \det(P_1^t \cdots P_{k-1}^t P_k^t) \\
&= \det(P_1 \cdots P_{k-1} P_k) \\
&= \det(P_1) \cdots \det(P_{k-1})\det(P_k) \\
&= \det(P_k)\det(P_{k-1}) \cdots \det(P_k) \\
&= \det(P_k P_{k-1} \cdots P_1) \\
&= \det(P),
\end{aligned}
$$

which proves (2.8) for general permutations and finally (2.4) for general $n \times n$ matrices.

We can apply (2.4) to orthogonal matrices. Recall that an $n \times n$ matrix U is orthogonal if and only if $U^t U = I$. In particular, every permutation matrix is an orthogonal matrix.

If U is orthogonal, then

$$1 = \det(U^t U) = \det(U^t)\det(U) = (\det(U))^2.$$

The only numbers satisfying the equation $x^2 = 1$ are $x = 1$ and $x = -1$, so if U is orthogonal, $\det(U) = \pm 1$.

We summarize.

THEOREM 5   *For any $n \times n$ matrix A,*

$$\det(A^t) = \det(A). \tag{2.9}$$

Consequently, if U is an $n \times n$ orthogonal matrix, then $\det(U) = \pm 1$.

Here is another important consequence of (2.9): Since the rows of A^t are the columns of A, $\det(A) = \det(A^t)$ is a linear function of the columns of A. This consequence deserves emphasis.

* *The determinant function is linear in the columns as well as the rows.*

To see how this fact can be used, consider the general 3×3 matrix

$$A = \begin{bmatrix} A_{1,1} & A_{1,2} & A_{1,3} \\ A_{2,1} & A_{2,2} & A_{2,3} \\ A_{3,1} & A_{3,2} & A_{3,3} \end{bmatrix}. \tag{2.10}$$

The first column is

$$\begin{bmatrix} A_{1,1} \\ A_{2,1} \\ A_{3,1} \end{bmatrix} = A_{1,1}\mathbf{e}_1 + A_{2,1}\mathbf{e}_2 + A_{3,1}\mathbf{e}_3.$$

Therefore, by linearity in the columns,

$$\det(A) = A_{1,1}\det\left(\begin{bmatrix} 1 & A_{1,2} & A_{1,3} \\ 0 & A_{2,2} & A_{2,3} \\ 0 & A_{3,2} & A_{3,3} \end{bmatrix}\right) + A_{2,1}\det\left(\begin{bmatrix} 0 & A_{1,2} & A_{1,3} \\ 1 & A_{2,2} & A_{2,3} \\ 0 & A_{3,2} & A_{3,3} \end{bmatrix}\right)$$

$$+ A_{3,1}\det\left(\begin{bmatrix} 0 & A_{1,2} & A_{1,3} \\ 0 & A_{2,2} & A_{2,3} \\ 1 & A_{3,2} & A_{3,3} \end{bmatrix}\right).$$

Now

$$\begin{bmatrix} 1 & A_{1,2} & A_{1,3} \\ 0 & A_{2,2} & A_{2,3} \\ 0 & A_{3,2} & A_{3,3} \end{bmatrix}$$

is a block matrix, and we can easily compute its determinant using Theorem 2:

$$\det\left(\begin{bmatrix} 1 & A_{1,2} & A_{1,3} \\ 0 & A_{2,2} & A_{2,3} \\ 0 & A_{3,2} & A_{3,3} \end{bmatrix}\right) = A_{2,2}A_{3,3} - A_{2,3}A_{3,2}.$$

The other two matrices are not block diagonal, but with row swaps we can make them block diagonal and then compute them. The result is a formula for $\det(A)$:

$$\begin{aligned} \det(A) = \; &A_{1,1}A_{2,2}A_{3,3} + A_{2,1}A_{3,2}A_{1,3} + A_{3,1}A_{1,2}A_{2,3} \\ &- A_{2,1}A_{1,2}A_{3,3} - A_{1,1}A_{3,2}A_{2,3} - A_{3,1}A_{2,2}A_{1,3}. \end{aligned} \tag{2.11}$$

This is the analog of the formula

$$\det\left(\begin{bmatrix} a & b \\ c & d \end{bmatrix}\right) = ad - bc.$$

As you see, it is quite a lot more complicated. Now that we have a formula for the 3×3 case, we could check that it does indeed have the properties we require, so we are in a position to justify our definition in the 3×3 case. However, that will become still easier later on, so we postpone doing so for now.

It is easier to use (2.11) if we introduce a better notation. Let us think of A as the matrix whose first row is $\mathbf{a}$, whose second row is $\mathbf{b}$, and whose third row is $\mathbf{c}$:

$$A = \begin{bmatrix} \mathbf{a} \\ \mathbf{b} \\ \mathbf{c} \end{bmatrix} = \begin{bmatrix} a_1 & a_2 & a_3 \\ b_1 & b_2 & b_3 \\ c_1 & c_2 & c_3 \end{bmatrix}.$$

In this notation, the result becomes

$$\det(A) = a_1 b_2 c_3 + b_1 c_2 a_3 + c_1 a_2 b_3 - b_1 a_2 c_3 - a_1 c_2 b_3 - c_1 b_2 a_3. \tag{2.12}$$

There are several simple ways to remember this. Here is one: Form the array

$$\begin{array}{ccc} a_1 & a_2 & a_3 \\ b_1 & b_2 & b_3 \\ c_1 & c_2 & c_3 \\ a_1 & a_2 & a_3 \\ b_1 & b_2 & b_3 \end{array}$$

where we have just repeated the first two rows at the bottom. Now take the products of the terms on the "diagonals" indicated by the arrows:

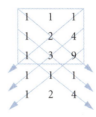

$$
\begin{array}{ccc}
a_1 & a_2 & a_3 \\
b_1 & b_2 & b_3 \\
c_1 & c_2 & c_3 \\
a_1 & a_2 & a_3 \\
b_1 & b_2 & b_3
\end{array}
$$

The three products entering (2.12) with a positive sign are the products of the terms on the three diagonals going down and to the right starting from the a_1, b_1, and c_1, respectively. We call these the "positive diagonals." The three products entering (2.12) with a negative sign are the products of the terms on the three diagonals going down and to the left starting from the a_3, b_3, and c_3, respectively. We call these the "negative diagonals."

EXAMPLE 11 **(A 3×3 determinant)** For example, consider the matrix

$$
A = \begin{bmatrix} 1 & 1 & 1 \\ 1 & 2 & 4 \\ 1 & 3 & 9 \end{bmatrix}, \tag{2.13}
$$

and let us use this device to compute its determinant. Form the array

$$
\begin{array}{ccc}
1 & 1 & 1 \\
1 & 2 & 4 \\
1 & 3 & 9 \\
1 & 1 & 1 \\
1 & 2 & 4
\end{array}
$$

Add the products of the terms on the positive diagonals, and subtract the products of the terms on the negative diagonals. The result is

$$
\det(A) = 18 + 3 + 4 - 2 - 12 - 9 = 2.
$$

The method we used to reduce the computation of a 3×3 determinant to the computation of three 2×2 determinants could be used to reduce the computation of a 4×4 determinant by using linearity in the first column to reduce to the computation of four 3×3 matrices. Since the formula (2.11) for 3×3 determinants has 6 terms, the corresponding formula would have 24 terms. But there are only 8 "diagonals" in a 4×4 matrix, so the "diagonals method" used in Example 11 *does not* generalize to larger matrices.

Still, we can use expansion in the first column—or even any other column or row—to reduce the computation of the determinant of an $n \times n$ matrix to the computation of n determinants of $(n - 1) \times (n - 1)$ matrices, and by recursion, we can work our way down to $n!/2$ determinants of 2×2 matrices. This is called the *Laplace expansion method*. It has uses (see Exercises 2.6 and 2.7), but since we can row reduce an $n \times n$ matrix using fewer than n^2 row operations and hence n^3 arithmetic operations, which is much less than $n!/2$ for large n, the expansion method is relatively inefficient in general.

2.2 Determinants, area, and volume

In Chapter 1 we saw, by direct calculation, that when A is any 2×2 matrix, the area of the image of the unit square under the linear transformation of $\mathbb{R}^2$ induced by A is $|\det(A)|$. We can now give a more sophisticated explanation that has the virtue of easily extending to higher dimensions. Here is how it goes in $\mathbb{R}^2$:

Let A be any 2×2 matrix. If $\text{rank}(A) = 0$, A is the zero matrix, and there is nothing to explain.

If $\text{rank}(A) = 1$, then $\det(A) = 0$, and $\text{Img}(A)$ is a line, so the image of the unit square under linear transformation of $\mathbb{R}^2$ induced by A is contained in this line, and its area is zero.

Therefore, we can suppose that $\text{rank}(A) = 2$. This means that the columns of A are linearly independent, so A has a QR factorization

$$A = QR$$

where Q is a 2×2 orthogonal matrix and R is an invertible 2×2 upper-triangular matrix. Hence, R has the form

$$R = \begin{bmatrix} a & b \\ 0 & d \end{bmatrix},$$

and neither a nor d is zero since R is invertible. Therefore, we can factor R as

$$R = TD$$

where

$$T = \begin{bmatrix} 1 & b/a \\ 0 & 1 \end{bmatrix} \quad \text{and} \quad D = \begin{bmatrix} a & 0 \\ 0 & d \end{bmatrix}.$$

This gives us

$$A = QTD.$$

Now, Q is orthogonal, so $|\det(Q)| = 1$ by Theorem 5. Also, T is upper triangular, and both of its diagonal entries are 1, so $\det(T) = 1$. Therefore,

$$|\det(A)| = |\det(QTD)| = |\det(Q)\det(T)\det(D)| = |\det(D)|.$$

The decomposition of A as the product of the three matrices Q, T, and D enables us to see what effect A has on the area of the unit square, since the linear transformations corresponding to each of Q, T, and D are easy to understand geometrically.

First of all, D simply stretches the unit square into the rectangle whose base has width $|a|$ height $|d|$, and area $|ad| = |\det(D)| = |\det(A)|$. Next we claim that neither T nor Q has any effect on the area of sets in the plane. For Q this claim is particularly clear. Since Q is an isometry, it does not change the distances or angles, and so it does not affect area: All it does is reflect or rotate.

It takes a bit more thought to see that T does not affect area. Let us consider an example. Suppose that

$$T = \begin{bmatrix} 1 & 1/2 \\ 0 & 1 \end{bmatrix}.$$

Then

$$T\mathbf{e}_1 = \mathbf{e}_1 + \frac{1}{2}\mathbf{e}_2 \quad \text{and} \quad T\mathbf{e}_2 = \mathbf{e}_2.$$

Therefore, the image of the unit square under T is

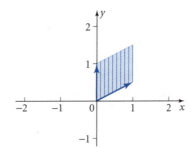

Any matrix of the type

$$\begin{bmatrix} 1 & \alpha \\ 0 & 1 \end{bmatrix}$$

for any number α is called a *shear transformation*. Look at the diagram to see why. If you "cut" the unit square into thin vertical strips—this is the "shearing"—and then "slide the strips upward" by an amount that is proportional to their distance from the y-axis, you can duplicate the effect of the transformation. This rearrangement of the strips has no effect on their total area, so the total area is the same after the transformation as it was before. Since the rearrangement has nothing to do with the particular value of α, we conclude that T has no effect on the area of the unit square.

Recall from Chapter 1 that a linear transformation magnifies* the area of any polygon by the same factor that it magnifies the area of the unit square. This number is called the *magnification factor* of a linear transformation. Since Q and T have magnification factor 1 and D has the magnification factor $|ad| = |\det(D)| = |\det(A)|$, the total magnification factor of $A = QTD$ is $1 \cdot 1 \cdot |\det(D)| = |\det(A)|$. In summary, we have the following theorem, which is essentially a restatement of Theorem 16 of Chapter 1.

THEOREM 6   **(Area and determinants)** *Let A be any 2×2 matrix. Let R be a region in the plane with a well-defined area. Let $\tilde{R}$ be the image of R under A. Then $\tilde{R}$ has a well-defined area, and the area of $\tilde{R}$ is $|\det(A)|$ times the area of R.*

The derivation of Theorem 6, based on the QR decomposition and the product formula, can be adapted to higher dimensions.

THEOREM 7   **(Volume and determinants)** *Let A be any 3×3 matrix. Let R be a region in $\mathbb{R}^3$ with a well-defined volume. Let $\tilde{R}$ be the image of R under A. Then $\tilde{R}$ has a well-defined volume, and the volume of $\tilde{R}$ is $|\det(A)|$ times the volume of R.*

We will prove this theorem at the end of the section, but first we give an application and example.

DEFINITION

(Parallelepiped) Given two vectors $\mathbf{v}_1$ and $\mathbf{v}_2$ in $\mathbb{R}^2$, the *parallelepiped spanned by $\mathbf{v}_1$ and $\mathbf{v}_2$* is the set of vectors of the form

$$x\mathbf{v}_1 + y\mathbf{v}_2 \qquad \text{with } 0 \le x, y \le 1.$$

Given three vectors $\mathbf{v}_1$, $\mathbf{v}_2$, and $\mathbf{v}_3$ in $\mathbb{R}^3$, the *parallelepiped spanned by $\mathbf{v}_1$, $\mathbf{v}_2$, and $\mathbf{v}_3$* is the set of vectors of the form

$$x\mathbf{v}_1 + y\mathbf{v}_2 + z\mathbf{v}_3 \qquad \text{with } 0 \le x, y, z \le 1.$$

---

*We can be inclusive and think of "shrinking" as magnification by a multiple less than one.

It is clear from the definition that if $A = [\mathbf{v}_1, \mathbf{v}_2, \mathbf{v}_3]$, the 3×3 matrix whose jth column is $\mathbf{v}_j$, then

$$x\mathbf{v}_1 + y\mathbf{v}_2 + z\mathbf{v}_3 = A \begin{bmatrix} x \\ y \\ z \end{bmatrix},$$

so the parallelepiped spanned by $\mathbf{v}_1$, $\mathbf{v}_2$, and $\mathbf{v}_3$ is the image under A of the unit cube. Since the unit cube has unit volume, it follows immediately from Theorem 7 that the volume of the parallelepiped spanned by $\mathbf{v}_1$, $\mathbf{v}_2$, and $\mathbf{v}_3$ is $|\det([\mathbf{v}_1, \mathbf{v}_2, \mathbf{v}_3])|$. The analogous result holds in two dimensions, and we record the following theorem.

THEOREM 8   **(Volumes of parallelepipeds)** *The area of the parallelepiped spanned by $\mathbf{v}_1$ and $\mathbf{v}_2$ in $\mathbb{R}^2$ is $|\det([\mathbf{v}_1, \mathbf{v}_2])|$. The volume of the parallelepiped spanned by $\mathbf{v}_1$, $\mathbf{v}_2$, and $\mathbf{v}_3$ in $\mathbb{R}^3$ is $|\det([\mathbf{v}_1, \mathbf{v}_2, \mathbf{v}_3])|$.*

EXAMPLE 12   **(Volume of a parallelepiped)** Consider

$$A = \begin{bmatrix} 2 & -2 \\ 1 & 1 \end{bmatrix}.$$

Let $\mathbf{v}_1$ and $\mathbf{v}_2$ be its columns so that

$$\mathbf{v}_1 = \begin{bmatrix} 2 \\ 1 \end{bmatrix}, \qquad \mathbf{v}_1 = \begin{bmatrix} -2 \\ 1 \end{bmatrix}, \qquad \text{and} \qquad A = [\mathbf{v}_1, \mathbf{v}_2]. \tag{2.14}$$

The unit square in the x, y plane is the set of all vectors

$$\mathbf{x} = \begin{bmatrix} x \\ y \end{bmatrix} = x\mathbf{e}_1 + y\mathbf{e}_2 \qquad \text{with } 0 \leq x, y \leq 1. \tag{2.15}$$

The vertices (or "corners") are 0, $\mathbf{e}_1$, $\mathbf{e}_2$, and $\mathbf{e}_1 + \mathbf{e}_2$. Here is a diagram:

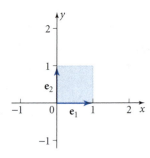

Since $A\mathbf{x} = x\mathbf{v}_1 + y\mathbf{v}_2$, the image of the unit square under A is the set of all linear combinations

$$x\mathbf{v}_1 + y\mathbf{v}_2 \qquad \text{with } 0 \leq x, y \leq 1. \tag{2.16}$$

This parallelogram has the vertices 0, $A\mathbf{e}_1 = \mathbf{v}_1$, $A\mathbf{e}_2 = \mathbf{v}_2$, and $A(\mathbf{e}_1 + \mathbf{e}_2) = \mathbf{v}_1 + \mathbf{v}_2$. Here is a diagram:

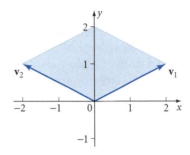

As you can see from the diagram, the area of the image is four units where, by definition, a "unit" is the area of the unit square. You also see that $\det(A) = 4$, which agrees with Theorem 5 since $\det(A) = 4$.

A useful related result is also a consequence of the QR decomposition and the product formula.

THEOREM 9

(Hadamard's inequality) *Let $A = [\mathbf{v}_1, \mathbf{v}_2, \ldots, \mathbf{v}_n]$ be any $n \times n$ matrix. Then*

$$|\det(A)| \le |\mathbf{v}_1||\mathbf{v}_2| \cdots |\mathbf{v}_n|, \tag{2.17}$$

and there is equality in (2.17) if and only if the columns of A are all orthogonal to one another.

The theorem says that $|\det(A)|$ is never larger than the product of the lengths of its columns. By Theorem 6, $\det(A) = \det(A^t)$, and since the columns of A^t are the rows of A, we also see that $|\det(A)|$ is never larger than the product of the lengths of its rows.

PROOF: We may suppose that $\det(A) \ne 0$ or there is nothing to prove. In this case, the columns of A are linearly independent, so

$$A = QR$$

where Q and R are $n \times n$. Then Q is orthogonal, so $|\det(Q)| = 1$. Let $Q = [\mathbf{u}_1, \mathbf{u}_2, \ldots, \mathbf{u}_n]$, Then R is upper triangular, and by Theorem 15 of Chapter 3, its jth diagonal entry is $\mathbf{u}_j \cdot \mathbf{v}_j$. Therefore,

$$\det(R) = (\mathbf{u}_1 \cdot \mathbf{v}_1)(\mathbf{u}_2 \cdot \mathbf{v}_2) \cdots (\mathbf{u}_n \cdot \mathbf{v}_n).$$

By the Schwarz inequality and the fact that each $\mathbf{u}_j$ is a unit vector,

$$|\mathbf{u}_j \cdot \mathbf{v}_j| \le |\mathbf{u}_j||\mathbf{v}_j| = |\mathbf{v}_j|. \tag{2.18}$$

Therefore,

$$|\det(R)| \le |\mathbf{v}_1||\mathbf{v}_2| \cdots |\mathbf{v}_n|. \tag{2.19}$$

By the product formula and the fact that $|\det(Q)| = 1$,

$$|\det(A)| = |\det(QR)| = |\det(Q)||\det(R)| = |\det(R)|.$$

Combining this equation with (2.17) gives (2.19).

To see that there is equality only when the $\mathbf{v}_j$ are orthogonal, recall that there is equality in (2.18) if and only if $\mathbf{v}_j$ and $\mathbf{u}_j$ are proportional. ∎

Finally, we give the proof of Theorem 7. It is an excellent example of the strategy of analyzing problems in linear algebra by factoring the matrices in question into simple factors.

PROOF OF THEOREM 7: If A is a 3×3 matrix with rank 3, it has a QR decomposition $A = QR$, where Q is an orthogonal matrix and R is an invertible upper-triangular matrix. Since R is invertible, none of its diagonal entries is zero. Therefore, we can divide each column through by its diagonal entry to produce a unit upper-triangular matrix:

$$T = \begin{bmatrix} 1 & \alpha & \beta \\ 0 & 1 & \gamma \\ 0 & 0 & 1 \end{bmatrix}.$$

If a, b, and c were the diagonal entries of R, R would have the factorization

$$R = TD = \begin{bmatrix} 1 & \alpha & \beta \\ 0 & 1 & \gamma \\ 0 & 0 & 1 \end{bmatrix} \begin{bmatrix} a & 0 & 0 \\ 0 & b & 0 \\ 0 & 0 & c \end{bmatrix}.$$

It is clear what D does to volumes by considering its effect on the unit cube: It magnifies volumes by a factor of $|abc| = |\det(D)|$.

This time T is a bit more complicated; there are three entries above the diagonal. But T itself can be factored at

$$T = \begin{bmatrix} 1 & \alpha & 0 \\ 0 & 1 & 0 \\ 0 & 0 & 1 \end{bmatrix} \begin{bmatrix} 1 & 0 & 0 \\ 0 & 1 & \gamma \\ 0 & 0 & 1 \end{bmatrix} \begin{bmatrix} 1 & 0 & \beta - \alpha\gamma \\ 0 & 1 & 0 \\ 0 & 0 & 1 \end{bmatrix},$$

as you can check. Each of the factors on the right is a *simple shear transformation* in $\mathbb{R}^3$.

Consider the first one. Let us call it S:

$$S = \begin{bmatrix} 1 & \alpha & 0 \\ 0 & 1 & 0 \\ 0 & 0 & 1 \end{bmatrix}.$$

We have

$$S \begin{bmatrix} x \\ y \\ z \end{bmatrix} = \begin{bmatrix} x + \alpha y \\ y \\ z \end{bmatrix}.$$

Only the x coordinate is affected, and it is shifted by an amount proportional to the y coordinate. Therefore, if you slice up the unit cube into a stack of sheets, cutting perpendicular to the y-axis, and then shift a sheet at height y in the x direction by the amount αy, you duplicate the effect of this transformation.

To see that this has no affect on volume, it may help to think of a deck of cards. If you place the deck on a table, and then push the cards in some direction so that each card is moved an amount proportional to its height, you are applying a shear transformation to the deck. Clearly it has no effect on the volume of the deck—an important observation known as *Cavalieri's principle.*

Therefore, S has no effect on volumes, which tells us that the magnification factor of S is 1. The same reasoning applies to each factor of T, so the total magnification factor of T is $1 \cdot 1 \cdot 1 = 1$ and the determinant of T is 1.

As before, since Q is orthogonal, it preserves both distances and angles. Therefore, it transforms cubes into cubes of the same size, which preserves volume, so the magnification factor of Q is 1, and we know $\det(Q) = \pm 1$, so $|\det(Q)| = 1$. Also, since T is a unit upper-triangular matrix, $|\det(T)| = 1$. Therefore,

$$|\det(A)| = |\det(Q)||\det(T)||\det(D)| = |\det(D)| = |abc|,$$

and the total magnification factor of $A = QTD$ is $1 \cdot 1 \cdot |abc| = |\det(A)|$. ∎

Exercises

2.1 Let v_1 and v_2 be the vectors $v_1 = \begin{bmatrix} 1 \\ -2 \end{bmatrix}$ and $v_2 = \begin{bmatrix} 3 \\ 1 \end{bmatrix}$. Find the area of the parallelepiped spanned by v_1 and v_2.

2.2 Consider the three points $(2, 3)$, $(3, 5)$, and $(-1, -1)$. What is the area of the triangle with these vertices?

2.3 Let $v_1 = \begin{bmatrix} 1 \\ -2 \\ 1 \end{bmatrix}$, $v_2 = \begin{bmatrix} 0 \\ -1 \\ 2 \end{bmatrix}$, and $v_3 = \begin{bmatrix} 3 \\ -1 \\ 0 \end{bmatrix}$. Find the volume of the parallelepiped spanned by v_1, v_2, and v_3.

2.4 Consider the four points $(0, 0, 0)$, $(2, 3, 1)$, $(3, 0, 1)$, and $(-1, -1, 0)$. What is the area of the simplex with these vertices? (Note that the simplex determined by four points is the set of all convex combinations of these points or, in other words, all weighted averages of these points.)

2.5 Prove that the determinant changes sign when two *columns* are swapped.

Exercises on Laplace's expansion and Cramer's formula

We derived a formula, (2.12), for the determinant of a 3×3 matrix by using the fact that the determinant of a matrix is linear in the columns. This idea can be developed to give a recursive formula expressing the determinant of any $n \times n$ matrix as a linear combination of n determinants of $(n - 1) \times (n - 1)$ matrices. We say that the formula is recursive because, unless $n - 1$ is 2 or 3, we would next use it again to express each of the $(n - 1) \times (n - 1)$ determinants as a linear combination of $n - 1$ determinants of $(n - 2) \times (n - 2)$ matrices. We would then keep on going until we had reduced to 2×2 or 3×3 matrices.

This is a very labor-intensive way to compute determinants and is not recommended for that purpose. However, it does have some theoretical uses, and it leads to a general formula for the inverse of a square matrix, known as Cramer's formula. The following exercises develop this circle of ideas. First, we fix some notation that will be used in both: For any $n \times n$ matrix A and any $1 \le i, j \le n$,

$$A^{[i,j]} = \text{the } (n - 1) \times (n - 1) \text{ matrix obtained by deleting row } i \text{ and column } j.$$

Make sure that you clearly understand the derivation of (2.12) before starting these exercises.

2.6 (a) Let A be an $n \times n$ matrix, and suppose that the first row of A is e_j. Show that

$$\det(A) = (-1)^{j-1} \det(A^{[1,j]}).$$

(b) Let A be an $n \times n$ matrix, and suppose that the ith row of A is e_j. Show that

$$\det(A) = (-1)^{i+j} \det(A^{[i,j]}).$$

Then use linearity in the ith row to show that in general, without any assumption on the ith row of A,

$$\det(A) = \sum_{j=1}^{n} A_{i,j}(-1)^{i+j} \det(A^{[i,j]}).$$

This is the *Laplace expansion formula*; notice that by taking the transpose, we get another version in which we can expand along any column.

(c) Suppose that $A(t)$ is an $n \times n$ matrix such that each of its entries is a differentiable function of a real variable t. Define a function $g(t)$ by $g(t) = \det(A(t))$. Prove that $g(t)$ is differentiable by first using the formula for $\det(A(t))$ when $n = 2$, and then use induction and the Laplace expansion to deal with the general case. (This is an example of the theoretical utility of the Laplace expansion. Because of the geometric significance of determinants, it is useful in geometry.)

2.7 (a) Let A be an $n \times n$ matrix, and define the $n \times n$ matrix B by

$$B_{j,i} = (-1)^{i+j} \det(A^{[i,j]}).$$

(Notice the transpose here: j is the row index and i is the column index for B.) The result of part (b) from the previous exercise can now be written

$$\sum_{j=1}^{n} A_{i,j} B_{j,i} = \det(A) \qquad \text{for all } i = 1. \ldots, n.$$

Show that for any $\ell \neq i$, $\sum_{j=1}^{n} A_{\ell,j}(-1)^{i+j} \det(A^{[i,j]})$ is the determinant of a matrix with two identical rows—the ith and the ℓth—and hence is zero. Conclude that

$$A^{-1} = \frac{1}{\det(A)} B,$$

with B defined as earlier in the exercise. This is *Cramer's formula* for the inverse of A.

(c) Suppose that $A(t)$ is an $n \times n$ matrix such that each of its entries is a differentiable function of a real variable t. Suppose that $A(0)$ is invertible. Prove that all the entries of $(A(t))^{-1}$ are differentiable at $t = 0$. (This is an example of the theoretical utility of Cramer's formula.)

SECTION 3 | Area, Volume, and Cross Product

3.1 The cross product

Let

$$\mathbf{a} = \begin{bmatrix} a_1 \\ a_2 \\ a_3 \end{bmatrix} \qquad \text{and} \qquad \mathbf{b} = \begin{bmatrix} b_1 \\ b_2 \\ b_3 \end{bmatrix}$$

be any two vectors in $\mathbb{R}^3$. Given another vector

$$\mathbf{x} = \begin{bmatrix} x \\ y \\ z \end{bmatrix}$$

in $\mathbb{R}^3$, form the matrix

$$A = \begin{bmatrix} \mathbf{a} \\ \mathbf{b} \\ \mathbf{x} \end{bmatrix},$$

and compute its determinant using (2.12):

$$\det(A) = (a_2 b_3 - a_3 b_2)x + (a_3 b_1 - a_1 b_3)y + (a_1 b_2 - a_2 b_1)z,$$

which we can write in the form

$$\begin{bmatrix} a_2b_3 - a_3b_2 \\ a_3b_1 - a_1b_3 \\ a_1b_2 - a_2b_1 \end{bmatrix} \cdot \begin{bmatrix} x \\ y \\ z \end{bmatrix}. \tag{3.1}$$

The vector on the left-hand side is the *cross product* of **a** and **b**.

DEFINITION

(Cross product) The *cross product* of vectors

$$\mathbf{a} = \begin{bmatrix} a_1 \\ a_2 \\ a_3 \end{bmatrix} \quad \text{and} \quad \mathbf{b} = \begin{bmatrix} b_1 \\ b_2 \\ b_3 \end{bmatrix}$$

in $\mathbb{R}^3$ is the vector $\mathbf{a} \times \mathbf{b}$ defined by

$$\mathbf{a} \times \mathbf{b} = \begin{bmatrix} a_2b_3 - a_3b_2 \\ a_3b_1 - a_1b_3 \\ a_1b_2 - a_2b_1 \end{bmatrix}. \tag{3.2}$$

The definition of $\mathbf{a} \times \mathbf{b}$ has been made so that for any vector $\mathbf{v}$,

$$(\mathbf{a} \times \mathbf{b}) \cdot \mathbf{v} = \det\left(\begin{bmatrix} \mathbf{a} \\ \mathbf{b} \\ \mathbf{v} \end{bmatrix}\right). \tag{3.3}$$

Moreover, since $\det(A^t) = \det(A)$ for any square matrix A, it is also the case that

$$(\mathbf{a} \times \mathbf{b}) \cdot \mathbf{v} = \det\left([\mathbf{a}, \mathbf{b}, \mathbf{v}]\right). \tag{3.4}$$

You may find it easier to remember (3.2) in the form

$$\mathbf{a} \times \mathbf{b} = \det\left(\begin{bmatrix} a_1 & a_2 & a_3 \\ b_1 & b_2 & b_3 \\ \mathbf{e}_1 & \mathbf{e}_2 & \mathbf{e}_3 \end{bmatrix}\right)$$
$$= (a_2b_3 - a_3b_2)\mathbf{e}_1 + (a_3b_1 - a_1b_3)\mathbf{e}_2 + (a_1b_2 - a_2b_1)\mathbf{e}_3. \tag{3.5}$$

This may look like a strange sort of determinant since the entries of the bottom row are vectors, not numbers. But we can multiply vectors and numbers, and since each of the products in the definition of the determinant contains only one entry from each row, and hence, only one vector, the formula makes sense—and works.

EXAMPLE 13 **(A cross product)** Let

$$\mathbf{a} = \begin{bmatrix} 1 \\ 2 \\ 3 \end{bmatrix} \quad \text{and} \quad \mathbf{b} = \begin{bmatrix} 3 \\ 2 \\ 1 \end{bmatrix}.$$

Then from (3.2),

$$\mathbf{a} \times \mathbf{b} = \begin{bmatrix} 2 - 6 \\ 9 - 1 \\ 2 - 6 \end{bmatrix} = \begin{bmatrix} -4 \\ 8 \\ -4 \end{bmatrix}.$$

EXAMPLE 14 **(Some special cross products)** Let $\mathbf{a} = \mathbf{e}_1$ and $\mathbf{b} = \mathbf{e}_2$. Then from (3.2) you find $\mathbf{a} \times \mathbf{b} = \mathbf{e}_3$. In fact in the same way, you find

$$\mathbf{e}_1 \times \mathbf{e}_2 = -\mathbf{e}_2 \times \mathbf{e}_1 = \mathbf{e}_3, \qquad \mathbf{e}_2 \times \mathbf{e}_3 = -\mathbf{e}_3 \times \mathbf{e}_2 = \mathbf{e}_1, \qquad \text{and} \qquad \mathbf{e}_3 \times \mathbf{e}_1 = -\mathbf{e}_1 \times \mathbf{e}_3 = \mathbf{e}_2 \tag{3.6}$$

and

$$\mathbf{e}_1 \times \mathbf{e}_1 = 0, \qquad \mathbf{e}_2 \times \mathbf{e}_2 = 0, \qquad \text{and} \qquad \mathbf{e}_3 \times \mathbf{e}_3 = 0. \tag{3.7}$$

In particular,

$$(\mathbf{e}_1 \times \mathbf{e}_2) \times \mathbf{e}_2 = \mathbf{e}_3 \times \mathbf{e}_2 = -\mathbf{e}_1,$$

while

$$\mathbf{e}_1 \times (\mathbf{e}_2 \times \mathbf{e}_2) = \mathbf{e}_1 \times 0 = 0.$$

Hence, the cross product *does not* have the associative property.

The next theorem records some important properties of the cross product.

THEOREM 10 **(Properties of the cross product)** *The cross product of two vectors* $\mathbf{a}$ *and* $\mathbf{b}$ *in* $\mathbb{R}^3$ *has the following properties:*

1. *It is anticommutative; that is,*

$$\mathbf{a} \times \mathbf{b} = -\mathbf{b} \times \mathbf{a}. \tag{3.8}$$

2. *It is linear in both* $\mathbf{a}$ *and* $\mathbf{b}$*. In particular, for any numbers* s *and* t *and any vectors* $\mathbf{a}_1$*,* $\mathbf{a}_2$*, and* $\mathbf{b}$*,*

$$(s\mathbf{a}_1 + t\mathbf{a}_2) \times \mathbf{b} = s(\mathbf{a}_1 \times \mathbf{b}) + t(\mathbf{a}_2 \times \mathbf{b}). \tag{3.9}$$

3. *The cross product* $\mathbf{a} \times \mathbf{b}$ *is orthogonal to both* $\mathbf{a}$ *and* $\mathbf{b}$*. That is,*

$$(\mathbf{a} \times \mathbf{b}) \cdot \mathbf{a} = 0 \qquad and \qquad (\mathbf{a} \times \mathbf{b}) \cdot \mathbf{b} = 0. \tag{3.10}$$

4. *The length of* $\mathbf{a} \times \mathbf{b}$ *is the area of the parallelogram with edges along* $\mathbf{a}$ *and* $\mathbf{b}$*. That is,*

$$|\mathbf{a} \times \mathbf{b}| = |\mathbf{a}||\mathbf{b}| \sin(\theta) \tag{3.11}$$

where θ *is the angle between* $\mathbf{a}$ *and* $\mathbf{b}$*. In particular,* $\mathbf{a} \times \mathbf{b} = 0$ *if* $\mathbf{b}$ *is a multiple of* $\mathbf{a}$*.*

PROOF: The anticommutivity (3.8) follows directly from (3.5) and the fact that the determinant changes sign when two rows are swapped. The linearity (3.9) follows directly from (3.5) and the fact that the determinant is a linear function of each of its rows. The fact that (3.10) is true follows from (3.3) and the fact that a determinant with two identical rows is zero.

Finally, let $\mathbf{n}$ be given by

$$\mathbf{n} = \frac{1}{|\mathbf{a} \times \mathbf{b}|} \mathbf{a} \times \mathbf{b}.$$

This is a unit vector pointing in the direction of $\mathbf{a} \times \mathbf{b}$, so $|\mathbf{a} \times \mathbf{b}| = \mathbf{n} \cdot \mathbf{a} \times \mathbf{b}$. By (3.3),

$$|\mathbf{a} \times \mathbf{b}| = \det\left(\begin{bmatrix} \mathbf{a} \\ \mathbf{b} \\ \mathbf{n} \end{bmatrix}\right).$$

We know that this determinant is the volume of the parallelepiped spanned by the vectors $\mathbf{a}$, $\mathbf{b}$, and $\mathbf{n}$, which in turn is the area of the parallelogram spanned by $\mathbf{a}$ and $\mathbf{b}$ times the height of the parallelepiped in the direction rising out of the plane spanned by $\mathbf{a}$ and $\mathbf{b}$. Since $\mathbf{n}$ is a unit vector in this direction, the height is one, so $|\mathbf{a} \times \mathbf{b}|$ equals the area of the parallelogram spanned by $\mathbf{a}$ and $\mathbf{b}$. By elementary planar geometry, this area is the product of the lengths of the sides and the sine of the angle θ between them, which is to say, $|\mathbf{a}||\mathbf{b}| \sin(\theta)$. ■

Here is a diagram representing the orthogonality property of the cross product:

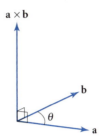

The fact that $|\mathbf{a} \times \mathbf{b}| = |\mathbf{a}||\mathbf{b}||\sin(\theta)|$ is at least as important as the fact that $\mathbf{a} \times \mathbf{b}$ is orthogonal to both $\mathbf{a}$ and $\mathbf{b}$, as we shall see. We have derived this in the proof just given by considering volume, but it can also be seen from (3.2) by direct computation. From (3.2) we have

$$
\begin{aligned}
|\mathbf{a} \times \mathbf{b}|^2 =& (a_2b_3 - a_3b_2)^2 + (a_3b_1 - a_1b_3)^2 + (a_2b_1 - a_1b_2)^2 \\
=& (a_2b_3)^2 + (a_3b_2)^2 + (a_3b_1)^2 + (a_1b_3)^2 + (a_2b_1)^2 + (a_1b_2)^2 \\
& - 2a_2b_3a_3b_2 - 2a_3b_1a_1b_3 - 2a_2b_1a_1b_2 \\
=& |\mathbf{a}|^2|\mathbf{b}|^2 - (\mathbf{a} \cdot \mathbf{b})^2 \\
=& |\mathbf{a}|^2|\mathbf{b}|^2 - |\mathbf{a}|^2|\mathbf{b}|^2 \cos^2(\theta) \\
=& |\mathbf{a}|^2|\mathbf{b}|^2 \sin^2(\theta),
\end{aligned}
$$

where in the next to the last line we have used the fact that $\mathbf{a} \cdot \mathbf{b} = |\mathbf{a}||\mathbf{b}| \cos(\theta)$.

The cross product gives us another way to compute the equation of the plane in $\mathbb{R}^3$ spanned by two linearly independent vectors.

EXAMPLE 15   **(Cross products and planes)** Let us find the equation for the plane spanned by

$$
\mathbf{a} = \begin{bmatrix} 1 \\ 2 \\ 3 \end{bmatrix} \quad \text{and} \quad \mathbf{b} = \begin{bmatrix} 3 \\ 2 \\ 1 \end{bmatrix}.
$$

We have already computed in Example 13 that

$$
\mathbf{a} \times \mathbf{b} = \begin{bmatrix} -4 \\ 8 \\ -4 \end{bmatrix}.
$$

Since $\{\mathbf{a}, \mathbf{b}\}$ is a basis for the plane, and since $\mathbf{a} \times \mathbf{b}$ is orthogonal to each of these vectors, $\{\mathbf{a} \times \mathbf{b}\}$ is a basis for the orthogonal complement of the plane, so the equation of the plane is

$$
\begin{bmatrix} x \\ y \\ z \end{bmatrix} \cdot \begin{bmatrix} -4 \\ 8 \\ -4 \end{bmatrix} = 0, \quad \text{or} \quad x - 2y + z = 0.
$$

We already have two other ways to do this; the cross product gives us a third. Let us review the other two.

Form the matrix

$$
\begin{bmatrix} \mathbf{a} \\ \mathbf{b} \end{bmatrix} = \begin{bmatrix} 1 & 2 & 3 \\ 3 & 2 & 1 \end{bmatrix}.
$$

The vectors in its kernel are orthogonal to each of the rows, namely, **a** and **b**. So finding a nonzero vector in the kernel gives us the orthogonal vector we seek.

To find it, row reduce the matrix. In one step,

$$\begin{bmatrix} 1 & 2 & 3 \\ 3 & 2 & 1 \end{bmatrix} \rightarrow \begin{bmatrix} 1 & 2 & 3 \\ 0 & -4 & -8 \end{bmatrix}.$$

The variable z is nonpivotal, so we set $z = 1$ and solve for x and y, finding $y = -2$ and then $x = 1$. Hence,

$$\mathbf{v} = \begin{bmatrix} 1 \\ -2 \\ 1 \end{bmatrix}$$

is the vector we seek. Notice that it is $-1/4$ times $\mathbf{a} \times \mathbf{b}$. Hence, we find the equation

$$\begin{bmatrix} x \\ y \\ z \end{bmatrix} \cdot \begin{bmatrix} 1 \\ -2 \\ 1 \end{bmatrix} = 0 \text{ or } x - 2y + z = 0,$$

as before.

In both these approaches we found the equation by finding a nonzero vector that was orthogonal to both **a** and **b**, first, by computing $\mathbf{a} \times \mathbf{b}$ and then by using row reduction to find a basis for the kernel of

$$\begin{bmatrix} \mathbf{a} \\ \mathbf{b} \end{bmatrix}.$$

There is yet another way that we already know: Row reduce

$$\begin{bmatrix} 1 & 3 & x \\ 2 & 2 & y \\ 3 & 1 & z \end{bmatrix}.$$

In two steps we find

$$\begin{bmatrix} 1 & 3 & x \\ 2 & 2 & y \\ 3 & 1 & z \end{bmatrix} \rightarrow \begin{bmatrix} 1 & 3 & x \\ 0 & -4 & -2x + y \\ 0 & -8 & -3x + z \end{bmatrix} \rightarrow \begin{bmatrix} 1 & 3 & x \\ 0 & -4 & -2x + y \\ 0 & 0 & x - 2y + z \end{bmatrix}.$$

Evidently,

$$\begin{bmatrix} x \\ y \\ z \end{bmatrix}$$

is in the image of [**a**, **b**] exactly when $x - 2y + z = 0$.

As far as computing equations of planes goes, the cross product gives us another way to do something we already knew how to do. In fact, it is not even faster in most cases; usually finding the kernel, as in our second approach, involves the least amount of computation.

The real value of the cross product $\mathbf{a} \times \mathbf{b}$ is not that it gives us *some* vector orthogonal to both **a** and **b**: The length of $\mathbf{a} \times \mathbf{b}$ is the area of the parallelogram spanned by **a** and **b**. This fact is very useful in multivariable calculus when one has to compute the area of a *curved* two-dimensional surface in $\mathbb{R}^3$: You break up the surface into a large number of parallelograms and then use the cross product to compute each of their areas. Then you add up these areas and taking an appropriate limit, the sum becomes an integral.

Pursuing this line of inquiry would take us outside the field of linear algebra. However, we will illustrate the use of the cross product to compute the areas of parallelograms in $\mathbb{R}^3$.

EXAMPLE 16   **(Cross products and area)** Let

$$\mathbf{a} = \begin{bmatrix} 1 \\ -1 \\ 1 \end{bmatrix} \quad \text{and} \quad \mathbf{b} = \begin{bmatrix} 3 \\ 2 \\ 1 \end{bmatrix}.$$

What is the area of the parallelogram spanned by **a** and **b**? From (3.2),

$$\mathbf{a} \times \mathbf{b} = \begin{bmatrix} -1-2 \\ 3-1 \\ 2+3 \end{bmatrix} = \begin{bmatrix} -3 \\ 2 \\ 5 \end{bmatrix},$$

so

$$|\mathbf{a} \times \mathbf{b}| = \sqrt{38},$$

the area of the parallelogram. (Though it is not required to answer the question, it is a good idea to check that $\mathbf{a} \times \mathbf{b} \cdot \mathbf{a}$ and $\mathbf{a} \times \mathbf{b} \cdot \mathbf{b}$ are both zero, as they must be.)

EXAMPLE 17   **(Cross products and area)** Let us compute the total surface area of the parallelepiped spanned by the three vectors

$$\mathbf{a} = \begin{bmatrix} 1 \\ -1 \\ 1 \end{bmatrix}, \qquad \mathbf{b} = \begin{bmatrix} 3 \\ 2 \\ 1 \end{bmatrix}, \qquad \text{and} \qquad \mathbf{c} = \begin{bmatrix} 1 \\ 2 \\ 3 \end{bmatrix}.$$

Any such parallelepiped in $\mathbb{R}^3$ has six sides, like the cube, which come in three parallel pairs. There is the parallelogram spanned by **a** and **b**, another parallelogram of the same area on the opposite side, and so on for each of the three pairs of vectors. Hence, the total area is

$$2|\mathbf{a} \times \mathbf{b}| + 2|\mathbf{a} \times \mathbf{c}| + 2|\mathbf{b} \times \mathbf{c}|.$$

In Examples 13 and 16 we have already computed $\mathbf{a} \times \mathbf{b}$ and $\mathbf{b} \times \mathbf{c}$ (using different names for the same vectors). In Example 15 we found that

$$\mathbf{a} \times \mathbf{b} = \begin{bmatrix} -3 \\ 2 \\ 5 \end{bmatrix} \qquad \text{and} \qquad |\mathbf{a} \times \mathbf{b}| = \sqrt{38}.$$

In Example 13 we found that

$$\mathbf{b} \times \mathbf{c} = \begin{bmatrix} -4 \\ 8 \\ -4 \end{bmatrix}, \qquad \text{so} \qquad \mathbf{b} \times \mathbf{c} = \sqrt{96}.$$

Finally,

$$\mathbf{a} \times \mathbf{c} = \begin{bmatrix} -3-2 \\ 1-3 \\ 2+1 \end{bmatrix} = \begin{bmatrix} -5 \\ -2 \\ 3 \end{bmatrix},$$

so

$$|\mathbf{a} \times \mathbf{c}| = \sqrt{38}.$$

Hence, the total surface area of the parallelepiped is $4\sqrt{38} + 2\sqrt{86}$.

3.2 The right-hand rule

Theorem 10 tells us that for linearly independent vectors **a** and **b**, $\mathbf{a} \times \mathbf{b}$ is perpendicular to the plane they span. But to which side of the plane does it point? To answer this question, we use the *right-hand rule*.

Let $\{\mathbf{u}_1, \mathbf{u}_2, \mathbf{u}_3\}$ be an orthonormal basis for $\mathbb{R}^3$. The 3×3 matrix $[\mathbf{u}_1, \mathbf{u}_2, \mathbf{u}_3]$ is an orthogonal matrix, which from Theorem 5 means

$$\det([\mathbf{u}_1, \mathbf{u}_2, \mathbf{u}_3]) = \pm 1.$$

We use this dichotomy to distinguish two classes of orthonormal bases for $\mathbb{R}^3$.

DEFINITION   **(Right- and left-handed bases)** An orthonormal basis $\{\mathbf{u}_1, \mathbf{u}_2, \mathbf{u}_3\}$ is called a *right-handed basis* if $\det([\mathbf{u}_1, \mathbf{u}_2, \mathbf{u}_3]) = 1$, and it is called a *left-handed basis* if and only if $\det([\mathbf{u}_1, \mathbf{u}_2, \mathbf{u}_3]) = -1$.

By the definition of the cross product, $\det([\mathbf{u}_1, \mathbf{u}_2, \mathbf{u}_3]) = (\mathbf{u}_1 \times \mathbf{u}_2) \cdot \mathbf{u}_3$. Since $\mathbf{u}_1 \times \mathbf{u}_2$ is a unit vector orthogonal to both $\mathbf{u}_1$ and $\mathbf{u}_2$, it must be one of $\pm\mathbf{u}_3$. If $\mathbf{u}_1 \times \mathbf{u}_2 = \mathbf{u}_3$, then $\det([\mathbf{u}_1, \mathbf{u}_2, \mathbf{u}_3]) = 1$, and otherwise, $\det([\mathbf{u}_1, \mathbf{u}_2, \mathbf{u}_3]) = -1$. This gives us a second equivalent formulation of the definition:

- *An orthonormal basis $\{\mathbf{u}_1, \mathbf{u}_2, \mathbf{u}_3\}$ is right handed if and only if $\mathbf{u}_3 = \mathbf{u}_1 \times \mathbf{u}_2$, left handed if and only if $\mathbf{u}_3 = -\mathbf{u}_1 \times \mathbf{u}_2$.*

There is a third, equivalent definition that explains the terminology. It is based on the following diagram:

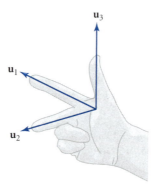

- *If you can arrange your right hand so that the index finger points along $\mathbf{u}_1$, the middle finger along $\mathbf{u}_2$, and the thumb along $\mathbf{u}_3$, then $\{\mathbf{u}_1, \mathbf{u}_2, \mathbf{u}_3\}$ is a right-handed basis. If you can do it with your left hand, $\{\mathbf{u}_1, \mathbf{u}_2, \mathbf{u}_3\}$ is a left-handed basis.*

It is not immediately clear why the new definition is equivalent to the first two, but the next theorem provides the link.

THEOREM 11   **(Orthogonal transformations and the cross product)** *Let U be any 3×3 orthogonal matrix. If $\det(U) = 1$, then for any two vectors $\mathbf{a}$ and $\mathbf{b}$ in $\mathbb{R}^3$*

$$U\mathbf{a} \times U\mathbf{b} = U(\mathbf{a} \times \mathbf{b}), \tag{3.12}$$

while if $\det(U) = -1$, then for any three vectors $\mathbf{a}$, $\mathbf{b}$, and $\mathbf{v}$ in $\mathbb{R}^3$,

$$U\mathbf{a} \times U\mathbf{b} = -U(\mathbf{a} \times \mathbf{b}).$$

Theorem 5 tells us that there are two types of $n \times n$ orthogonal matrices U: those with $\det(U) = 1$ and those with $\det(U) = -1$. Theorem 11 highlights an important special property of orthogonal matrices with a positive determinant.

In fact, the linear transformation represented by a 3×3 orthogonal matrix U with $\det(U) = 1$ is a rotation of $\mathbb{R}^3$ about some line through the origin. Exercises at the end of this section will lead you through the reasoning behind this fact, which will justify the following definition.

DEFINITION   **(Rotation matrices)** An $n \times n$ matrix U is a *rotation matrix* if and only if U is orthogonal, and moreover, $\det(U) = 1$.

PROOF OF THEOREM 11: Let $\mathbf{v}$ be any vector in $\mathbb{R}^3$. Then

$$U\mathbf{a} \times U\mathbf{b} \cdot U\mathbf{v} = \det([U\mathbf{a}, U\mathbf{b}, U\mathbf{v}])$$
$$= \det(U[\mathbf{a}, \mathbf{b}, \mathbf{v}])$$
$$= \det(U)\det([\mathbf{a}, \mathbf{b}, \mathbf{v}])$$
$$= \det([\mathbf{a}, \mathbf{b}, \mathbf{v}])$$
$$= (\mathbf{a} \times \mathbf{b}) \cdot \mathbf{v}.$$

Hence by **V.I.F.5**,

$$U^t(U\mathbf{a} \times U\mathbf{b}) \cdot \mathbf{v} = (\mathbf{a} \times \mathbf{b}) \cdot \mathbf{v}$$

for all $\mathbf{v}$. Taking $\mathbf{v} = \mathbf{e}_j$ for $j = 1, 2$, and 3 successively, we see that $U^t(U\mathbf{a} \times U\mathbf{b})$ and $\mathbf{a} \times \mathbf{b}$ have the same entries. Hence, $U^t(U\mathbf{a} \times U\mathbf{b}) = \mathbf{a} \times \mathbf{b}$. Multiplying on the left by U, we have the result. ∎

This proof has an important consequence. Let $\{\mathbf{u}_1, \mathbf{u}_2, \mathbf{u}_3\}$ be a right-handed orthonormal basis. Let $U^t = [\mathbf{u}_1, \mathbf{u}_2, \mathbf{u}_3]$. Since U^t is an orthogonal matrix, so is $U = (U^t)^t$.

For any vector $\mathbf{a}$ in $\mathbb{R}^3$, $U^t\mathbf{a}$ is the coordinate vector of $\mathbf{a}$ with respect to the basis $\{\mathbf{u}_1, \mathbf{u}_2, \mathbf{u}_3\}$. Indeed, for each i, by **V.I.F.2**,

$$\mathbf{u}_i \cdot \mathbf{a} = (U^t\mathbf{a})_i.$$

Define $\mathbf{x}$ and $\mathbf{y}$ by

$$\mathbf{x} = U^t\mathbf{a} \quad \text{and} \quad \mathbf{y} = U^t\mathbf{b} \tag{3.13}$$

so that

$$\mathbf{a} = U\mathbf{x} \quad \text{and} \quad \mathbf{b} = U\mathbf{y}.$$

By Theorem 11,

$$\mathbf{a} \times \mathbf{b} = U\mathbf{x} \times U\mathbf{y}$$
$$= U(\mathbf{x} \times \mathbf{y}), \tag{3.14}$$

so $\mathbf{x} \times \mathbf{y}$ is the coordinate vector of $\mathbf{a} \times \mathbf{b}$ for the $\{\mathbf{u}_1, \mathbf{u}_2, \mathbf{u}_3\}$ basis. That is,

- *You can compute the cross product of $\mathbf{a}$ and $\mathbf{b}$ by taking the cross products of the coordinate vectors of $\mathbf{a}$ and $\mathbf{b}$ using any right-handed orthonormal coordinate system.*

We can now answer the question raised at the beginning of this subsection. Let $\mathbf{a}$ and $\mathbf{b}$ be any two linearly independent vectors in $\mathbb{R}^3$. Let $\mathbf{u}_1$ be the unit vector pointing in the direction of $\mathbf{a}$. Let $\mathbf{u}_3$ be either unit vector that is orthogonal to the plane spanned by $\{\mathbf{a}, \mathbf{b}\}$. Finally, let $\mathbf{u}_2$ be the unit vector that is orthogonal to the plane spanned by $\{\mathbf{u}_1, \mathbf{u}_3\}$, so that $\{\mathbf{u}_1, \mathbf{u}_2, \mathbf{u}_3\}$ is a right-handed basis. (One of the two unit vectors orthogonal to this plane gives a right-handed basis, the other a left-handed basis. Pick the "right" one.)

We now compute $\mathbf{a} \times \mathbf{b}$ in this basis. Let $U^t = [\mathbf{u}_1, \mathbf{u}_2, \mathbf{u}_3]$ as earlier, and define $\mathbf{x}$ and $\mathbf{y}$ as in (3.13). Then, since $\mathbf{u}_1$ points along $\mathbf{a}$,

$$\mathbf{x} = \begin{bmatrix} \mathbf{a} \cdot \mathbf{u}_1 \\ \mathbf{a} \cdot \mathbf{u}_2 \\ \mathbf{a} \cdot \mathbf{u}_3 \end{bmatrix} = \begin{bmatrix} |\mathbf{a}| \\ 0 \\ 0 \end{bmatrix}.$$

Also, since $\mathbf{u}_3$ is orthogonal to $\mathbf{b}$,

$$\mathbf{y} = \begin{bmatrix} \mathbf{b} \cdot \mathbf{u}_1 \\ \mathbf{b} \cdot \mathbf{u}_2 \\ \mathbf{b} \cdot \mathbf{u}_3 \end{bmatrix} = \begin{bmatrix} \mathbf{b} \cdot \mathbf{u}_1 \\ \mathbf{b} \cdot \mathbf{u}_2 \\ 0 \end{bmatrix}.$$

It is easy to compute the cross product of these two vectors:

$$\begin{bmatrix} |\mathbf{a}| \\ 0 \\ 0 \end{bmatrix} \times \begin{bmatrix} \mathbf{b} \cdot \mathbf{u}_1 \\ \mathbf{b} \cdot \mathbf{u}_2 \\ 0 \end{bmatrix} = \begin{bmatrix} 0 \\ 0 \\ |\mathbf{a}|\mathbf{b} \cdot \mathbf{u}_2 \end{bmatrix}.$$

Therefore, by (3.14),

$$\mathbf{a} \times \mathbf{b} = U \begin{bmatrix} 0 \\ 0 \\ |\mathbf{a}|\mathbf{b} \cdot \mathbf{u}_2 \end{bmatrix} = (|\mathbf{a}|\mathbf{b} \cdot \mathbf{u}_2)\,\mathbf{u}_3.$$

We see that $\mathbf{a} \times \mathbf{b}$ points in the $\mathbf{u}_3$ direction if and only if

$$\mathbf{b} \cdot \mathbf{u}_2 \geq 0.$$

Since $\mathbf{b} \cdot \mathbf{u}_2 = |\mathbf{b}| \cos(\theta)$, where θ is the angle between $\mathbf{b}$ and $\mathbf{u}_2$, we see that $\mathbf{b} \cdot \mathbf{u}_2 \geq 0$ if and only if $\mathbf{b}$ and $\mathbf{u}_2$ are both on the same side of the $\{\mathbf{u}_1, \mathbf{u}_3\}$ plane.

Now imagine grasping the $\mathbf{u}_3$ axis with your right hand so that your extended thumb points in the $\mathbf{u}_3$ direction and your extended index finger points along the direction of $\mathbf{a}$. If you can bend your index finger so that it points along the direction of $\mathbf{b}$ without otherwise moving your hand, either you are very double jointed, or else $\mathbf{b}$ is on the same side of the $\{\mathbf{u}_1, \mathbf{u}_3\}$ plane as $\mathbf{u}_2$. In this case, $\mathbf{a} \times \mathbf{b}$ points in the $\mathbf{u}_3$ direction. Otherwise, it points in the $-\mathbf{u}_3$ direction.

Although we used the basis $\{\mathbf{u}_1, \mathbf{u}_2, \mathbf{u}_3\}$ to deduce the rule, we can express it more briefly if we leave out the basis in the end:

- *Imagine grasping the axis orthogonal to the plane spanned by $\{\mathbf{a}, \mathbf{b}\}$ with your right hand and curling your fingers around it so that they wrap in the sense of rotation that would bring $\mathbf{a}$ parallel to $\mathbf{b}$ using an angle of rotation less than π. (The other sense would take a larger angle.) Extend your thumb along the axis. It points in the direction of $\mathbf{a} \times \mathbf{b}$. This statement is independent of the coordinates, and is called the "right-hand rule" for determining the direction of $\mathbf{a} \times \mathbf{b}$.*

Exercises

3.1 Let $\mathbf{u} = \begin{bmatrix} 2 \\ 0 \\ 1 \end{bmatrix}$ $\mathbf{v} = \begin{bmatrix} -1 \\ 1 \\ 1 \end{bmatrix}$ and $\mathbf{w} = \begin{bmatrix} 2 \\ 0 \\ -2 \end{bmatrix}.$

(a) Compute the cross products $\mathbf{u} \times \mathbf{v}$, $\mathbf{u} \times \mathbf{w}$, and $\mathbf{v} \times \mathbf{w}$.

(b) Compute the total surface area of the parallelepiped spanned by $\mathbf{u}$, $\mathbf{v}$, and $\mathbf{w}$.

3.2 Let $\mathbf{u} = \begin{bmatrix} 1 \\ 1 \\ 1 \end{bmatrix}$, $\mathbf{v} = \begin{bmatrix} 1 \\ -1 \\ 1 \end{bmatrix}$,   and $\mathbf{w} = \begin{bmatrix} 1 \\ 2 \\ 4 \end{bmatrix}.$

(a) Compute the cross products $\mathbf{u} \times \mathbf{v}$, $\mathbf{u} \times \mathbf{w}$, and $\mathbf{v} \times \mathbf{w}$.

(b) Compute the total surface area of the parallelepiped spanned by $\mathbf{u}$, $\mathbf{v}$, and $\mathbf{w}$.

3.3 Let $\mathbf{a} = \begin{bmatrix} 1 \\ 3 \\ 2 \end{bmatrix}$ and $\mathbf{b} = \begin{bmatrix} 1 \\ 2 \\ -1 \end{bmatrix}.$

(a) Compute the cross product $\mathbf{a} \times \mathbf{b}$, and use this computation to find the equation of the plane spanned by $\mathbf{a}$ and $\mathbf{b}$.

(b) Use row reduction to find a nonzero vector in Ker $\left(\begin{bmatrix} a \\ b \end{bmatrix} \right)$, and use this computation to find the equation of the plane spanned by **a** and **b**.

(c) What is the area of the parallelogram spanned by **a** and **b**?

3.4 Let **a** and **b** be the two vectors $\mathbf{a} = \begin{bmatrix} 1 \\ 2 \\ 4 \end{bmatrix}$ and $\mathbf{b} = \begin{bmatrix} 1 \\ -2 \\ 4 \end{bmatrix}$.

(a) Compute the cross product $\mathbf{a} \times \mathbf{b}$, and use this computation to find the equation of the plane spanned by **a** and **b**.

(b) Use row reduction to find a nonzero vector in Ker $\left(\begin{bmatrix} a \\ b \end{bmatrix} \right)$, and use this computation to find the equation of the plane spanned by **a** and **b**.

(c) What is the area of the parallelogram spanned by **a** and **b**?

3.5 Let B be a 3×3 matrix. What conditions on B guarantee that

$$(B\mathbf{u} \times B\mathbf{v}) \cdot B\mathbf{w} = (\mathbf{u} \times \mathbf{v}) \cdot \mathbf{w}$$

for all vectors **u**, **v**, and **w** in $\mathbb{R}^3$? (Do not give just the trivial ones like $B = I$ or $B = 0$; there are other more interesting cases.)

3.6 Let B be a 3×3 matrix. What conditions on B guarantee that

$$(B\mathbf{u} \times B\mathbf{v}) = B(\mathbf{u} \times \mathbf{v})$$

for all vectors **u** and **v** in $\mathbb{R}^3$? (Do not give just the trivial ones like $B = I$ or $B = 0$; there are other, more interesting cases.)

3.7 Given any three vectors $\mathbf{v}_1, \mathbf{v}_2$, and $\mathbf{v}_3$ in $\mathbb{R}^3$, define a *modified cross product* $\mathbf{a} \bowtie \mathbf{b}$ by requiring that

$$\mathbf{a} \bowtie \mathbf{b} = -\mathbf{b} \bowtie \mathbf{a}$$

for all **a** and **b**, and that $\mathbf{a} \bowtie \mathbf{b}$ be linear in both **a** and **b**. Finally, we specify that

$$\mathbf{e}_1 \bowtie \mathbf{e}_2 = \mathbf{v}_3, \qquad \mathbf{e}_2 \bowtie \mathbf{e}_3 = \mathbf{v}_1, \qquad \text{and} \qquad \mathbf{e}_3 \bowtie \mathbf{e}_1 = \mathbf{v}_2.$$

(a) Explain why this is enough information to compute $\mathbf{a} \bowtie \mathbf{b}$ for any **a** and **b** in $\mathbb{R}^3$. (If you are not sure how to explain why, consider part (b) first.)

(b) A bit more concretely, let

$$\mathbf{v}_1 = \begin{bmatrix} 1 \\ 0 \\ 0 \end{bmatrix}, \qquad \mathbf{v}_2 = \begin{bmatrix} 1 \\ 1 \\ 0 \end{bmatrix}, \qquad \text{and} \qquad \mathbf{v}_3 = \begin{bmatrix} 1 \\ 1 \\ 1 \end{bmatrix},$$

and

$$\mathbf{a} = \begin{bmatrix} 1 \\ 2 \\ 4 \end{bmatrix} \quad \text{and} \quad \mathbf{b} = \begin{bmatrix} 1 \\ -2 \\ 4 \end{bmatrix}.$$

Compute $\mathbf{a} \bowtie \mathbf{b}$.

(c) Let B be the matrix $B = [\mathbf{v}_1, \mathbf{v}_2, \mathbf{v}_3]$. How are $\mathbf{a} \bowtie \mathbf{b}$ and $B(\mathbf{a} \times \mathbf{b})$ related? (Answer in general, not just for the specific case considered in part (b).)

3.8 Define a generalized "triple cross product" of *three* vectors in $\mathbb{R}^4$, and explain why you consider your definition to be a natural generalization.

The correspondence between 3×3 antisymmetric matrices and cross products

Exercises 3.9 through 3.11 develop some important properties of 3×3 antisymmetric matrices A, that is, those with $A^t = -A$. It turns out that they are closely connected with cross products, and there is an important one-to-one correspondence between 3×3 antisymmetric matrices and vectors in $\mathbb{R}^3$. Try working through these exercises in sequence, using what you learn in the first one to help you solve the second, and so on.

3.9 Let $\mathbf{a} = \begin{bmatrix} a \\ b \\ c \end{bmatrix}$ be any nonzero vector in $\mathbb{R}^3$, and define a transformation $f_{\mathbf{a}}$ from $\mathbb{R}^3$ onto itself by

$$f_{\mathbf{a}}(\mathbf{x}) = \mathbf{a} \times \mathbf{x}.$$

By Theorem 10, this transformation is linear. Therefore, there is a uniquely determined 3×3 matrix $A_{\mathbf{a}}$ such that for all $\mathbf{x}$ in $\mathbb{R}^3$,

$$A_{\mathbf{a}}\mathbf{x} = \mathbf{a} \times \mathbf{x}.$$

Show that

$$A_{\mathbf{a}} = \begin{bmatrix} 0 & -c & b \\ c & 0 & -a \\ -b & a & 0 \end{bmatrix}.$$

Conversely, show that if A is any nonzero 3×3 antisymmetric matrix, then there is a nonzero vector $\mathbf{a}$ in $\mathbb{R}^3$ so that for all $\mathbf{x}$ in $\mathbb{R}^3$,

$$A\mathbf{x} = \mathbf{a}_A \times \mathbf{x}.$$

This establishes a one-to-one correspondence

$$A_{\mathbf{a}} = \begin{bmatrix} 0 & -c & b \\ c & 0 & -a \\ -b & a & 0 \end{bmatrix} \quad \longleftrightarrow \quad \begin{bmatrix} a \\ b \\ c \end{bmatrix} = \mathbf{a}$$

between 3×3 antisymmetric matrices and vectors in $\mathbb{R}^3$.

3.10 Let A be a 3×3 antisymmetric matrix, and let $\mathbf{a}$ be the corresponding vector in $\mathbb{R}^3$. (See the previous exercise.)

(a) Show that unless $A = 0$, $\{\mathbf{a}\}$ is a basis for $\mathrm{Ker}(A)$ and that $\mathrm{rank}(A) = 2$. *In particular, no 3×3 antisymmetric matrix is invertible.*

(b) Show that $\mathrm{Img}(A)$ and $\mathrm{Ker}(A)$ are orthogonal complements in $\mathbb{R}^3$. *In particular, $\mathrm{Img}(A)$ is a plane through the origin, and $\mathrm{Ker}(A)$ is the normal line to this plane.*

(c) Show that there are no 3×3 rotation matrices that are antisymmetric.

3.11 Let A be a nonzero 3×3 antisymmetric matrix. Let $\mathbf{a}$ be the vector in $\mathbb{R}^3$ so that for all $\mathbf{x}$ in $\mathbb{R}^3$, $A\mathbf{x} = \mathbf{a} \times \mathbf{x}$. Define an orthonormal basis $\{\mathbf{u}_1, \mathbf{u}_2, \mathbf{u}_3\}$ of $\mathbb{R}^3$ as follows: Let $\mathbf{u}_3 = \dfrac{1}{|\mathbf{a}|}\mathbf{a}$. Let $\mathbf{u}_1$ be any unit vector that is orthogonal to $\mathbf{u}_3$, and define $\mathbf{u}_2 = \mathbf{u}_3 \times \mathbf{u}_1$. Let $U = [\mathbf{u}_1, \mathbf{u}_2, \mathbf{u}_3]$.

(a) Show that $\{\mathbf{u}_1, \mathbf{u}_2, \mathbf{u}_3\}$ is a right-handed orthonormal basis and that U is a rotation matrix.

(b) Show that

$$U^t A U = |\mathbf{a}| \begin{bmatrix} 0 & -1 & 0 \\ 1 & 0 & 0 \\ 0 & 0 & 0 \end{bmatrix}.$$

Rotation matrices

3.12 Let U_1 and U_2 be two rotation matrices. Show that $U_1 U_2$ and U_1^{-1} are also rotation matrices. (Recall that all orthogonal matrices are invertible.)

3.13 Let $U = [\mathbf{u}_1, \mathbf{u}_2]$ be a 2×2 orthogonal matrix. Since $\mathbf{u}_1$ is a unit vector, $\mathbf{u}_1 = \begin{bmatrix} \cos(\theta) \\ \sin(\theta) \end{bmatrix}$ for some θ with $0 \le \theta < 2\pi$. Since U is orthogonal, $\mathbf{u}_2$ is a unit vector that is orthogonal to $\mathbf{u}_1$. There are only two such unit vectors, namely $\mathbf{u}_1^{\perp}$ and $-\mathbf{u}_1^{\perp}$. Therefore

$$\text{either} \qquad U = \begin{bmatrix} \cos(\theta) & -\sin(\theta) \\ \sin(\theta) & \cos(\theta) \end{bmatrix} \qquad \text{or else} \qquad U = \begin{bmatrix} \cos(\theta) & \sin(\theta) \\ \sin(\theta) & -\cos(\theta) \end{bmatrix}.$$

Show that in the first case, the linear transformation represented by U is counterclockwise rotation in $\mathbb{R}^2$ through the angle θ, and in the second case, the linear transformation represented by U is reflection in $\mathbb{R}^2$ about the line through the origin and $\begin{bmatrix} \cos(\theta/2) \\ \sin(\theta/2) \end{bmatrix}$. (Drawing diagrams will help.) *In particular, every 2×2 orthogonal matrix U is either a rotation or a reflection, according to whether $\det(U) = 1$ or $\det(U) = -1$.*

3.14 Let $U_1 = \begin{bmatrix} \cos(2\theta) & \sin(2\theta) \\ \sin(2\theta) & -\cos(2\theta) \end{bmatrix}$ and $U_2 = \begin{bmatrix} \cos(3\theta) & \sin(3\theta) \\ \sin(3\theta) & -\cos(3\theta) \end{bmatrix}$. Compute the product $U_1 U_2$, and show that every 2×2 rotation matrix is the product of two 2×2 reflection matrices. Draw a diagram depicting a rotation as the product of two reflections.

3.15 Let $U_1 = \begin{bmatrix} \cos(3\theta/2) & \sin(3\theta/2) \\ \sin(3\theta/2) & -\cos(3\theta/2) \end{bmatrix}$ and $U_2 = \begin{bmatrix} \cos(\theta/2) & \sin(\theta/2) \\ \sin(\theta/2) & -\cos(\theta/2) \end{bmatrix}$. Show that U_1 and U_2 are reflection matrices. Compute their product, and give another proof that every 2×2 rotation matrix can be written as the product of two reflection matrices.

3.16 Let U be a 3×3 rotation matrix, and suppose that U is symmetric; that is $U^t = U$.

(a) Show that $(U + I)(U - I) = (U - I)(U + I) = 0$.

(b) Show that $\text{Ker}(U + I)$ and $\text{Ker}(U - I)$ are othogonal complements of one another as subspaces of $\mathbb{R}^3$.

(c) Suppose that U is not the identity matrix, so that $\text{Ker}(U - I)$ is not all of $\mathbb{R}^3$. Show that it is possible to build an orthonormal basis $\{v_1, v_2, v_3\}$ of $\mathbb{R}^3$ out of orthonormal bases for $\text{Ker}(U + I)$ and $\text{Ker}(U - I)$ such that with $V = [v_1, v_2, v_3]$,

$$V^t U V = \begin{bmatrix} -1 & 0 & 0 \\ 0 & -1 & 0 \\ 0 & 0 & 1 \end{bmatrix} .$$

(d) Note that the linear transformation of rotating a vector $\mathbf{v}$ in the plane through an angle π (counterclockwise or clockwise; it does not matter) sends $\mathbf{v}$ into $-\mathbf{v}$. Conclude that the linear transformation corresponding to

$$\begin{bmatrix} -1 & 0 & 0 \\ 0 & -1 & 0 \\ 0 & 0 & 1 \end{bmatrix}$$

is rotation through an angle π in the x, y plane.

(e) Show that a 3×3 rotation matrix U is symmetric if and only if the linear transformation represented by U is rotation through the angle π about some line through the origin in $\mathbb{R}^3$.

3.17 Let Q be a 3×3 rotation matrix, and assume that Q is not symmetric. Define

$$S = \frac{1}{2}(Q + Q^t) \qquad \text{and} \qquad A = \frac{1}{2}(Q - Q^t)$$

so that $Q = S + A$, where S is symmetric and A is antisymmetric. By assumption, $A \neq 0$.

Define the *plane of rotation* for Q to be $\text{Img}(A)$ and the *axis of rotation* for Q to be $\text{Ker}(A)$. (Exercises 3.10 and 3.11 explain why $\text{Ker}(A)$ is a line).

The goal of this exercise is to show that the linear transformation of $\mathbb{R}^3$ that Q represents is the rotation of $\mathbb{R}^3$ about the axis of rotation through some angle θ. This justifies our use of the term rotation matrices.

(a) Let $\{\mathbf{v}\}$ be either unit vector in $\text{Ker}(A)$. Show that either $Q\mathbf{v} = \mathbf{v}$ or else $Q\mathbf{v} = -\mathbf{v}$ and that either $Q^t\mathbf{v} = \mathbf{v}$ or else $Q^t\mathbf{v} = -\mathbf{v}$.

(b) Let $\{\mathbf{u}_1, \mathbf{u}_2\}$ be an orthonormal basis of $\text{Img}(A)$. By changing the sign of $\mathbf{u}_2$, if necessary, we may assume that

$$\mathbf{u}_2 \cdot Q\mathbf{u}_1 \geq 0 ,$$

and we do so. Define an angle θ by

$$\theta = \cos^{-1}(\mathbf{u}_1 \cdot Q\mathbf{u}_1) .$$

Show that

$$Q\mathbf{u}_1 = \cos(\theta)\mathbf{u}_1 + \sin(\theta)\mathbf{u}_2 .$$

(c) Let $U = [\mathbf{u}_1, \mathbf{u}_2, \mathbf{v}]$, and define

$$V = U^t Q U .$$

Show that V is a rotation matrix and that

$$V = \begin{bmatrix} \cos(\theta) & * & 0 \\ \sin(\theta) & * & 0 \\ 0 & * & \pm 1 \end{bmatrix}.$$

(d) Show that since Q is not a symmetric matrix (by hypothesis), neither is V. Using this fact and the fact that the second column of V is orthogonal to the first and third, show that in fact

$$V = \begin{bmatrix} \cos(\theta) & -\sin(\theta) & 0 \\ \sin(\theta) & \cos(\theta) & 0 \\ 0 & 0 & 1 \end{bmatrix},$$

with $0 < \theta < \pi$.

(e) Show that the linear transformation represented by V is a counterclockwise rotation through the angle θ, fixing the z-axis.

(f) Show that the linear transformation represented by Q is a rotation through the angle θ about the axis of rotation, that is, the line through the origin and $\mathbf{v}$.

3.18 Let Q be a 3×3 rotation matrix that is not the identity matrix.

(a) Show that $\text{Ker}(Q - I)$ is always one dimensional and that for every vector $\mathbf{v}$ in $\text{Ker}(Q - I)$, $Q\mathbf{v} = \mathbf{v}$. *The line through the origin and $\mathbf{v}$, which is kept fixed by Q, is the axis of rotation of Q.*

(b) The plane that is orthogonal to the axis of rotation is called the *plane of rotation*. Show that for all unit vectors $\mathbf{w}$ in the plane of rotation, $\mathbf{w} \cdot Q\mathbf{w}$ is the same number. Define the angle of rotation θ by

$$\theta = \cos^{-1}(\mathbf{w} \cdot Q\mathbf{w})$$

for any unit vector $\mathbf{w}$ in the plane of rotation.

(c) The angle of rotation θ, by definition, satisfies $0 < \theta \leq \pi$. Suppose that $\theta < \pi$, and define $\mathbf{u}$ by

$$\mathbf{u} = \frac{1}{\sin(\theta)} \mathbf{w} \times Q\mathbf{w}$$

where $\mathbf{w}$ is any unit vector in the plane of rotation. Show that $\mathbf{u}$ is a unit vector lying in the axis of rotation, and that $\mathbf{u}$ is independent of the choice of the unit vector $\mathbf{w}$ in the plane of rotation. The unit vector $\mathbf{u}$ is called the *direction of rotation*.

(d) Again suppose that the angle of rotation θ satisfies $\theta < \pi$ so that the direction of rotation, $\mathbf{u}$, is defined. Think of grasping the axis of rotation with your right hand so that your thumb extends along the direction specified by $\mathbf{u}$. Show that the linear transformation represented by Q is a rotation through the angle θ about the axis of rotation and that the sense of the rotation is the one in which the fingers of your right hand curl about the axis of rotation.

In Exercises 3.19 through 3.22, let Q_1, Q_2, Q_3 and Q_4 be the following rotation matrices:

$$Q_1 = \frac{1}{9} \begin{bmatrix} 1 & 8 & 4 \\ 4 & -4 & 7 \\ 8 & 1 & -4 \end{bmatrix} \qquad Q_2 = \frac{1}{75} \begin{bmatrix} -23 & -14 & 70 \\ -14 & 73 & 10 \\ -70 & -10 & -25 \end{bmatrix}$$

$$Q_3 = \frac{1}{45} \begin{bmatrix} 35 & -20 & -20 \\ 4 & 35 & -28 \\ 28 & 20 & 29 \end{bmatrix} \qquad Q_4 = \frac{1}{117} \begin{bmatrix} 77 & -68 & -56 \\ 4 & 77 & -88 \\ 88 & 56 & 53 \end{bmatrix}.$$

3.19 Check that Q_1 is a rotation matrix. Then find the angle of rotation θ and the direction of rotation **u** for Q_1. Also, find an equation for the plane of rotation.

3.20 Check that Q_2 is a rotation matrix. Then find the angle of rotation θ and the direction of rotation **u** for Q_2. Also, find an equation for the plane of rotation.

3.21 Check that Q_3 is a rotation matrix. Then find the angle of rotation θ and the direction of rotation **u** for Q_3. Also, find an equation for the plane of rotation.

3.22 Check that Q_4 is a rotation matrix. Then find the angle of rotation θ and the direction of rotation **u** for Q_4. Also, find an equation for the plane of rotation.

3.23 Let $\{\mathbf{u}_1, \mathbf{u}_2, \mathbf{u}_3\}$ be a right-handed orthonormal basis of $\mathbb{R}^3$.

(a) Show that not only does $\mathbf{u}_1 \times \mathbf{u}_2 = \mathbf{u}_3$, but also

$$\mathbf{u}_2 \times \mathbf{u}_3 = \mathbf{u}_1 \qquad \text{and} \qquad \mathbf{u}_3 \times \mathbf{u}_1 = \mathbf{u}_2 \,.$$

(b) Show that for any **x** in $\mathbb{R}^3$,

$$\mathbf{u}_1 \times \mathbf{x} = (\mathbf{u}_2 \cdot \mathbf{x})\mathbf{u}_3 - (\mathbf{u}_3 \cdot \mathbf{x})\mathbf{u}_2$$
$$\mathbf{u}_2 \times \mathbf{x} = (\mathbf{u}_3 \cdot \mathbf{x})\mathbf{u}_1 - (\mathbf{u}_1 \cdot \mathbf{x})\mathbf{u}_3$$
$$\mathbf{u}_3 \times \mathbf{x} = (\mathbf{u}_1 \cdot \mathbf{x})\mathbf{u}_2 - (\mathbf{u}_2 \cdot \mathbf{x})\mathbf{u}_1$$

3.24 There is a formula due to Euler for the matrix of the rotation with angle θ and direction vector **u**. Euler's formula is

$$Q = \cos(\theta)I + (1 - \cos(\theta))\mathbf{u}\mathbf{u}^t + \sin(\theta)A_{\mathbf{u}} \,.$$

Compute the right-hand side for each of the four pairs of rotation angles and directions found in Exercises 3.19–3.22, thus checking Euler's formula in these examples. (Euler's formula is derived in the next exercise.)

3.25 Let Q be a 3×3 rotation matrix, and let θ be the angle of rotation of Q.

(a) Suppose that $\theta = \pi$. Let **u** be either unit vector in the axis of rotation. Show that

$$Q = 2\mathbf{u}\mathbf{u}^t - I \,.$$

(b) Suppose that $\theta < \pi$. Let **u** be the uniquely specified direction of rotation. Let $A_{\mathbf{u}}$ be the antisymmetric matrix corresponding to **u**; that is,

$$A = \begin{bmatrix} 0 & -u_3 & u_2 \\ u_3 & 0 & -u_1 \\ -u_2 & u_1 & 0 \end{bmatrix} \qquad \text{where } \mathbf{u} = \begin{bmatrix} u_1 \\ u_2 \\ u_3 \end{bmatrix} \,.$$

Show that

$$Q = \cos(\theta)I + (1 - \cos(\theta))\mathbf{u}\mathbf{u}^t + \sin(\theta)A_{\mathbf{u}} \,.$$

3.26 Let f be a transformation from $\mathbb{R}^3$ to $\mathbb{R}^3$ such that for all $\mathbf{x}$ and $\mathbf{y}$ in $\mathbb{R}^3$,

$$|f(\mathbf{x}) - f(\mathbf{y})| = |\mathbf{x} - \mathbf{y}| \, .$$

Such a transformation is called a *rigid body transformation*: Under the transformation, the distance between any two points is unchanged; the two points may as well be connected by an iron bar.

We are *not* assuming that f is linear. Indeed, for any fixed nonzero vector $\mathbf{a}$ in $\mathbb{R}^3$, the *translation transformation* f defined by $f(\mathbf{x}) = \mathbf{x} + \mathbf{a}$ has the required property, but it is not linear. (Note that in general $f(2\mathbf{x}) \neq 2f(\mathbf{x})$). However, it turns out that apart from a translation, any rigid body transformation is given by an orthogonal transformation U. That is, if f is a rigid body transformation, then there is a vector $\mathbf{a}$ in $\mathbb{R}^3$ and a 3×3 orthogonal matrix U so that

$$f(\mathbf{x}) = \mathbf{a} + U\mathbf{x} \, .$$

The goal of this exercise is to make it clear why every rigid body transformation has this simple form.

(a) Let f be a rigid body transformation of $\mathbb{R}^3$. Define a new transformation g by

$$g(\mathbf{x}) = f(\mathbf{x}) = f(0) \, .$$

Show that g is also a rigid body transformation.

(b) Show that for any two vectors $\mathbf{x}$ and $\mathbf{y}$ in $\mathbb{R}^3$,

$$g(\mathbf{x}) \cdot g(\mathbf{y}) = \mathbf{x} \cdot \mathbf{y} \, .$$

(c) Define vectors $\mathbf{u}_1, \mathbf{u}_2$ and $\mathbf{u}_3$ in $\mathbb{R}^3$ by $\mathbf{u}_j = g(\mathbf{e}_j)$ for $j = 1, 2, 3$. Show that $\{\mathbf{u}_1, \mathbf{u}_2, \mathbf{u}_3\}$ is an orthonormal basis of $\mathbb{R}^3$.

(d) Since $\{\mathbf{u}_1, \mathbf{u}_2, \mathbf{u}_3\}$ is an orthonormal basis of $\mathbb{R}^3$, it follows that for any vector $\mathbf{x}$ in $\mathbb{R}^3$,

$$g(\mathbf{x}) = \sum_{j=1}^{3} (\mathbf{u}_j \cdot g(\mathbf{x}))\mathbf{u}_j \, .$$

Show that, in fact,

$$g(\mathbf{x}) = \sum_{j=1}^{3} (\mathbf{e}_j \cdot \mathbf{x})\mathbf{u}_j$$

and thus that

$$g(\mathbf{x}) = U\mathbf{x}$$

where $U = [\mathbf{u}_1, \mathbf{u}_2, \mathbf{u}_3]$.

(e) Show that if f is any rigid body transformation of $\mathbb{R}^3$, then there is an orthogonal matrix U such that for all $\mathbf{x}$ in $\mathbb{R}^3$,

$$f(\mathbf{x}) = f(0) + U\mathbf{x} \, .$$

│ Permutations

4.1 Permutations and permutation matrices

In this section, we finally give a formula for the determinant of a square matrix. We will see from this formula that there really is a function on the set of square matrices that has the three properties we have used to define the determinant. Thus, the formula shows that our definition does make sense. Although the formula is not helpful for actually computing the determinant of a given square matrix A—the approach we have explained using the three properties is much much more efficient—it does have some theoretical uses, and becoming familiar with it is worthwhile.

For this purpose, it is necessary to delve a bit more deeply into permutation matrices and permutations in general—not so surprising since an essential property of the determinant is the sign change when two rows are swapped.

D E F I N I T I O N │ **(Permutation)** A *permutation* of $\{1, 2, \ldots, n\}$ is a function σ from this set *onto* itself.

Recall that "onto" means that for every j in $\{1, 2, \ldots, n\}$, there is an i with $\sigma(i) = j$. We can specify a permutation σ of $\{1, 2, \ldots, n\}$ by listing the assignments it makes:*

$$\sigma = \begin{array}{ccccc} 1 & 2 & 3 & \cdots & n \\ \downarrow & \downarrow & \downarrow & \cdots & \downarrow \\ \sigma(1) & \sigma(2) & \sigma(3) & \cdots & \sigma(n) \end{array}.$$

For example, if $n = 3$, and $\sigma(1) = 2$, $\sigma(2) = 3$, and $\sigma(3) = 1$,

$$\sigma = \begin{array}{ccc} 1 & 2 & 3 \\ \downarrow & \downarrow & \downarrow \\ 2 & 3 & 1 \end{array}.$$

The arrows are not really telling us much; we can remember that the top row is inputs and the bottom row is outputs. Let us shorten the notation to

$$\sigma = \begin{array}{ccc} 1 & 2 & 3 \\ 2 & 3 & 1 \end{array}.$$

The generalization of this way of writing permutations to higher values of n is plain, and we will use it freely.

There are exactly $n!$ permutations of $\{1, 2, \ldots, n\}$. Indeed, consider any permutation σ of $\{1, 2, \ldots, n\}$. There are n choices for the value of $\sigma(1)$. Make this choice, and then, $\sigma(1)$ being taken, there are $n - 1$ choices remaining for value of $\sigma(2)$. Next, there are $n - 2$ choices for

---

*The equal sign makes sense: σ is a function, and the right-hand side specifies a function. The equality is the equality of two functions.

$\sigma(3)$, the value to be assigned to 3. Continuing in this way, there are $n(n-1)(n-2)\cdots 1 = n!$ choices to make, and each one leads to a distinct permutation.

EXAMPLE 18   **(Permutations of {1,2,3})** There are six permutations of $\{1, 2, 3\}$:

$$\sigma_a = \begin{matrix} 1 & 2 & 3 \\ 1 & 2 & 3 \end{matrix}$$

$$\sigma_b = \begin{matrix} 1 & 2 & 3 \\ 2 & 1 & 3 \end{matrix} \qquad \sigma_c = \begin{matrix} 1 & 2 & 3 \\ 1 & 3 & 2 \end{matrix}$$

$$\sigma_d = \begin{matrix} 1 & 2 & 3 \\ 2 & 3 & 1 \end{matrix} \qquad \sigma_e = \begin{matrix} 1 & 2 & 3 \\ 3 & 1 & 2 \end{matrix}$$ (4.1)

$$\sigma_f = \begin{matrix} 1 & 2 & 3 \\ 3 & 2 & 1 \end{matrix}.$$

Since permutations of $\{1, 2, \ldots, n\}$ are functions from this set onto itself, we can compose them: If σ_1 and σ_2 are two permutations of $\{1, 2, \ldots, n\}$, then $\sigma_2 \circ \sigma_1$ is defined by

$$\sigma_2 \circ \sigma_1(i) = \sigma_2(\sigma_1(i)), \qquad \text{for each } i = 1, \ldots, n. \tag{4.2}$$

EXAMPLE 19   **(Composing permutations)** Let us compute $\sigma_d \circ \sigma_b$, where σ_d and σ_b are the permutations given in (4.1). From (4.1) we see that

$$\sigma_d \circ \sigma_b(1) = \sigma_d(\sigma_b(1)) = \sigma_d(2) = 3$$

$$\sigma_d \circ \sigma_b(2) = \sigma_d(\sigma_b(2)) = \sigma_d(1) = 2$$

$$\sigma_d \circ \sigma_b(2) = \sigma_d(\sigma_b(2)) = \sigma_d(3) = 1.$$

Thus,

$$\sigma_d \circ \sigma_b = \begin{matrix} 1 & 2 & 3 \\ 3 & 2 & 1 \end{matrix} = \sigma_f.$$

Notice that in this example, the composition product of two permutations gave another permutation. This closure under composition always happens. Indeed, consider any two permutations σ_2 and σ_1 of $\{1, 2, \ldots, n\}$. To see that $\sigma_2 \circ \sigma_1$ is also a permutation, we just need to check that for each j in $\{1, 2, \ldots, n\}$, there is an i such that $\sigma_2 \circ \sigma_1(i) = j$. But since σ_2 is a permutation, there is a k so that $\sigma_2(k) = j$. And since σ_1 is a permutation, there is an i so that $\sigma_1(i) = k$. Then $\sigma_2 \circ \sigma_1(i) = \sigma_2(\sigma_1(i)) = \sigma_2(k) = j$. We have found the i for which $\sigma_2 \circ \sigma_1(i) = j$, so $\sigma_2 \circ \sigma_1$ is a permutation.

The permutation at the top of the list in (4.1), σ_a, is called the *identity* permutation since it sends each element of $\{1, 2, 3\}$ to itself. This has an obvious generalization to other values of n. Moreover, every permutation σ has an inverse, σ^{-1}, which sends any j in back to the integer i in $\{1, 2, \ldots, n\}$ from whence it came.* The inverse also is a map of $\{1, 2, \ldots, n\}$ onto itself, and hence it is a permutation. (It is the original map "played in reverse.")

---

*Since $\{1, 2, \ldots, n\}$ is a finite set and since σ is onto σ is also one-to-one. Indeed, if $\sigma(i) = \sigma(j)$ for $i \neq j$, it would have spent two of n "shots" at hitting a single target, which would preclude hitting all n. So σ is necessarily one-to-one from $\{1, 2, \ldots, n\}$ onto itself, and hence it is invertible.

There is a connection between the composition product for permutations and the matrix product. Let σ_1 be a permutation of $\{1, 2, \ldots, n\}$. The corresponding permutation matrix P_σ is the $n \times n$ matrix given by

$$P_\sigma = [\, \mathbf{e}_{\sigma(1)}, \mathbf{e}_{\sigma(2)}, \ldots, \mathbf{e}_{\sigma(n)} \,]. \tag{4.3}$$

Since the jth column of an $n \times n$ matrix A is $A\mathbf{e}_j$, we also have $P_\sigma \mathbf{e}_j = \mathbf{e}_{\sigma(j)}$; that is, the jth column of P_σ is $\mathbf{e}_{\sigma(j)}$.

Consider any two permutations σ_2 and σ_1 of $\{1, 2, \ldots, n\}$. Then for any $1 \leq j \leq n$,

$$\left(P_{\sigma_1} P_{\sigma_2}\right) \mathbf{e}_j = P_{\sigma_1} \left(P_{\sigma_2} \mathbf{e}_j\right) = P_{\sigma_1} \mathbf{e}_{\sigma_2(j)} = \mathbf{e}_{\sigma_1 \circ \sigma_2(j)} = P_{\sigma_1 \circ \sigma_2} \mathbf{e}_j.$$

Since this statement holds for each $1 \leq j \leq n$,

$$P_{\sigma_1} P_{\sigma_2} = P_{\sigma_1 \circ \sigma_2}, \tag{4.4}$$

and the product on the left is matrix multiplication.

EXAMPLE 20   **(Composition products of permutations and matrix products)** Let σ_d and σ_b be the permutations given in (4.1). Then

$$P_{\sigma_d} = \begin{bmatrix} 0 & 0 & 1 \\ 1 & 0 & 0 \\ 0 & 1 & 0 \end{bmatrix} \quad \text{and} \quad P_{\sigma_b} = \begin{bmatrix} 0 & 1 & 0 \\ 1 & 0 & 0 \\ 0 & 0 & 1 \end{bmatrix},$$

and

$$P_{\sigma_d} P_{\sigma_b} = \begin{bmatrix} 0 & 0 & 1 \\ 1 & 0 & 0 \\ 0 & 1 & 0 \end{bmatrix} \begin{bmatrix} 0 & 1 & 0 \\ 1 & 0 & 0 \\ 0 & 0 & 1 \end{bmatrix} = \begin{bmatrix} 0 & 0 & 1 \\ 0 & 1 & 0 \\ 1 & 0 & 0 \end{bmatrix} = P_{\sigma_f}.$$

These are the same permutation matrices we encountered when studying the LU decomposition. There we got them by swapping rows of the identity matrix, and they served as records of the row swapping we did. For studying the LU decomposition, we did not need to know as much about them as we will to study determinants, so we did not go into detail back then. We did, however, make one observation that bears repeating:

- *The columns of a permutation matrix are orthonormal since they are just the standard basis vectors for $\mathbb{R}^n$ in a different order. Hence, every permutation matrix $P\sigma$ is an orthogonal matrix, so*

$$\left(P_\sigma\right)^{-1} = \left(P_\sigma\right)^t. \tag{4.5}$$

It follows from (4.4) that $\left(P_\sigma\right)^{-1} = P_{\sigma^{-1}}$, so another expression is

$$P_{\sigma^{-1}} = \left(P_\sigma\right)^t. \tag{4.6}$$

4.2 The degree of mixing of a permutation

Consider once more the list (4.1) of permutations. Except for σ_a, which is the identity permutation, all the permutations "mix things up" to some extent. In fact, we have arranged these permutations in an order that reflects a measure of "how much mixing" is involved in each one, starting from no mixing at the top, to the most mixing at the bottom. Here is the definition we use to quantify the mixing:

DEFINITION	**(Degree of mixing)** The *degree of mixing* of a permutation σ of $\{1, 2, \ldots, n\}$ is the number of pairs of integers (i, j) in $\{1, 2, \ldots, n\}$ with $\qquad\qquad i < j \qquad$ and $\qquad \sigma(i) > \sigma(j).$ $\qquad\qquad\qquad$ (4.7) This number is denoted $D(\sigma)$.

Evidently, the degree of mixing of a permutation is a count of the number of pairs in $\{1, 2, \ldots, n\}$ whose order is reversed by σ. The more "reversed" pairs, the more mixing.

EXAMPLE 21 **(Computing the degree of mixing)** Let us compute $D(\sigma)$ for each of the six permutations of $\{1, 2, 3\}$. There are three pairs (i, j) with $i < j$, namely,

$$(1, 2), \qquad (1, 3), \qquad \text{and} \qquad (2, 3).$$

To compute the degree of mixing of σ, we look at

$$(\sigma(1), \sigma(2)), \qquad (\sigma(1), \sigma(3)), \qquad \text{and} \qquad (\sigma(2), \sigma(3))$$

and count the number of times these pairs are "out of order." You can easily check that

$$D(\sigma_a) = 0, \qquad D(\sigma_b) = 1, \qquad D(\sigma_c) = 1, \qquad D(\sigma_d) = 2, \qquad D(\sigma_e) = 2, \qquad \text{and} \qquad D(\sigma_f) = 3.$$

The definition of $D(\sigma)$ is useful because of the way it interacts with the composition product. Indeed, consider the following question:

- *Given two permutations σ_1 and σ_2 of $\{1, 2, \ldots, n\}$, what can we say about $D(\sigma_2 \circ \sigma_1)$?*

First, in applying σ_1, we reverse the order of $D(\sigma_1)$ pairs. Then, applying σ_2 after that, we reverse the order of $D(\sigma_2)$ pairs. So the number of pairs that are reversed by $\sigma_2 \circ \sigma_1$ is no more than $D(\sigma_1) + D(\sigma_2)$.

However, some of the pairs that σ_2 reverses may have already been put out of order by σ_1. *In this case, σ_2 puts them back in order.* An extreme case is when $\sigma_2 = (\sigma_1)^{-1}$. Then σ_2 undoes all the mixing done by σ_1, and $D(\sigma_2 \circ \sigma_1) = 0$.

So we conclude that $0 \leq D(\sigma_2 \circ \sigma_1) \leq D(\sigma_1) + D(\sigma_2)$. We can say more: Suppose that when σ_2 is applied, c pairs that had been put out of order by σ_1 are "reordered" when we apply σ_2. Then,

- *Of the $D(\sigma_1)$ pairs reversed by σ_1, exactly $D(\sigma_1) - c$ are still reversed after applying σ_2.*
- *Of the $D(\sigma_2)$ pair reversals created by σ_2, c are "used up" undoing reversals created by σ_1, so exactly $D(\sigma_2) - c$ new reversals are created.*

Adding things up,

$$D(\sigma_2 \circ \sigma_1) = (D(\sigma_1) - c) + (D(\sigma_2) - c) = D(\sigma_1) + D(\sigma_2) - 2c. \qquad (4.8)$$

Whatever c is, $2c$ is always an even integer, so $(-1)^{2c} = 1$ and

$$(-1)^{D(\sigma_2 \circ \sigma_1)} = (-1)^{D(\sigma_1)}(-1)^{D(\sigma_1)}. \qquad (4.9)$$

DEFINITION	**(Character of a permutation)** The *character* $\chi(\sigma)$ of a permutation σ is defined by $$\chi(\sigma) = (-1)^{D(\sigma)}, \qquad (4.10)$$ where $D(\sigma)$ is given by (1.7). A permutation σ is called an *even permutation* if $\chi(\sigma) = 1$, and an *odd permutation* if $\chi(\sigma) = -1$.

In Example 18 σ_a, σ_d, and σ_e are even permutations, whereas σ_b, σ_c, and σ_f are odd permutations. The point of the definition is that $\chi(\sigma_2 \circ \sigma_1) = \chi(\sigma_2)\chi(\sigma_1)$. That is, *the character of a product equals the product of the characters*, which follows directly from (4.9).

If you want to determine $\chi(\sigma)$ for a given permutation σ, you do not always have to compute $D(\sigma)$ first and then apply the definition (4.10). There are some general rules for particular kinds of permutations.

DEFINITION	**(Pair permutations)** For each $i < j$ in $\{1, 2, \ldots, n\}$ the *pair permutation* $\sigma_{i,j}$ is defined by $$\sigma_{i,j}(i) = j, \qquad \sigma_{i,j}(j) = i, \qquad \text{and} \qquad \sigma_{i,j}(k) = k \qquad \text{for } k \neq i, j. \qquad (4.11)$$ It is called an *adjacent pair permutation* if and only if $j = i + 1$.

EXAMPLE 22   **(A pair permutation)**   For $n = 4$,

$$\sigma_{2,4} = \begin{matrix} 1 & 2 & 3 & 4 \\ 1 & 4 & 3 & 2 \end{matrix}.$$

Notice that each pair permutation is its own inverse; applying it twice swaps the reversed pair back into place.

Next notice that for each adjacent pair permutation $\sigma_{i,i+1}$, $D(\sigma_{i,i+1}) = 1$, and hence, $\chi(\sigma_{i,i+1}) = -1$. What about general pair permutations?

- *For any $i < j$, $\sigma_{i,j}$ can be written as the product of $2k - 1$ adjacent pair permutations where $k = j - i$.*

Therefore, since the character of a product is the product of the characters,

$$\chi(\sigma_{i,j}) = (-1)^{2k-1} = -1$$

for every pair permutation, adjacent or not.

To justify the claim about σ_{ij} with $i < j$, write $j = i + k$. Then we can "move" i to the right of j using k adjacent pair permutations. We can then move j back to the ith spot with $k - 1$ pair permutations. Only $k - 1$ are required, because the last pair permutation used to move i into the jth place already moved j one place to the left. The total used is $2k - 1$, an odd number.

We summarize the discussion in the following theorem.

THEOREM 12   **(Properties of the character)**   *For any two permutations σ_1 and σ_2 of $\{1, 2, \ldots, n\}$,*

$$\chi(\sigma_2 \circ \sigma_1) = \chi(\sigma_2)\chi(\sigma_1). \qquad (4.12)$$

Moreover, for any pair permutation $\sigma_{i,j}$,

$$\chi(\sigma_{i,j}) = -1. \qquad (4.13)$$

The theorem gives us a convenient way to compute $\chi(\sigma)$: Bring the sequence $(1, 2, \ldots, n)$ into the order $(\sigma(1), \sigma(2), \ldots, \sigma(n))$ by swapping pairs, that is, by pair permutations. Then $\chi(\sigma)$ is the product of the characters of these pair permutations, so it is $(-1)^N$, where N is the number of pair permutations you used.

EXAMPLE 23

(Computing $\chi(\sigma)$ by counting pair swaps) Consider

$$\sigma = \begin{matrix} 1 & 2 & 3 & 4 \\ 4 & 1 & 3 & 2 \end{matrix}.$$

We can transform $(1, 2, 3, 4)$ to $(4, 1, 3, 2)$ using pair permutations

$$(1, 2, 3, 4) \rightarrow (4, 2, 3, 1) \rightarrow (4, 1, 3, 2)$$

or as well by

$$(1, 2, 3, 4) \rightarrow (1, 2, 4, 3) \rightarrow (1, 4, 2, 3) \rightarrow (4, 1, 2, 3) \rightarrow (4, 1, 3, 2).$$

In the first case, we used $N = 2$ pair permutations and in the second case, $N = 4$. Either way, we see $\chi(\sigma) = (-1)^2 = (-1)^4 = 1$, so σ is even.

You might wonder why we did not just *define* $\chi(\sigma)$ to be $(-1)^N$ where N is the number of pair swaps required to produce σ. The point is this: Suppose you could write some σ as a product of 7 pair permutations and also 242 pair permutations. Then you would have $\chi(\sigma) = (-1)^7 = -1$ and $\chi(\sigma) = (-1)^{242} = 1$, and they cannot *both* be right. If this happened, $\chi(\sigma) = (-1)^N$ would not be a well-defined function.

Evidently, our preceding analysis implies that for any given permutation σ, if you can write σ as a product of an odd number of pair permutations, then *every* way of writing σ as a product of pair permutations uses an odd number of them. This fact is not obvious! We know it is true by Theorem 12 and (4.10), which defines $\chi(\sigma)$.

We are now ready to apply the idea to linear algebra.

4.3 The determinant formula

In this section we break down the formula for the determinant into building blocks. The building blocks will be simpler functions that we will combine to form the determinant function. Here is the first one:

Given an $n \times n$ matrix A and a permutation σ of $\{1, 2, \ldots, n\}$, define the number $\sigma(A)$ by

$$\sigma(A) = A_{\sigma(1),1} A_{\sigma(2),2} \cdots A_{\sigma(n),n}. \tag{4.14}$$

This is the product of n entries from A: There is one from each column, and the one taken from the jth column comes from row $\sigma(j)$. Because the permutation is one-to-one, there is exactly one term from each row.

EXAMPLE 24

(Computation of $\sigma(A)$) With $n = 3$, let

$$\sigma = \begin{matrix} 1 & 2 & 3 \\ 2 & 3 & 1 \end{matrix} \qquad \text{and} \qquad A = \begin{bmatrix} 1 & 2 & 3 \\ 4 & 5 & 6 \\ 7 & 8 & 9 \end{bmatrix}.$$

Then

$$\sigma(A) = A_{2,1} A_{3,2} A_{1,3} = 4 \times 8 \times 3 = 96.$$

There is an easy way to see which entries of A are included in the product defining $\sigma(A)$ by looking at the permutation matrix P_σ. You see from (4.14) that the entry $A_{k,\ell}$ is included in the product defining $\sigma(A)$ if and only if $k = \sigma(\ell)$. This is the case if and only if $(P_\sigma)_{k,\ell} = 1$, which gives us the following rule.

- *To compute $\sigma(A)$, take the product of the entries of A in places where P_σ has a 1. Note that there is exactly one entry from each column and that there is exactly one entry from each row in the product.*

EXAMPLE 25   **(Computation of $\sigma(A)$ using P_σ)** Let $\sigma = \begin{matrix} 1 & 2 & 3 & 4 \\ 4 & 3 & 1 & 2 \end{matrix}$ so that $P_\sigma = \begin{bmatrix} 0 & 0 & 1 & 0 \\ 0 & 0 & 0 & 1 \\ 0 & 1 & 0 & 0 \\ 1 & 0 & 0 & 0 \end{bmatrix}$. From where

the 1s are, we see that for any 4×4 matrix A,

$$\sigma(A) = A_{4,1}A_{3,2}A_{1,3}A_{2,4}.$$

Note that $A_{4,1}A_{3,2}A_{1,3}A_{2,4} = A_{\sigma(1),1}A_{\sigma(2),2}A_{\sigma(3),3}A_{\sigma(4),4}$, in agreement with (4.14).

There is one other building block that goes into the determinant function: the character function $\chi(\sigma)$. Since we are already familiar with χ, we are ready for our second definition of the determinant function.

DEFINITION

(Determinant) Let A be an $n \times n$ matrix. The *determinant* of A is the number $\det(A)$ defined by

$$\det(A) = \sum_\sigma \chi(\sigma)\sigma(A), \tag{4.15}$$

where $\sigma(A)$ is given by (4.14) and where the sum is taken over all of the $n!$ permutations of $\{1, 2, \ldots, n\}$.

Of course, we have to show that this definition is equivalent to the one we made in Section 1. We shall do this by showing that the function defined by (4.15) has the three properties listed in Theorem 3.

The present definition is a bit complicated, and it is hard to use directly for calculation. But it has the advantage that it is clearly well defined; it makes no reference to anything whose existence is in doubt. It is fully justified, and once we have shown that the two definitions are the same, our previous definition will finally be fully justified.

Of course, we have to be careful not to be circular in our logic. But because this definition is *so different* from the first one, we cannot even be tempted to use results based on the first definition to study the second. First, though, let us check the 2×2 case.

EXAMPLE 26   **(A 2×2 determinant)** Consider the general 2×2 matrix

$$A = \begin{bmatrix} a & b \\ c & d \end{bmatrix}.$$

There are only two permutations of $\{1, 2\}$ to consider, namely,

$$\sigma_1 = \begin{matrix} 1 & 2 \\ 1 & 2 \end{matrix} \quad \text{and} \quad \sigma_2 = \begin{matrix} 1 & 2 \\ 2 & 1 \end{matrix}.$$

Clearly $\chi(\sigma_1) = 1$ and $\chi(\sigma_2) = -1$. Hence $\det(A) = A_{1,1}A_{2,2} - A_{2,1}A_{1,2} = ad - bc$.

This is reassuring! Let us next check the 3×3 case.

EXAMPLE 27 **(A 3×3 case)** Consider a general 3×3 matrix A. We have already worked out a list of the six permutations of $\{1, 2, 3\}$ in (4.1) and computed the characters of each of them. In the 3×3 case, then, the definition (4.15) leads to

$$\det(A) = A_{1,1}A_{2,2}A_{3,3} + A_{2,1}A_{3,2}A_{1,3} + A_{3,1}A_{1,2}A_{2,3}$$

$$- A_{2,1}A_{1,2}A_{3,3} - A_{1,1}A_{3,2}A_{2,3} - A_{3,1}A_{2,2}A_{1,3}.$$

This, too, is reassuring—the formula (4.15) leads us to the formula (2.11) that we found before. We now go on to the general case.

THEOREM 13 **(Agreement of the determinant definitions)** *Let f be the numerically valued function on the $n \times n$ matrices defined by*

$$f(A) = \sum_{\sigma} \chi(\sigma)\sigma(A). \tag{4.16}$$

Then

1. $f(A)$ changes sign when any two rows of A are interchanged.

2. $f(A)$ is linear in each row of A.

3. $f(I) = 1$.

This theorem assures us that there is at least one function f with the properties (1), (2), and (3), and Theorem 3 assures us that there is at most one. Hence, it makes sense to refer to *the* function with these properties, and with f defined by (4.16),

$$f(A) = \det(A).$$

This final result closes our chapter on determinants.

PROOF: To prove property (1), suppose that B is obtained from A by interchanging the kth and ℓth rows of A. Then we have to show that $f(B) = -f(A)$.

Note that $B = P_{\sigma_{k,\ell}}A$, and hence, for any permutation σ,

$$\sigma(B) = \prod_{j=1}^{n} B_{\sigma(j),j} = \prod_{j=1}^{n}(P_{\sigma_{k,\ell}}A)_{\sigma(j),j} \,.$$

Now define a permuation ρ by $\rho = \sigma_{k,\ell} \circ \sigma$. Then the $\sigma(j)$th row of $P_{\sigma_{k,\ell}}A$ is the $\rho(j)$th row of A. Therefore,

$$\sigma(B) = \rho(A) \,.$$

Because of the additional pair permutation, if σ is even, ρ is odd, and vice versa:

$$\chi(\rho) = -\chi(\sigma).$$

Now, as σ ranges over the $n!$ permutations of $\{1, 2 \ldots, n\}$, so does ρ. The point is that if $\sigma_{k,\ell} \circ \sigma_1 = \sigma_{k,\ell} \circ \sigma_2$, then $\sigma_1 = \sigma_2$: Just multiply both sides by $\sigma_{k,\ell}$. So as σ ranges over all possible permutations, $\rho = \sigma_{k,\ell} \circ \sigma$ passes through each one exactly once. In other words, summing over ρ is the same as summing over σ; either way we are just summing over all $n!$ permutations. Hence:

$$f(B) = \sum_\sigma \chi(\sigma)\sigma(B) = \sum_\sigma \chi(\sigma)\rho(A)$$

$$= -\sum_\sigma \chi(\rho)\rho(A)$$

$$= -f(A).$$

To prove property (2), we have to show that if

$$\mathbf{r}_i = \alpha\mathbf{v} + \beta\mathbf{w},$$

then

$$f\left(\begin{bmatrix} \mathbf{r}_1 \\ \mathbf{r}_2 \\ \vdots \\ \alpha\mathbf{v} + \beta\mathbf{w} \\ \vdots \\ \mathbf{r}_n \end{bmatrix}\right) = \alpha f\left(\begin{bmatrix} \mathbf{r}_1 \\ \mathbf{r}_2 \\ \vdots \\ \mathbf{v} \\ \vdots \\ \mathbf{r}_n \end{bmatrix}\right) + \beta f\left(\begin{bmatrix} \mathbf{r}_1 \\ \mathbf{r}_2 \\ \vdots \\ \mathbf{w} \\ \vdots \\ \mathbf{r}_n \end{bmatrix}\right). \tag{4.17}$$

This is true since each product $\sigma(A) = A_{\sigma(1),1}A_{\sigma(2),2}\cdots A_{\sigma(n),n}$ contains exactly one factor coming from the ith row and hence is a linear function of the entries of the ith row. By definition, $f(A)$ is a linear combination of the $\sigma(A)$. A linear combination of linear functions is linear, so the determinant is a linear function of the entries of the ith row. Notice that the permutation characters $\chi(\sigma)$ did not play any role in this first property.

To prove property (3), observe that $\sigma(I)$ contains a zero factor unless σ is the identity. This means that unless σ is the identity, $\sigma(I) = 0$. Hence, when $A = I$, there is just one nonzero term in the sum (4.16), namely, the one coming from the identity permutation. When σ is the identity, clearly $\sigma(I) = 1$ and $\chi(\sigma) = 1$, so $f(I) = 1$. ∎

Exercises

4.1 Consider the following permutations:

$$\sigma_1 = \begin{matrix} 1 & 2 & 3 & 4 & 5 & 6 \\ \downarrow & \downarrow & \downarrow & \downarrow & \downarrow & \downarrow \\ 3 & 1 & 4 & 5 & 6 & 2 \end{matrix} \qquad \sigma_2 = \begin{matrix} 1 & 2 & 3 & 4 & 5 & 6 \\ \downarrow & \downarrow & \downarrow & \downarrow & \downarrow & \downarrow \\ 4 & 3 & 6 & 5 & 2 & 1 \end{matrix} \qquad \sigma_3 = \begin{matrix} 1 & 2 & 3 & 4 & 5 & 6 \\ \downarrow & \downarrow & \downarrow & \downarrow & \downarrow & \downarrow \\ 4 & 5 & 6 & 1 & 2 & 3 \end{matrix}.$$

(a) Compute $D(\sigma_j)$ for $j = 1, 2, 3$.

(b) For each $j = 1, 2, 3$, find a way to write σ_j as a product of pair permutations, as in Example 23.

(c) For each $j = 1, 2, 3$, compute $\chi(\sigma_j)$.

(d) Compute the value of $\chi(\sigma_1 \circ (\sigma_2 \circ \sigma_3)^{-1})$.

4.2 Consider the following permutations:

$$\sigma_1 = \begin{matrix} 1 & 2 & 3 & 4 & 5 & 6 \\ \downarrow & \downarrow & \downarrow & \downarrow & \downarrow & \downarrow \\ 2 & 4 & 6 & 1 & 3 & 5 \end{matrix} \qquad \sigma_2 = \begin{matrix} 1 & 2 & 3 & 4 & 5 & 6 \\ \downarrow & \downarrow & \downarrow & \downarrow & \downarrow & \downarrow \\ 5 & 1 & 6 & 4 & 2 & 3 \end{matrix} \qquad \sigma_3 = \begin{matrix} 1 & 2 & 3 & 4 & 5 & 6 \\ \downarrow & \downarrow & \downarrow & \downarrow & \downarrow & \downarrow \\ 4 & 1 & 5 & 2 & 6 & 3 \end{matrix}.$$

(a) Compute $D(\sigma_j)$ for $j = 1, 2, 3$.

(b) For each $j = 1, 2, 3$, find a way to write σ_j as a product of pair permutations, as in Example 23.

(c) For each $j = 1, 2, 3$, compute $\chi(\sigma_j)$.

(d) Compute the value of $\chi(\sigma_1 \circ (\sigma_2 \circ \sigma_3)^{-1})$.

4.3 Consider the following statements:

(a) For any n and any permutation τ of $\{1, 2 \ldots, n\}$, $\sigma(P_\tau) = 1$ for $\sigma = \tau$, and is zero otherwise.

(b) For any n and any permutation σ of $\{1, 2 \ldots, n\}$, $\det(P_\sigma) = \chi(\sigma)$.

Decide whether these statements are true or not, and justify your answer.

The Eigenvalue Problem

Overview of Chapter 5

Many methods in linear algebra are based on finding ways to "take matrices apart" into products of simpler matrices. The LU and QR decompositions are examples. Here we develop a new decomposition involving diagonal matrices. Diagonal matrices are particularly simple. All the questions we generally ask about matrices can be answered for them at a glance.

Unfortunately, matrices that come up in problems we would like to solve tend not to be diagonal. However, it turns out that *every* symmetric matrix A can be written as a product

$$A = UDU^t$$

where D is a diagonal matrix and U is an orthogonal matrix. In other words, any symmetric matrix A can be factored into a product of an orthogonal matrix U, a diagonal matrix D, and the transpose of U. As we shall see, this factorization is the key to a number of matrix computations. It is very useful since very often the matrices arising in problems we would like to solve are symmetric. Better yet, even when A is not symmetric but is at least square, it is often possible to find an invertible matrix V and a diagonal matrix D so that $A = VDV^{-1}$. As we shall see, this factorization renders many questions concerning A transparent.

Such factorizations come from finding bases that are well adapted to the matrix at hand. To see the connection, let $\{\mathbf{v}_1, \ldots, \mathbf{v}_n\}$ be *any* basis of $\mathbb{R}^n$, and let $V = [\mathbf{v}_1, \ldots, \mathbf{v}_n]$. Then V^{-1} is the change-of-basis matrix that gives the coordinates of a vector $\mathbf{x}$ with respect to the new basis. That is, if $\mathbf{y} = V^{-1}\mathbf{x}$, then $\mathbf{x} = y_1\mathbf{v}_1 + \cdots + y_n\mathbf{v}_n$.

How are these coordinates transformed under the action of an $n \times n$ matrix A? That is, how are the coordinate vectors $V^{-1}(A\mathbf{x})$ and $V^{-1}\mathbf{x}$ related? The answer is

$$V^{-1}(A\mathbf{x}) = (V^{-1}AV)V^{-1}\mathbf{x},$$

so that applying A to $\mathbf{x}$ corresponds to applying $V^{-1}AV$ to its coordinate vector, $\mathbf{y} = V^{-1}\mathbf{x}$. Using the new coordinates will simplify life if the matrix S defined by

$$S = V^{-1}AV$$

is a simple matrix.

How can we choose the basis $\{\mathbf{v}_1, \ldots, \mathbf{v}_n\}$ to make S simple? By **V.I.F.3**, the jth column of S is $V^{-1}(A\mathbf{v}_j)$, which is the coordinate vector of $A\mathbf{v}_j$. *So the jth column of S will be simple when $A\mathbf{v}_j$ has a simple expression as a linear combination of the vectors in the basis.* Here is one of the simplest things we could hope for:

$$A\mathbf{v}_j = \mu_j\mathbf{v}_j$$

for some number μ_j. Then the jth coordinate of $A\mathbf{v}_j$ is μ_j, and the others are all zero, so $V^{-1}(A\mathbf{v}_j) = \mu_j\mathbf{e}_j$. When this happens for each j, then S is a diagonal matrix. Nonzero vectors $\mathbf{v}$ such that $A\mathbf{v} = \mu\mathbf{v}$ for some number μ are called *eigenvectors* of A, and then μ is called an *eigenvalue* of A. (The fully translated versions of these hybrid German–English words would be "characteristic vector" and "characteristic value," which are sometimes used.)

Of course,

$$V^{-1}AV = S \quad \Longleftrightarrow \quad A = VSV^{-1}$$

and on the right-hand side we have a matrix factorization of A. On the left-hand side, we have the change of basis formula from which it came. Every factorization $A = VSV^{-1}$ comes from a change of basis, and vice versa. The factorization will be useful when the change of basis results in a simple matrix S expressing the action of A in the new coordinates. One good way to go about finding a nice basis for working with A is to find as many linearly independent eigenvectors as we can.

SECTION 1 | The Eigenvalue Problem

1.1 Eigenvectors and eigenvalues

> **DEFINITION**
>
> (**Eigenvectors and eigenvalues**) Let A be an $n \times n$ matrix. A number μ is an *eigenvalue* of A if there is a nonzero vector $\mathbf{v}$ such that
>
> $$A\mathbf{v} = \mu\mathbf{v}. \tag{1.1}$$
>
> Any such vector $\mathbf{v}$ is called an *eigenvector of A with eigenvalue μ*.

It is a crucial that the vector **v** in the definition is a nonzero vector. Since $A0 = \mu 0$ for *any* number μ, the definition of eigenvalue would be vacuous otherwise. Speaking geometrically, a nonzero vector **v** is an eigenvector if and only if $A\mathbf{v}$ points in the same direction as **v** or is zero.

Given an $n \times n$ matrix A, the *eigenvalue problem* for A is to find all the eigenvalues μ for A and all the eigenvectors.

Compare this problem with the problem of solving $A\mathbf{x} = \mathbf{b}$, which has been our major concern until now. The eigenvalue problem is quite different. Although only **x** is unknown when we are trying to solve $A\mathbf{x} = \mathbf{b}$, there are two unknowns—μ and **v**—to be solved for in the eigenvalue problem.

The following theorem relates the eigenvalue problem to something we have already studied: determining the kernel of a matrix.

THEOREM 1   **(Kernels, eigenvalues, and eigenvectors)** *Let A be an $n \times n$ matrix. Then μ is an eigenvalue of A if and only if $\mathrm{Ker}(A - \mu I) \neq 0$. Moreover, suppose that μ is an eigenvalue of A. Then a nonzero vector **v** is an eigenvector of A with eigenvalue μ if and only if **v** belongs to $\mathrm{Ker}(A - \mu I)$.*

PROOF: $A\mathbf{v} = \mu\mathbf{v} \iff (A - \mu I)\mathbf{v} = 0 \iff$ **v** belongs to $\mathrm{Ker}(A - \mu I)$. ∎

Since the kernel of any matrix is a subspace, the following definition is reasonable.

DEFINITION   **(Eigenspace)** If μ is an eigenvalue of A, the kernel of $A - \mu I$ is called the *eigenspace of A corresponding to the eigenvalue μ.* We denote it by writing E_μ when the context makes it clear what A is.*

In summary so far: Finding the eigenvalues of A amounts to finding the numbers μ so that $\mathrm{Ker}(A - \mu I) \neq 0$; we know how to do this. Once we know the eigenvalues μ of A, finding the eigenvectors is just the problem of determining the kernel of $(A - \mu I)$. We know how to do this, too.

The best way to go about finding eigenvalues and eigenvectors depends on the size of the matrix A, among other things. But when the size is small, computing the determinant of $A - \mu I$ is a convenient way to check whether or not $\mathrm{Ker}(A - \mu I) \neq 0$. By the Corollary to Theorem 7 of Chapter 2 and Theorem 1 of Chapter 4,

$$\mathrm{Ker}(A - \mu I) \neq 0 \iff (A - \mu I) \text{ is not invertible} \iff \det(A - \mu I) = 0. \quad (1.2)$$

EXAMPLE 1   **(Finding eigenvalues and eigenvectors)** Consider

$$A = \begin{bmatrix} 1 & 2 \\ 2 & 1 \end{bmatrix}.$$

In this case, for any number t, we have

$$A - tI = \begin{bmatrix} 1-t & 2 \\ 2 & 1-t \end{bmatrix}.$$

Then

$$\det(A - tI) = (1 - t)^2 - 4$$
$$= t^2 - 2t - 3$$
$$= (t - 3)(t + 1).$$

*All that is new about E_μ is notation; it is just a short way of writing $\mathrm{Ker}(A - \mu I)$.

Therefore, $\det(A - \mu I) = 0$ if and only if $\mu = 3$ or $\mu = -1$. Hence, there are exactly two eigenvalues of A, namely, 3 and -1.

Now for the eigenvectors. The eigenvectors of A with eigenvalue 3 are the nonzero vectors in the kernel of

$$A - 3I = \begin{bmatrix} -2 & 2 \\ 2 & -2 \end{bmatrix}.$$

This matrix row reduces to

$$\begin{bmatrix} 1 & -1 \\ 0 & 0 \end{bmatrix},$$

and hence, the kernel of $A - 3I$ consists of all multiples of

$$\mathbf{v}_1 = \begin{bmatrix} 1 \\ 1 \end{bmatrix}.$$

The eigenvectors of A with eigenvalue 3 are exactly the nonzero multiples of $\mathbf{v}_1$. In other words, E_3 is the span of $\{\mathbf{v}_1\}$.

Next consider $\mu = -1$. The corresponding eigenvectors are the nonzero vectors in the kernel of

$$A + I = \begin{bmatrix} 2 & 2 \\ 2 & 2 \end{bmatrix}.$$

You can see that the kernel of this matrix is all multiples of

$$\mathbf{v}_2 = \begin{bmatrix} -1 \\ 1 \end{bmatrix}.$$

Hence, the eigenvectors of A with eigenvalue -1 are exactly the nonzero multiples of $\mathbf{v}_2$. In other words, E_{-1} is the span of $\{\mathbf{v}_2\}$.

If we examine Example 1, we learn useful things. First, notice that $\det(A - tI) = t^2 - 2t - 3$ is a quadratic polynomial in t and that the eigenvalues of A are the roots of this polynomial. This is no accident; however, let us look at another example before drawing general conclusions.

EXAMPLE 2 **(Eigenvalues and eigenvectors for a 3×3 matrix)** Consider

$$A = \begin{bmatrix} 1 & 1 & 1 \\ 1 & 1 & 1 \\ 1 & 1 & 1 \end{bmatrix}.$$

In this case, for any number t, we have

$$A - tI = \begin{bmatrix} 1-t & 1 & 1 \\ 1 & 1-t & 1 \\ 1 & 1 & 1-t \end{bmatrix}.$$

Then computing,

$$\det(A - tI) = -t^3 + 3t^2 = t^2(3 - t).$$

Hence, $\det(A - \mu I) = 0$ if and only if $\mu = 0$ or $\mu = 3$, and A has exactly two eigenvalues, 0 and 3.

Now let us find the eigenvectors. The eigenspace E_0 is just the kernel of A. Since A row reduces to

$$\begin{bmatrix} 1 & 1 & 1 \\ 0 & 0 & 0 \\ 0 & 0 & 0 \end{bmatrix},$$

we see that $\text{rank}(A) = 1$, so $E_0 = \text{Ker}(A)$ is two-dimensional. We can find a basis $\{\mathbf{v}_1, \mathbf{v}_2\}$ for it using the standard procedure, with the result that

$$\mathbf{v}_1 = \begin{bmatrix} -1 \\ 0 \\ 1 \end{bmatrix} \quad \text{and} \quad \mathbf{v}_2 = \begin{bmatrix} -1 \\ 1 \\ 0 \end{bmatrix}.$$

The span of these two vectors, which is the plane in $\mathbb{R}^3$ with the equation $x + y + z = 0$, is the subspace E_0, and the eigenvectors with eigenvalue 0 are exactly the nonzero vectors in it.

The eigenspace E_3 is the kernel of

$$A - 3I = \begin{bmatrix} -2 & 1 & 1 \\ 1 & -2 & 1 \\ 1 & 1 & -2 \end{bmatrix}.$$

This row reduces to

$$\begin{bmatrix} -2 & 1 & 1 \\ 0 & -3 & 3 \\ 0 & 0 & 0 \end{bmatrix},$$

so we see that $\text{rank}(A - 3I) = 2$ and so $E_3 = \text{Ker}(A - 3I)$ is one-dimensional. The usual method gives us the basis consisting of the single vector

$$\mathbf{v}_3 = \begin{bmatrix} 1 \\ 1 \\ 1 \end{bmatrix},$$

so E_3 is the line spanned by this vector.

Notice that in Example 2, the eigenspaces E_0 and E_3 had different dimensions. The dimensions of the eigenspaces turn out to be important, so we give them a name.

DEFINITION

(Geometric multiplicity) If μ is an eigenvalue of a square matrix A, then the *geometric multiplicity* of μ is the dimension of the corresponding eigenspace E_μ.

Using this new terminology for the matrix A of Example 2, the eigenvalue 0 has geometric multiplicity 2, and the eigenvalue 3 has geometric multiplicity 1.

In our first two examples, we found the eigenvalues of the given matrices by finding the roots of a polynomial. In general for any $n \times n$ matrix A, from the permutation formula for the determinant,*

$$\det(A - tI) = \sum_\sigma \chi(\sigma)(A - tI)_{\sigma(1),1}(A - tI)_{\sigma(2),2} \cdots (A - tI)_{\sigma(n),n}. \tag{1.3}$$

For each permutation σ,

$$(A - tI)_{\sigma(1),1}(A - tI)_{\sigma(2),2} \cdots (A - tI)_{\sigma(n),n}$$

is a polynomial of degree at most n in t, as in the examples. Adding up any number of polynomials of degree at most n in t just gives us another such polynomial, so $\det(A - tI)$ is a polynomial of degree n in t.

DEFINITION

(Characteristic polynomial) For any $n \times n$ matrix A, the nth-degree polynomial $p_A(t)$ defined by

$$p_A(t) = \det(A - tI)$$

is called the *characteristic polynomial* of A.

Theorem 1 tells us that the eigenvalues of A are exactly the roots of the characteristic polynomial of A.

---

*You do not really need the determinant formula to see that $\det(A - tI)$ is a polynomial of degree n in t. You could also use the Laplace expansion method, for example, which is developed in Exercise 2.6 of Chapter 4.

The fundamental theorem of algebra says that every such nth-degree polynomial can be completely factored in the complex plane as

$$p(t) = C(\mu_1 - t)(\mu_2 - t) \cdots (\mu_n - t), \tag{1.4}$$

where the complex numbers $\{\mu_1, \mu_2, \ldots, \mu_n\}$ are the roots of $p(t)$ in the complex plane and C is a constant. Of course, these roots do not have to be distinct; they can be repeated. Also, they do not have to be complex; sometimes we get lucky and can factor our characteristic polynomial using only real numbers. Both things happened in Example 2. In that case, $p(t)$ factored as

$$p(t) = (0 - t)(0 - t)(3 - t) = (0 - t)^2(3 - t).$$

Also, as we saw in the examples and shall show in the next subsection, the coefficient of t^n in is always $(-1)^n$ for any $n \times n$ matrix A so that $C = 1$.

As you can see, finding the eigenvectors of A amounts to factoring the characteristic polynomial of A. For each different root, we can find nonzero eigenvectors. These vectors may have complex entries, but that will not mean they are useless, *even in a problem where we expect that the final answer will involve only real numbers*. We will see examples of this soon.

How many times a given root of a characteristic polynomial is repeated turns out to be significant, so we give this number a name.

DEFINITION

(Algebraic multiplicity) If A is any square matrix and μ is any eigenvalue of A, the *algebraic multiplicity* of μ is the number of times μ is repeated as a root of $p(t)$, the characteristic polynomial of A. In other words, it is the largest integer k so that $(\mu - t)^k$ divides $p(t)$.

For the matrix A of Example 2, the algebraic multiplicity of 0 is 2, which happens to be equal to the geometric multiplicity of 0. Likewise, the algebraic multiplicity of 3 is 1, which happens to be equal to the geometric multiplicity of 1. This *does not* always happen.

EXAMPLE 3 **(Different algebraic and geometric multiplicities)** Consider

$$A = \begin{bmatrix} 2 & 1 \\ 0 & 2 \end{bmatrix}.$$

In this case, for any number t, we have

$$A - tI = \begin{bmatrix} 2 - t & 1 \\ 0 & 2 - t \end{bmatrix}.$$

Then $\det(A - tI) = (2 - t)^2$, so there is just one eigenvalue, namely, 2. Its algebraic multiplicity is 2, as we see from the characteristic polynomial, which, nicely enough, is already factored.

Next,

$$E_2 = \text{Ker}(A - 2I) \quad \text{and} \quad A - 2I = \begin{bmatrix} 0 & 1 \\ 0 & 0 \end{bmatrix}$$

—already row reduced. Evidently, $\text{rank}(A - 2I) = 1$, so

$$\dim(E_2) = \dim(\text{Ker}(A - 2I)) = 2 - \text{rank}(A - 2I) = 1.$$

Hence, the geometric multiplicity is 1, while the algebraic multiplicity is 2. As we will see, this is significant, though we are not yet quite ready to explain why.

1.2 Trace, determinant, and characteristic polynomials

Let A be an $n \times n$ matrix, and let $p(t) = \det(A - tI)$ be the characteristic polynomial of A. Let

$$p(t) = a_0 + a_1 t + a_2 t^2 + \cdots + a_n t^n.$$

This subsection focuses on the following question:

- *How are the coefficients of p related to the entries of A?*

There are a few simple answers, and they turn out to be very useful. They are summarized in the following formula:

$$p(t) = (-1)^n t^n + (-1)^{n-1} \left(\sum_{j=1}^{n} A_{j,j} \right) t^{n-1} + \cdots + \det(A). \tag{1.5}$$

That is, the coefficient of t^n is $(-1)^n$ no matter what A is, and the constant term is $\det(A)$. Finally, the coefficient of t^{n-1} is

$$(-1)^{n-1} \left(\sum_{j=1}^{n} A_{j,j} \right).$$

This quantity will come up again and again; it deserves a name.

DEFINITION | **(Trace of a square matrix)** Let A be any $n \times n$ matrix. Then the *trace* of A, $\mathrm{tr}(A)$, is the number

$$\mathrm{tr}(A) = A_{1,1} + A_{2,2} + \cdots A_{n,n}.$$

EXAMPLE 4   **(Computing a trace)** Let

$$A = \begin{bmatrix} 1 & 2 \\ 2 & 1 \end{bmatrix},$$

as in Example 1. Then

$$\mathrm{tr}(A) = 1 + 1 = 2.$$

That is easy! Computing a trace is *much* easier than computing a determinant. There is no need to do any row reduction at all; just add up the diagonal entries.

With this definition, we can formulate the following theorem.

THEOREM 2   **(Coefficients of the characteristic polynomial)** *Let A be any n × n matrix, and let p(t) be the characteristic polynomial of A. Then*

$$p(t) = (-1)^n t^n + (-1)^{n-1} (\mathrm{tr}(A)) t^{n-1} + \cdots + \det(A). \tag{1.6}$$

At the end of this subsection, we give a simple proof of Theorem 2 based on the permutation formula for the determinant. The consequences of Theorem 2 are more important than its proof, so we turn to them first.*

---

*Though their justification relies on the formula for the determinant, if you are willing to accept the theorems, you can easily understand and apply them and get on with the rest of this chapter. If, on the other hand, you really want to know why these theorems are true, you will have to study the somewhat subtle final section of Chapter 4, if you have not done so already.

First of all, since the coefficient of t^n is always $(-1)^n$, it follows that the constant C in (1.4) is always 1, as claimed. This means that the characteristic polynomial of an $n \times n$ matrix A always has the factorization

$$p(t) = \prod_{j=1}^{n} (\mu_j - t),$$

where $\mu_1, \mu_2, \ldots, \mu_n$ are the eigenvalues of A repeated according to their algebraic multiplicity.

Expanding the product that defines $p(t)$,

$$p(t) = (-1)^n t^n + (-1)^{n-1} t^{n-1} \left(\sum_{j=1}^{n} \mu_j \right) + \cdots + \prod_{j=1}^{n} \mu_j. \tag{1.7}$$

Comparing the coefficients of t^{n-1} and t^0 in (1.6) and (1.7) gives us the following result.

THEOREM 3   **(Trace, determinant, and eigenvalues)** *Let A be an $n \times n$ matrix, and suppose that the eigenvalues of A, repeated according to their algebraic multiplicities, are $\mu_1, \mu_2, \ldots, \mu_n$. Then*

$$\mathrm{tr}(A) = \sum_{j=1}^{n} \mu_j \quad \text{and} \quad \det(A) = \prod_{j=1}^{n} \mu_j. \tag{1.8}$$

EXAMPLE 5   **(Trace, determinant, and eigenvalues)** Let

$$A = \begin{bmatrix} 1 & 2 \\ 2 & 1 \end{bmatrix},$$

as in Example 1. Then

$$\mathrm{tr}(A) = 2 \quad \text{and} \quad \det(A) = -3.$$

By Theorem 3, the eigenvalues add up to 2, and their product is -3. This checks with Example 1, where we found that the eigenvalues are -1 and 3.

PROOF OF THEOREM 3: First of all, $p(0) = \det(A)$ from the definition $p(t) = \det(A - tI)$ and evaluating any polynomial at 0 gives the constant term. Thus, $a_0 = \det(A)$.

At the other end of the polynomial, both terms involving t^n and t^{n-1} come from the product of the diagonal entries $\prod_{j=1}^{n} (A_{j,j} - t)$, which enters the formula (1.3) when σ is the identity permutation. All other terms in (1.3) are of degree at most $n-2$ in t. The key observation is that if a permutation σ is not the identity permutation, then there are at least two values of j with $\sigma(j) \neq j$.

Indeed, by definition, if σ is not the identity, there is *some* k so that $\sigma(k) \neq k$. Let $\sigma(k) = \ell$. But then $\sigma(\ell) \neq \ell$ since $\sigma(k) = \ell$, and permutations are one-to-one.

Next, since $I_{i,j} = 0$ when $i \neq j$, $(A - tI)_{\sigma(j),j} = A_{\sigma(j),j}$ or both $j = k$ and $j = \ell$. This is just a constant, independent of t. Thus, when σ is not the identity permutation, the product $\prod_{j=1}^{n} (A - tI)_{\sigma(j),j}$ contains at least two constant terms and is therefore a polynomial of degree $n - 2$ at most. Hence, the coefficients of t^n and t^{n-1} in $p(t)$ all come from the single product $\prod_{j=1}^{n} (A_{j,j} - t)$ corresponding to the identity permutation.

- *The only term in the sum defining $\det(A - tI)$ that contributes to degree n or $n - 1$ in the characteristic polynomial is the diagonal product $\prod_{j=1}^{n} (A_{j,j} - t)$.*

With the sum over permutations out of the way, the coefficients of t^n and t^{n-1} are easily computed. Expanding $\prod_{j=1}^{n}(A_{j,j} - t)$, the terms of order n and $n-1$ are

$$\prod_{j=1}^{n}(A_{j,j} - t) = (-1)^n t^n + (-1)^{n-1}\left(\sum_{j=1}^{n} A_{j,j}\right) t^{n-1} + \cdots,$$

and hence,

$$p(t) = (-1)^n t^n + (-1)^{n-1}\left(\sum_{j=1}^{n} A_{j,j}\right) t^{n-1} + \cdots + \det(A).$$

Therefore,

$$a_n = (-1)^n \quad \text{and} \quad a_{n-1} = (-1)^{n-1}(A_{1,1} + A_{2,2} + \cdots + A_{n,n}). \qquad \blacksquare$$

1.3 Similar matrices

DEFINITION

(Similar matrices) Two $n \times n$ matrices A and B are *similar* if and only if there is an invertible $n \times n$ matrix C so that
$$A = CBC^{-1}.$$

Notice that the condition is symmetric in A and B: if $A = CBC^{-1}$, then $B = C^{-1}AC$.

THEOREM 4

(Similar matrices and eigenvalues) *If A and B are similar matrices, then they have the same eigenvalues with the same geometric and algebraic multiplicities.*

PROOF: We first show that A and B have the same characteristic polynomial, which means that they have the same eigenvalues with the same algebraic multiplicities. The key is that since $A = CBC^{-1}$,

$$A - tI = CBC^{-1} - tI = C(B - tI)C^{-1}. \qquad (1.9)$$

Then, by the multiplicative property of determinants,

$$\det(A - tI) = \det(C(B - tI)C^{-1}) = \det(C)\det(B - tI)\det(C^{-1})$$
$$= \det(C)\det(B - tI)(\det(C))^{-1} = \det(B - tI).$$

Next, suppose that μ is an eigenvalue of A and therefore of B also. By definition, the geometric multiplicity of μ as an eigenvalue of A is $\text{nullity}(A - \mu I)$, and the geometric multiplicity of μ as an eigenvalue of B is $\text{nullity}(B - \mu I)$. Hence, we have to show that

$$\text{nullity}(A - \mu I) = \text{nullity}(B - \mu I).$$

by the dimension formula, it suffices to show that

$$\text{rank}(A - \mu I) = \text{rank}(B - \mu I). \qquad (1.10)$$

But multiplying a matrix by an invertible matrix does not change its rank, so (1.10) follows from (1.9). $\qquad \blacksquare$

1.4 Complex eigenvectors and the geometry of $\mathbb{C}^n$

By the fundamental theorem of algebra, every polynomial has roots in the complex plane. So as long as we allow complex eigenvalues and complex eigenvectors, every matrix A has at least one eigenvalue.

EXAMPLE 6 **(Complex eigenvalues)** Let

$$A = \begin{bmatrix} 0 & 1 \\ -1 & 0 \end{bmatrix}.$$

The characteristic polynomial $p_A(t)$ is

$$p_A(t) = \det\left(\begin{bmatrix} -t & 1 \\ -1 & -t \end{bmatrix} \right) = t^2 + 1.$$

Evidently, there are no real solutions to $p_A(t) = 0$. However, we do have the two complex solutions, $\mu_1 = i$ and $\mu_2 = -i$, where i denotes $\sqrt{-1}$. These are the eigenvalues of A, and to find the eigenvectors with eigenvalue i, we form

$$A - iI = \begin{bmatrix} -i & 1 \\ -1 & -i \end{bmatrix}.$$

The kernel of this matrix is spanned by $\mathbf{v}_1 = \begin{bmatrix} -i \\ 1 \end{bmatrix}$, and hence this complex vector satisfies $A\mathbf{v}_1 = \mu_1 \mathbf{v}_1$. Likewise,

$$A + iI = \begin{bmatrix} i & 1 \\ -1 & i \end{bmatrix},$$

so the kernel of this matrix is spanned by $\mathbf{v}_2 = \begin{bmatrix} i \\ 1 \end{bmatrix}$, and hence this complex vector satisfies $A\mathbf{v}_2 = \mu_2 \mathbf{v}_2$. So as long as we are willing to admit complex eigenvalues and eigenvectors with complex entries, then every $n \times n$ matrix has at least one eigenvalue.

As we saw in Example 6, even if all the entries of an $n \times n$ matrix A are real numbers, all the eigenvalues may be complex, or even purely imaginary. The corresponding eigenvectors will therefore have complex entries.

DEFINITION

($\mathbb{C}^n$, complex vectors) The set of all vectors

$$\mathbf{z} = \begin{bmatrix} z_1 \\ z_2 \\ \vdots \\ z_n \end{bmatrix},$$

where $z_1, z_2, \ldots, z_n$ are complex numbers, is denoted $\mathbb{C}^n$ and called *complex n-dimensional space.*

The algebra of vectors in $\mathbb{C}^n$ is just the same as in $\mathbb{R}^n$, with one difference: Now we allow the entries of vectors to be complex numbers, and we allow multiplication of vectors by complex numbers. We do this entrywise, just as before, and still add the vectors entry by entry, just as before.

While we are at it, we may as well consider $m \times n$ matrices with complex entries. Again, all the algebra is the same as before—all our formulas generalize to the case in which the entries are complex numbers, and the proofs are the same since addition and multiplication of complex numbers have the same associative, commutative, and distributive properties as do the addition and multiplication of real numbers.

For exactly the same reason, all the theorems we have proved solving linear systems of equations by row reduction still hold—with the same proofs—if we allow the coefficients and variables to take on complex values as well.

The geometry of $\mathbb{C}^n$ is another matter. We cannot define the length of a vector $\mathbf{z}$ in $\mathbb{C}^n$ as the positive square root of $\mathbf{z} \cdot \mathbf{z}$ since in general, $\mathbf{z} \cdot \mathbf{z}$ will be a complex number and there will not be any positive square root.

When $z = x + iy$ is a single complex number, the *modulus* of z, also called the *length* of z, is denoted by $|z|$ and is defined by

$$|z|^2 = x^2 + y^2.$$

If, as usual, we identify the set of complex numbers with the plane $\mathbb{R}^2$ so that $z = x + iy$ is identified with the vector

$$\begin{bmatrix} x \\ y \end{bmatrix},$$

then $|z|$ coincides with the length of the vector

$$\begin{bmatrix} x \\ y \end{bmatrix}.$$

Recall that the complex conjugate $\bar{z}$ of $z = x + iy$ is defined by $\bar{z} = x - iy$, and

$$\bar{z}z = x^2 + y^2 = |z|^2.$$

We extend the complex conjugate to $\mathbb{C}^n$ as follows: If

$$\mathbf{z} = \begin{bmatrix} z_1 \\ z_2 \\ \vdots \\ z_n \end{bmatrix},$$

then $\bar{\mathbf{z}}$ is the complex vector

$$\bar{\mathbf{z}} = \begin{bmatrix} \bar{z}_1 \\ \bar{z}_2 \\ \vdots \\ \bar{z}_n \end{bmatrix}.$$

With this definition,

$$\bar{\mathbf{z}} \cdot \mathbf{z} = \bar{z}_1 z_1 + \bar{z}_2 z_2 + \cdots + \bar{z}_n z_n$$
$$= (x_1^2 + y_1^2) + (x_2^2 + y_2^2) + \cdots (x_n^2 + y_n^2).$$

This *is* a positive number, and we can therefore define

$$|\mathbf{z}| = \sqrt{\bar{\mathbf{z}} \cdot \mathbf{z}}.$$

We can then define the *distance* between two vectors $\mathbf{w}$ and $\mathbf{z}$ to be the length of their difference, $|\mathbf{w} - \mathbf{z}|$, as before. Of course, if we are going to refer to this quantity as "distance," we must be able to prove that it has the properties of a distance function, as explained in Section 4 of Chapter 1. In particular, we must be able to prove that it satisfies the triangle inequality. This is the case, and as with $\mathbb{R}^n$, the key is the Schwarz inequality. First, we provide a definition.

DEFINITION

(Inner product in $\mathbb{C}^n$) Given two vectors $\mathbf{w}$ and $\mathbf{z}$ in $\mathbb{C}^n$, we define their *inner product*, denoted $\langle \mathbf{w}, \mathbf{z} \rangle$, by

$$\langle \mathbf{w}, \mathbf{z} \rangle = \bar{\mathbf{w}} \cdot \mathbf{z}.$$

Notice that $\langle \mathbf{z}, \mathbf{w} \rangle$ is the complex conjugate of $\langle \mathbf{w}, \mathbf{z} \rangle$; order matters in inner products! The Schwarz inequality then is

$$|\langle \mathbf{w}, \mathbf{z} \rangle| \le |\mathbf{w}||\mathbf{z}|.$$

It is left as an exercise to adapt the proofs of the Schwarz inequality and then the triangle inequality from $\mathbb{R}^n$ to $\mathbb{C}^n$.

One final definition is required to extend all the theorems to the complex case. Recall the key identity involving the transpose:

$$\mathbf{x} \cdot A\mathbf{y} = (A^t \mathbf{x}) \cdot \mathbf{y}.$$

The identity is still true for complex vectors and matrices, but it no longer has geometric meaning in the complex case since there it is the inner product, not the dot product, that has geometric meaning. Here is the geometric analog of the transpose in the complex case.

DEFINITION

(Adjoint of a complex matrix) Given an $m \times n$ matrix A with complex entries, its *adjoint* A^* is defined by

$$A_{i,j}^* = \overline{A_{j,i}}.$$

That is, the entries of A^* are the complex conjugates of the entries of A^t, the transpose of A.

To see why this definition is "right,"

$$\begin{aligned}
\langle \mathbf{w}, A\mathbf{z} \rangle &= \overline{\mathbf{w}} \cdot A\mathbf{z} \\
&= A^t \overline{\mathbf{w}} \cdot \mathbf{z} \\
&= \overline{A^* \mathbf{w}} \cdot \mathbf{z} \\
&= \langle A^* \mathbf{w}, \mathbf{z} \rangle.
\end{aligned}$$

The relation

$$\langle \mathbf{w}, A\mathbf{z} \rangle = \langle A^* \mathbf{w}, \mathbf{z} \rangle \tag{1.11}$$

is the complex analog of the fundamental property of the transpose **V.I.F.5**.

With the changes we have described here, every theorem we have proved for real vectors and matrices extends to complex vectors and matrices. In particular, any collection of orthonormal vectors in $\mathbb{C}^n$ is linearly independent. We close with one final definition; the generalization of orthogonal matrices to the complex setting.

DEFINITION

(Unitary matrix) A complex $n \times n$ matrix is called *unitary* in case its columns are orthonormal.

Exercises

1.1 Let $A = \begin{bmatrix} 1 & 1 & 0 \\ 1 & 0 & 1 \\ 0 & 1 & 1 \end{bmatrix}$.

(a) Find the characteristic polynomial of A, and find all its roots.

(b) Find the eigenvalues of A, and for each eigenvalue give a basis for the corresponding eigenspace.

(c) What are the algebraic and geometric multiplicities of each of the eigenvalues?

(d) Compute the trace of A.

1.2 Let $A = \begin{bmatrix} -3 & 1 & 1 \\ 1 & -3 & 1 \\ 1 & 1 & -1 \end{bmatrix}$.

(a) Find the characteristic polynomial of A, and find all its roots.

(b) Find the eigenvalues of A, and for each eigenvalue give a basis for the corresponding eigenspace.

(c) What are the algebraic and geometric multiplicities of each of the eigenvalues?

(d) Compute the trace of A.

1.3 Let $A = \begin{bmatrix} 1 & 2 \\ 2 & 4 \end{bmatrix}$.

(a) Find the characteristic polynomial of A, and find all its roots.

(b) Find the eigenvalues of A, and for each eigenvalue give a basis for the corresponding eigenspace.

(c) What are the algebraic and geometric multiplicities of each of the eigenvalues?

(d) Compute the trace of A.

1.4 Let $A = \begin{bmatrix} 1 & 1 \\ -1 & 3 \end{bmatrix}$.

(a) Find the characteristic polynomial of A, and find all its roots.

(b) Find the eigenvalues of A, and for each eigenvalue give a basis for the corresponding eigenspace.

(c) What are the algebraic and geometric multiplicities of each of the eigenvalues?

(d) Compute the trace of A.

1.5 Consider $A = \begin{bmatrix} 1 & -2 & 0 \\ 1 & 2 & 0 \\ 0 & 2 & -1 \end{bmatrix}$ and $B = \begin{bmatrix} 1 & 2 & 3 \\ 0 & 3 & 3 \\ 0 & 0 & -3 \end{bmatrix}$.

(a) Compute the traces of A and B.

(b) Compute the determinants of A and B.

(c) Are A and B similar?

1.6 Let A be an $n \times n$ matrix. Suppose there is a nonzero vector $\mathbf{v}$ and a number μ so that $(A - \mu I)^2 \mathbf{v} = 0$. Show that μ is an eigenvalue of A.

1.7 Let A be an $n \times n$ matrix. A is called *idempotent* in case $A^2 = A$. Orthogonal projections are an important example. Show that if A is idempotent and $A \neq I$ and $A \neq 0$, then A has exactly two eigenvalues, namely, 0 and 1.

1.8 Let R be the 2×2 matrix $R = \begin{bmatrix} \cos(\theta) & -\sin(\theta) \\ \sin(\theta) & \cos(\theta) \end{bmatrix}$, and let $0 < \theta < 2\pi, \theta \neq \pi$. This is the matrix corresponding to the counterclockwise rotation through the angle θ in the plane $\mathbb{R}^2$.

(a) R has no eigenvectors in $\mathbb{R}^2$. Give a geometric explanation of this fact.

(b) According to part (a), R cannot have any real eigenvalues. Compute the eigenvalues of R.

1.9 Let A and B be $n \times n$ matrices such that $AB = BA$, that is, such that A and B commute. Show that if $\mathbf{v}$ is an eigenvector of B and $A\mathbf{v} \neq 0$, then $A\mathbf{v}$ is an eigenvector of B.

1.10 Let A be any $n \times n$ matrix.

(a) Show that $\operatorname{tr}(A) = \sum_{j=1}^{n} \mathbf{e}_j \cdot A\mathbf{e}_j$.

(b) Let $\{\mathbf{u}_1, \mathbf{u}_2, \ldots, \mathbf{u}_n\}$ be any orthonormal basis for $\mathbb{R}^n$, and let $U = [\mathbf{u}_1, \mathbf{u}_2, \ldots, \mathbf{u}_n]$. Explain why $U^t A U$ and A have the same trace, and use this to derive the formula $\operatorname{tr}(A) = \sum_{j=1}^{n} \mathbf{u}_j \cdot A\mathbf{u}_j$.

(c) Let $\mathbf{u}$ be any unit vector in $\mathbb{R}^n$, and let M be the reflection matrix

$$M = I - 2\mathbf{u}\mathbf{u}^t.$$

Show that $\operatorname{tr}(M) = n - 2$.

1.11 A 2×2 matrix of the form $\begin{bmatrix} a & b \\ b & a \end{bmatrix}$ is called *doubly symmetric*, since not only is A the same when you flip it about the main diagonal (upper left to lower right), it is also symmetric when you flip it about the other diagonal. For such matrices you can write down the eigenvectors and eigenvalues very easily. Show that no matter what the values of a and b are, $\mathbf{v}_1 = \begin{bmatrix} 1 \\ 1 \end{bmatrix}$ is an eigenvector with eigenvalue $a + b$, and $\mathbf{v}_2 = \begin{bmatrix} 1 \\ -1 \end{bmatrix}$ is an eigenvector with eigenvalue $a - b$.

1.12 (a) Use the fact that $\operatorname{rank}(B) = \operatorname{rank}(B^t)$ for any matrix B to show that for any square matrix A, A and A^t have the same eigenvalues with the same geometric multiplicities.

(b) Use the fact that $\det(B) = \det(B^t)$ for any square matrix B to show that for any square matrix A, A and A^t have the same eigenvalues with the same algebraic multiplicities.

1.13 If an $n \times n$ matrix A with nonnegative entries has the property that for each i,

$$\sum_{j=1}^{n} A_{i,j} = 1,$$

then A is called a *stochastic matrix*. Such matrices arise in many problems in probability theory.

In other words, a matrix is stochastic if its entries are all nonnegative and the sum of the entries in each row is one. More generally, an $n \times n$ matrix with arbitrary entries is called a *constant row sum with row sum c* matrix if and only if

$$\sum_{j=1}^{n} A_{i,j} = c$$

for each $i = 1, 2, \ldots, n$.

(a) Show that if A is a constant row sum matrix with row sum c, then the vector $\mathbf{v} = \begin{bmatrix} 1 \\ 1 \\ \vdots \\ 1 \end{bmatrix}$ is an eigenvector with eigenvalue c.

(b) A matrix A is a *constant column sum matrix with column sum c* if and only if A^t is a constant row sum matrix with row sum c. Show that in this case c is an eigenvalue of A.

1.14 Compute $|\mathbf{z}|$ for $\mathbf{z} = \begin{bmatrix} 1 + i \\ -3 + 2i \\ 1 - 4i \end{bmatrix}$.

1.15 Compute $|\mathbf{z}|$ for $\mathbf{z} = \begin{bmatrix} 5 - i \\ i \\ 5 \\ 1 - 3i \end{bmatrix}$.

1.16 Show that if $\mathbf{z} = \begin{bmatrix} z_1 \\ z_2 \end{bmatrix}$ is any vector in $\mathbb{C}^2$, then $\mathbf{z}^{\perp} = \begin{bmatrix} -\bar{z}_2 \\ \bar{z}_1 \end{bmatrix}$ is such that $|\mathbf{z}^{\perp}| = |\mathbf{z}|$ and $\langle \mathbf{z}, \mathbf{z}^{\perp} \rangle = 0$. Hence, given any unit vector $\mathbf{z}$ in $\mathbb{C}^2$, $\{\mathbf{z}, \mathbf{z}^{\perp}\}$ is an orthonormal basis of $\mathbb{C}^2$.

1.17 Let $\mathbf{z} = \begin{bmatrix} 1 + 2i \\ i \\ 3 - i \end{bmatrix}$ and $\mathbf{w} = \begin{bmatrix} 3 + 4i \\ 4 + 3i \\ 2 + 11i \end{bmatrix}$. Compute $\langle \mathbf{w}, \mathbf{z} \rangle$.

1.18 Let $\mathbf{z} = \begin{bmatrix} 2 + 3i \\ i \\ 1 - i \end{bmatrix}$ and $\mathbf{w} = \begin{bmatrix} 1 + 4i \\ 4 + i \\ 3 + 3i \end{bmatrix}$. Compute $\langle \mathbf{w}, \mathbf{z} \rangle$.

1.19 Find a vector $\mathbf{w}$ in $\mathbb{C}^2$ that is orthogonal to $\mathbf{u} = \begin{bmatrix} 2 + 4i \\ 4i \end{bmatrix}$. Then find a 2×2 unitary matrix whose first column is a real multiple of $\mathbf{u}$.

1.20 Find a vector $\mathbf{w}$ in $\mathbb{C}^2$ that is orthogonal to $\mathbf{u} = \begin{bmatrix} 7 + 6i \\ 6 \end{bmatrix}$. Then find a 2×2 unitary matrix whose first column is a real multiple of $\mathbf{u}$.

1.21 Compute the adjoint of $A = \begin{bmatrix} 1 + i & i \\ 2 - i & 3 + i \end{bmatrix}$.

1.22 Compute the adjoint of $A = \begin{bmatrix} 3 + 2i & 5 + 2i \\ 1 - 7i & 2 - i \end{bmatrix}$.

SECTION 2 | Diagonalization

2.1 Diagonalization

Now let us go back to Example 1 and look more closely at the eigenvectors we found:

$$\mathbf{v}_1 = \begin{bmatrix} 1 \\ 1 \end{bmatrix} \quad \text{and} \quad \mathbf{v}_2 = \begin{bmatrix} -1 \\ 1 \end{bmatrix}.$$

Notice that $\mathbf{v}_1 \cdot \mathbf{v}_2 = 0$. We know that nonzero orthogonal vectors are linearly independent; hence, $\{\mathbf{v}_1, \mathbf{v}_2\}$ is a basis for $\mathbb{R}^2$. Better yet, if we define

$$\mathbf{u}_1 = \frac{1}{\sqrt{2}}\mathbf{v}_1 \quad \text{and} \quad \mathbf{u}_2 = \frac{1}{\sqrt{2}}\mathbf{v}_2,$$

we have in $\{\mathbf{u}_1, \mathbf{u}_2\}$ an orthonormal basis of $\mathbb{R}^2$ consisting of eigenvectors of A.

Now let U be the matrix whose first column is $\mathbf{u}_1$ and whose second column is $\mathbf{u}_2$; that is, $U = [\mathbf{u}_1, \mathbf{u}_2]$. Now we know that $A\mathbf{u}_1 = 3\mathbf{u}_1$ and $A\mathbf{u}_2 = -1\mathbf{u}_2$, so by **V.I.F.3**,

$$A[\mathbf{u}_1, \mathbf{u}_2] = [A\mathbf{u}_1, A\mathbf{u}_2] = [3\mathbf{u}_1, -\mathbf{u}_2] = [\mathbf{u}_1, \mathbf{u}_2]\begin{bmatrix} 3 & 0 \\ 0 & -1 \end{bmatrix}.$$

If we define

$$D = \begin{bmatrix} 3 & 0 \\ 0 & -1 \end{bmatrix},$$

we can rewrite the equation for AU as

$$AU = UD \quad \text{or} \quad A = UDU^{-1}.$$

Notice that not only is U invertible, but because its columns are orthonormal, U is an orthogonal matrix, and hence $U^{-1} = U^t$. *Thus, our solution of the eigenvalue problem for A has led to a way to write A as the product of an orthogonal matrix U, a diagonal matrix D, and U^{-1}.*

Diagonal matrices are "nice," and orthogonal matrices are "nice." Any time we "take a matrix apart into nice pieces," we accomplish something of considerable computational value. Was there something very special about this example that enabled us to take the matrix apart into nice pieces, or is there something more general going on here? Let us investigate.

The next theorem says that eigenvectors corresponding to different eigenvalues are *guaranteed* to be linearly independent.

THEOREM 5

(Distinct eigenvalues and independence) *Let A be an $n \times n$ matrix, and let $\{v_1, v_2, \ldots, v_m\}$ be m eigenvectors of A with distinct eigenvalues $\{\mu_1, \mu_2, \ldots, \mu_m\}$. Then $\{v_1, v_2, \ldots, v_m\}$ is linearly independent.*

PROOF: Clearly, the theorem is true for $m = 1$. Now make the inductive hypothesis that it is true for $m - 1$ eigenvectors with distinct eigenvalues.

Suppose that

$$a_1 v_1 + a_2 v_2 + \cdots + a_m v_m = 0. \tag{2.1}$$

Then

$$\begin{aligned}
0 &= (A - \mu_m I)(a_1 v_1 + a_2 v_2 + \cdots + a_{m-1} v_{m-1} + a_m v_m) \\
&= a_1(A - \mu_m I)v_1 + \cdots + a_{m-1}(A - \mu_m I)v_{m-1} + a_m(A - \mu_m I)v_m \\
&= a_1(\mu_1 - \mu_m)v_1 + \cdots + a_{m-1}(\mu_{m-1} - \mu_m)v_{m-1} + a_m(\mu_m - \mu_m)v_m \\
&= a_1(\mu_1 - \mu_m)v_1 + \cdots + a_{m-1}(\mu_{m-1} - \mu_m)v_{m-1}.
\end{aligned} \tag{2.2}$$

By the inductive hypothesis, $a_j(\mu_j - \mu_m) = 0$ for each $j = 1, 2, \ldots, m-1$. Since the eigenvalues are distinct, $(\mu_j - \mu_m) \neq 0$ for any $j < m$, and hence, $a_j = 0$ for each $j < m$.

Going back to (2.1), $a_m v_m = 0$, so $a_m = 0$, too. Hence, if (2.1) is true, then $a_j = 0$ for each j, so $\{v_1, v_2, \ldots, v_m\}$ is linearly independent. ∎

Theorem 5 implies that if A is an $n \times n$ real matrix with n distinct real eigenvalues, then there is a basis $\{v_1, v_2, \ldots, v_n\}$ of $\mathbb{R}^n$ consisting of eigenvectors of A. All we have to do to produce it is to find an eigenvector for each of the eigenvalues. This gives us n linearly independent vectors in $\mathbb{R}^n$ and therefore a basis of $\mathbb{R}^n$.

However, not every matrix has enough linearly independent eigenvectors to make a basis, even if complex numbers are admitted.

EXAMPLE 7

(There is not always a basis of eigenvectors) Consider the matrix

$$A = \begin{bmatrix} 0 & 1 \\ 0 & 0 \end{bmatrix}.$$

Notice that

$$A^2 = \begin{bmatrix} 0 & 0 \\ 0 & 0 \end{bmatrix}.$$

When we multiply numbers, we never get zero when we square a nonzero number. But evidently it can happen with matrices.

Computing

$$\det(A - \mu I) = \det\left(\begin{bmatrix} -\mu & 1 \\ 0 & -\mu \end{bmatrix}\right) = \mu^2,$$

we see that the only eigenvalue μ of A is $\mu = 0$. This means that $\mathbf{v}$ is an eigenvector of A if and only if $\mathbf{v}$ is a nonzero vector in the kernel of A. But the kernel of A consists of all nonzero multiples of

$$\mathbf{v} = \begin{bmatrix} 1 \\ 0 \end{bmatrix}.$$

Hence, the kernel of A is a one-dimensional subspace of $\mathbb{R}^2$, and there cannot be a basis of $\mathbb{R}^2$ consisting of eigenvectors of A.

So for an $n \times n$ matrix A, there may or may not be a basis $\{\mathbf{v}_1, \mathbf{v}_2, \ldots, \mathbf{v}_n\}$ of $\mathbb{R}^n$ consisting of eigenvectors of A. But when there is, we can *diagonalize* the matrix. Here is what we mean.

Let $\{\mathbf{v}_1, \mathbf{v}_2, \ldots, \mathbf{v}_n\}$ be a basis of $\mathbb{R}^n$ consisting of eigenvectors of A, and form the matrix

$$V = [\mathbf{v}_1, \mathbf{v}_2, \ldots, \mathbf{v}_n] \tag{2.3}$$

whose jth column is $\mathbf{v}_j$. Then, if $A\mathbf{v}_j = \mu_j \mathbf{v}_j$, from **V.I.F.3**,

$$A[\mathbf{v}_1, \mathbf{v}_2, \ldots, \mathbf{v}_n] = [A\mathbf{v}_1, A\mathbf{v}_2, \ldots, A\mathbf{v}_n] = [\mu_1 \mathbf{v}_1, \mu_2 \mathbf{v}_2, \ldots, \mu_n \mathbf{v}_n]. \tag{2.4}$$

There is another way to get the matrix on the right side in (2.4): Let D be the diagonal matrix given by

$$D = \begin{bmatrix} \mu_1 & 0 & 0 & \cdots & 0 \\ 0 & \mu_2 & 0 & \cdots & 0 \\ \vdots & \vdots & \vdots & \ddots & \vdots \\ 0 & 0 & 0 & \cdots & \mu_n \end{bmatrix}. \tag{2.5}$$

Then

$$[\mathbf{v}_1, \mathbf{v}_2, \ldots, \mathbf{v}_n]D = [\mu_1 \mathbf{v}_1, \mu_2 \mathbf{v}_2, \ldots, \mu_n \mathbf{v}_n]. \tag{2.6}$$

Comparing (2.4) and (2.6), we have the fundamental identity

$$AV = VD, \tag{2.7}$$

where V and D are given by (2.3) and (2.5), respectively.

By Theorem 7 of Chapter 2 and Theorem 5 of Chapter 3, since the columns of V are independent, V is invertible. Multiplying on the right by V^{-1}, we get

$$A = VDV^{-1}. \tag{2.8}$$

DEFINITION | **(Diagonalizable)** An $n \times n$ matrix A is *diagonalizable* in case there is an invertible $n \times n$ matrix V and a diagonal $n \times n$ matrix D so that $A = VDV^{-1}$.

The next theorem tells us when we can diagonalize a matrix and how to do it.

THEOREM 6 **(Diagonalization and eigenvectors)** *An $n \times n$ matrix A is diagonalizable if and only if there is a basis of $\mathbb{R}^n$ consisting of eigenvectors of A. In this case, if $\{\mathbf{v}_1, \mathbf{v}_2, \ldots, \mathbf{v}_n\}$ is such a basis and $V = [\mathbf{v}_1, \mathbf{v}_2, \ldots, \mathbf{v}_n]$, then $V^{-1}AV$ is the diagonal matrix D whose jth diagonal entry is μ_j, the eigenvalue of A corresponding to $\mathbf{v}_j$.*

PROOF: We have seen that if there is such a basis, then A is diagonalizable in the manner described in the theorem.

For the "only if" part, suppose that A is diagonalizable, and $A = VDV^{-1}$. Then $AV = VD$. Then from (2.4) and (2.6), we see that $A\mathbf{v}_j = \mu_j\mathbf{v}_j$ for each j. ∎

Combining Theorems 5 and 6, we get a very useful criterion for diagonalizability:

- *Any $n \times n$ matrix A with n distinct eigenvalues is diagonalizable.*

To see this, let $\{\mu_1, \mu_2, \ldots, \mu_n\}$ be the n distinct eigenvalues. For each of them, choose a corresponding eigenvector. By Theorem 5, $\{\mathbf{v}_1, \mathbf{v}_2, \ldots, \mathbf{v}_n\}$ is linearly independent. Any n linearly independent vectors in $\mathbb{R}^n$ or $\mathbb{C}^n$ are a basis for $\mathbb{R}^n$ or $\mathbb{C}^n$, as the case may be. Hence, by Theorem 6, A is diagonalizable.

What is diagonalization good for? Many things! Here is just one: computing powers of matrices.

THEOREM 7 **(Diagonalization and matrix powers)** *Let A be a diagonalizable $n \times n$ matrix, and let $A = VDV^{-1}$ with D diagonal. Then for all positive integers k,*

$$A^k = VD^k V^{-1}. \tag{2.9}$$

PROOF: Notice that

$$A^2 = (VDV^{-1})(VDV^{-1}) = VD(V^{-1}V)DV^{-1} = VD^2V^{-1}.$$

Now make the inductive assumption that $A^{k-1} = VD^{k-1}V^{-1}$. Then

$$
\begin{aligned}
A^k = AA^{k-1} &= (VDV^{-1})(VD^{k-1}V^{-1}) \\
&= VD(V^{-1}V)D^{k-1}V^{-1} = VDD^{k-1}V^{-1} \\
&= VD^k V^{-1}.
\end{aligned}
$$

∎

EXAMPLE 8 **(Diagonalization and matrix powers)** Let

$$A = \begin{bmatrix} 4 & 3 \\ 2 & 3 \end{bmatrix},$$

and let us find a formula for A^k.

The first step is to diagonalize A. We compute

$$\det(A - \mu I) = (4 - \mu)(3 - \mu) - 6 = \mu^2 - 7\mu + 6 = (\mu - 6)(\mu - 1).$$

The eigenvalues are $\mu_1 = 6$ and $\mu_2 = 1$. Since there are two distinct eigenvalues, there will be two linearly independent eigenvectors, so A is diagonalizable.

Now let us find the eigenvectors.

$$A - 6I = \begin{bmatrix} -2 & 3 \\ 2 & -3 \end{bmatrix}.$$

Evidently

$$\mathbf{v}_1 = \begin{bmatrix} 3 \\ 2 \end{bmatrix}$$

is orthogonal to the rows and is in the kernel. So that is one eigenvector.

Next,

$$A - I = \begin{bmatrix} 3 & 3 \\ 2 & 2 \end{bmatrix}.$$

Evidently

$$\mathbf{v}_2 = \begin{bmatrix} -1 \\ 1 \end{bmatrix}$$

is orthogonal to the rows and is in the kernel. So that is the other eigenvector. Hence, we have

$$V = \begin{bmatrix} 3 & -1 \\ 2 & 1 \end{bmatrix} \quad \text{and} \quad D = \begin{bmatrix} 6 & 0 \\ 0 & 1 \end{bmatrix}.$$

We easily compute that $V^{-1} = \dfrac{1}{5}\begin{bmatrix} 1 & 1 \\ -2 & 3 \end{bmatrix}$. Now it is easy to find a formula for A^k, because

$$D^k = \begin{bmatrix} 6^k & 0 \\ 0 & 1 \end{bmatrix}.$$

Therefore,

$$A^k = VD^kV^{-1}$$
$$= \frac{1}{5}\begin{bmatrix} 3 & -1 \\ 2 & 1 \end{bmatrix}\begin{bmatrix} 6^k & 0 \\ 0 & 1 \end{bmatrix}\begin{bmatrix} 1 & 1 \\ -2 & 3 \end{bmatrix}$$
$$= \frac{1}{5}\begin{bmatrix} 2 + 3 \times 6^k & -3 + 3 \times 6^k \\ -2 + 2 \times 6^k & 3 + 2 \times 6^k \end{bmatrix}.$$

Such a formula would not be so easy to deduce without diagonalization. We will see what this kind of formula is good for in the next section. For now, we point out that we are not restricted to integer powers! We can also use diagonalization to compute fractional powers, such as square roots.

EXAMPLE 9   **(Diagonalization and matrix square roots)** Let

$$A = \begin{bmatrix} 4 & 3 \\ 2 & 3 \end{bmatrix}.$$

Consider the problem of finding a matrix B with

$$B^2 = A.$$

Such a matrix is a "square root" of A; that is, $B = A^{1/2}$. How can we solve for B?
Diagonalization! As in the previous example, we compute that $A = VDV^{-1}$ with

$$V = \begin{bmatrix} 3 & -1 \\ 2 & 1 \end{bmatrix} \quad \text{and} \quad D = \begin{bmatrix} 6 & 0 \\ 0 & 1 \end{bmatrix}.$$

Since D is diagonal, it is easy to find a square root of D:

$$\begin{bmatrix} \sqrt{6} & 0 \\ 0 & 1 \end{bmatrix}\begin{bmatrix} \sqrt{6} & 0 \\ 0 & 1 \end{bmatrix} = \begin{bmatrix} 6 & 0 \\ 0 & 1 \end{bmatrix},$$

and hence,

$$V\begin{bmatrix} \sqrt{6} & 0 \\ 0 & 1 \end{bmatrix}V^{-1}V\begin{bmatrix} \sqrt{6} & 0 \\ 0 & 1 \end{bmatrix}V^{-1} = V\begin{bmatrix} 6 & 0 \\ 0 & 1 \end{bmatrix}V^{-1} = A.$$

Therefore, we define

$$B = V\begin{bmatrix} \sqrt{6} & 0 \\ 0 & 1 \end{bmatrix}V^{-1} = \frac{1}{5}\begin{bmatrix} 3 & -1 \\ 2 & 1 \end{bmatrix}\begin{bmatrix} \sqrt{6} & 0 \\ 0 & 1 \end{bmatrix}\begin{bmatrix} 1 & 1 \\ -2 & 3 \end{bmatrix}$$

$$= \frac{1}{5}\begin{bmatrix} 3\sqrt{6}+2 & 3\sqrt{6}-3 \\ 2\sqrt{6}-2 & 2\sqrt{6}+3 \end{bmatrix}.$$

You can multiply B by itself and verify that indeed $B^2 = A$.

So now we know how to take the square roots of diagonalizable matrices. There are other functions, such as the exponential function, that we can handle the same way. This is very useful for solving systems of linear differential equations, as we shall see in the next section.

Exercises

2.1 Let $A = \begin{bmatrix} 1 & 1 & 0 \\ 1 & 0 & 1 \\ 0 & 1 & 1 \end{bmatrix}$. Find an invertible matrix V and a diagonal matrix D so that $A = VDV^{-1}$, or explain why this cannot be done.

2.2 Let $A = \begin{bmatrix} -3 & 1 & 1 \\ 1 & -3 & 1 \\ 1 & 1 & -1 \end{bmatrix}$. Find an invertible matrix V and a diagonal matrix D so that $A = VDV^{-1}$, or explain why this cannot be done.

2.3 Let $A = \begin{bmatrix} 1 & 2 & 3 \\ 0 & 1 & 2 \\ 0 & 0 & 1 \end{bmatrix}$. Find an invertible matrix V and a diagonal matrix D so that $A = VDV^{-1}$, or explain why this cannot be done.

2.4 Let $A = \begin{bmatrix} 5 & 8 \\ -2 & -3 \end{bmatrix}$. Find an invertible matrix V and a diagonal matrix D so that $A = VDV^{-1}$, or explain why this cannot be done.

2.5 Let $A = \begin{bmatrix} 1 & 2 \\ 2 & 4 \end{bmatrix}$. Find an invertible matrix V and a diagonal matrix D so that $A = VDV^{-1}$, or explain why this cannot be done.

2.6 Consider $A = \begin{bmatrix} 1 & -2 & 0 \\ 1 & 2 & 0 \\ 0 & 2 & -1 \end{bmatrix}$ and $B = \begin{bmatrix} 1 & 2 & 3 \\ 0 & 3 & 3 \\ 0 & 0 & -3 \end{bmatrix}$.

(a) Find all the eigenvalues of A.

(b) Find all the eigenvalues of B.

(c) Is B diagonalizable? Justify your answer.

(d) What is the largest eigenvalue of B^4, and what is the dimension of the corresponding eigenspace?

2.7 Consider

$$A = \begin{bmatrix} 0 & 1 \\ 0 & 0 \end{bmatrix}, \qquad B = \begin{bmatrix} 1 & 1/2 \\ 2 & 1 \end{bmatrix}, \qquad C = \begin{bmatrix} 2 & 1 \\ 3 & 3 \end{bmatrix} \quad \text{and} \quad D = \begin{bmatrix} 2 & 3 \\ 1 & 0 \end{bmatrix}.$$

(a) Which of these matrices, if any, have only a single eigenvalue, and which have two eigenvalues?

(b) Which of these matrices, if any, can be diagonalized by a change of basis?

(c) Compute $D^2 - 2D - 3I$, where I is the 2×2 identity matrix.

(d) Compute B^{15}.

2.8 Consider $A = \begin{bmatrix} 2 & 1 & 0 & 0 \\ 1 & 2 & 0 & 0 \\ 0 & 0 & 3 & 1 \\ 0 & 0 & 1 & 3 \end{bmatrix}$ and $B = \begin{bmatrix} 1 & 0 & 0 \\ 2 & 3 & 0 \\ 1 & 3 & 3 \end{bmatrix}$.

(a) Find all the eigenvalues and eigenvectors of A.

(b) Find all the eigenvalues and eigenvectors of B.

(c) Is A diagonalizable? Justify your answer.

(d) Is B diagonalizable? Justify your answer.

2.9 Consider $A = \begin{bmatrix} 5 & -8 \\ 2 & 5 \end{bmatrix}$. Compute a closed-form formula for A^k, as in Example 8. The eigenvalues of A are complex, so your computations will involve complex numbers, but you should be able to eliminate them from your final answer.

2.10 Consider $A = \begin{bmatrix} 1 & -2 \\ 1 & 1 \end{bmatrix}$. Compute a closed-form formula for A^k, as in Example 8. The eigenvalues of A are complex, so your computations will involve complex numbers, but you should be able to eliminate them from your final answer.

2.11 Here is a very useful fact about eigenvalues: If A is any $n \times n$ matrix, k is any positive integer, and μ is an eigenvalue of A, then μ^k is an eigenvalue of A^k.

(a) Explain why this fact is true.

(b) Is every eigenvector of A necessarily an eigenvector of A^k?

(c) Is every eigenvector of A^k necessarily an eigenvector of A?

2.12 Here is a useful fact about diagonalization: If A is any diagonalizable $n \times n$ matrix and k is any positive integer, then A^k is also diagonalizable.

(a) Explain why this fact is true.

(b) Give an example of a 2×2 matrix A such that A^2 is diagonalizable but A is not.

2.13 Let A be any 3×3 matrix with real entries.

(a) Show that A has at least one real eigenvalue.

(b) Show that if A has at least one complex eigenvalue, then A is diagonalizable; that is, there is a basis of $\mathbb{C}^n$ consisting of eigenvectors of A. (Recall that complex roots of real polynomials come in complex-conjugate pairs.)

2.14 Let R be the 2×2 matrix $R = \begin{bmatrix} \cos(\theta) & -\sin(\theta) \\ \sin(\theta) & \cos(\theta) \end{bmatrix}$, and let $0 < \theta < 2\pi$. This is the matrix corresponding to the counterclockwise rotation through the angle θ in the plane $\mathbb{R}^2$. Find a basis of $\mathbb{C}^2$ consisting of eigenvectors of R.

2.15 (a) Show that if A is a diagonal $n \times n$ matrix, the eigenvalues of A are exactly the diagonal entries of A. More generally, show that if A is an upper- or lower-triangular $n \times n$ matrix, the eigenvalues of A are exactly the diagonal entries of A.

(b) Show that every upper-triangular matrix A with distinct diagonal entries is diagonalizable.

2.16 Let $A = \begin{bmatrix} 1 & -2 \\ 0 & 9 \end{bmatrix}$. Compute a square root of A. That is, find a matrix B so that $B^2 = A$.

| ## Application to Differential and Difference Equations

3.1 Differentiating vector functions

Consider a vector-valued function $\mathbf{x}(t)$ of the real variable t with values in $\mathbb{R}^n$. For example, let us consider $n = 3$ and

$$\mathbf{x}(t) = \begin{bmatrix} \cos(t) \\ \sin(t) \\ 1/t \end{bmatrix}. \tag{3.1}$$

Here is a three-dimensional plot of the curve traced out by $\mathbf{x}(t)$ as t varies from $t = 1$ to $t = 20$:

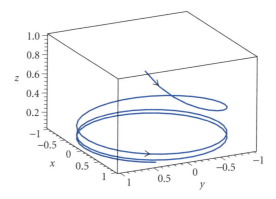

Such vector-valued functions arise whenever we need to describe the position of a particle as a function of time. But more generally, we might have any sort of system that is described by n parameters. These could be, for example, the voltages across n points in an electric circuit. We can arrange these data into a vector, and if the data are varying with time, as is often the case in applications, we then have a time-dependent vector $\mathbf{x}(t)$ in $\mathbb{R}^n$.

When quantities are varying in time, it is often useful to consider their rates of change.

DEFINITION

(Derivatives of vector-valued functions) Let $\mathbf{x}(t)$ be a vector-valued function of the variable t. We say that $\mathbf{x}(t)$ is *differentiable* at $t = t_0$ with derivative $\mathbf{x}'(t_0)$ if

$$\lim_{h \to 0} \frac{1}{h} (\mathbf{x}(t_0 + h) - \mathbf{x}(t_0)) = \mathbf{x}'(t_0)$$

in the sense that this limit exists for each of the n entries separately. A vector-valued function is *differentiable* in some interval (a, b) if it is differentiable for each t_0 in (a, b).

There is nothing really new going on here. To compute the derivative of $\mathbf{x}(t)$, you differentiate it entry by entry in the usual way.

Consider a t-dependent vector

$$\mathbf{x}(t) = \begin{bmatrix} x(t) \\ y(t) \end{bmatrix} \text{ in } \mathbb{R}^2.$$

Then, by the rules for vector subtraction and scalar multiplication,

$$\frac{1}{h}\left(\mathbf{x}(t+h) - \mathbf{x}(t)\right) = \frac{1}{h}\left(\begin{bmatrix} x(t+h) \\ y(t+h) \end{bmatrix} - \begin{bmatrix} x(t) \\ y(t) \end{bmatrix}\right)$$
$$= \begin{bmatrix} (x(t+h) - x(t))/h \\ (y(t+h) - y(t))/h \end{bmatrix}.$$

Taking the limits on the right, entry by entry, we see that

$$\mathbf{x}'(t) = \begin{bmatrix} x'(t) \\ y'(t) \end{bmatrix},$$

provided $x(t)$ and $y(t)$ are both differentiable. The same reduction to single-variable differentiation clearly extends to any number of entries.

EXAMPLE 10   **(Computing the derivative of a vector-valued function of t)** Let $\mathbf{x}(t)$ be given by (3.1). Then for any $t \neq 0$,

$$\mathbf{x}'(t) = \begin{bmatrix} -\sin(t) \\ \cos(t) \\ -1/t^2 \end{bmatrix}.$$

Because we differentiate vectors entry by entry without mixing the entries up in any way, familiar rules for differentiating numerically valued functions hold for vector-valued functions as well. In particular, the derivative of a sum is still the sum of the derivatives.

$$(\mathbf{x}(t) + \mathbf{y}(t))' = \mathbf{x}'(t) + \mathbf{y}'(t). \tag{3.2}$$

Things are only slightly more complicated with the product rule, because now we have several types of products to consider. Here is an example that we shall need soon: a "product rule" for the dot product.

THEOREM 8   **(Differentiating dot products)** *Suppose that $\mathbf{v}(t)$ and $\mathbf{w}(t)$ are differentiable vector-valued functions for t in (a, b) with values in $\mathbb{R}^n$. Then $\mathbf{v}(t) \cdot \mathbf{w}(t)$ is differentiable for t in (a, b), and*

$$\frac{d}{dt}\mathbf{v}(t) \cdot \mathbf{w}(t) = \mathbf{v}'(t) \cdot \mathbf{w}(t) + \mathbf{v}(t) \cdot \mathbf{w}'(t). \tag{3.3}$$

PROOF: For each i, we have by the usual product rule

$$\frac{d}{dt}v_i(t)w_i(t) = v_i'(t)w_i(t) + v_i(t)w_i'(t).$$

Summing on i now gives us (3.3).   ∎

3.2 Systems of differential equations and the superposition principle

Consider the system of differential equations

$$\begin{aligned} x'(t) &= 4x(t) + 3y(t) \\ y'(t) &= 2x(t) + 3y(t) \end{aligned} \tag{3.4}$$

with the initial conditions

$$x(0) = 1 \quad\text{and}\quad y(0) = 4. \tag{3.5}$$

The four equations in (3.4) and (3.5) specify, as we shall see, a pair of functions $x(t)$ and $y(t)$. You can think of these as the x and y coordinates at time t of a point particle moving in the plane $\mathbb{R}^2$. Then the equations in (3.4) relate the velocity of the particle at time t to the position of the particle at time t. The equations in (3.5) say that at time $t = 0$, the particle is at the point $(1, 4)$.

Very often, a physical principle, or some other scientific principle, leads to a description of motion in terms of such a system of equations. In most cases, though, this is not good enough. Often we want a *formula* for $x(t)$ and $y(t)$. How do we find that?

The reason we raise this question here is that we can use what we know about eigenvectors to find a pair of functions satisfying all these conditions, which is what it means to *solve this system of differential equations for the given initial data.* Here is how:

The first step is to rewrite (3.4) in vector and matrix form, so that it becomes a single equation for a time dependent vector in $\mathbb{R}^2$. This is easy: Introduce the vector

$$\mathbf{x}(t) = \begin{bmatrix} x(t) \\ y(t) \end{bmatrix}$$

and the matrix

$$A = \begin{bmatrix} 4 & 3 \\ 2 & 3 \end{bmatrix}.$$

Then we can rewrite (3.4) as

$$\mathbf{x}'(t) = A\mathbf{x}(t). \tag{3.6}$$

The point of rewriting (3.4) in the form (3.6) is that we can combine the eigenvectors and eigenvalues of A to produce two very simple solutions of (3.6).

As we saw in Examples 8 and 9, $A\mathbf{v}_1 = 6\mathbf{v}_1$ and $A\mathbf{v}_2 = \mathbf{v}_2$, where

$$\mathbf{v}_1 = \begin{bmatrix} 3 \\ 2 \end{bmatrix} \quad\text{and}\quad \mathbf{v}_2 = \begin{bmatrix} -1 \\ 1 \end{bmatrix}.$$

Here are the simple solutions built out of the eigenvectors and eigenvalues:

$$\mathbf{x}_1(t) = e^{6t}\mathbf{v}_1 \quad\text{and}\quad \mathbf{x}_2(t) = e^t\mathbf{v}_2. \tag{3.7}$$

Notice how simple this is: Just multiply the eigenvector $\mathbf{v}_j$ by $e^{t\mu_j}$ to give a solution of (3.6). This fact is worth recording as a general theorem.

THEOREM 9   **(Eigenvectors, eigenvalues, and special solutions)** *Let A be any $n \times n$ matrix, and suppose that* **v** *is an eigenvector of A with* $A\mathbf{v} = \mu\mathbf{v}$. *Define* $\mathbf{x}(t) = e^{t\mu}\mathbf{v}$. *Then* $\mathbf{x}'(t) = A\mathbf{x}(t)$.

PROOF: On the one hand,

$$\mathbf{x}'(t) = (e^{t\mu})'\mathbf{v} = (\mu e^{t\mu})\mathbf{v} = \mu(e^{t\mu}\mathbf{v}) = \mu\mathbf{x}(t).$$

On the other hand,

$$A\mathbf{x}(t) = A(e^{t\mu}\mathbf{v}) = e^{t\mu}A\mathbf{v} = e^{t\mu}\mu\mathbf{v} = \mu(e^{t\mu}\mathbf{v}) = \mu\mathbf{x}(t).$$

∎

According to Theorem 9, (3.7) defines two solutions to (3.6). However, neither solution satisfies the vector version of (3.5), which is

$$\mathbf{x}(0) = \begin{bmatrix} 1 \\ 4 \end{bmatrix}. \tag{3.8}$$

But all is not lost! All we need do is to find a linear combination $a\mathbf{v}_1 + b\mathbf{v}_2$ of $\mathbf{v}_1$ and $\mathbf{v}_2$ with

$$a\mathbf{v}_1 + b\mathbf{v}_2 = \begin{bmatrix} 1 \\ 4 \end{bmatrix}. \tag{3.9}$$

Then the solution we seek is the *superposition*

$$\mathbf{x}(t) = a\mathbf{x}_1(t) + b\mathbf{x}_2(t). \tag{3.10}$$

To see why this works, observe that at $t = 0$,

$$a\mathbf{x}_1(0) + b\mathbf{x}_2(0) = a\mathbf{v}_1 + b\mathbf{v}_2 = \begin{bmatrix} 1 \\ 4 \end{bmatrix} \tag{3.11}$$

since $\mathbf{x}_1(0) = \mathbf{v}_1$ and $\mathbf{x}_2(0) = \mathbf{v}_2$. Moreover, since differentiation is a linear operation,

$$\begin{aligned}
\mathbf{x}'(t) &= (a\mathbf{x}_1(t) + b\mathbf{x}_2(t))' \\
&= a\mathbf{x}_1'(t) + b\mathbf{x}_2'(t) \\
&= aA\mathbf{x}_1(t) + bA\mathbf{x}_2(t) \\
&= A(a\mathbf{x}_1(t) + b\mathbf{x}_2(t)) = A\mathbf{x}(t).
\end{aligned}$$

Therefore, the function $\mathbf{x}(t)$ defined by (3.10) satisfies both the equation in (3.6) and the initial condition in (3.8). This is called the *superposition principle*.

Finally, since $\{\mathbf{v}_1, \mathbf{v}_2\}$ is a basis of $\mathbb{R}^2$, there is a unique choice of the numbers a and b in (3.10). To find them, introduce the matrix

$$V = [\mathbf{v}_1, \mathbf{v}_2] = \begin{bmatrix} 3 & -1 \\ 2 & 1 \end{bmatrix}$$

and the vector

$$\mathbf{a} = \begin{bmatrix} a \\ b \end{bmatrix}.$$

Then by **V.I.F.1**, we can write (3.9) as

$$V\mathbf{a} = \begin{bmatrix} 1 \\ 4 \end{bmatrix}.$$

Since

$$V^{-1} = \frac{1}{5}\begin{bmatrix} 1 & 1 \\ -2 & 3 \end{bmatrix},$$

we easily find

$$\mathbf{a} = \begin{bmatrix} 1 \\ 2 \end{bmatrix}.$$

Hence $a = 1$ and $b = 2$, and our solution is

$$\mathbf{x}(t) = a\mathbf{v}_1(t) + b\mathbf{v}_2(t) = \mathbf{v}_1(t) + 2\mathbf{v}_2(t) = e^{6t}\begin{bmatrix} 3 \\ 2 \end{bmatrix} + 2e^t\begin{bmatrix} -1 \\ 1 \end{bmatrix}. \tag{3.12}$$

Writing out $\mathbf{x}(t)$ explicitly,

$$\mathbf{x}(t) = \begin{bmatrix} 3e^{6t} - 2e^t \\ 2e^{6t} + 2e^t \end{bmatrix}.$$

Hence, the pair of functions

$$x(t) = 3e^{6t} - 2e^t \qquad \text{and} \qquad y(t) = 2e^{6t} + 2e^t$$

solves (3.4) and the initial conditions $x(0) = 1$ and $y(0) = 4$. In fact, this is the *only* solution,* and we have succeeded in going from the differential equation description of the "particle motion" given by (3.4) and (3.5) to explicit formulas for $x(t)$ and $y(t)$.

When can we use this method to solve systems of differential equations? Whenever the corresponding matrix A is diagonalizable. The point of diagonalizability is that an $n \times n$ matrix A is diagonalizable if and only if there is a basis $\{\mathbf{v}_1, \mathbf{v}_2, \ldots, \mathbf{v}_n\}$ of $\mathbb{R}^n$ consisting of eigenvectors of A, as we know from Theorem 6.

The fact that there is a basis of eigenvectors ensures that *any* vector in $\mathbb{R}^n$, in particular $\mathbf{x}_0$, has a unique expression as a linear combination of eigenvectors of A. Therefore, there are uniquely determined numbers $a_1, a_2, \ldots, a_n$ so that

$$\mathbf{x}_0 = a_1\mathbf{v}_1 + a_2\mathbf{v}_2 + \cdots + a_n\mathbf{v}_n.$$

These numbers are simply the coordinates of $\mathbf{x}_0$ with respect to the basis $\{\mathbf{v}_1, \mathbf{v}_2, \ldots, \mathbf{v}_n\}$. To find them, just solve

$$V\mathbf{a} = \mathbf{x}_0,$$

where

$$V = [\mathbf{v}_1, \mathbf{v}_2, \ldots, \mathbf{v}_n] \qquad \text{and} \qquad \mathbf{a} = \begin{bmatrix} a_1 \\ a_2 \\ \vdots \\ a_n \end{bmatrix}.$$

Then, if μ_j is the eigenvalue corresponding to $\mathbf{v}_j$, the solution to $\mathbf{x}'(t) = A\mathbf{x}(t)$ with initial condition $\mathbf{x}(0) = \mathbf{x}_0$ is

$$\mathbf{x}(t) = a_1 e^{t\mu_1}\mathbf{v}_1 + a_2 e^{t\mu_2}\mathbf{v}_2 + \cdots + a_n e^{t\mu_n}\mathbf{v}_n.$$

We summarize:

- **The superposition principle** *If A is a diagonalizable $n \times n$ matrix, then the solution of $\mathbf{x}'(t) = A\mathbf{x}(t)$ with $\mathbf{x}(0) = \mathbf{x}_0$ can be found by expanding $\mathbf{x}_0$ in a basis of eigenvectors of A and then forming the corresponding superposition, that is, linear combination, of the special eigenvector solutions.*

*We shall soon show that there is just one solution.

In using the superposition principle to solve $x'(t) = Ax(t)$ with the initial condition $x(0) = x_0$, we are emphasizing the *coordinate change* aspect of diagonalization. The point is that written in coordinates that are based on a basis $\{v_1, \ldots, v_n\}$ of eigenvectors of A, the system of differential equations $x'(t) = Ax(t)$ *decouples* into n independent equations, one for each coordinate.

Indeed, define $y(t) = V^{-1}x(t)$, which is the coordinate vector of $x(t)$. Then since V^{-1} is constant in t,

$$y'(t) = V^{-1}x'(t) = V^{-1}Ax(t) = (V^{-1}AV)V^{-1}x(t) = (V^{-1}AV)y(t).$$

But since $V^{-1}AV$ is diagonal, with the jth diagonal entry being μ_j, this equation reduces to

$$y_j'(t) = \mu_j y_j(t) \qquad j = 1, \ldots, n.$$

For each j, this is solved—uniquely—by $y_j(t) = e^{t\mu_j}y_j(0)$. (To see the uniquness, note that if $y_j'(t) = \mu_j y_j(t)$, then the derivative of $e^{-t\mu_j}y_j(t)$ is zero. Hence, this function is constant, so $e^{-t\mu_j}y_j(t) = y_j(0)$.)

But then having found the coordinate vector $y(t)$, we can "reassemble" $x(t)$:

$$x(t) = Vy(t) = e^{t\mu_1}y_1(0)v_1 + \cdots + e^{t\mu_n}y_n(0)v_n.$$

An n-variable calculus problem has been reduced—by a judicious change of coordinates—to n single-variable calculus problems, and we see that the solution provided by the superposition principle is unique.

This chapter and much of the rest of the book is a further development of the ideas we first encountered at the end of Chapter 3 about simplifying linear algebra problems by making a judicious change of coordinates. Since coordinate systems are based on bases, this means making a judicious change of bases.

EXAMPLE 11 **(The superposition principle in action)** Consider the system of differential equations

$$x'(t) = x(t) + y(t) + z(t)$$
$$y'(t) = x(t) + y(t) + z(t)$$
$$z'(t) = x(t) + y(t) + z(t)$$

with the initial condition $x(0) = 2$, $y(0) = 1$, and $y(0) = 0$. Let us find $x(t)$, $y(t)$, and $z(t)$ using the superposition principle.

The first step is to introduce the matrix

$$A = \begin{bmatrix} 1 & 1 & 1 \\ 1 & 1 & 1 \\ 1 & 1 & 1 \end{bmatrix}$$

and the initial data vector

$$x_0 = \begin{bmatrix} 2 \\ 1 \\ 0 \end{bmatrix}.$$

Then our problem is to solve

$$x'(t) = Ax(t) \qquad \text{and} \qquad x(0) = x_0.$$

To proceed, we have to find the eigenvalues and eigenvectors of A. We have already done this back in Example 2. We found the eigenvalues

$$\mu_1 = 0, \qquad \mu_2 = 0, \qquad \text{and} \qquad \mu_3 = 3$$

with the corresponding eigenvectors

$$\mathbf{v}_1 = \begin{bmatrix} -1 \\ 0 \\ 1 \end{bmatrix}, \qquad \mathbf{v}_2 = \begin{bmatrix} -1 \\ 1 \\ 0 \end{bmatrix}, \qquad \text{and} \qquad \mathbf{v}_3 = \begin{bmatrix} 1 \\ 1 \\ 1 \end{bmatrix}.$$

We have a basis of eigenvectors, and so we can apply the superposition principle. Letting $V = [\mathbf{v}_1, \mathbf{v}_2, \mathbf{v}_3]$ and solving $V\mathbf{a} = \mathbf{x}_0$,

$$\mathbf{a} = \begin{bmatrix} -1 \\ 0 \\ 1 \end{bmatrix},$$

so

$$\mathbf{x}_0 = -\mathbf{v}_1 + \mathbf{v}_3.$$

The solution then is

$$\mathbf{x}(t) = -\mathbf{v}_1 + e^{3t}\mathbf{v}_3 = \begin{bmatrix} e^{3t} + 1 \\ e^{3t} \\ e^{3t} - 1 \end{bmatrix}.$$

That is,

$$x(t) = e^{3t} + 1, \qquad y(t) = e^{3t}, \qquad \text{and} \qquad z(t) = e^{3t} - 1.$$

We can even apply the superposition principle when the eigenvalues of A are complex.

EXAMPLE 12   **(Superposition with complex eigenvalues)** Consider the system of differential equations

$$x'(t) = y(t) \qquad \text{and} \qquad y'(t) = -x(t)$$

with the initial conditions $x(0) = 3$ and $y(0) = 4$. Let us find $x(t)$ and $y(t)$ using the superposition principle.

The first step is to introduce the matrix

$$A = \begin{bmatrix} 0 & 1 \\ -1 & 0 \end{bmatrix}$$

and the initial data vector

$$\mathbf{x}_0 = \begin{bmatrix} 3 \\ 4 \end{bmatrix}.$$

Then our problem is to solve

$$\mathbf{x}'(t) = A\mathbf{x}(t) \qquad \text{and} \qquad \mathbf{x}(0) = \mathbf{x}_0.$$

To proceed, we have to find the eigenvalues and eigenvectors of A. We already did this back in Example 6. The eigenvalues are imaginary: $\mu_1 = i$ and $\mu_2 = -i$. In Example 6, we found the corresponding eigenvectors

$$\mathbf{v}_1 = \begin{bmatrix} -i \\ 1 \end{bmatrix} \qquad \text{and} \qquad \mathbf{v}_2 = \begin{bmatrix} i \\ 1 \end{bmatrix}.$$

The vector $\mathbf{x}_0$ is a vector in $\mathbb{C}^2$ as well as $\mathbb{R}^2$—just as the real numbers are a subset of the complex numbers—and $\{\mathbf{v}_1, \mathbf{v}_2\}$ is a basis for $\mathbb{C}^2$. To find the coordinates of $\mathbf{x}_0$ with respect to this basis, we solve $V\mathbf{a} = \mathbf{x}_0$, where

$$V = \begin{bmatrix} -i & i \\ 1 & 1 \end{bmatrix}.$$

Using the usual formula for 2×2 inverses, we find

$$V^{-1} = \frac{1}{2} \begin{bmatrix} i & 1 \\ -i & 1 \end{bmatrix}.$$

Hence,

$$\begin{bmatrix} a_1 \\ a_2 \end{bmatrix} = \frac{1}{2} \begin{bmatrix} i & 1 \\ -i & 1 \end{bmatrix} \begin{bmatrix} 3 \\ 4 \end{bmatrix} = \frac{1}{2} \begin{bmatrix} 4 + 3i \\ 4 - 3i \end{bmatrix}.$$

Our solution then is

$$\mathbf{x}(t) = \frac{4 + 3i}{2} e^{it} \begin{bmatrix} -i \\ 1 \end{bmatrix} + \frac{4 - 3i}{2} e^{-it} \begin{bmatrix} i \\ 1 \end{bmatrix} = \frac{1}{2} \begin{bmatrix} (3 - 4i)e^{it} + (3 - 4i)e^{-it} \\ (4 + 3i)e^{it} + (4 - 3i)e^{-it} \end{bmatrix}.$$

This is the answer, but it would be unsatisfactory to leave it in this form. We are expecting a motion in $\mathbb{R}^2$, not $\mathbb{C}^2$, and in fact we have one: If we use Euler's identity

$$e^{i\theta} = \cos(\theta) + i \sin(\theta)$$

so that

$$\cos(\theta) = \frac{e^{i\theta} + e^{-i\theta}}{2} \qquad \text{and} \qquad \sin(\theta) = \frac{e^{i\theta} - e^{-i\theta}}{2i},$$

we find

$$\mathbf{x}(t) = \begin{bmatrix} 3 \cos(t) + 4 \sin(t) \\ 4 \cos(t) - 3 \sin(t) \end{bmatrix}.$$

You can easily check that this is the solution. The point $\mathbf{x}(t)$ is moving clockwise at unit angular speed along the centered circle of radius 5.

Imaginary eigenvalues come up whenever one uses the superposition principle to find a motion that involves rotation, even if they disappear in the final answer.

What happens when A is not diagonalizable? You might get lucky and still be able to express your given initial data $\mathbf{x}_0$ as a linear combination of eigenvalues of A. But this will not work for every $\mathbf{x}_0$, or else there would be a spanning set and hence a basis of eigenvectors.

To treat general initial data using the superposition principle, one needs to introduce "generalized" eigenvectors. It turns out that *every* matrix has a basis of generalized eigenvectors, and a simple extension of Theorem 9 gives special solutions for generalized eigenvectors. However, there is more to be said about the eigenvalue problem before generalizing it.

3.3 Fibonacci's rabbits

Some problems are best considered in discrete time instead of in continuous time. They often involve *difference equations* instead of differential equations. Here is a classic example.

Fibonacci wrote a mathematics text in the thirteenth century that posed the following problem: Suppose that in January you are given a newborn pair of rabbits for breeding. Suppose that it takes one month for rabbits to mature to breeding age, and that each month, as soon as they are able, each breeding pair produces another pair of rabbits. How many rabbits will you have after one year?

Let us figure out what happens during the first few months and find the pattern. At the beginning of February, you would still have just one pair, but that pair would now be mature.

At the beginning of March, you would have 2 pairs, the original pair, plus the new one produced by the original breeding pair.

At the beginning of April, there will be 3 pairs; the 2 we had the previous month, plus a new one produced by the only mature pair.

However, during April, the pair born at the end of January is mature, so there are two breeding pairs. Hence, two new pairs will be added at the end of April so that at the beginning of May there are 5 pairs, of which 3 are mature.

Likewise, at the beginning of June there will be 8 pairs—the 5 present at the beginning of the month plus the 3 new ones produced by the 3 mature pairs.

Let a_n be the number of pairs of rabbits at the beginning of the nth month. As we have deduced,

$$a_1 = 1, \qquad a_2 = 1, \qquad a_3 = 2, \qquad a_4 = 3, \qquad a_5 = 5, \qquad \text{and} \qquad a_6 = 8.$$

From here you can see and understand the pattern: For $n \geq 2$,

$$a_{n+1} = a_n + a_{n-1}. \tag{3.13}$$

This is because the a_{n-1} pairs that were present in month $n - 1$ are the mature ones during month n, so they produce a_{n-1} new pairs during the nth month. Of course you also still have the a_n pairs originally present at the beginning of the nth month, and the result is $a_n + a_{n-1}$ pairs at the beginning of month $n + 1$.

To answer Fibonacci's question, we need to compute a_{13}. But let us make a further hypothesis, namely, that rabbits are immortal, and let us produce an infinite sequence $\{a_n\}$ from the recursive formula

$$a_{n+1} = a_n + a_{n-1} \qquad \text{with} \quad a_1 = a_2 = 1.$$

This is a "difference equation" in disguise. If we rewrite it as

$$(a_{n+1} - a_n) = a_{n-1} \qquad \text{with} \quad a_1 = a_2 = 1,$$

it *is* a difference equation. That is, it is an equation giving the change made in passing from a_n to a_{n+1} in terms of the known values of the sequence at time n. However, in applying matrix methods, the first form of the equation is preferable.

Let us now find an explicit formula for the Fibonacci numbers using eigenvectors and eigenvalues. Here is how: Introduce the vectors

$$\mathbf{x}_n = \begin{bmatrix} a_n \\ a_{n-1} \end{bmatrix}$$

and the matrix

$$A = \begin{bmatrix} 1 & 1 \\ 1 & 0 \end{bmatrix}.$$

Then (3.13) is equivalent to

$$\mathbf{x}_{n+1} = A\mathbf{x}_n.$$

It follows that

$$\mathbf{x}_n = A^{n-2}\mathbf{x}_2. \tag{3.14}$$

To get an explicit formula for $\mathbf{x}_n$ and hence the nth Fibonacci number a_n, all we need to do is to get an explicit formula for A^{n-2}. This is something we know how to do, as we did it back in Example 8 for a different matrix. To do for the present matrix A, we just need to diagonalize it.

Proceeding as usual, we compute the characteristic polynomial of A, which is

$$p(t) = t^2 - t - 1 = (t - \gamma_+)(t - \gamma_-),$$

where

$$\gamma_\pm = \frac{1 \pm \sqrt{5}}{2}.$$

The number γ_+ is called the *golden mean*. There are two distinct eigenvectors, so A is diagonalizable.

Let us find the eigenvectors. The eigenspace corresponding to γ_+ is $\mathrm{Ker}(A - \gamma_+ I)$, and

$$A - \gamma_+ I = \begin{bmatrix} 1 - \gamma_+ & 1 \\ 1 & -\gamma_+ \end{bmatrix}.$$

Any eigenvector $\mathbf{v}$ of A with eigenvalue γ_+ must be in the kernel of this matrix and therefore must be orthogonal to both rows of $A - \gamma_+ I$.

The bottom row is simpler; let us work with that: Any eigenvector $\mathbf{v}$ of A with eigenvalue γ_+ must be orthogonal to

$$\begin{bmatrix} 1 \\ -\gamma_+ \end{bmatrix}.$$

In $\mathbb{R}^2$, finding orthogonal vectors is easy. The vectors in $\mathbb{R}^2$ that are orthogonal to

$$\begin{bmatrix} a \\ b \end{bmatrix}$$

are the multiples of

$$\begin{bmatrix} a \\ b \end{bmatrix}^\perp = \begin{bmatrix} -b \\ a \end{bmatrix}.$$

Therefore,

$$\mathbf{v}_1 = \begin{bmatrix} \gamma_+ \\ 1 \end{bmatrix}$$

is an eigenvector of A with eigenvalue γ_+. (You could have found it by solving for $\mathrm{Ker}(A - \gamma_+ I)$ in the usual way, but in $\mathbb{R}^2$, taking advantage of orthogonality is expeditious.) Doing the same for γ_-, we find that

$$\mathbf{v}_2 = \begin{bmatrix} \gamma_- \\ 1 \end{bmatrix}$$

is an eigenvector with eigenvalue γ_-.

We now form

$$V = [\mathbf{v}_1, \mathbf{v}_2] = \begin{bmatrix} \gamma_+ & \gamma_- \\ 1 & 1 \end{bmatrix} \quad \text{and} \quad D = \begin{bmatrix} \gamma_+ & 0 \\ 0 & \gamma_- \end{bmatrix}.$$

We therefore have $A = VDV^{-1}$, and hence

$$\mathbf{x}_n = A^{n-2}\mathbf{x}_2 = VD^{n-2}V^{-1}\mathbf{x}_2.$$

Since

$$V^{-1} = \frac{1}{\sqrt{5}} \begin{bmatrix} 1 & -\gamma_- \\ -1 & \gamma_+ \end{bmatrix},$$

the product $VD^{n-2}V^{-1}\mathbf{x}_2$ is not hard to work out: The result is *Binet's formula* for a_n:

$$a_n = \frac{1}{\sqrt{5}} \left(\left(\frac{1 + \sqrt{5}}{2} \right)^n - \left(\frac{1 - \sqrt{5}}{2} \right)^n \right).$$

In doing the computations, you can make use of the relation $\gamma_+ \gamma_- = -1$.

This striking formula for the Fibonacci numbers was not discovered until many centuries after Fibonacci. The method used here can be used to find formulas for the solution of other so-called *difference equations*.

Here is a class of difference equations that generalizes the case considered by Fibonacci. Let α and β be any two numbers and consider

$$a_{n+1} - a_n = \alpha a_n + \beta a_{n-1}. \tag{3.15}$$

Taking $\alpha = 0$ and $\beta = 1$, we have Fibonacci's equation. In the following exercises, you will apply the method we used to solve Fibonacci's problem to deal with various cases of (3.15) and even more general difference equations.

Exercises

3.1 Let $\mathbf{x}(t)$ be the t-dependent vector $\mathbf{x}(t) = \begin{bmatrix} t^2 \\ t \\ t^3 \end{bmatrix}$, and let $\mathbf{y}(t)$ be the t-dependent vector $\mathbf{y}(t) = \begin{bmatrix} t^{-1} \\ t \\ t^{-2} \end{bmatrix}$, $t \neq 0$.

(a) Compute $\mathbf{x}'(t)$.

(b) Compute $\mathbf{y}'(t)$.

(c) Compute $\dfrac{d}{dt}|\mathbf{x}(t)|$.

(d) Compute $\dfrac{d}{dt}|\mathbf{y}(t)|$.

(e) For which values of t, if any, are $\mathbf{x}'(t)$ and $\mathbf{y}'(t)$ orthogonal?

3.2 Let $\mathbf{x}(t)$ be the t-dependent vector $\mathbf{x}(t) = \begin{bmatrix} \cos(2t) \\ \sin(2t) \\ t \end{bmatrix}$, and let $\mathbf{y}(t)$ be the t-dependent vector $\mathbf{y}(t) = \begin{bmatrix} t \\ \cos(t) \\ \sin(t) \end{bmatrix}$, $t \neq 0$.

(a) Compute $\mathbf{x}'(t)$.

(b) Compute $\mathbf{y}'(t)$.

(c) Compute $\dfrac{d}{dt}|\mathbf{x}(t)|$.

(d) Compute $\dfrac{d}{dt}|\mathbf{y}(t)|$.

(e) For which values of t, if any, are $\mathbf{x}'(t)$ and $\mathbf{y}'(t)$ orthogonal?

3.3 Use the superposition principle to solve the system

$$x'(t) = x(t) - 2y(t)$$
$$y'(t) = -2x(t) + y(t),$$

subject to the initial conditions $x(0) = 1$ and $y(0) = -3$.

3.4 Use the superposition principle to solve the system

$$x'(t) = x(t) - 2y(t)$$
$$y'(t) = 3y(t),$$

subject to the initial conditions $x(0) = 1$ and $y(0) = 1$.

3.5 Use the superposition principle to solve the system

$$x'(t) = x(t) + y(t)$$
$$y'(t) = x(t) + z(t)$$
$$z'(t) = y(t) + z(t),$$

subject to the initial conditions $x(0) = 1$, $y(0) = 2$, and $z(0) = 3$.

3.6 Use the superposition principle to solve the system

$$x'(t) = x(t) + y(t) + 2z(t)$$
$$y'(t) = x(t) + 2y(t) + z(t)$$
$$z'(t) = 2x(t) + y(t) + z(t),$$

subject to the initial conditions $x(0) = 1$, $y(0) = 2$, and $z(0) = 1$.

3.7 Use the superposition principle to solve the system

$$x'(t) = 5x(t) - 8y(t)$$
$$y'(t) = 2x(t) + 5y(t),$$

subject to the initial conditions $x(0) = 1$ and $y(0) = 1$.

3.8 Use the superposition principle to solve the system

$$x'(t) = x(t) - 2y(t)$$
$$y'(t) = x(t) + y(t),$$

subject to the initial conditions $x(0) = 1$ and $y(0) = 1$.

3.9 As we have seen, Fibonacci's difference equation is $(a_{n+1} - a_n) = a_{n-1}$.

The similar-looking equation $(a_{n+1} - a_n) = a_n$ is rather trivial; it amounts to $a_{n+1} = 2a_n$, so $a_{n+1} = 2^n a_1$. In between these two is the equation

$$(a_{n+1} - a_n) = \frac{a_n + a_{n-1}}{2}.$$

Solve this equation subject to $a_1 = a_2 = 1$.

3.10 Here is a three-term difference equation. Consider the sequence $\{a_n\}$ defined recursively by

$$a_{n+1} - a_n = \frac{a_{n-1} + a_{n-2} - 2a_n}{3},$$

subject to the initial data $a_3 = a_2 = a_1 = 1$. Introduce the vector $\mathbf{x}_n = \begin{bmatrix} a_n \\ a_{n-1} \\ a_{n-2} \end{bmatrix}$.

(a) Find a 3×3 matrix A so that $\mathbf{x}_{n+1} = A\mathbf{x}_n$.

(b) Find a closed-form formula for a_n.

3.11 Let A be the matrix $A = \begin{bmatrix} 1 & -2 \\ -2 & 1 \end{bmatrix}$. Let S denote the set of all initial data vectors $\mathbf{x}_0$ for which the solution of $\mathbf{x}'(t) = A\mathbf{x}(t)$ with $\mathbf{x}(0) = \mathbf{x}_0$ has the property

$$\lim_{t \to \infty} \mathbf{x}(t) = 0$$

for all $t > 0$. Is S a subspace of $\mathbb{R}^2$? If so, find a basis. If not, explain why not.

SECTION 4 | # Diagonalizing Symmetric Matrices

4.1 Eigenvectors and eigenvalues of symmetric matrices

There is a particularly important case in which we can be sure that there is a basis of eigenvectors—the case in which $A = A^t$. Matrices that are their own transposes are called *symmetric*.

The key to this is the very important formula $\mathbf{x} \cdot (A\mathbf{y}) = (A^t\mathbf{x}) \cdot \mathbf{y}$ for all $\mathbf{x}$ and $\mathbf{y}$, which in the symmetric case reduces to

$$\mathbf{x} \cdot (A\mathbf{y}) = (A\mathbf{x}) \cdot \mathbf{y}. \tag{4.1}$$

Now, if A is an $n \times n$ symmetric matrix with real entries, we also have $A^* = A^t = A$; the complex conjugation in the definition of the adjoint A^* of A has no effect if all the entries are real. Therefore, we have from the generalization of (4.1) to $\mathbb{C}^n$, namely (1.11), that

$$\langle \mathbf{w}, A\mathbf{z} \rangle = \langle A\mathbf{w}, \mathbf{z} \rangle \tag{4.2}$$

for all vectors $\mathbf{w}$ and $\mathbf{z}$ in $\mathbb{C}^n$,

The identity (4.2) holds for all $n \times n$ matrices A with $A^* = A$. Such matrices are called *self-adjoint*. In all the examples considered here, our matrices will be real, in which case $A^* = A$ reduces to $A^t = A$. Therefore, we discuss only the symmetric case. However, once you understand it, you will have no problem extending all the results to the case of complex, self-adjoint matrices.

Here is our first use of (4.2).

THEOREM 10   **(Distinct eigenvalues and orthogonality)** *Let A be a symmetric $n \times n$ matrix with real entries. If $\mathbf{v}$ and $\mathbf{w}$ are eigenvectors of A with distinct eigenvalues, then $\mathbf{v}$ and $\mathbf{w}$ are orthogonal. Moreover, all the eigenvalues of A are real.*

PROOF: We begin by showing that all the eigenvalues are real and that therefore we can restrict our attention to real eigenvectors.

Suppose that $\mathbf{v}$ is an eigenvector with eigenvalue μ so that $A\mathbf{v} = \mu\mathbf{v}$. For all we know now, μ and $\mathbf{v}$ might be complex. Therefore, we will use the inner product in $\mathbb{C}^n$:

$$\langle \mathbf{v}, A\mathbf{v} \rangle = \langle \mathbf{v}, \mu\mathbf{v} \rangle = \mu \langle \mathbf{v}, \mathbf{v} \rangle = \mu|\mathbf{v}|^2.$$

On the other hand, since the entries of A are real and A is symmetric, $A^* = A$. Therefore, by (4.2),

$$\langle \mathbf{v}, A\mathbf{v} \rangle = \langle A^*\mathbf{v}, \mathbf{v} \rangle = \langle \mu\mathbf{v}, \mathbf{v} \rangle = \bar{\mu}\langle \mathbf{v}, \mathbf{v} \rangle = \bar{\mu}|\mathbf{v}|^2.$$

Comparing results from our two computations of $\langle \mathbf{v}, A\mathbf{v} \rangle$, we conclude that

$$\mu|\mathbf{v}|^2 = \bar{\mu}|\mathbf{v}|^2,$$

so $\mu = \bar{\mu}$, which means that μ is real.

We see that when working with symmetric matrices, we do not need to concern ourselves with complex numbers or vectors in $\mathbb{C}^n$. We will be working in $\mathbb{R}^n$, as before.

Now let us return to the first part. Consider two eigenvectors $\mathbf{v}$ and $\mathbf{w}$ of A corresponding to distinct eigenvalues μ and v, respectively. Then, working in $\mathbb{R}^n$ again,

$$v\mathbf{v} \cdot \mathbf{w} = \mathbf{v} \cdot (A\mathbf{w}) = (A^t\mathbf{v}) \cdot \mathbf{w} = (A\mathbf{v}) \cdot \mathbf{w} = \mu\mathbf{v} \cdot \mathbf{w},$$

and hence

$$(v - \mu)\mathbf{v} \cdot \mathbf{w} = 0.$$

By hypothesis, $v - \mu \neq 0$, and hence $\mathbf{v} \cdot \mathbf{w} = 0$. ∎

The next theorem gives another special property of symmetric matrices, the key to the fact that symmetric matrices can always be diagonalized.

THEOREM 11   **(Symmetric matrices with only zero as an eigenvalue)** *If A is any $n \times n$ symmetric matrix such that zero is the only eigenvalue of A, then A is the zero matrix.*

Before giving the proof, we point out that the theorem would be false if we did not require A to be symmetric. The simplest example is

$$A = \begin{bmatrix} 0 & 1 \\ 0 & 0 \end{bmatrix}.$$

Indeed, any strictly upper-triangular $n \times n$ matrix has the characteristic polynomial $(-t)^n$, so 0 is the only eigenvalue.

PROOF: Let A be a symmetric $n \times n$ matrix, and assume that A is not the zero matrix. We will show that A has a nonzero eigenvalue.

Let r be the rank of A, and let $\{\mathbf{u}_1, \mathbf{u}_2, \ldots, \mathbf{u}_r\}$ be an orthonormal basis for $\text{Img}(A)$. Let U be the $n \times r$ matrix given by $U = [\mathbf{u}_1, \mathbf{u}_2, \ldots, \mathbf{u}_r]$, and let B be the $r \times r$ matrix given by

$$B = U^t A U.$$

Notice that B is symmetric. Every matrix has at least one eigenvalue, and since B is symmetric, Theorem 10 ensures that there is a real number μ and a nonzero vector $\mathbf{w}$ in $\mathbb{R}^r$ with

$$B\mathbf{w} = \mu\mathbf{w}.$$

Multiplying through on the left by U, and recalling the definition of B,

$$UU^t A(U\mathbf{w}) = \mu(U\mathbf{w}). \tag{4.3}$$

Now recall that UU^t is the orthogonal projection onto $\text{Img}(A)$, so $UU^tA = A$. Therefore, defining $\mathbf{v}$ by $\mathbf{v} = U\mathbf{w}$, we have

$$A\mathbf{v} = \mu\mathbf{v}. \tag{4.4}$$

Since U is an isometry and $\mathbf{w} \neq 0$, $\mathbf{v} \neq 0$. Therefore, $\mathbf{v}$ is an eigenvector of A with eigenvalue μ.

It remains to show that $\mu \neq 0$. But if it were the case that $\mu = 0$, we would have from (4.4) that $A\mathbf{v} = A(U\mathbf{w}) = 0$, which would mean that $U\mathbf{w}$ belongs to $\text{Ker}(A)$. But since $\{\mathbf{u}_1, \mathbf{u}_2, \ldots, \mathbf{u}_r\}$ is a basis for $\text{Img}(A)$, $U\mathbf{w}$ belongs to $\text{Img}(A)$. Now we use that symmetry of A. Since $A = A^t$,

$$(\text{Img}(A))^{\perp} = \text{Ker}(A^t) = \text{Ker}(A).$$

Since $\text{Img}(A)$ and $\text{Ker}(A)$ are orthogonal complements, the only vector belonging to both of them is the zero vector. Since $U\mathbf{w} \neq 0$, $\mu = 0$ is impossible. Hence, $\mu \neq 0$, and A has a nonzero eigenvalue. ∎

4.2 The spectral theorem

We can put together the things we have learned so far to prove an important theorem.

THEOREM 12   **(The spectral theorem for symmetric matrices)** *Let A be any $n \times n$ symmetric matrix. Then there is an orthonormal basis of $\mathbb{R}^n$ consisting of eigenvectors of A so that A is diagonalizable. In other words, there is a factorization $A = UDU^t$ where U is orthogonal and D is diagonal.*

Moreover, suppose that A has k distinct nonzero eigenvalues $v_1, v_2, \ldots, v_k$. For each $j = 1, 2 \ldots, k$, let P_j denote the orthogonal projection onto $E_j = \text{Ker}(A - v_jI)$, the eigenspace of A corresponding to v_j. Then

$$A = v_1P_1 + v_2P_2 + \cdots + v_kP_k. \tag{4.5}$$

First we prove the following lemma, and then we use it to prove the theorem.

LEMMA   **(Removal of eigenvalues)** *Let A be any $n \times n$ symmetric matrix. Suppose that A has k distinct nonzero eigenvalues $v_1, v_2, \ldots, v_k$. Let P_k denote the orthogonal projection onto $E_k = \text{Ker}(A - v_kI)$, the eigenspace of A corresponding to v_k, and let*

$$B = A - v_kP_k. \tag{4.6}$$

Then B is a symmetric matrix, the nonzero eigenvalues of B are $v_1, v_2, \ldots, v_{k-1}$, and for $j < k$, $\text{Ker}(B - v_jI) = \text{Ker}(A - v_jI) = E_j$.

The lemma says that the subtraction in (4.6) "cancels off" one eigenvalue. Theorem 11 says that when we have canceled off all the nonzero eigenvalues in this way, what is left is just the zero matrix. Therefore, inductive application of the lemma, together with Theorem 11 will then easily yield (4.5), and the rest will follow.

PROOF OF THE LEMMA: Let $\mathbf{v}$ be any vector in $E_k = \text{Img}(P_k)$. Then since $A\mathbf{v} = v_k\mathbf{v}$ and $P_k\mathbf{v} = \mathbf{v}$,

$$B\mathbf{v} = A\mathbf{v} - v_kP_k\mathbf{v} = v_k\mathbf{v} - v_k\mathbf{v} = 0.$$

Therefore, $\mathbf{v}$ is in the kernel of B, so

$$E_k \subset \text{Ker}(B). \tag{4.7}$$

Taking orthogonal complements,

$$(\text{Ker}(B))^\perp \subset E_k^\perp . \tag{4.8}$$

Theorem 10 tells us that if $\mathbf{v}$ is an eigenvector of B with a nonzero eigenvalue μ, then $\mathbf{v}$ is orthogonal to the kernel of B, which is the eigenspace with eigenvalue 0. Thus, (4.8) tells us that $\mathbf{v}$ belongs to $E_k^\perp$, so $P_k\mathbf{v} = 0$. Then by the definition of B,

$$B\mathbf{v} = (A - v_k P_k)\mathbf{v} = A\mathbf{v} - v_k P_k \mathbf{v} = A\mathbf{v},$$

so $B\mathbf{v} = \mu\mathbf{v}$ becomes $A\mathbf{v} = \mu\mathbf{v}$. Therefore, μ is one of the nonzero eigenvalues of A. It cannot be v_k, since if $A\mathbf{v} = v_k\mathbf{v}$, we would have $\mu\mathbf{v} = B\mathbf{v} = A\mathbf{v} - v_k P_k\mathbf{v} = v_k\mathbf{v} - v_k\mathbf{v} = 0$, and by hypothesis, $\mu \neq 0$. Hence, μ is one of the numbers $v_1, v_2, \ldots, v_{k-1}$. ∎

PROOF OF THE SPECTRAL THEOREM: We will prove (4.5) first. We do so by induction on the number of nonzero eigenvalues of A. Let k denote the number of nonzero eigenvalues of A, and note that by Theorem 11, (4.5) holds if $k = 0$ since then both sides are zero.

We now assume that (4.5) is true for all symmetric matrices with $k-1$ distinct eigenvalues. Then, by the lemma, $B = A - v_k P_k$ is a symmetric matrix with $k-1$ eigenvalues $v_1, v_2, \ldots, v_{k-1}$, which also happen to be eigenvalues of A.

By the inductive hypothesis,

$$B = v_1 P_1 + v_2 P_2 + \cdots + v_{k-1} P_{k-1}.$$

Combining this equation with (4.6) yields (4.5).

From here it is easy to see that there is a basis of $\mathbb{R}^n$ consisting of eigenvectors of A so that A is diagonalizable.

For each $j = 1, 2, \ldots, k$, pick an orthonormal basis for E_j, the eigenspace corresponding to v_j. Combine all the unit vectors into one set

$$\{\mathbf{u}_1, \mathbf{u}_2, \ldots, \mathbf{u}_r\}. \tag{4.9}$$

We claim that this set is an orthonormal basis for $\text{Img}(A)$. First, it is an orthonormal set, since any two vectors in it are orthogonal, since either they belong to the same E_j and hence the same orthonormal basis or they belong to different eigenspaces and are orthogonal by Theorem 10.

Any orthonormal set is linearly independent, so now we only need to show that (4.9) spans $\text{Img}(A)$. By definition, $\mathbf{v}$ is in $\text{Img}(A)$ if and only if $\mathbf{v} = A\mathbf{x}$ for some $\mathbf{x}$ in $\mathbb{R}^n$. But then from (4.5),

$$\mathbf{v} = A\mathbf{x} = v_1 P_1 \mathbf{x} + v_2 P_2 \mathbf{x} + \cdots + v_k P_k \mathbf{x}.$$

Each vector $v_j P_j \mathbf{x}$ belongs to E_j, and (4.9) includes an orthonormal basis for each E_j. Therefore, each vector $v_j P_j \mathbf{x}$ is a linear combination of vectors in (4.9), and hence so is $\mathbf{v}$.

Now we know that (4.9) is an orthonormal basis for $\text{Img}(A)$, the dimension of which is $\text{rank}(A)$, so $r = \text{rank}(A)$.

Next, let $\{\mathbf{u}_{r+1}, \mathbf{u}_{r+2}, \ldots, \mathbf{u}_n\}$ be an orthonormal basis for $\text{Ker}(A)$, which has dimension $n - r$, as indicated. Each of these vectors is an eigenvector of A with eigenvalue 0 and, by Theorem 10, is orthogonal to each of the vectors in (4.9). Hence, the combined set

$$\{\mathbf{u}_1, \mathbf{u}_2, \ldots, \mathbf{u}_r, \mathbf{u}_{r+1}, \mathbf{u}_{r+2}, \ldots, \mathbf{u}_n\}$$

is an orthonormal set of n vectors in $\mathbb{R}^n$ and thus a basis for $\mathbb{R}^n$. Each vector is an eigenvector of A. ∎

Formula (4.5) gives an *additive* decomposition of a symmetric matrix into simple pieces. All the decompositions we have considered up to now have been multiplicative: $A = LU$, $A = QR$, $A = VDV^{-1}$. Another particularly useful form of the additive decomposition (4.5) carries it a bit further.

THEOREM 13   **(The spectral decomposition for symmetric matrices)** *Let A be any $n \times n$ symmetric matrix. Let $\{\mathbf{u}_1, \mathbf{u}_2, \ldots, \mathbf{u}_n\}$ be an orthonormal basis of $\mathbb{R}^n$ consisting of eigenvectors of A, and let μ_j be the jth eigenvalue: $A\mathbf{u}_j = \mu_j\mathbf{u}_j$. Then*

$$A = \sum_{j=1}^{n} \mu_j \mathbf{u}_j \mathbf{u}_j^t. \tag{4.10}$$

PROOF: Let $B = \sum_{j=1}^{n} \mu_j \mathbf{u}_j \mathbf{u}_j^t$. We will show that for all $\mathbf{v}$ in $\mathbb{R}^n$, $B\mathbf{v} = A\mathbf{v}$, which means that $B = A$.

We first compute $B\mathbf{v}$. By the definition of B,

$$B\mathbf{v} = \left(\sum_{j=1}^{n} \mu_j \mathbf{u}_j \mathbf{u}_j^t \right) \mathbf{v}$$

$$= \sum_{j=1}^{n} \mu_j (\mathbf{u}_j \cdot \mathbf{v}) \mathbf{u}_j.$$

Next, the jth component of $\mathbf{v}$ in the orthonormal basis $\{\mathbf{u}_1, \mathbf{u}_2, \ldots, \mathbf{u}_n\}$ is $\mathbf{u}_j \cdot \mathbf{v}$ so that

$$\mathbf{v} = \sum_{j=1}^{n} (\mathbf{u}_j \cdot \mathbf{v}) \mathbf{u}_j.$$

Applying A, we find

$$A\mathbf{v} = A \left(\sum_{j=1}^{n} (\mathbf{u}_j \cdot \mathbf{v}) \mathbf{u}_j \right)$$

$$= \sum_{j=1}^{n} (\mathbf{u}_j \cdot \mathbf{v}) A\mathbf{u}_j$$

$$= \sum_{j=1}^{n} \mu_j (\mathbf{u}_j \cdot \mathbf{v}) \mathbf{u}_j.$$

Comparing the result of the two computations, we see that $B\mathbf{v} = A\mathbf{v}$.   ∎

In the sum (4.10) it is customary to choose the indices so that

$$\mu_1 \le \mu_2 \le \cdots \le \mu_n.$$

The numbers are called the *spectrum* of A.

The spectral theorem has many important implications and applications. The next section describes one of them.

4.3 The norm of a matrix

Here is a natural problem to consider: Given any $m \times n$ matrix A, we ask how large $|A\mathbf{v}|$ can get, as $\mathbf{v}$ ranges over all unit vectors. A maximum value for $|A\mathbf{v}|$ with $|\mathbf{v}| = 1$ would be a measure of how much "stretching" A does. The maximum amount of stretching has a name.

DEFINITION

(Norm of a matrix) For any $m \times n$ matrix A, the norm of A, denoted $\|A\|$, is defined by

$$\|A\| = \text{least upper bound of } \big\{|A\mathbf{v}| : \mathbf{v} \text{ in } R^n, \quad |\mathbf{v}| = 1\big\}. \tag{4.11}$$

The definition says in particular that if $\mathbf{v}$ is any unit vector,

$$|A\mathbf{v}| \leq \|A\|. \tag{4.12}$$

What about vectors that are not unit vectors? If $\mathbf{w}$ is any nonzero vector, define the unit vector $\mathbf{v}$ by $\mathbf{v} = (1/|\mathbf{w}|)\mathbf{w}$, so that $\mathbf{w} = |\mathbf{w}|\mathbf{v}$. Then (4.12) becomes

$$|A\mathbf{w}| = |A(|\mathbf{w}|\mathbf{v})| = ||\mathbf{w}|(A\mathbf{v})| = |\mathbf{w}||A\mathbf{v}| \leq |\mathbf{w}|\|A\|,$$

or, more briefly,

$$|A\mathbf{w}| \leq \|A\||\mathbf{w}|. \tag{4.13}$$

Moreover, since $\|A\|$ is defined as a *least* upper bound, $\|A\|$ is the smallest number so that (4.13) is true for all vectors $\mathbf{w}$. Because of (4.13), we say that the norm of $\|A\|$ measures the maximum amount of "stretching" that A does.

Computing least upper bounds is often a challenge. In this case, however, it reduces to an eigenvalue problem.

THEOREM 14

(Eigenvalues and norms) *Let A any $m \times n$ matrix with real entries. Then the norm of A is equal to the square root of the largest eigenvalue of $A^t A$.*

In particular, the norm of an $n \times n$ matrix is always a finite number.

EXAMPLE 13

(Computing the norm of A) Compute the norm of

$$A = \begin{bmatrix} 1 & 2 \\ 3 & 4 \end{bmatrix}.$$

We compute

$$A^t A = \begin{bmatrix} 10 & 14 \\ 14 & 20 \end{bmatrix},$$

which has the characteristic polynomial $\mu^2 - 3 - \mu + 4$. The two eigenvalues are then $15 \pm \sqrt{221}$, and hence,

$$\|A\| = \sqrt{15 + \sqrt{221}}.$$

PROOF OF THEOREM 14: Let $\mathbf{v}$ be any unit vector in $\mathbb{R}^n$. Then

$$|A\mathbf{v}|^2 = A\mathbf{v} \cdot A\mathbf{v} = \mathbf{v} \cdot A^t A\mathbf{v}. \tag{4.14}$$

The matrix $A^t A$ is an $n \times n$ symmetric matrix. Moreover, if μ is any eigenvalue of $A^t A$ and $A^t A\mathbf{w} = \mu\mathbf{w}$, then

$$\mu|\mathbf{w}|^2 = \mathbf{w} \cdot (\mu\mathbf{w}) = \mathbf{w} \cdot (A^t A\mathbf{w}) = A\mathbf{w} \cdot A\mathbf{w} = |A\mathbf{w}|^2 \geq 0,$$

so $\mu \geq 0$

Let $\{\mathbf{u}_1, \mathbf{u}_2, \ldots, \mathbf{u}_n\}$ be an orthonormal basis of eigenvectors of $A^t A$, and let μ_j be the jth eigenvalue so that $A^t A\mathbf{u}_j = \mu_j\mathbf{u}_j$. Choose the indexing so that

$$0 \leq \mu_1 \leq \mu_2 < \cdots \leq \mu_n.$$

Then by Theorem 13,

$$A^t A = \sum_{j=1}^{n} \mu_j\mathbf{u}_j\mathbf{u}_j^t.$$

Note that $\sum_{j=1}^{n} (\mathbf{u}_j \cdot \mathbf{v})^2 = |\mathbf{v}|^2 = 1$. Therefore,

$$
\begin{aligned}
\mathbf{v} \cdot (A^t A)\mathbf{v} &= \sum_{j=1}^{n} \mu_j(\mathbf{u}_j \cdot \mathbf{v})^2 \\
&\leq \mu_n \left(\sum_{j=1}^{n} (\mathbf{u}_j \cdot \mathbf{v})^2 \right) \\
&= \mu_n|\mathbf{v}|^2 = \mu_n.
\end{aligned}
\tag{4.15}
$$

Combining (4.14) and (4.15), we have

$$|A\mathbf{v}|^2 \leq \mu_n.$$

This shows that μ_n is an upper bound for $|A\mathbf{v}|^2$. To see that it is the least upper bound, choose $\mathbf{v} = \mathbf{u}_n$. Then $|A\mathbf{v}|^2 = \mathbf{u}_n \cdot A^t A\mathbf{u}_n = \mu_n\mathbf{u}_n \cdot \mathbf{u}_n = \mu_n$. Hence, no smaller bound exists. ∎

COROLLARY   *Let A be any symmetric matrix. Then all the eigenvalues μ of A satisfy*

$$-\|A\| \leq \mu \leq \|A\|,$$

and at least one of $\|A\|$ or $-\|A\|$ is an eigenvalue of A.

PROOF: If μ is an eigenvalue of A, then μ^2 is an eigenvalue of $A^2 = A^t A$. By Theorem 14, $\|A\|^2$ is the largest eigenvalue of $A^t A$, so $\mu^2 \leq \|A\|^2$, implying that $-\|A\| \leq \mu \leq \|A\|$.

Next,

$$(A^t A - \|A\|^2 I) = (A^2 - \|A\|^2 I) = (A - \|A\|I)(A + \|A\|I).$$

If both $(A - \|A\|I)$ and $(A + \|A\|I)$ were invertible, $(A^t A - \|A\|^2 I)$ would be invertible. By Theorem 14, $\|A\|^2$ is an eigenvalue of $A^t A$, so $(A^t A - \|A\|^2 I)$ is not invertible. Hence, at least one of $(A - \|A\|I)$ or $(A + \|A\|I)$ is not invertible, which means that at least one of $\|A\|$ or $-\|A\|$ is an eigenvalue of A. ∎

The corollary has important implications. It says that for symmetric matrices, the eigenvalue problem is a min-max problem, which is important since for large matrices it is difficult to compute the characteristic polynomial, let alone factor it to find the eigenvalues. For large symmetric matrices, the corollary provides an alternative.

4.4 The eigenvalue problem as a min-max problem

We have seen that for any symmetric matrix A, either $\|A\|$ or $-\|A\|$ is an eigenvalue of A. This leads to a way to find the eigenvalues and eigenvectors of A without computing and factoring the characteristic polynomial of A. The same ideas apply to the $n \times n$ case, but the key ideas are particularly clear in the 2×2 case, which we now explain.

Consider a 2×2 symmetric matrix

$$A = \begin{bmatrix} a & b \\ b & d \end{bmatrix}, \tag{4.16}$$

and consider the time-dependent vector

$$\mathbf{v}(t) = \begin{bmatrix} \cos(t) \\ \sin(t) \end{bmatrix}. \tag{4.17}$$

Notice that for each t, $|\mathbf{v}(t)| = 1$. As t varies, $\mathbf{v}(t)$ traces out the unit circle.

Now consider the function $f(t)$ given by

$$f(t) = \mathbf{v}(t) \cdot A\mathbf{v}(t). \tag{4.18}$$

We ask the following question:

- *For what values of t does $f(t)$ take on its maximum and minimum values?*

The question is properly formulated, since computing $f(t)$, we find that

$$\begin{aligned} f(t) &= a \cos^2(t) + d \sin^2(t) + 2b \sin(t) \cos(t) \\ &= \left(\frac{a+d}{2} \right) + \left(\frac{a-d}{2} \right) (\cos^2(t) - \sin^2(t)) + 2b \sin(t) \cos(t) \\ &= \left(\frac{a+d}{2} \right) + \left(\frac{a-d}{2} \right) \cos(2t) + b \sin(2t). \end{aligned}$$

Evidently, $f(t)$ is a continuous function of t that is periodic with period π, which means that $f(t + \pi) = f(t)$ for all t. Under these circumstances, $f(t)$ takes on both its maximum and minimum somewhere in the interval $[0, \pi)$.

To find out *where*, we differentiate—

$$f'(t) = -(a - d) \sin(2t) + 2b \cos(2t)$$

—so for $b \neq 0$,

$$f'(t_0) = 0 \quad \Longleftrightarrow \quad \tan(2t_0) = \frac{a - d}{2b}. \tag{4.19}$$

(If $b = 0$, A is diagonal.) If we graph the function $y = \tan(2t)$ for $0 \leq t < \pi$, we get

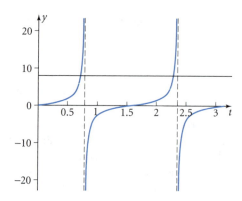

The horizontal line shown here at $y = 8$ represents the value $(a - d)/2b$. No matter what this value is, you see that the horizontal line crosses the graph of $y = \tan(2t)$ exactly twice for t in $[0, \pi)$, once in $[0, \pi/2)$, and once again in $[\pi/2, \pi)$. (The two vertical lines are asymptotes, not part of the graph, and of course, we have to exclude the case $b = 0$, in which case A is already diagonal.)

By (4.19), the two values of t at which the graph of $y = \tan(2t)$ crosses the line $y = (a - d)/2b$ are the two solutions to $f'(t) = 0$. One gives the minimum of $f(t)$, and the other gives the maximum. Let t_0 be the first solution; that is, the one in $[0, \pi/2)$. Then, since $\tan(2t)$ is periodic with period $\pi/2$, the second solution occurs at $t_0 + \pi/2$. We have

$$f'(t_0) = f'(t_0 + \pi/2) = 0. \tag{4.20}$$

Now by (3.3),

$$f'(t_0) = \mathbf{v}'(t_0) \cdot A\mathbf{v}(t_0) + \mathbf{v}(t_0) \cdot A\mathbf{v}'(t_0). \tag{4.21}$$

By the symmetry of A, $\mathbf{v}(t_0) \cdot A\mathbf{v}'(t_0) = A\mathbf{v}(t_0) \cdot \mathbf{v}'(t_0)$, and by (4.20) and (4.21),

$$\mathbf{v}'(t_0) \cdot A\mathbf{v}(t_0) = 0, \tag{4.22}$$

so $\mathbf{v}'(t_0)$ and $A\mathbf{v}(t_0)$ are orthogonal.

Next notice that

$$\mathbf{v}'(t) \cdot \mathbf{v}(t) = 0,$$

and that $\mathbf{v}'(t)$ is also a unit vector for each t. Indeed, by direct computation,

$$\mathbf{v}'(t) = \begin{bmatrix} -\sin(t) \\ \cos(t) \end{bmatrix} = \begin{bmatrix} \cos(t + \pi/2) \\ \sin(t + \pi/2) \end{bmatrix} = \mathbf{v}(t_0 + \pi/2). \tag{4.23}$$

It follows that if we define the pair of vectors $\{\mathbf{u}_1, \mathbf{u}_2\}$ by

$$\mathbf{u}_1 = \mathbf{v}(t_0) \qquad \text{and} \qquad \mathbf{u}_2 = \mathbf{v}'(t_0),$$

we get an orthonormal basis for $\mathbb{R}^2$. Moreover, we can rewrite (4.22) as $\mathbf{u}_2 \cdot A\mathbf{u}_1 = 0$, and since $A = A^t$, this means $A\mathbf{u}_2 \cdot \mathbf{u}_1 = 0$. Hence,

$$\mathbf{u}_2 \cdot A\mathbf{u}_1 = 0 \qquad \text{and} \qquad \mathbf{u}_1 \cdot A\mathbf{u}_2 = 0. \tag{4.24}$$

We know how to work out coordinates with respect to an orthonormal basis, so let us work out the coordinates for $A\mathbf{u}_1$:

$$A\mathbf{u}_1 = (\mathbf{u}_1 \cdot A\mathbf{u}_1)\mathbf{u}_1 + (\mathbf{u}_2 \cdot A\mathbf{u}_1)\mathbf{u}_2. \tag{4.25}$$

Combining this equation with (4.24), we see that

$$A\mathbf{u}_1 = (\mathbf{u}_1 \cdot A\mathbf{u}_1)\mathbf{u}_1, \tag{4.26}$$

which says that $\mathbf{u}_1$ is an eigenvector of A with eigenvalue

$$\mu_1 = \mathbf{u}_1 \cdot A\mathbf{u}_1 = \mathbf{v}(t_0) \cdot A\mathbf{v}(t_0) = f(t_0). \tag{4.27}$$

In the same way,

$$A\mathbf{u}_2 = (\mathbf{u}_1 \cdot A\mathbf{u}_2)\mathbf{u}_1 + (\mathbf{u}_2 \cdot A\mathbf{u}_2)\mathbf{u}_2, \tag{4.28}$$

and by (4.24) this equation simplifies to

$$A\mathbf{u}_2 = (\mathbf{u}_2 \cdot A\mathbf{u}_2)\mathbf{u}_2, \tag{4.29}$$

which says that $\mathbf{u}_2$ is an eigenvector of A with eigenvalue

$$\mu_2 = \mathbf{u}_2 \cdot A\mathbf{u}_2 = \mathbf{v}(t_0 + \pi/2) \cdot A\mathbf{v}(t_0 + \pi/2) = f(t_0 + \pi/2). \tag{4.30}$$

We have found an orthonormal basis of eigenvectors of A by seeking the maximum and minimum of the function $f(t)$ defined in (4.17). *Moreover, the eigenvalues are exactly the maximum and minimum values of $f(t)$, namely, $f(t_0)$ and $f(t_0 + \pi/2)$.* With a few small twists, the same approach works in n dimensions.

Exercises

4.1 Let A be the matrix $A = \begin{bmatrix} 2 & 3 \\ 3 & 4 \end{bmatrix}$.

(a) Find the eigenvalues of A and an orthonormal basis consisting of eigenvectors of A.

(b) Find a diagonal matrix D and an orthogonal matrix U so that $A = UDU^t$.

4.2 Let A be the matrix $A = \begin{bmatrix} 3 & 1 \\ 1 & 5 \end{bmatrix}$.

(a) Find the eigenvalues of A and an orthonormal basis consisting of eigenvectors of A.

(b) Find a diagonal matrix D and an orthogonal matrix U so that $A = UDU^t$.

4.3 Let A be the matrix $A = \begin{bmatrix} 1 & 1 & 2 \\ 1 & 2 & 1 \\ 2 & 1 & 1 \end{bmatrix}$.

(a) Find the eigenvalues of A and an orthonormal basis consisting of eigenvectors of A.

(b) Find a diagonal matrix D and an orthogonal matrix U so that $A = UDU^t$.

4.4 Let A be the matrix $A = \begin{bmatrix} 1 & 1 & 1 \\ 1 & 1 & 1 \\ 1 & 1 & 1 \end{bmatrix}$.

(a) Find the eigenvalues of A and an orthonormal basis consisting of eigenvectors of A.

(b) Find a diagonal matrix D and an orthogonal matrix U so that $A = UDU^t$.

4.5 Compute the norms of the matrices A and B in Exercise 1.5

4.6 (a) Let $A = \begin{bmatrix} 1 & 3 \\ 0 & 1 \end{bmatrix}$, which is invertible. Compute $\|A\|$ and $\|A^{-1}\|$.

(b) Let $B = \begin{bmatrix} 1 & 3 \\ 3 & 1 \end{bmatrix}$, which is invertible. Compute $\|B\|$ and $\|B^{-1}\|$.

(c) Let C be any invertible $n \times n$ matrix. Show that $\|C^{-1}\| \geq (\|C\|)^{-1}$.

4.7 Let A be the matrix $A = \begin{bmatrix} 0 & 1 & 2 \\ 0 & 0 & 1 \\ 0 & 0 & 0 \end{bmatrix}$.

(a) Find the unit vectors $\mathbf{u}$ such that $|A\mathbf{u}| \geq |A\mathbf{v}|$ for any other unit vector $\mathbf{v}$.

(b) Find the unit vectors $\mathbf{u}$ such that $|A\mathbf{u}| \leq |A\mathbf{v}|$ for any other unit vector $\mathbf{v}$.

4.8 Let A be the matrix $A = \begin{bmatrix} 2 & 1 \\ 0 & 1 \end{bmatrix}$.

(a) Find the unit vectors $\mathbf{u}$ such that $|A\mathbf{u}| \geq |A\mathbf{v}|$ for any other unit vector $\mathbf{v}$.

(b) Find the unit vectors $\mathbf{u}$ such that $|A\mathbf{u}| \leq |A\mathbf{v}|$ for any other unit vector $\mathbf{v}$.

4.9 Define a *distance function* $d(A, B)$ on the set of $m \times n$ matrices by $d(A, B) = \|A - B\|$. Show that

(i) $d(A, B) = 0$ if and only if $A = B$.

(ii) $d(A, B) = d(B, A)$ for all A and B.

(iii) $d(A, C) \leq d(A, B) + d(B, C)$ for all A, B, and C.

4.10 An $n \times n$ matrix A is called *positive* if and only if A is symmetric and $\mathbf{v} \cdot A\mathbf{v} \geq 0$ for all vectors $\mathbf{v}$ in $\mathbb{R}^n$. (See Section 5.5 of Chapter 2.) Show that A is positive if and only if there is a positive $n \times n$ matrix C so that $A = C^2$.

4.11 An $n \times n$ matrix A is called *antisymmetric* if and only if $A^t = -A$. For example, $\begin{bmatrix} 0 & 1 \\ -1 & 0 \end{bmatrix}$ is antisymmetric.

(a) Compute the eigenvalues of $\begin{bmatrix} 0 & 1 \\ -1 & 0 \end{bmatrix}$.

(b) Show that every eigenvalue of any antisymmetric matrix with real entries is purely imaginary.

4.12 An $n \times n$ matrix A is called *nilpotent* if and only if for some n, $A^n = 0$.

(a) Show that zero is the only eigenvalue of a nilpotent matrix.

(b) Can a matrix possibly be both nilpotent and symmetric? Explain why not, or give an example.

Eigenvectors, eigenvalues, and rotation matrices

The following exercises lead you through an important topic: the structure of rotation matrices in $\mathbb{R}^3$. What we have learned about eigenvalues, eigenvectors, and diagonalization sheds useful light on this topic—though we have already treated many aspects of it in Chapter 4. (See Exercises 3.12 through 3.26 of Chapter 4.)

4.13 Let Q be an $n \times n$ rotation matrix, and suppose that Q is symmetric.

(a) Show that if μ is an eigenvalue of Q, then either $\mu = 1$ or $\mu = -1$.

(b) Show that the algebraic multiplicity of the eigenvalue -1 is even.

(c) Show that if $n = 3$ and Q is not the identity matrix, then there is a right-handed orthonormal basis $\{\mathbf{u}_1, \mathbf{u}_2, \mathbf{u}_3\}$ such that with $U = [\mathbf{u}_1, \mathbf{u}_2, \mathbf{u}_3]$,

$$U^t Q U = \begin{bmatrix} -1 & 0 & 0 \\ 0 & -1 & 0 \\ 0 & 0 & 1 \end{bmatrix}.$$

(d) Note the the linear transformation of rotating a vector $\mathbf{v}$ in the plane through an angle π (counterclockwise or clockwise; it does not matter) sends $\mathbf{v}$ into $-\mathbf{v}$. Conclude that the linear transformation corresponding to

$$\begin{bmatrix} -1 & 0 & 0 \\ 0 & -1 & 0 \\ 0 & 0 & 1 \end{bmatrix}$$

is rotation through an angle π in the x, y plane.

(e) Show that a 3×3 rotation matrix Q is symmetric if and only if the linear transformation represented by Q is rotation through the angle π about some line through the origin in $\mathbb{R}^3$.

4.14 Let $\mathbf{u}$ be a unit vector in $\mathbb{R}^3$. Show that

$$Q = 2\mathbf{u}\mathbf{u}^t - I$$

is a 3×3 rotation matrix and that in fact the linear transformation it describes is a rotation through the angle π about the line through the origin and $\mathbf{u}$.

4.15 Let Q be an $n \times n$ rotation matrix, and suppose further that Q is *not* symmetric; that is, $Q \neq Q^t$.

(a) Show that if n is odd, then $\det(Q-I) = 0$, so 1 is an eigenvalue of 1. Hint: Note that $Q^t(Q-I) = (I = Q^t)$. Then use Theorems 4 and 5 of Chapter 4.

(b) Let Q be a 3×3 rotation matrix. Let S and A be the symmetric and antisymmetric parts of Q; that is,

$$S = \frac{1}{2}(Q + Q^t) \quad \text{and} \quad A = \frac{1}{2}(Q - Q^t).$$

Using the identities $Q^t Q = I = QQ^t$, show that

$$S^2 - A^2 + (SA - AS) = I = S^2 - A^2 - (SA - AS)$$

and that hence $AS = SA$. *That is, S and A commute.* Show also that $QA = AQ$ and $QS = SQ$.

(c) Let $\mathbf{v}$ be a unit vector with $Q\mathbf{v} = \mathbf{v}$. Show that $Q(A\mathbf{v}) = A\mathbf{v}$ and that $A\mathbf{v}$ is orthogonal to $\mathbf{v}$, so if $A\mathbf{v} \neq 0$, there are at least two linearly independent eigenvectors with eigenvalue 1.

(d) Show that if the eigenspace E_1 of Q is at least two-dimensional, then it is three-dimensional, making $Q = I$. Conclude that when Q is not symmetric and hence not the identity, the geometric multiplicity of the eigenvalue 1 is 1. Show then that if $Q\mathbf{v} = \mathbf{v}$, then $A\mathbf{v} = 0$ and so $S\mathbf{v} = \mathbf{v}$.

(e) Let $\{\mathbf{u}_1, \mathbf{u}_2, \mathbf{u}_3\}$ be a right-handed orthonormal basis of eigenvectors of S with $S\mathbf{u}_3 = \mathbf{u}_3$. Show that $\mathbf{u}_1$ and $\mathbf{u}_2$ belong to the same eigenspace E_μ; that is,

$$S\mathbf{u}_1 = \mu\mathbf{u}_1 \quad \text{and} \quad S\mathbf{u}_2 = \mu\mathbf{u}_2 .$$

Show that $-1 < \mu < 1$, and define an angle θ in the range $0 < \theta < \pi$ by $\theta = \cos^{-1}(\mu)$. Changing $\mathbf{u}_2$ into $-\mathbf{u}_2$ if need be, we may arrange that $\mathbf{u}_2 \cdot Q\mathbf{u}_1 \geq 0$. (If we need to do so, we also change $\mathbf{u}_2$ into $-\mathbf{u}_2$ so that our basis is still right-handed.) Show that $\mathbf{u}_2 \cdot Q\mathbf{u}_1 = \sin(\theta)$.

(f) Let $U = [\mathbf{u}_1, \mathbf{u}_2, \mathbf{u}_3]$, using the basis from part (e). Show that

$$U^t A U = \sin(\theta) \begin{bmatrix} 0 & -1 & 0 \\ 1 & 0 & 0 \\ 0 & 0 & 0 \end{bmatrix}$$

and then that

$$U^t Q U = \begin{bmatrix} \cos(\theta) & -\sin(\theta) & 0 \\ \sin(\theta) & \cos(\theta) & 0 \\ 0 & 0 & 1 \end{bmatrix}.$$

Conclude that the linear transformation represented by Q is a rotation through the angle θ about the line through the origin and $\mathbf{u}_3$, which is the axis of rotation.

SECTION 5 | The Exponential of a Square Matrix

5.1 The matrix function e^{tA}

Consider the differential equation

$$x'(t) = ax(t) \qquad x(0) = x_0, \tag{5.1}$$

where a is a number. The function

$$x(t) = e^{at} x_0 \tag{5.2}$$

satisfies (5.1). That is, if we differentiate, we find

$$\frac{d}{dt} e^{at} x_0 = a e^{at} x_0,$$

and $x(0) = x_0$, which is what it means for $x(t) = \exp(at) x_0$ to satisfy (5.1).

In Section 3, we studied systems of differential equations such as

$$x'(t) = ax(t) + by(t) \quad \text{and} \quad y'(t) = cx(t) + dy(t) \tag{5.3}$$

subject to the initial conditions

$$x(0) = x_0 \quad \text{and} \quad y(0) = y_0. \tag{5.4}$$

Just as in Section 3, introduce

$$\mathbf{x}(t) = \begin{bmatrix} x(t) \\ y(t) \end{bmatrix} \quad \text{and} \quad \mathbf{x}_0 = \begin{bmatrix} x_0 \\ y_0 \end{bmatrix}.$$

Then (5.3) can be rewritten

$$\mathbf{x}'(t) = A\mathbf{x}(t) \quad \text{and} \quad \mathbf{x}(0) = \mathbf{x}_0. \tag{5.5}$$

This form brings out the similarity to (5.1). In Section 3, we used the superposition principle to solve such equations. We saw that this solution strategy will succeed whenever A is diagonalizable, although we did not actually diagonalize A—we just used the basis of eigenvectors.

It turns out that there is a way to use diagonalization to produce a matrix version of (5.2). This formula is very useful in applications.

Recall that for numbers a,

$$e^{ta} = \sum_{k=0}^{\infty} \frac{1}{k!} (ta)^k = \sum_{k=0}^{\infty} \frac{1}{k!} t^k a^k.$$

This formula suggests that for an $n \times n$ matrix A, we should define

$$e^{tA} = \sum_{k=0}^{\infty} \frac{1}{k!} t^k A^k. \tag{5.6}$$

We interpret the infinite sum on the right-hand side as

$$\lim_{N \to \infty} \sum_{k=0}^{N} \frac{1}{k!} t^k A^k. \tag{5.7}$$

For each fixed N, the matrix on the right-hand side of (5.7) is a well-defined $n \times n$ matrix. We say that a sequence of matrices converges to a limit if for *each* i and j, the corresponding numerical sequence of the entries converges. It is the familiar issue of convergence for numerical sequences, with the matrix considered one entry at a time.

At the end of this section, we show that the limit in (5.7) exists for every $n \times n$ matrix for every t. Therefore we can use it, or what is the same thing, (5.6), to define the matrix-valued function of t, e^{tA}.

The exponential notation for this function is justified by the properties it shares with the usual exponential function. The first of these is

$$\frac{d}{dt} e^{tA} = A e^{tA}. \tag{5.8}$$

The left-hand side requires a word of explanation: When we differentiate a matrix function of t, we differentiate it entry by entry, just as with vector functions of t. You can see that (5.8) is true by differentiating (5.6) term by term, just as you would in the numerical case. The justification, which is a matter of analysis, not algebra, is essentially the same (see the end of the section for more details).

Notice also that for any A,

$$e^{0A} = I, \tag{5.9}$$

since only the very first term in (5.6) is not zero.

Now we can give the matrix version of the formula (5.2):

- *The solution of (5.5) is given by*

$$\mathbf{x}(t) = e^{tA}\mathbf{x}_0. \tag{5.10}$$

To see that (5.10) defines a solution, we need another product rule:

- *For any differentiable $m \times n$ matrix-valued function $B(t)$ and any differentiable n-dimensional vector-valued function $\mathbf{v}(t)$, let $\mathbf{x}(t)$ denote the product: $\mathbf{x}(t) = B(t)\mathbf{v}(t)$. Then $\mathbf{x}(t)$ is differentiable, and*

$$\mathbf{x}'(t) = B'(t)\mathbf{v}(t) + B(t)\mathbf{v}'(t). \tag{5.11}$$

The formula (5.11) is an easy consequence of Theorem 10, the product rule for dot products, and **V.I.F.2**. The details are left as an exercise.

Applying (5.11) to $\mathbf{x}(t) = e^{tA}\mathbf{x}_0$, since $\mathbf{x}_0$ is constant,

$$\mathbf{x}'(t) = \left(e^{tA}\right)' \mathbf{x}_0 = \left(Ae^{tA}\right)\mathbf{x}_0 = A\left(e^{tA}\mathbf{x}_0\right) = A\mathbf{x}(t).$$

Also, since $e^{0A} = I$, $\mathbf{x}(0) = \mathbf{x}_0$. Therefore, (5.10) gives a solution of (5.5).

Moreover, we can use the exponential function to show that $\mathbf{x}(tt) = e^{tA}\mathbf{x}_0$ is the *only* solution. To do so, we need to know that

$$(e^{tA})^{-1} = e^{-tA}, \tag{5.12}$$

familiar when A is a number and a special case of what we shall prove in Theorem 16.

Accepting (5.12) for the time being, let $\mathbf{y}(t)$ be any vector-valued function satisfying $\mathbf{y}'(t) = A\mathbf{y}(t)$ and $\mathbf{y}'(0) = \mathbf{x}_0$. Define

$$\mathbf{z}(t) = e^{-tA}\mathbf{y}(t).$$

Then by (5.11),

$$\mathbf{z}'(t) = \left(-Ae^{-tA}\right)\mathbf{y}(t) + e^{-tA}(A\mathbf{y}(t)) = 0,$$

which means that every entry of $\mathbf{z}(t)$ is constant, so for all t,

$$\mathbf{z}(t) = \mathbf{z}(0) = \mathbf{x}_0.$$

But then, by the definition of $\mathbf{z}(t)$,

$$\mathbf{y}(t) = e^{tA}\mathbf{x}_0,$$

proving that $\mathbf{x}(t) = e^{tA}\mathbf{x}_0$ is the *only* solution.

An advantage of using e^{tA} instead of the superposition principle is that once e^{tA} is computed, the system $\mathbf{x}'(t) = A\mathbf{x}(t)$ can be solved for a thousand different initial conditions alone by matrix multiplication.

We come to a pressing issue:

- *Given an $n \times n$ matrix A, how can we compute a closed-form formula for e^{tA}?*

The point is that the formula we have, namely, (5.6), is an *infinite sum* involving infinitely many algebraic operations. We need a formula that can be evaluated in only finitely many steps if we are going to make practical use of (5.11).

If A is diagonalizable, arriving at such a formula is easy. Recall that when A is diagonalizable, there is an invertible matrix V whose columns are eigenvectors of A and a diagonal matrix D whose diagonal entries are the corresponding eigenvalues of A so that $A = VDV^{-1}$. We have also seen that in this case,

$$A^k = VD^kV^{-1}.$$

Now, if

$$D = \begin{bmatrix} \mu_1 & 0 & 0 & \ldots & 0 \\ 0 & \mu_2 & 0 & \ldots & 0 \\ \vdots & \vdots & \vdots & \ddots & \vdots \\ 0 & 0 & 0 & \ldots & \mu_n \end{bmatrix},$$

then

$$D^k = \begin{bmatrix} \mu_1^k & 0 & 0 & \ldots & 0 \\ 0 & \mu_2^k & 0 & \ldots & 0 \\ \vdots & \vdots & \vdots & \ddots & \vdots \\ 0 & 0 & 0 & \ldots & \mu_n^k \end{bmatrix}.$$

But then for any N,

$$\sum_{k=0}^{N} \frac{1}{k!} t^k A^k = \sum_{k=0}^{N} \frac{1}{k!} t^k VD^kV^{-1} = V\left(\sum_{k=0}^{N} \frac{1}{k!} t^k D^k\right) V^{-1} = Ve^{tD}V^{-1}.$$

Since D is diagonal, computing e^{tD} is simple enough:

$$\sum_{k=0}^{N} \frac{1}{k!} t^k D^k = \begin{bmatrix} \sum_{k=0}^{N}(t^k\mu_1^k/k!) & 0 & 0 & \ldots & 0 \\ 0 & \sum_{k=0}^{N}(t^k\mu_2^k/k!) & 0 & \ldots & 0 \\ \vdots & \vdots & \vdots & \ddots & \vdots \\ 0 & 0 & 0 & \ldots & \sum_{k=0}^{N}(t^k\mu_n^k/k!) \end{bmatrix}.$$

Taking the limit entry by entry,

$$\lim_{N\to\infty} \sum_{k=0}^{N}(t^k\mu_1^k/k!) = e^{t\mu_1}.$$

The result is that

$$e^{tD} = \sum_{k=0}^{\infty} \frac{1}{k!} t^k D^k = \lim_{N\to\infty} \sum_{k=0}^{N} \frac{1}{k!} t^k D^k = \begin{bmatrix} e^{t\mu_1} & 0 & 0 & \ldots & 0 \\ 0 & e^{t\mu_2} & 0 & \ldots & 0 \\ \vdots & \vdots & \vdots & \ddots & \vdots \\ 0 & 0 & 0 & \ldots & e^{t\mu_n} \end{bmatrix}.$$

THEOREM 15   **(Matrix exponentials)** *Let A be an $n \times n$ diagonalizable matrix. Then for any t, e^{tA}, defined as in (5.6), exists, and moreover, if $A = VDV^{-1}$, where*

$$D = \begin{bmatrix} \mu_1 & 0 & 0 & \ldots & 0 \\ 0 & \mu_2 & 0 & \ldots & 0 \\ \vdots & \vdots & \vdots & \ddots & \vdots \\ 0 & 0 & 0 & \ldots & \mu_n \end{bmatrix},$$

then

$$e^{tA} = V \begin{bmatrix} e^{t\mu_1} & 0 & 0 & \dots & 0 \\ 0 & e^{t\mu_2} & 0 & \dots & 0 \\ \vdots & \vdots & \vdots & \ddots & \vdots \\ 0 & 0 & 0 & \dots & e^{t\mu_n} \end{bmatrix} V^{-1}. \tag{5.13}$$

The formula (5.13) is easily used to compute closed-form formulas for the exponentials of diagonalizable matrices.

EXAMPLE 14 (**Computing e^{tA}**) Consider the matrix

$$A = \begin{bmatrix} 1 & 2 \\ 2 & 1 \end{bmatrix}$$

from Example 1. We found that $A = VDV^{-1}$, where

$$V = \frac{1}{\sqrt{2}} \begin{bmatrix} 1 & -1 \\ 1 & 1 \end{bmatrix} \quad \text{and} \quad D = \begin{bmatrix} 3 & 0 \\ 0 & -1 \end{bmatrix}.$$

Since V is orthogonal, V^{-1} is the transpose of V, and we have

$$e^t A = \frac{1}{2} \begin{bmatrix} 1 & -1 \\ 1 & 1 \end{bmatrix} \begin{bmatrix} e^{3t} & 0 \\ 0 & e^{-t} \end{bmatrix} \begin{bmatrix} 1 & 1 \\ -1 & 1 \end{bmatrix}$$

$$= \frac{1}{2} \begin{bmatrix} 1 & -1 \\ 1 & 1 \end{bmatrix} \begin{bmatrix} e^{3t} & e^{3t} \\ -e^{-t} & e^{-t} \end{bmatrix}$$

$$= \frac{1}{2} \begin{bmatrix} e^{3t} + e^{-t} & e^{3t} - e^{-t} \\ e^{3t} - e^{-t} & e^{3t} + e^{-t} \end{bmatrix}.$$

5.2 Solving differential equations with matrix exponentials

Now that we can compute matrix exponentials, we can put (5.10) to work.

EXAMPLE 15 (**Solving $x'(t) = Ax(t)$ using matrix exponentials**) Let

$$A = \begin{bmatrix} 1 & 2 \\ 2 & 1 \end{bmatrix}$$

as in Example 1. Let $\mathbf{x}_0 = \begin{bmatrix} 2 \\ 1 \end{bmatrix}$, and let us use (5.10) to solve

$$\mathbf{x}'(t) = A\mathbf{x}(t) \quad \text{and} \quad \mathbf{x}(0) = \mathbf{x}_0.$$

Using (5.10) and the result of Example 14, we compute

$$e^{tA}\mathbf{x}_0 = \frac{1}{2} \begin{bmatrix} e^{3t} + e^{-t} & e^{3t} - e^{-t} \\ e^{3t} - e^{-t} & e^{3t} + e^{-t} \end{bmatrix} \begin{bmatrix} 2 \\ 1 \end{bmatrix} = \frac{1}{2} \begin{bmatrix} 3e^{3t} + e^{-t} \\ 3e^{3t} - e^{-t} \end{bmatrix}.$$

Writing the right-hand side as $\begin{bmatrix} x(t) \\ y(t) \end{bmatrix}$ so that

$$x(t) = \frac{1}{2}(3e^{3t} + e^{-t}) \quad \text{and} \quad y(t) = \frac{1}{2}(3e^{3t} - e^{-t}),$$

$$x'(t) = x(t) + 2y(t) \quad \text{and} \quad y'(t) = 2x(t) + y(t).$$

It is then straightforward to do the computations to check that

$$x(0) = 2 \quad \text{and} \quad y(0) = 1.$$

Thus, we have solved this system of linear differential equations. As t varies, $\mathbf{x}(t) = e^{tA}\mathbf{x}_0$ traces out a curve in the x, y plane. Here is a graph of that curve for $-2 \le t \le 1/4$.

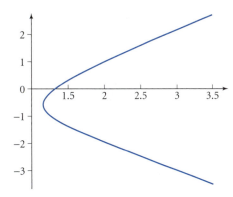

As you can see, the curve passes through the point $(2, 1)$. Also, when t is positive and large, e^{-t} is negligible compared to e^{3t}, so for such t, we have $x(t) \approx y(t) \approx e^{3t}$. Therefore, we expect the curve to be asymptotic to the line $y = x$ for large t. Indeed, we see this already near $t = 1/4$. (Notice the different scales on the x- and y-axes.) Also, when t is negative and large, e^{3t} is negligible compared to e^{-t}, so for such t, $x(t) \approx -y(t) \approx e^{-t}$. Therefore, we expect the curve to be asymptotic to the line $y = -x$ for such t. Again this is what we see.

EXAMPLE 16   **(Solving x'(t) = Ax(t) using matrix exponentials)** Let us find the solution to

$$x'(t) = x(t) + 3y(t), \qquad x(0) = -4$$
$$y'(t) = 2x(t), \qquad\qquad y(0) = 5.$$

In matrix form the differential equation is $\mathbf{x}'(t) = A\mathbf{x}(t)$, where

$$A = \begin{bmatrix} 1 & 3 \\ 2 & 0 \end{bmatrix}.$$

This matrix differs from the matrix in Example 4 in Section 1 by a multiple of the identity, and hence it has the same eigenvectors. From the results of that example, we have $A = VDV^{-1}$, where

$$V = \begin{bmatrix} 3 & -1 \\ 2 & 1 \end{bmatrix}, \qquad D = \begin{bmatrix} 3 & 0 \\ 0 & -2 \end{bmatrix}, \qquad \text{and} \qquad V^{-1} = \frac{1}{5}\begin{bmatrix} 1 & 1 \\ -2 & 3 \end{bmatrix}.$$

(Notice that this time, since A was not symmetric, the inverse was not simply the transpose.) Therefore,

$$
\begin{aligned}
e^{tA} &= \frac{1}{5}\begin{bmatrix} 3 & -1 \\ 2 & 1 \end{bmatrix}\begin{bmatrix} e^{3t} & 0 \\ 0 & e^{-2t} \end{bmatrix}\begin{bmatrix} 1 & 1 \\ -2 & 3 \end{bmatrix} \\
&= \frac{1}{5}\begin{bmatrix} 3 & -1 \\ 2 & 1 \end{bmatrix}\begin{bmatrix} e^{3t} & e^{3t} \\ -2e^{-2t} & 3e^{-2t} \end{bmatrix} \\
&= \frac{1}{5}\begin{bmatrix} 3e^{3t} + 2e^{-2t} & 3e^{3t} - 3e^{-2t} \\ 2e^{3t} - 2e^{-2t} & 2e^{3t} + 3e^{-2t} \end{bmatrix}.
\end{aligned}
$$

The data for $x(0)$ and $y(0)$ tell us that

$$\mathbf{x}_0 = \begin{bmatrix} -4 \\ 5 \end{bmatrix},$$

and our solution is

$$
\begin{aligned}
\mathbf{x}(t) = e^{tA}\mathbf{x}_0 &= \frac{1}{5}\begin{bmatrix} 3e^{3t} + 2e^{-2t} & 3e^{3t} - 3e^{-2t} \\ 2e^{3t} - 2e^{-2t} & 2e^{3t} + 3e^{-2t} \end{bmatrix}\begin{bmatrix} -4 \\ 5 \end{bmatrix} \\
&= \frac{1}{5}\begin{bmatrix} 3e^{3t} - 23e^{-2t} \\ 2e^{3t} + 23e^{-2t} \end{bmatrix}.
\end{aligned}
$$

The solution, therefore, is

$$x(t) = \frac{1}{5}(3e^{3t} - 23e^{-2t}) \quad \text{and} \quad y(t) = \frac{1}{5}(2e^{3t} + 23e^{-2t}).$$

For $-1/2 \le t \le 1$, this solution traces out the following curve:

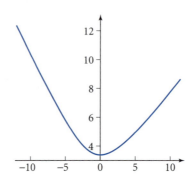

Again we see the negative eigenvalue dominating for negative times and the positive eigenvalue dominating for positive times, with a transition between the two asymptotic behaviors near $t = 0$.

We can use these ideas to solve an even more important class of systems of linear differential equations. Let A be an $n \times n$ matrix, and consider the vector differential equation

$$\mathbf{x}'(t) = A\mathbf{x}(t) + \mathbf{r}(t), \tag{5.14}$$

where $\mathbf{r}(t)$ is a given time-dependent vector. Suppose there is a solution $\mathbf{x}(t)$ that satisfies the initial condition $\mathbf{x}(0) = \mathbf{x}_0$, and try to find a formula for it. Introduce a new variable $\mathbf{z}(t)$ by

$$\mathbf{z}(t) = e^{-tA}\mathbf{x}(t). \tag{5.15}$$

Then

$$\mathbf{z}'(t) = -Ae^{-tA}\mathbf{x}(t) + e^{-tA}\mathbf{x}'(t). \tag{5.16}$$

Using (5.14) to eliminate $\mathbf{x}'(t)$ from (5.16),

$$
\begin{aligned}
\mathbf{z}'(t) &= -Ae^{-tA}\mathbf{x}(t) + e^{-tA}(A\mathbf{x}(t) + \mathbf{r}(t)) \\
&= -Ae^{-tA}\mathbf{x}(t) + Ae^{-tA}\mathbf{x}(t) + e^{-tA}\mathbf{r}(t) \\
&= e^{-tA}\mathbf{r}(t).
\end{aligned} \tag{5.17}
$$

It follows that

$$\mathbf{z}(t) - \mathbf{z}(0) = \int_0^t e^{-sA}\mathbf{r}(s)\, ds.$$

By (5.15), this equation is the same as

$$e^{-tA}\mathbf{x}(t) = \mathbf{x}_0 + \int_0^t e^{-sA}\mathbf{r}(s)\, ds$$

or

$$
\begin{aligned}
\mathbf{x}(t) &= e^{tA}\mathbf{x}_0 + e^{tA}\left(\int_0^t e^{-sA}\mathbf{r}(s)\, ds\right) \\
&= e^{tA}\mathbf{x}_0 + \int_0^t e^{(t-s)A}\mathbf{r}(s)\, ds.
\end{aligned} \tag{5.18}
$$

It is now easy to differentiate and verify that the vector-valued function $\mathbf{x}(t)$ defined by (5.18) does indeed satisfy (5.14) with the initial condition $\mathbf{x}(0) = \mathbf{x}_0$.

EXAMPLE 17 **(Solving x′(t) = Ax(t) + r(t) using matrix exponentials)** Let us solve the system of differential equations

$$x'(t) = -x(y) + y(t) + \sin(2t) \quad \text{and} \quad y(t) = -x(y) - y(t). \quad (5.19)$$

with the initial conditions $x(0) = 3$ and $y(0) = 2$.

We let

$$\mathbf{x}(t) = \begin{bmatrix} x(t) \\ y(t) \end{bmatrix} \quad \text{and} \quad \mathbf{x}_0 = \begin{bmatrix} 3 \\ 2 \end{bmatrix}$$

as usual and define

$$A = \begin{bmatrix} -1 & 1 \\ -1 & -1 \end{bmatrix} \quad \text{and} \quad \mathbf{r}(t) = \begin{bmatrix} \sin(2t) \\ 0 \end{bmatrix}.$$

With these definitions, (5.19) has the form (5.14) and hence is solved by (5.18).

We leave it as an exercise to compute that in this case

$$e^{tA} = e^{-t} \begin{bmatrix} \cos(t) & \sin(t) \\ -\sin(t) & \cos(t) \end{bmatrix}.$$

(This is a case in which the eigenvalues and eigenvectors of A are complex. However, our final expression for e^{tA} is real, as it must be.) In this case

$$\int_0^t e^{(t-s)A} \mathbf{r}(s)\,ds = \begin{bmatrix} \displaystyle\int_0^t e^{s-t}\cos(s-t)\sin(2s)\,ds \\[2mm] \displaystyle\int_0^t e^{s-t}\sin(s-t)\sin(2s)\,ds \end{bmatrix}.$$

The integrals may look a bit messy but are not so bad, and in any case they can be done on a computer. The results are

$$\int_0^t e^{s-t}\cos(s-t)\sin(2s)\,ds = \frac{2}{5} - \frac{4}{5}\cos^2(t) + \frac{3}{10}\sin(2t) + \frac{2}{5}e^{-t}\cos(t) - \frac{1}{5}e^{-t}\sin(t)$$

and

$$\int_0^t e^{s-t}\sin(s-t)\sin(2s)\,ds = -\frac{1}{5} + \frac{2}{5}\cos^2(t) + \frac{1}{10}\sin(2t) - \frac{1}{5}e^{-t}\cos(t) - \frac{2}{5}e^{-t}\sin(t).$$

It is now easy to add on $e^{tA}\mathbf{x}_0$, and we obtain the solutions

$$x(t) = 3e^{-t}\cos(t) + \frac{2}{5} - \frac{4}{5}\cos^2(t) + \frac{3}{10}\sin(2t) + \frac{2}{5}e^{-t}\cos(t) - \frac{1}{5}e^{-t}\sin(t)$$

$$y(t) = -2e^{-t}\sin(t) - \frac{1}{5} + \frac{2}{5}\cos^2(t) + \frac{1}{10}\sin(2t) - \frac{1}{5}e^{-t}\cos(t) - \frac{2}{5}e^{-t}\sin(t).$$

Notice that for large t, all the terms containing a factor of e^{-t} are negligible, so we have

$$x(t) \approx \frac{2}{5} - \frac{4}{5}\cos^2(t) + \frac{3}{10}\sin(2t) \quad \text{and} \quad y(t) \approx -\frac{1}{5} + \frac{2}{5}\cos^2(t) + \frac{1}{10}\sin(2t). \quad (5.20)$$

Here is a graph if the curve traced out by $\mathbf{x}(t)$ for $0 \le t \le 20$:

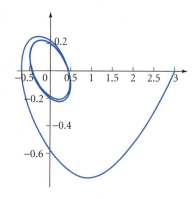

And here is a graph for $40 \leq t \leq 50$:

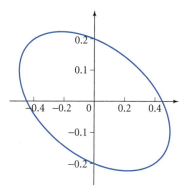

Notice in the last example that the solution seems to have settled into a "steady state" behavior of running around a fixed ellipse for large t. In fact, it is easy to see that $e^{tA}\mathbf{x}_0$ tends rapidly to zero as t increases, so the dependence of the solution on the initial data is quickly washed out. The curve you see in the second graph represents the limiting "steady state" behavior *independent of the initial data*. You should consider this fundamental example very carefully. The computations themselves are not too bad, especially if you use a computer to do them. But make sure you understand what computations you need to ask your computer to do if you need to solve a problem like this.

5.3 The general case: Generalized eigenvectors

We now know how to explicitly compute e^{tA} if A is diagonalizable. But what if A is not diagonalizable?

Recall that an $n \times n$ matrix A is diagonalizable if and only if there is a basis of $\mathbb{C}^n$ consisting of eigenvectors of A. (Recall also that even if A has real entries, the eigenvectors may be complex.)

Although not every matrix has a basis of eigenvectors, every matrix has a basis of something almost as good—generalized eigenvectors. Using them, we can compute e^{tA} in closed form.

Recall that a nonzero vector $\mathbf{v}$ is an eigenvector of the $n \times n$ matrix A if and only if μ is an eigenvalue of A and

$$(A - \mu I)\mathbf{v} = 0.$$

Here is the generalization.

DEFINITION

(Generalized eigenvectors) Let A be an $n \times n$ matrix, and let μ be an eigenvalue of A. A vector $\mathbf{v}$ in $\mathbb{C}^n$ is a *generalized eigenvector of A with eigenvalue μ* if and only if there is some positive integer p so that

$$(A - \mu I)^p \mathbf{v} = 0. \tag{5.21}$$

The smallest integer p for which (5.21) is true is called the *degree* of $\mathbf{v}$.

Notice that a generalized eigenvector is a true eigenvector if and only if its degree is 1. To make use of the generalization, we need to know one more thing about matrix exponentials.

THEOREM 16   **(Matrix exponentials and addition to multiplication)** *Let B and C be two $n \times n$ matrices such that*

$$BC = CB. \tag{5.22}$$

Then

$$e^{B+C} = e^B e^C. \tag{5.23}$$

PROOF: The only difference in the rules for the algebra of matrices and the algebra of numbers is that, in general, matrices do not commute; that is, in general, (5.22) is not true. But when it is true, any formula involving addition and multiplication of two real numbers also applies to the matrices A and B. In particular, the *binomial formula*

$$(A + B)^k = \sum_{j=0}^{k} \frac{k!}{(k-j)!j!} A^j B^{k-j} \tag{5.24}$$

is true when (5.22) is true. Since the binomial formula is what one uses to show, starting from the power series, that for any two numbers a and b,

$$e^{a+b} = e^a e^b,$$

the same proof extends directly to matrices A and B, satisfying (5.22).   ■

To apply Theorem 16, let $B = t\mu I$ and $C = (A - \mu I)$. Clearly, (5.22) is satisfied, and since $tA = B + C$,

$$e^{tA} = e^{t\mu I} e^{t(A-\mu I)} = e^{t\mu} e^{t(A-\mu I)}. \tag{5.25}$$

Now let $\mathbf{v}$ be a generalized eigenvector of $\mathbf{v}$ with eigenvalue μ and degree d. Then for all $k \geq d$, $(A - \mu I)^k \mathbf{v} = 0$, so

$$\sum_{k=0}^{\infty} \frac{t^k}{k!}(A - \mu I)^k \mathbf{v} = \sum_{k=0}^{d-1} \frac{t^k}{k!}(A - \mu I)^k \mathbf{v}.$$

This is the point of the definition: When **v** is a generalized eigenvector, the infinite sum on the left side of the foregoing summation reduces to a finite sum. Combining with (5.25),

$$e^{tA}\mathbf{v} = e^{t\mu}\left(\sum_{k=0}^{d-1} \frac{t^k}{k!}(A - \mu I)^k\right)\mathbf{v}, \tag{5.26}$$

Now, suppose that there is a basis of $\{\mathbf{v}_1, \mathbf{v}_2, \ldots, \mathbf{v}_n\}$ of $\mathbb{C}^n$ consisting of generalized eigenvectors of A. Form the $n \times n$ matrix $V = [\mathbf{v}_1, \mathbf{v}_2, \ldots, \mathbf{v}_n]$. Then, by **V.I.F.3**,

$$e^{tA}V = [e^{tA}\mathbf{v}_1, e^{tA}\mathbf{v}_2, \ldots, e^{tA}\mathbf{v}_n]. \tag{5.27}$$

We can use (5.26) to compute each column of $e^{tA}\mathbf{v}$, and therefore we can compute a closed-form formula for $e^{tA}V$. But then since

$$e^{tA} = \left(e^{tA}V\right)V^{-1}, \tag{5.28}$$

we can compute a closed-form formula for e^{tA}. The next theorem says that this gives us a general method for computing e^{tA}.

THEOREM 17   **(Every matrix has a basis of generalized eigenvectors)** *Let A be any $n \times n$ matrix. Then there is a basis of $\mathbb{C}^n$ consisting of generalized eigenvectors of A.*

Moreover, if μ is an eigenvalue of A and d is the algebraic multiplicity of μ, then every generalized eigenvector of A with eigenvalue μ belongs to $\mathrm{Ker}((A - \mu I)^d)$.

Theorem 17 can be proved using only the ideas we have developed in Chapters 3 and 4, but such a proof would be somewhat intricate. The abstract point of view developed in Chapter 7 really helps clarify the explanation of why Theorem 17 is true, so we postpone detailing this proof until then. Here, let us simply give an example instead.

EXAMPLE 18   **(Using generalized eigenvectors to exponentiate)** Let

$$A = \begin{bmatrix} 1 & 1 & 0 \\ -2 & 3 & 1 \\ 0 & 1 & 1 \end{bmatrix}.$$

Then $\det(A - tI) = (2 - t)^2(1 - t)$, so that 1 and 2 are the eigenvalues of A. To find the eigenspaces, we compute the kernels of $A - I$ and $A - 2I$, as usual.
    We find that

$$A - I = \begin{bmatrix} 0 & 1 & 0 \\ -2 & 2 & 1 \\ 0 & 1 & 0 \end{bmatrix}.$$

It is not hard to see that $A - I$ has rank 2 and that

$$\mathbf{v}_1 = \begin{bmatrix} 1 \\ 0 \\ 2 \end{bmatrix}$$

is in the kernel. The line spanned by this vector is the eigenspace E_1 of A.

In the same way, we find that

$$A - 2I = \begin{bmatrix} -1 & 1 & 0 \\ -2 & 1 & 1 \\ 0 & 1 & -1 \end{bmatrix}.$$

It is not hard to see that $A - 2I$ has rank 2 and that

$$\mathbf{v}_2 = \begin{bmatrix} 1 \\ 1 \\ 1 \end{bmatrix}$$

is in the kernel. The line spanned by this vector is the eigenspace E_2 of A.

There is no basis consisting of eigenvectors of A, so A is not diagonalizable. We need to seek generalized eigenvalues. Since the algebraic multiplicity of the eigenvalue 1 is 1, Theorem 17 says that we do not get anything new for this eigenvalue. But the algebraic multiplicity of the eigenvalue 2 is 2, so Theorem 17 says we should look for the kernel of $(A - 2I)^2$.

Computing in the usual way, we find that

$$(A - 2I)^2 = \begin{bmatrix} -1 & 0 & 1 \\ 0 & 0 & 0 \\ -2 & 0 & 2 \end{bmatrix}.$$

It is not hard to see that $(A - 2I)^2$ has rank 1 and that

$$\mathbf{v}_3 = \begin{bmatrix} 0 \\ 1 \\ 0 \end{bmatrix}$$

belongs to the kernel of A, along with $\mathbf{v}_2$. Therefore, $\mathbf{v}_3$ is a generalized eigenvector of A of second degree.

We now have the basis $\{\mathbf{v}_1, \mathbf{v}_2, \mathbf{v}_3\}$ consisting of generalized eigenvectors and form

$$V = [\mathbf{v}_1, \mathbf{v}_2, \mathbf{v}_3] = \begin{bmatrix} 1 & 1 & 0 \\ 0 & 1 & 1 \\ 2 & 1 & 0 \end{bmatrix}$$

and compute

$$e^{tA} V = [e^{tA}\mathbf{v}_1, e^{tA}\mathbf{v}_2, e^{tA}\mathbf{v}_3].$$

This computation is easy. Since $A^k \mathbf{v}_1 = \mathbf{v}_1$ for all k,

$$e^{tA}\mathbf{v}_1 = \sum_{k=0}^{\infty} \frac{t^k}{k!} A^k \mathbf{v}_1 = \sum_{k=0}^{\infty} \frac{t^k}{k!} \mathbf{v}_1 = e^t \mathbf{v}_1,$$

so

$$e^{tA}\mathbf{v}_1 = e^t \mathbf{v}_1 = \begin{bmatrix} e^t \\ 0 \\ 2e^t \end{bmatrix}.$$

Likewise, since $A^k \mathbf{v}_2 = 2^k \mathbf{v}_2$ for all k,

$$e^{tA}\mathbf{v}_2 = e^{2t}\mathbf{v}_2 = \begin{bmatrix} e^{2t} \\ e^{2t} \\ e^{2t} \end{bmatrix}.$$

To compute $e^{tA}\mathbf{v}_3$, we use (5.26) and

$$\sum_{k=0}^{\infty} \frac{t^k}{k!} (A - 2I)^k \mathbf{v}_3 = I\mathbf{v}_3 + t(A - 2I)\mathbf{v}_3.$$

It is easy to see that $(A - 2I)\mathbf{v}_3 = \mathbf{v}_2$, so

$$e^{tA}\mathbf{v}_2 = e^{2t}(\mathbf{v}_3 + t\mathbf{v}_2) = \begin{bmatrix} e^{2t}t \\ (1+t)e^{2t} \\ e^{2t}t \end{bmatrix},$$

giving

$$e^{tA}V = [e^{tA}\mathbf{v}_1, e^{tA}\mathbf{v}_2, e^{tA}\mathbf{v}_3] = \begin{bmatrix} e^t & e^{2t} & te^{2t} \\ 0 & e^{2t} & e^{2t}(1+t) \\ 2e^t & e^{2t} & te^{2t} \end{bmatrix}.$$

We find that

$$V^{-1} = \begin{bmatrix} -1 & 0 & 1 \\ 2 & 0 & -1 \\ -2 & 1 & 1 \end{bmatrix},$$

and therefore $e^{tA} = \left(e^{tA}V\right)V^{-1}$ works out to be

$$e^{tA} = \begin{bmatrix} -e^t + 2e^{2t}(1-t) & e^{2t}t & e^t + e^{2t}(t-1) \\ -2e^{2t}t & e^{2t}(1+t) & e^{2t}t \\ -2e^t + 2e^{2t}(1-t) & e^{2t}t & 2e^t + 2e^{2t}(t-1) \end{bmatrix}.$$

Notice that when generalized eigenvectors are present, e^{tA} involves powers of t as well as exponentials.

This example works out nicely since A has real eigenvalues. The procedure is the same if the eigenvalues are complex, and on the basis of this example, you should be able to compute e^{tA} for any square matrix A as long as you can find the eigenvalues. (When A is larger than 4×4, say, this may not be so easy. Computing and then factoring the characteristic polynomial can be difficult to do, even numerically, for large matrices.)

The general procedure is very simple: For each eigenvalue μ of A, let d be the algebraic multiplicity of μ. Then find a basis for $\mathrm{Ker}((A - \mu I)^d)$. Lump all these bases together into one set of vectors; this is the basis $\{\mathbf{v}_1, \mathbf{v}_2, \ldots, \mathbf{v}_n\}$ of generalized eigenvectors.

The method explained in this section enables us to solve *any* system of n linear differential equations in n unknowns.

EXAMPLE 19   **(Solving systems in general)** Consider the system of differential equation

$$\begin{aligned} x'(t) &= x(t) + y(t) \\ y'(t) &= -2x(t) + 3y(t) + z(t) \\ z'(t) &= y(t) + z(t) \end{aligned}$$

subject to the initial conditions

$$x(0) = 1, \qquad y(0) = 1, \qquad \text{and} \qquad z(0) = 0.$$

Letting

$$\mathbf{x}(t) = \begin{bmatrix} x(t) \\ y(t) \\ z(t) \end{bmatrix},$$

the system becomes $\mathbf{x}'(t) = A\mathbf{x}(t)$ with $\mathbf{x}(0) = \mathbf{x}_0$ where A is the 3×3 matrix from Example 18 and

$$\mathbf{x}_0 = \begin{bmatrix} 1 \\ 1 \\ 0 \end{bmatrix}.$$

Since we have computed e^{tA} in Example 18, we have the solution

$$\mathbf{x}(t) = e^{tA}\mathbf{x}_0 = \begin{bmatrix} -e^t + e^{2t}(2-t) \\ e^{2t}(1-t) \\ -2e^t + e^{2t}(2-t) \end{bmatrix}.$$